COLLEGE OF MARIN LIBRARY
KENTFIELD, CALIFORNIA

Orang-utan Biology

Orang-utan Biology

EDITED BY

JEFFREY H. SCHWARTZ

University of Pittsburgh

New York Oxford

Oxford University Press

1988

Oxford University Press

Oxford New York Toronto
Delhi Bombay Calcutta Madras Karachi
Petaling Jaya Singapore Hong Kong Tokyo
Nairobi Dar es Salaam Cape Town
Melbourne Auckland

and associated companies in
Berlin Ibadan

Copyright © 1988 by Oxford University Press, Inc.

Published by Oxford University Press, Inc.,
200 Madison Avenue, New York, New York 10016

Oxford is a registered trademark of Oxford University Press

All rights reserved. No part of this publication may be reproduced,
stored in a retrieval system, or transmitted, in any form or by any means,
electronic, mechanical, photocopying, recording, or otherwise,
without the prior permission of Oxford University Press.

Library of Congress Cataloging-in-Publication Data
Orang-utan biology.
Includes bibliographies and index.
1. Orang-utan. I. Schwartz, Jeffrey H.
II. Title: Orang-utan biology.
QL737.P96073 1988 599.88′42 87-20440
ISBN 0-19-504371-5

Printing (last digit): 9 8 7 6 5 4 3 2 1
Printed in the United States of America
on acid-free paper

Preface

As is evident from the scope and diversity of contributions to this volume, the orang-utan has become of more than tangential scientific interest to an array of disciplines—again. I say again because this hominoid had for a long time commanded the almost singular attention of both scholar and lay person alike. But with increasing focus during the 20th century on the African apes and, in particular, on the chimpanzee, the orang-utan fell into the backwaters of research, in large part because of the common notion that we cannot learn very much about ourselves from study of a presumed distant relative. But this attitude is certainly changing and the orang-utan has emerged as a hominoid of intrinsic, if not phylogenetic, interest in and of itself. To this we may add the historically unexpected: fossil hominoids that were considered dentally and gnathically hominid—i.e., *Sivapithecus* (= ?*Ramapithecus*)—have been found to have features of the facial skeleton otherwise considered unique to the orang-utan; and morphologically, it appears that modern humans share a greater number of unique features with the orang-utan than with any other extant hominoid. Thus, the orang-utan finds itself in the midst of debate and analysis critical to the understanding of hominoid phylogeny in general, and even human origins.

In response to the growing interest in the evolutionary biology of the orang-utan, I organized a small symposium on the subject for the 1985 annual meeting of the American Association of Physical Anthropologists. The participants were drawn from colleagues I knew had been actively engaged in research problems surrounding the orang-utan and who thought they could compose a detailed abstract in time for the submission deadline and were also planning to attend the meetings. Presentations were given by D. R. Swindler, J. D. Swarts, L. A. Winkler, B. T. Shea, M. D. Rose, W. L. Jungers and N. M. Czekala, S. E. Shideler, and B. L. Lasley on topics that appear in this volume.

The success of this symposium led to the development of the idea that a volume specifically on aspects of the evolutionary biology of the extant hominoid—in contradistinction to the emphases on paleontological themes of recent symposia and conferences on hominoid and hominid evolution—would be timely and of general interest. From a diversity of sources (e.g., recommendations of the participants, inquiries from colleagues, recent publications centering on the orang-utan) emerged an extensive list of potential contributors to such a project which eventually got sorted out by such means as incompatibility of scheduling or redundancy or narrowness of subject matter.

Given the topics represented in the original symposium, it seemed reasonable to use these as areas of research to expand upon in the proposed volume. The opening section deals with the history of study of the orang-utan (Röhrer-Ertl), subspecific taxonomy (Courtenay et al.), ecology and behavior (Rodman), and evolution as interpreted biomolecularly (Marks) and morphologically (Schwartz). Following are sections on reproductive physiology, behavior, and anatomy (Graham, Nadler, Czekala et al., Kingsley, Dahl, and Soma); neuroanatomy (Zilles and Rehkämper, Armstrong and Frost); craniofacial anatomy, growth, and functional morphology (Winkler, Röhrer-Ertl, Winkler et al., Shea, Brown and Ward); dental morphology (Swarts, Swindler and Olshan); and postcranial anatomy, growth, and functional morphology (Morbeck and Zihlman, Rose, Tuttle and Cortright, Anderton, Jungers and Hartman).

Since I have provided summaries of the appropriate chapters at the beginning of each section (based on abstracts contributed by the authors themselves), I shall not repeat such information here. Rather, I would like to thank all the contributors for participating in this venture, making the final product exciting, interesting, and even provocative. Disagreement and controversy are not absent in this collection. But while the growth of any science relies in part on elements of disagreement and controversy, because of the refinement of ideas that naturally occurs, no discipline can survive without a continual infusion of new ideas, novel approaches, valuable syntheses, and the very basic collection of data. These qualities are characteristic of all of the contributors' efforts.

In addition to thanking all authors for participating in this venture, I would like to extend our collective gratitude to William F. Curtis of Oxford University Press, who responded so positively to the proposed project and continued to offer support, even when an extension of the deadline was requested. The successful completion of this project also derives from logistical support provided to me by the Department of Anthropology, University of Pittsburgh. Without the unlimited use of telephone, express mailing, duplication of materials and manuscripts, and secretarial and computer facilities made available by my department, I doubt that the preparation of this volume could have been accomplished as expeditiously.

Obviously, no volume of collected works, regardless of how many authors it has, can ever claim complete representation of the subject matter. But the offerings presented here do, I think, contribute substantially to our understanding of the orang-utan, from the broader evolutionary perspective to the details of what makes this animal not just unique among hominoids, but distinctive among primates in general.

Pittsburgh
June 1987

J.H.S.

Contents

Contributors

John C. Anderton
8714 Gateshead Road
Alexandria, VA 22309

Peter Andrews
Department of Palaeontology
British Museum (Natural History)
Cromwell Road
London SW7 5BD
England

Este Armstrong*
Department of Anatomy
Louisiana State University
Medical Center
1901 Perdido Street
New Orleans, LA 70112-1393

Barbara Brown
Department of Sociology and Anthropology
Kent State University
Kent, OH 44240

Glenn C. Conroy
Department of Anatomy/Neurobiology
Washington University
School of Medicine
660 S. Euclid Avenue
St. Louis, MO 63110

Gerald Cortright
Section of Morphology
Division of Biology and Medicine
Brown University
Providence, RI 02912

Jackie Courtenay
Department of Zoology
Australian National University
Canberra, A.C.T. 2601
Australia

N. M. Czekala
The San Diego Zoo
Research Department
P.O. Box 551
San Diego, CA 92112-0551

Jeremy F. Dahl
Yerkes Regional Primate Research Center
Emory University
Atlanta, GA 30322

Olav Röhrer-Ertl
Zoologische Staatssammlung München
c/o Triesterstrasse 61-D
8000 Munich 80
Federal Republic of Germany

G. Thomas Frost
Department of Anatomy
Louisiana State University Medical Center
1901 Perdido Street
New Orleans, LA 70112-1393

Charles E. Graham
Reproductive Biology Division
Primate Research Institute
New Mexico State University
Holloman Air Force Base
New Mexico 88330

Colin Groves
Department of Prehistory and Anthropology
Australian National University
Canberra, A.C.T. 2601
Australia

Steve E. Hartman
Department of Anthropology
Northwestern University
Evanston, IL 60201

William L. Jungers
Department of Anatomical Sciences
Health Sciences Center
State University of New York at Stony Brook
Stony Brook, NY 11794

Susan R. Kingsley
Wellcome Research Laboratories
Langley Court
Beckenham Kent BR3 3BS
England

*Present address: Yakovlev Collection, Armed Forces Institute of Pathology, Washington, D.C. 20306

B. L. Lasley
L.E.H.R.
University of California, Davis
Davis, CA 95616

Jon Marks
Department of Anthropology
P.O. Box 2114
Yale University
New Haven, CT 06520

Mary Ellen Morbeck
Department of Anthropology
University of Arizona
Tucson, AZ 85721

Ronald D. Nadler
Yerkes Regional Primate
Research Center
Emory University
Atlanta, GA 30322

Andrew F. Olshan
Department of Epidemiology
University of Washington
Seattle, WA 98195

Gerd Rehkämper
Anatomisches Institut
Universität Köln
Joseph-Stelzmann Strasse 9
5000 Cologne 41
Federal Republic of Germany

Peter S. Rodman
Department of Anthropology and
California Primate Research Center
University of California, Davis
Davis, CA 95616

M. D. Rose
Department of Anatomy
New Jersey Medical School
100 Bergen Street
Newark, NJ 07103

Jeffrey H. Schwartz
Department of Anthropology
University of Pittsburgh
Pittsburgh, PA 15260

Brian T. Shea
Department of Anthropology
Northwestern University
Evanston, IL 60201

S. E. Shideler
L.E.H.R.
University of California, Davis
Davis, CA 95616

Hiroaki Soma
Department of Obstetrics and Gynecology
Tokyo Medical College Hospital
1-7 Nishi-shinjuku 6
Sinjuku-ku
Tokyo 160
Japan

J. Douglas Swarts
Department of Anthropology
University of Pittsburgh
Pittsburgh, PA 15260

Daris R. Swindler
Department of Anthropology
University of Pittsburgh
Pittsburgh, PA 15260

Russell H. Tuttle
Department of Anthropology
University of Chicago
Chicago, IL 60637

Michael W. Vannier
Department of Radiology
Washington University
Medical Center
St. Louis, MO 63110

Steven C. Ward
Human Anatomy Program
Northeastern Ohio Universities
College of Medicine
Rootstown, OH 44272
and
Department of Sociology and Anthropology
Kent State University
Kent, OH 44240

Linda A. Winkler
Department of Biology and Anthropology
University of Pittsburgh
Titusville, PA 16354

Adrienne L. Zihlman
Department of Anthropology
Clark Kerr Hall
University of California
Santa Cruz, CA 95064

Karl Zilles
Anatomisches Institut
Universität Köln
Joseph-Stelzmann Strasse 9
5000 Cologne 41
Federal Republic of Germany

Orang-utan Biology

I

OVERVIEW: HISTORY, TAXONOMY, BEHAVIOR AND ECOLOGY, AND EVOLUTION

The intent of this opening section is to try to place the orang-utan in nature broadly, as viewed from the synthetic perspectives of history, taxonomy, behavior and ecology, and evolution from biomolecular as well as morphological standpoints. And it is also here, from the beginning, that we must realize the lack of consensus (even if it is a majority versus a minority) that occurs around this particular hominoid, from something seemingly as simple as the proper species and subspecies names of the orang-utan, to the behavioral and ecological reconstruction of the orang-utan's closest known fossil relative, *Sivapithecus,* to the evolutionary relations of the orang-utan to the other large hominoids.

Olav Röhrer-Ertl's contribution on the research history, nomenclature, and taxonomy of the orang-utan begins this section. Here, he summarizes first the early references to the orang-utan, or at least to "Pongo," a name that came to refer not just to the southeast Asian great ape, but to the African apes as well as humans, real or imagined. Then, in the same detective-like fashion, Röhrer-Ertl traces the taxonomic history of the orang-utan, following the first specimen of an orang-utan from its arrival at the British Museum (Natural History) (then called the Sloane Museum) to the study of its skull by Edwards and then by Hoppius and Linnaeus, and concluding quite untraditionally that the proper species name of the orang-utan is *satyrus,* not *pygmaeus.* Thus, we find *Pongo satyrus* used by Röhrer-Ertl (as well as by Zilles and Rehkämper) amidst the more expected references to *Pongo pygmaeus.* Röhrer-Ertl's argument is certain to keep the taxonomic fires ablaze for some time to come.

Röhrer-Ertl's revisions do not stop at the level of the species. Indeed, he has pursued the anatomical, especially cranial, details down to the subspecies and, after providing diagnoses by which one can distinguish the Sumatran from the Bornean variety, he then argues that the subspecies of the former should be *Pongo satyrus satyrus* (Linnaeus, 1758) and the latter *P. s. borneensis* von Wurmb, 1784. The subfossil orang-utan is designated as *P. s. weidenreichii* (Hooijer, 1948).

But, needless to say for those who have followed the taxonomic literature on the orang-utan, Jackie Courtenay, Colin Groves, and Peter Andrews (adopting the position developed in other publications by Groves) disagree with Röhrer-Ertl's proposed species and subspecies names, at least with regard to the extant taxa, preferring instead *Pongo pygmaeus pygmaeus* Linnaeus, 1760 for the Bornean variety and *P. p. abelii* Lesson, 1827 for the Sumatran orang-utan. This taxonomic conclusion is not, however, the initial focus of Courtenay and co-workers' assessment of the differences between Bornean and Sumatran orang-utans. Rather, these authors' intent was to investigate the degree to which inter- and intra-island variation among orang-utans warrants their separation at the specific, not subspecific, level. And it is only after a review of the more obvious aspects of general appearance (e.g., hair color and form, throat sac and cheek flange development, and presence or absence of hallucal nail) and reproductive and social behavior, as well as the details of cranial morphology and biochemical and chromosomal criteria, that Courtenay et al. conclude that their original suspicions of "difference" were not substantiated at what they considered to be the species level, even though the impression that this should be the case remains strong.

Whatever species and subspecies names are eventually settled upon as valid nomenclaturally, there still exist two varieties of orang-utan, and it is the differences and similarities in behavior and ecology between the Bornean and Sumatran orang-utans that have been pursued with increasing vigor and detail over the last 15 years. It is the diversity and consistency in ecology and behavior of the orang-utan that Peter Rodman has focused upon in his comprehensive review.

Living orang-utans of Borneo and Sumatra feed predominantly on the ripe flesh of fruits. They travel and feed arboreally, moving over short (300–900 m) distances each day as they proceed from resting to feeding site. Populations consist of dispersed adult females with young, whose home ranges vary from 0.4 to 3 km^2 and overlap extensively. Observations at four sites indicate that range size and travel distance per day are dependent on the dispersion of local resources. Young females remain near their mothers' ranges as they reach sexual maturity, whereas young males apparently leave their mothers' ranges to travel more widely. Some males may persist as "subadults" without maturing physically, whereas others develop striking secondary sexual characteristics, including large body size (see Susan Kingsley's and Mary Ellen Morbeck and Adrienne Zihlman's chapters in other sections). Some of these mature males reside for months if not years in local areas where they attempt to exclude other males from mating, using both indirect communication (loud calls) and direct aggression to do so. After arguing that the behavior and ecology of the orang-utan would also have characterized fossil forms known to be similar morphologically to the extant taxon, Rodman rejects the suggestion made elsewhere by Peter Andrews that the fossil *Sivapithecus* would have been a terrestrial, group-living herbivore with no extant analog among primates.

Jon Marks next provides what is to date one of the most complete reviews of biomolecular and chromosomal studies relevant to understanding the evolutionary relationships of the orang-utan among the extant large hominoids. It is also one of the first attempts to apply a cladistic methodological approach to the interpretation of these data, which have traditionally been analyzed phenetically. Marks comes to reject many of the sources of biomolecular information (e.g., immunologic reactivity, DNA hybridization) and interpretation (e.g., via maximum parsimony, immunologic distances) as providing the details of relationship among the extant hominoids. Thus Marks would retreat from the conclusions of more recent studies on, for example, hemoglobins, mitochondrial and η-globin DNA, and DNA hybridization that indicate the chimpanzee is more closely related to the orang-utan than it is to the gorilla. Rather, Marks finds that, while there may be derived features uniting humans and the African apes via a common ancestor (e.g., via fibrinopeptide B [two substitutions], β-globin [one substitution], myoglobin [one substitution], chromosomal Q-brilliance, and mitochondrial DNA [29 substitutions],) the trichotomous relationship cannot be refined further.

Marks's theoretical conclusions and questioning of commonly accepted practice among biomolecular and chromosomal systematists may raise the hackles of some of them, but many would disagree with the conclusions that I reach in my chapter. There I point out not only that there has been little morphological evidence offered since Darwin in support of the relatedness of humans and one or both of the African apes, but that, when the relationships among the extant hominoids are investigated cladistically in an evaluation of more than 200 morphological features, there are well over twice as many morphological synapomorphies in support of uniting humans with the orang-utan (36) as with an African ape clade (14) and very few (5) in support of uniting the chimpanzee more closely with humans than with the gorilla. The latter arrangement is, however, consistent with more recent biomolecular and chromosomal analyses, while the association of humans with the African apes is consistent with Marks's interpretations.

I am at present inclined to accept the morphologically supported sister groups human–orang-utan and chimpanzee–gorilla, but I realize that, by doing so, I am also suggesting that apparent biomolecular synapomorphies in support of alternative phylogenetic schemes must be interpreted as primitive retentions or parallelisms. Obviously, whatever phylogenetic scheme among the large hominoids one favors, that arrangement is only one of the available, competing theories of relatedness. Although Marks certainly makes apparent his preferred phylogeny, he also provides the following caveat:

> [G]iven the fact that the genome is still *terra incognita,* and the complexities of its evolutionary dynamics are only beginning to be grasped, it is surely unwise for traditional systematics to yield absolute hegemony to molecular genetics. Genomic data constitute a valuable set of tools, but are not yet the ultimate tool-set.

The era of traditional systematics is surely coming to an end, and the field of molecular genetics is at the same time approaching the end of its initial phase of innocence. At the risk of falling prey to jargon, I shall take the liberty of suggesting that a "new" systematics will be forthcoming only when the urge to be "right" takes a back seat to the possibility that one is wrong.

1

Research History, Nomenclature, and Taxonomy of the Orang-utan

OLAV RÖHRER-ERTL

During the study of the morphology of the orang-utan, I became aware of problems regarding the recognition of subspecies of this hominoid (Röhrer-Ertl, 1982). But while I was pursuing an analysis of subspecific characters of the orang-utan, it also became clear not only that this level of research on the orang-utan had been only superficially understood, but also that there were fundamental problems with the nomenclature of this hominoid (Röhrer-Ertl, 1983a, 1984a). Such taxonomic confusion had been unexpected, since the history of research on the orang-utan spans nearly 300 years.

It would seem that interest in accurate nomenclature and taxonomy of the orang-utan took a back seat to other systematic concerns. Perhaps an emphasis on the African apes contributed to this situation. Furthermore, in disregard of what I define as "scientific pedantry," recent authors have increasingly dealt with the more speculative of evolutionary postulates and, in many cases, have incorrectly referenced original source material, even though the correct data are available (cf. Scheidegger and Wendnagel, 1944–1950). Such inaccuracies occur often and can be traced throughout the literature.

I shall here attempt to unravel these inaccuracies and confusions regarding the history of research on the orang-utan.

RESEARCH HISTORY

Up to 1758

Modern science can be limited to the "modern era"—the period from the late Middle Ages to the present. From the beginning, science has described process, a procedure that can be traced to the European view of history in the Middle Ages.

During the Renaissance the essence of what later came to be regarded as the taxon Hominidae (Gray, 1870) was seen as a "link" between Man and the animal world. This interpretation was derived from earlier authors, among them Vesalius (1542), and subsumed such creatures as "satyrs" and "pygmies." The notion of there being such a link between humans and the animal world was reinforced especially by Dutch seafarers' stories, in which sightings of unusual creatures were reported. One account, in particular, by an English seafarer, Andrew Battell (Purchas, 1613), had far-reaching consequences. Battell, who had been a prisoner of the Portuguese in Loango, north of Angola for some 30 years during the late 16th century, reported seeing two "Pongo" (= "Ponginus"), of different sizes and with black fur, in the areas of Fiote (today, the northern coast of Congo-Brazzaville). Such animals were not unknown, at least in the realm of myth, to the native Fiotes, who called the larger creature "M'Pungu" and the smaller one "M'Geko."

The first Pongo or Ponginus to reach Europe, via a Portuguese trader, was a juvenile chimpanzee, which had been preserved in alcohol and was subsequently dissected and described by the famous Dutch anatomist and physician Nicolaas Tulpius (1641); the country of origin of this specimen was given as "Angola" (Luanda Harbor). It was again through trade with the Portuguese that the next recorded Pongo, an infant chimpanzee that had also been preserved in alcohol, reached the hands of the brilliant British comparative anatomist, Edward Tyson, in 1699. But whereas Tulpius had published his dissection as a contribution to animal comparative anatomy, Tyson referred to his chimpanzee as a "pygmie" and even as a "satyr" and placed this individual in the line of ascension of the "great chain of being." In no small way because of Tyson's efforts, Pongo became ensconced in the line of succession that proceeded from Monkey, to Ape, to Pygmy (or Satyr), to Man. (It should be mentioned that a pygmy was regarded as a primitive human, whereas a satyr, a "wild man of the woods," perhaps such as Esau in the Bible, was a more primitive individual.) Thus Pongo, in the guise of a pygmie or satyr, became the intermediate between the animal world and the world of human and near-human beings.

Since the name Pongo in one form or another came to include not only the southeast Asian hominoid but also the African apes, some remarks on the history of colonization and trade during the period are appropriate. During the 16th century Portugal completely dominated the East India trade route. In the face of severe competition from the Spanish, the Portuguese also maintained a hold on the China and Japan routes, but they did so while suffering considerable losses. In the 17th century, the Dutch succeeded in wresting away from the Portuguese many important localities. As a result, the Portuguese dominated only the west coast of Africa south of the Guinea Coast (approximately 4° S), and did so into the 19th century by forming a coalition with England. One of the chief Portuguese slave ports, Luanda, was also one of the strongest colonial citadels of the period and continues in strategic importance (Röhrer-Ertl, 1983a). If a trade item, such as a slave or a Pongo, were to pass through this port, the place of origin would be given as "Angola," even though the individuals concerned might not themselves have come from that area.

Although the existence of pygmies and satyrs had been "authenticated" and was important in the literature preceding and following 1758, most of the serious reports were devoted to humans, not Pongo. Of particular interest were sick and phenotypically aberrant individuals. Those that were excessively hairy were regarded as satyrs (cf. Bontius, 1658). It would appear that most of the individuals described were suffering from hypertrichosis lanuginosa, since hypertrichosis vera is a much rarer condition. None of the individuals described survived the overseas transport.

Linnaeus (1758) correctly ascribed this latter "form" to the genus *Homo* but then referred individuals to the species *troglodytes,* calling them "Ourang-Outangs." The term Ourang-Outang or orang-utan, although commonly translated from Malay as "(wild) man of the forest," actually means "debtor," referring to individuals who were taken by seamen in payment of debts.

Much confusion in the earlier literature about the true identity of the Pongo being discussed derives from the fact that varieties of this "animal" were thought to live in Asia as well as in Africa. Portuguese traders had apparently verified this fact. However, it was a misinterpretation of classical authors, especially of Pliny the Elder, which led to this confusion and which was exacerbated and proliferated through an increase of specimens of poor quality and a disregard for their actual places of origin. The material was simply fitted into the notion of Pongo, which, in turn, was fitted into the prevailing philosophy. And this is what faced Linnaeus in 1735 and later Buffon in 1745 when they set to work on their principal treatises.

1758–1800

Linnaeus appears to have been more critical of his sources than were his contemporaries, and he seems to have made a concerted effort to describe correctly the different types of large animals. Thus Linnaeus included only the chimpanzee in the tenth edition of the *Systema naturae* (1758), since this was the only hominoid known to any extent. Following Tulpius, and departing from his customary use of binomial nomenclature, Linnaeus named the chimpanzee *Simia satyrus indicus.* Linnaeus's correspondence with Edwards indicates that he was aware of the latter's study on the "Dermoplastik" of a young orang-utan that was published in 1758 (see Röhrer-Ertl, 1984a). Edwards sent the cranium of this specimen to Linnaeus in 1759; this skull has since been lost.

During the 18th century, reports by Dutch researchers (e.g., Beeckman, 1714) were relatively precise in detail, and until approximately 1760–1770 most were based on Pongo from Sumatra. Reports thereafter came almost exclusively from Borneo. This geographical shift is related to the colonial history of northern Sumatra.

The British, who had tried to take the China route from the Dutch, eventually succeeded in conquering almost all of the Dutch trading posts on the Malacca Peninsula and in 1786 acquired the harbor. However, until 1811–1812, the Dutch held onto the southern exit of the Malacca Strait—through what is now the island of Singapore—and continued to compete for Chinese trade. The Spanish trade lines, which ran from Manila to Cadiz and Seville via Acapulco and Vera Cruz, were competitive for only a few luxury items.

Northern Sumatra, especially Atjeh (Aceh), was a British client state until 1870; thus, only ships from countries friendly toward England were admitted into the harbor for provisioning and repairs. Trade was controlled completely by the British. Even though Madan remained Dutch, Dutch ships could not easily call to port there because of the presence of the British fleet. Therefore Dutch trade was carried out by way of Batavia (present-day Djakarta) and the Sunda Strait, which was also less dangerous for large sailing ships to navigate than the Malacca Strait. This change in Dutch trade routes also shifted scientific focus from Sumatra to Borneo.

During the period 1758–1800 the Academy of Batavia—the oldest of its kind outside Europe—initiated research specifically in Insulinde (the western Malayan Archipelago), which resulted, for example, in the gathering of correct descriptions of the "satyrs"—correct, that is, in terms of the attitudes of the time. In 1776, the first living orang-utan arrived in Europe; other individuals were sent preserved in alcohol. Although attempts were constantly being made to ship live animals to Europe, the long sea voyages caused many losses. In 1779, Willem Adriaan van Palm shot a mature male orang-utan in Sukadana, southwest Borneo. This individual was correctly called an orang-utan, but at the same time, it was thought to belong to a taxon different from previously caught individuals. In 1784, Baron Friedrich von Wurmb, Secretary of the Academy of Batavia, described this orang-utan as "Groote Borneoosche Pongo" and "Pongo . . . van Borneo." Von Wurmb thus referred this individual to *Pongo borneensis,* citing the work of Buffon (1750) (see below). However, since Linnaeus had died in 1778, von Wurmb's work could not be included in the *Systema.* Linnaeus had, however, emended his 1758 description of the chimpanzee, first in the twelfth (1766) and then again in the thirteenth (1767) editions of the *Systema.* In the latter editions, Linnaeus proposed the following:

(a) *Simia satyrus,* the rust-red (= "ferruginea") (sub)species, is a closer relative of Man. (There was no definite or more detailed description of this form.)
(b) *Simia satyrus indicus,* the black (= "nigra") subspecies; a subordinate form. (Linnaeus used a full description taken from Tulpius. In the later editions, he referred primarily to Edwards [1758], rather than to Hoppius [1763], for the more precise distinctions of both types.)

Although the correct distinction between the orang-utan and the chimpanzee was realized as early as 1750, the confusion of the two hominoids persisted. Vosmaer (1778) and later Camper (1782) formally restricted the term orang-utan to Linnaeus's Asian *Simia satyrus;* this taxonomic move was followed by Gmelin (1788).

During the latter half of the 18th century the orang-utan was believed to represent the "real link" between Man and Animal. Nevertheless, and although this opinion was common among the scientists of the day, this notion was, according to Immanuel Kant (1775, 1785, 1786, 1798), not precisely articulated, even among the scientifically tolerant Dutch. (Kant himself, however, not only defined the term "race" (i.e., species) in a sense presaging notions of population genetics, but he also emphasized a monophyletic origin of the genus *Homo* from the Animal Kingdom).

1800–1900

The Napoleonic Wars also altered the history of research on the orang-utan. The collection of orang-utans under the aegis of the governor of the Netherlands and all other important European collections were transferred to Paris, where they remained. Live animals arriving in Europe were also sent to Paris; such was the case, for example, in 1808, with an orang-utan belonging to Josephine Beauhernais. Of course, this turn of events greatly improved French research on these and other primates.

During the Congress of Vienna of 1814–1815, monopoly by colonial trade was condemned and a declaration concerning "freedom of the seas" propounded. Prior to 1800, only those scientists connected with specific colonial powers could legally study orang-utans (as was the case, for example, with the Dutch scientists). But after the Congress of Vienna, prepared specimens could be acquired for any collection: Dumortier (1838) studied some 39 crania from Borneo in the collections in Brussels; Rudolphi (1826) had 20 crania from Borneo available to him in Berlin, while Mayer (1849) saw a similar number throughout various German collections; in London, Owen (1830–31ff) dissected more than three individuals, of which one had earlier been described by Abel (1825), who got hold of it as it was being transported via Calcutta; and Tiedemann (1821), in Heidelberg, produced the first sections of an orang-utan's brain.

With an increase of scientific activity in the field and, consequently, more precise observations, there was bound to be a shift in research as well as a reevaluation of previous reports. For example, Zimmermann (1777) and subsequently Tilesius (1813) had surmised that all animals reaching Europe had died at a subadult age. But when Rudolphi (1826) checked this against the specimens, he found that this was not the case, a conclusion that became inescapable because of the work of Dumortier (1838), Schwartze (1839), Schlegel and Müller (1839–44), Müller (1845), Mayer (1849), and Alessandrini (1854). This led eventually to a rejection of Buffon's classification of orang-utans according to body size, even though this scheme was accepted by various prominent scholars (e.g., Cuvier, 1829; Owen, 1839; cf. Röhrer-Ertl, 1983a, 1984a).

Prior to 1893 adult orang-utans were found captive only in the Calcutta Zoo. Subsequently, however, they were kept in large numbers in zoos in Europe and North America. Unfortunately, these animals died quickly in captivity. But because the bodies were sent to anatomy laboratories, there was an explosion around the turn of the century of anatomical work on the orang-utan. Fick (1895ff), for example, presented the first complete anatomical description since Camper's appeared in 1782. However, anatomical studies diminished drastically as the 20th century progressed.

Interest in anatomical as well as other studies shifted from the orang-utan to the chimpanzee at the turn of the century, in part as a result of Selenka's (1899) assertions that, of the three large apes, the orang-utan had diverged from the "line" leading to humans prior to the African apes. Although von Wurmb (1784), for example, had earlier argued that the behavior of the orang-utan was most similar to that of humans, Selenka took the "apathy" and "disinterest" of captive animals to reflect their more distant relations. It now appears that zoo orang-utans' lethargy ("apathy" and "disinterest") is the result of their tendency to develop hyperparathyroidism (Röhrer-Ertl and Frey, 1987).

1900–1985

The study of the order Primates developed during the early 1900s into a separate discipline, particularly in the United States (e.g., Hrdlička, 1907; Huber, 1931; Schultz, 1916ff). However, by 1945, published studies specifically on the orang-utan had become virtually nonexistent and the animal was often only briefly treated in comparative studies (cf. Chasen, 1939–40; Kleinschmidt, 1933). This lack of research on the orang-utan is surprising in light of the fact that, due largely to the activity of the importer Ruhe (who operated out of Alfeld, near Hanover), some 200 orang-utans were shipped during the 1920s from Atjeh (northern Sumatra) to all parts of the world. Reports on the orang-utan that one did see were almost exclusively from zoologists working in zoos (e.g., Aulmann, 1932; Brandes, 1939), although there was the occasional taxonomic endeavor (e.g., Stiles and Orleman, 1927).

Things have changed, however, during the last 25 years. For the first time, ethologists have dealt with the species, basing their work on the initial efforts of Brandes (1939), Portelje (1939), and Prey (1950). The orang-utan has been studied in the wild (e.g., Galdikas, 1985; Harrisson, 1963; MacKinnon, 1974; Rijksen, 1978; Rodman, 1979, and this volume) and in the laboratory (e.g., Becker, 1984; Lethmate, 1976; Linke, 1973; Rensch and Dücker, 1966). The original impulse for this research came from efforts beginning around 1960 in Indonesia and Malaysia to preserve the orang-utan, which had become virtually decimated and on the verge of extinction. There has also been a resurgence of research on morphological aspects of the orang-utan, either for their own sake or in connection with questions about human evolution (e.g., Abbie, 1963; Röhrer-Ertl, 1982; Schwartz, 1983, 1984; Wegner, 1966). In fact, scientific interest regarding the orang-utan has reached the same level it was at some 200 years ago—prior to Selenka—although this time the reasons and motivations are somewhat different.

NOMENCLATURE

Although the history presented here has been brief and somewhat simplified, it should be clear that such a complicated history could easily affect the resolution of proper nomenclatural usage.

In the period prior to 1758, news about the Pongo came to Europe from southeast Asia (Atjeh) and Africa (Loango). Regardless of their original places of origin, all Pongo transported came via the Portuguese trade lines, as is indicated by the place name Angola associated with specimens. This assignation appears to have been part of a general policy at the time of deliberately providing false information. Because the trade routes from Africa to Europe were relatively short, most of the Pongo delivered to European collections and anatomy laboratories were chimpanzees. During this period only one orang-utan, an infant represented by pelt and skeleton in the Sloane Museum (now the British Museum), can definitely be certified as having been imported; this individual is now lost. It was this orang-utan that Edwards (1758) studied and about which he corresponded with Linnaeus. And it was this individual that Linnaeus (1758) combined with the chimpanzee as variants (= subspecies?) of a common genus and species—*Simia satyrus*—in Africa as well as in Asia. The name *satyrus* (wild man of the woods) was chosen to reflect Linnaeus's belief that the species represented a connecting link between Man and the Animal Kingdom. Linnaeus used *Simia satyrus indicus* to refer to the typical representative of this taxon, the chimpanzee.

In 1759 Edwards sent the infant orang-utan's skull to Linnaeus. This skull was then included in the thesis of Linnaeus's student Hoppius. The thesis (defended in 1760 and published in 1763) was entitled "Anthropomorpha," and in it reference was made to a new taxon, *Simia pygmaeus.* Although Linnaeus rejected Hoppius's species in his classifications of 1766 and 1767, he nonetheless did incorporate Edwards's and Hoppius's recognition of the orang-utan as a distinct entity and, because of this, separated *Simia satyrus* into two forms (see above): the first was a red form, which he did not describe or designate (and which, by implication, is alpha); the second, the black form (beta), he designated as *Simia satyrus indicus* and described. Gmelin (1788), following Linnaeus's lead, separated the orang-utan and the chimpanzee at the level of the species as *Simia satyrus* and *Simia troglodytes,* respectively. Gmelin thus replaced Linnaeus's ternary nomenclatural reference to these hominoids with a binomial taxonomic designation.

Linnaeus's reference was as clear as could be expected under the circumstances. Prior to 1840, book printing was financed through sales, and the most expensive part of a book was the paper. Thus Linnaeus as well as many other authors writing before 1840 was forced to abbreviate, at times to the extreme. With the advent after 1840 of cheaper paper, due to the industrial revolution, these cost-saving measures were no longer a penalizing priority. Those authors who published before 1840 and were supported by private or other aid were not constrained by the price of paper: e.g., since von Wurmb (1784) received private support and Buffon's work was generously financed by the state, there was little concern for saving space. It is therefore obvious why Buffon, especially, had such influence on the research of his day.

Next to Linnaeus's classification, Buffon's was the most important of the period, influencing nomenclatural usage until 1900, especially as it applied to the apes. Buffon was influential not only because his books were inexpensive, but also because he interspersed his scientific reports with excerpts from original travelers' stories, which made for interesting reading even for the layperson.

Buffon's most important criterion for distinguishing species (and even subspecies) from one another was body size. Thus smaller individuals were ascribed to Linnaeus's *Simia satyrus* (= the orang-utan) and *Simia satyrus indicus* (= *Simia troglodytes*) (= the chimpanzee). However, when describing a large, mature male orang-utan from Sukadana (southwest Borneo), von Wurmb (1784) invoked Buffon's terminology ("Borneoosche Pongo" and "Pongo van Borneo") and referred the specimen to the genus and species *Pongo borneensis.* Perhaps one reason that Buffon's insistence on delineating species on the basis of size remained popular procedure was that mature orang-utans do not look as "human" as subadult individuals.

Throughout the 19th century an enormous number of taxonomic names were generated to receive various specimens of orang-utan (see reviews by Röhrer-Ertl, 1983a; Simonetta, 1957). By the turn of the century, however, *Simia satyrus* emerged as the most frequent taxon of choice (cf. Selenka, 1896, 1898), and inquiry had shifted to the nature and identification of subspecies. But the taxonomic identity of the orang-utan changed in 1929 when the International Commission on Zoological Nomenclature suppressed the genus name *Simia* in favor of *Pongo.* In this action, published as Opinion 114, reference was to the nomenclatural usage of Stiles and Orleman (1927), who had relied on the earlier publication of Rothschild (1904), who had, in turn, excerpted and translated freely, without quotation, from Matschie (1904). And it was Matschie who used the binomen *Pongo pygmaeus* (*Pongo* Lacepede, 1799, and *pygmaeus* Linnaeus, 1763) to refer to the orang-utan. Matschie reserved the species name *satyrus* for the chimpanzee, although, to add to the confusion, he made reference in the same work to the chimpanzee as *pygmaeus* as well (see Röhrer-Ertl, 1983a; Simonetta, 1957). Needless to say, and even with the Opinion of 1929, taxonomic confusion surrounding the chimpanzee and orang-utan abounded, and as late as 1945 most attempts at settling upon definite names were doomed to failure.

Both Matschie and Rothschild felt that, since Linnaeus had offered two revisions in his later works, one could choose a solution to the taxonomic problem that best fit one's own opinion—and they did not accept the proposals of the *Systema* of 1758–1766/67, anyway. (Linnaeus had, however, proposed a further change of the name *pygmaeus* in the *Amoenitates Academicae*

[Hoppius, 1763].) Matschie and Rothschild chose to base their nomenclature in large part, but not in all details, on the work of Sherborn (1902, referred to in Palmer, 1904). But Sherborn apparently knew only about the second edition of *Amoenitates Academicae* (edited by Schreber in 1789), which might be the source of his citation of the wrong date of publication for the first edition: 1760 instead of 1763, the former being the year in which Hoppius defended his dissertation and received his degree. This confusion could have arisen as a result of Sherborn's not knowing when an "author" rather than an "editor" was being referenced, because British librarians (even as late as 1945; Röhrer-Ertl, 1983a) would list collected works both by author and editor (cf. Wheatley, 1884). Other errors of citation by Sherborn have been detailed elsewhere (Röhrer-Ertl, 1983a).

Given this brief review, it should not be surprising that the orang-utan has been referred to in various ways in the literature. The following are the most common names applied to this animal, listed in order of descending frequency of use, for (1) species: orang-utan, *Pongo, Pongo pygmaeus, Pongo pygmaeus* (Hoppius, 1763), *Pongo pygmaeus* (Linnaeus, 1760), *Pongo pygmaeus* (Linnaeus), *Simia satyrus;* and (2) genus: *Pongo* Lacépède, 1799, and *Pongo* (Linnaeus, 1760) (see Röhrer-Ertl, 1983a). However, as far as I have been able to determine, and in accordance with the International Rules of Zoological Nomenclature, the correct assignation of the orang-utan is

Genus *Pongo* von Wurmb, 1784
species *satyrus* (Linnaeus, 1758).

The objective synonym of the species is *pygmaeus* (Hoppius, 1763).

SUBSPECIFIC TAXONOMY

Confusion surrounding the correct nomenclature of the orang-utan has been due not only to historical factors but also to difficulties in resolving the taxonomy of this hominoid at the subspecific level. This is because virtually all phenotypic characteristics of the orang-utan have an extremely broad range of variation, similar to that seen in modern humans (cf. Röhrer-Ertl, Cranial Growth, this volume). I (1982, 1983a,b) have tried to present evidence for the existence of two extant subspecies. Most of the relevant fossils and subfossils can be accommodated by one or the other of these subspecies, but others require the recognition of a third subspecies.

Although the Bornean variety of orang-utan has been used in the more recent literature as the nominative form for the taxon, I think this has been due to a misunderstanding of 18th- and 19th-century sources. For example, Abel (1825) mentioned that the first mature individual he had examined was from Sumatra (Atjeh), and von Wurmb (1784) denied the existence of such a variety outside of Borneo. In order to clarify the situation, I (1984a) attempted to locate the place of origin of Edwards's (1758) orang-utan. In all probability, this individual came from Sumatra (Atjeh), for which the nominative form should be reserved. We may thus subdivide *Pongo* von Wurmb, 1784, into the following subspecies:

Pongo satyrus satyrus (Linnaeus, 1758)
Habitat: Sumatra, recent; Sumatra and Java, fossil-subfossil (Pleistocene-Holocene); Figs. 1–1 and 1–3.
Pongo satyrus borneensis von Wurmb, 1784
Habitat: Borneo, recent and subfossil (Holocene); Figs. 1–2 and 1–3.
Pongo satyrus weidenreichii (Hooijer, 1948)
Habitat: southern China, northern Laos, northern Vietnam, fossil-subfossil (Pleistocene-Holocene); Fig. 1–3.

Description of Recent Subspecies

Pongo satyrus satyrus (Linnaeus, 1758), Sumatra

Figure 1–1 illustrates the skull of a young adult male (ZSM 1981/248) from Atjeh which served as the basis of the description (cf. Röhrer-Ertl, 1984a). The profile is straight and obliquely oriented, the anterior teeth protrude (especially in the mandible), a suborbital fossa is lacking, and the malar-maxillary region (the "cheek") is convex and its zygomaticomaxillary suture courses diagonally, being symmetrical on the right and left sides. The piriform aperture is not completely rounded (= intermediate) because the nasal bones do not extend sufficiently downward to "close up" the maxillary aspect of the nasal region. This configuration is noted in the majority of individuals studied, but some do have a more ovoid nasal aperture, of which the superior border is slightly narrower. The living animal is characterized by a long, oval face with convex cheeks and a distally and basally oriented snout (cf. MacKinnon, 1974; Röhrer-Ertl, 1984a).

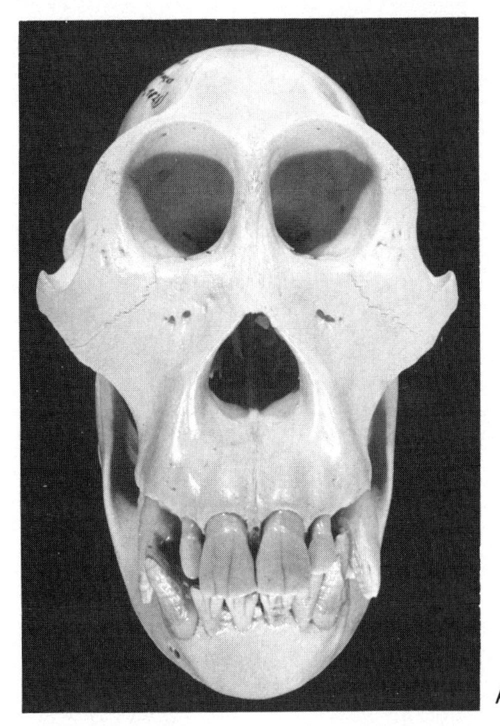

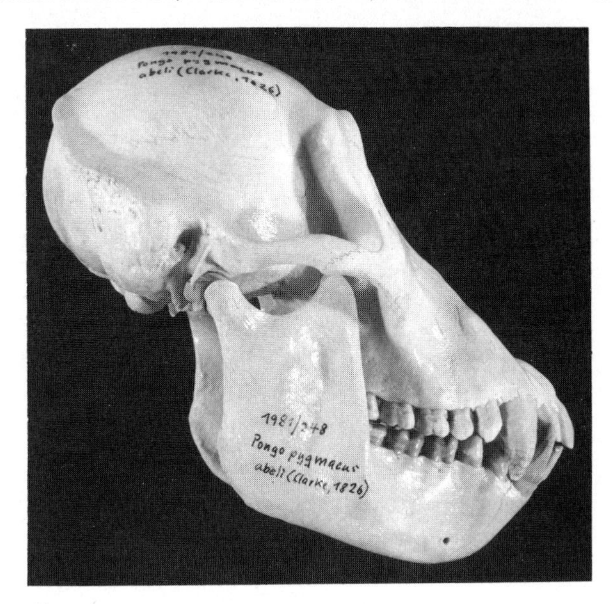

B

A

FIG. 1–1 Articulated skull and mandible of a young adult male orang-utan, *Pongo satyrus satyrus,* from Sumatra (IX) (Atjeh); (ZSM 1981/248). (A) Norma frontalis (OAE). (B) Norma lateralis dextra (OAE) (not to scale). OAE = Frankfort Horizontal. Further explanations are given in the text.

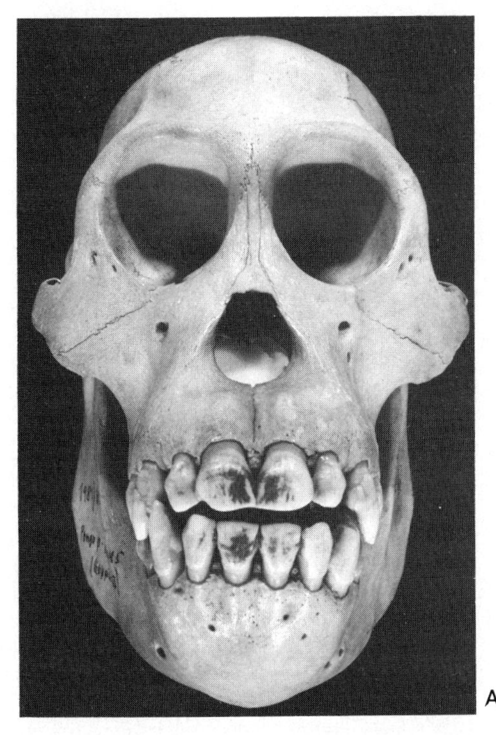

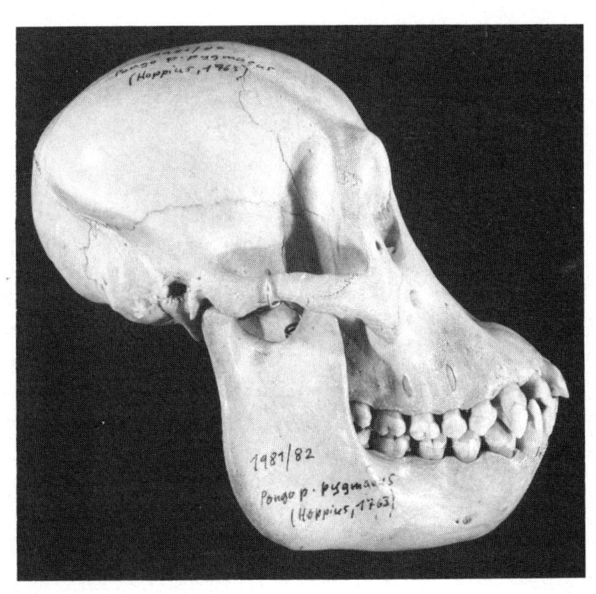

B

A

FIG. 1–2. Articulated skull and mandible of young adult female orang-utan; *Pongo satyrus borneensis,* from Borneo (III) (Skalau); (ZSM 1981/082). (A) Norma frontalis (OAE). (B) Norma lateralis dextra (OAE) (not to scale). Further explanations are given in the text.

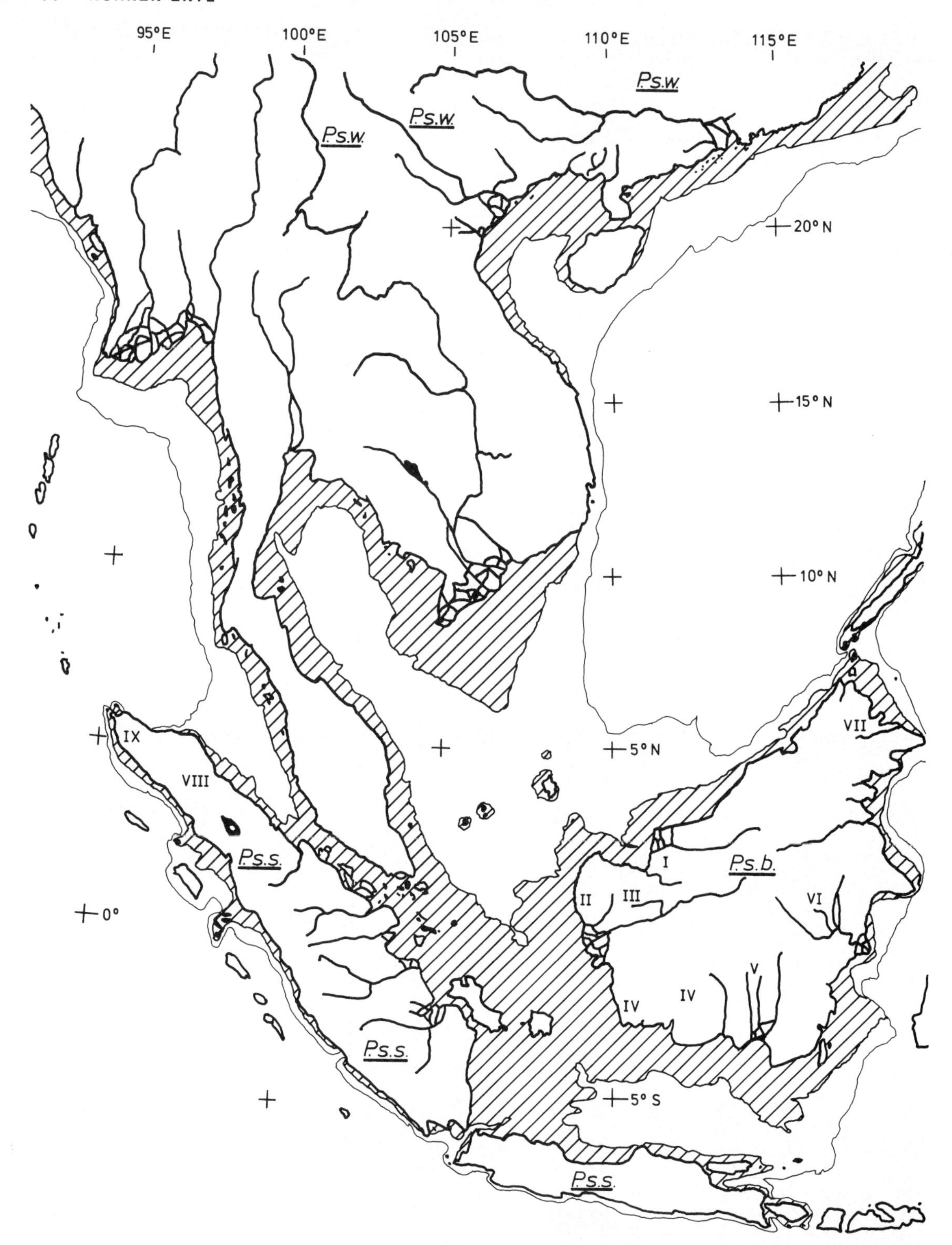

FIG. 1–3. Distribution map of *Pongo satyrus* in Indochina and the islands, showing important rivers, bathymetric depths of the Sunda Shelf in meters (m), and the known distribution of the orang-utan in recent and geological periods. *P.s.s.* = *Pongo satyrus satyrus; P.s.b.* = *Pongo satyrus borneensis; P.s.w.* = *Pongo satyrus weidenreichii. Hatched areas:* Sunda Shelf from ± 0 to 50 m depth. *Thin line:* Sunda Shelf from 50 to 200 m depth. Pleistocene underwater river valleys run from WNW and NW to SE and E. From sea level to 50 m in depth, there is recent filling and erosion, i.e., the Sunda Strait. Known Northern populations: (I) Batang Bara (West Sarawak, north of the Klingkang Mountains to north of the

Pongo satyrus borneensis von Wurmb, 1784, Borneo

Figure 1–2 illustrates the skull of a young adult female (ZSM 1981/082) from Skalau, Borneo. Although the "cheek" region is concave as in the Sumatran subspecies, the facial profile of this variety is also concave, not straight. In addition, the suborbital fossae are distinct and the anterior teeth, especially those of the lower jaw, are labially positioned. The piriform aperture is clearly "trumpet-shaped" (= triangular in cross section through the region of the nasal tubercle); this configuration is so far only known for Bornean orang-utans. The aperture arcs around the maxillary and premaxillary regions. The living Bornean orang-utan is characterized by a somewhat pentagonal face with concave cheeks and a clearly defined, distally and horizontally inclined snout.

In general, *P. s. borneensis* appears to be more specialized than *P. s. satyrus.* Among the features in which the Bornean subspecies became different from its Sumatran relative are increase in body size and mass, lengthening of the arms, enlargement of the adult's snout, and marked distension of the laryngeal sac. The amount of variation one sees in dental morphology among various populations of Bornean orang-utan (e.g., from Skalau) prevents its use at the subspecific level.

CONCLUSION

From the very beginning, research on the orang-utan has been interconnected with research on other apes as well as humans. From the earliest reports—Vesalius (1542) to Linnaeus (1758ff), Hunter (1775), Kant (1775ff), Soemmerring (1785), Camper (1782), and others—the anthropomorphous appearance of infant and juvenile orang-utans has received much attention. This is what led to the notion of the existence of satyrs, wild men of the woods—or orang-utans. In the 18th and 19th centuries Buffon and his followers apparently strove to retain the favorable comparison between humans and orang-utans by relegating the larger, mature (less human-looking) individuals to different species. Interest in the orang-utan and its possible connections with humans began to wane as a result of Selenka's (1899) influence on the field.

But the orang-utan has once again emerged as important in considerations of hominoid evolution: e.g., Schwartz (1983, 1984) has argued that there exist a variety of similarities shared exclusively by humans and orang-utans that attest to their being closely related, and Pilbeam (1984) has presented the case that the Miocene hominoid *Sivapithecus* is not an early hominid but is, instead, an ancestor of *Pongo.* Although I (1984a) might differ with the conclusions of these authors—that the morphological similarities of *Sivapithecus* and especially the Bornean orang-utan are more properly considered to be convergences—I do think we must try to understand the orang-utan better in order to resolve present issues in hominoid systematics. With regard to the more recent evolutionary history of the orang-utan, there is reason to believe (Röhrer-Ertl, 1984a,b) that the ancestor of modern orang-utans migrated from a region north of Indo-Pakistan following the late Neogene emergence of the mountains to the southeast, which extended to the islands (Fig. 1–3). Thus, rather than seeking the ancestry of the orang-utan in an open, flat countryside, one should expect such origins to have occurred in an elevated tropical rain forest environment.

Batang-Lupar River, a line which originally extended north through Brunei). Malaya: Sarawak. Contemporary: National Park Bako. (II) Landak (western West Borneo, north of the Kapuas River, south of the Klingkang Mountains). Indonesia: Kalimantan Barat. (III) Skalau (eastern West Borneo, north of the Kapuas River, south of the Klingkang Mountains). Indonesia: Kalimantan Barat. The landscape was relatively densely populated; by the 19th century, orang-utans had disappeared from south of the Kapuas River. Known Southern populations: (IV) Sampit; Sandokan (Southwest Borneo). Indonesia: Kalimantan Barat. (Migration during the Pleistocene probably took place through Sampit [Southern Borneo, south of the Schwaner Mountains] and Indonesia [Kalimantan Tengah].) Contemporary: Tanjung Puting Reserve. (V) Barito (eastern South Borneo, south of the Schwaner and Muller Mountains). Indonesia: Kalimantan Tengah and Salatan. (VI) Kuteilama (East Borneo, southeast of the Pegununga Mountains). Indonesia: Kalimantan Timur. Contemporary: Kutei Reserve. (Only individuals from the southern area could be authenticated; an additional population can be postulated for the area south of the border to North Borneo.) (VII) Kinabatangan (North Borneo). Malaya: Sabah. Contemporary: Ulu Segama Reserve. Sumatran populations: (VIII) Deli (Northeast Sumatra). Indonesia: Sumatra utara. Contemporary: Gunung-Leuser Reserve. (IX) Atjeh (Aceh) (Northwest Sumatra). Indonesia: Atjeh. (see Röhrer-Ertl, 1984a.)

REFERENCES

Abbie, A. A. 1963. The cranial centre. *Zietschrift für Morphologie und Anthropologie, 53:*6–11.

Abel, C. 1825. Some accounts of an ourang-outang of remarkable height found on the island of Sumatra, together with a description of certain remains of this animal, presented to the Asiatic Society by Capt. Cornfoot, and at present contained in its museum. *Asiatic Researches, 15:*489–498.

Alessandrini, A. 1854. Brevi note illustrative di uno scheletro di giovine orang-outan *Pithecus satyrus* Geoffr.—*Simia satyrus* Linn. *Nuovi Annali delle Scienze Naturali, 9, 3 ser.:*353–363.

Aulmann, G. 1932. Geglückte Nachzucht eines Orang-Utan im Düsseldorfer Zoo. *Zoologische Garten, 5:*83–90.

Becker, C. 1984. Orang-Utans and Bonobos im Speil. Untersuchungen zum Spielverhalten von Menschenaffen. Profil, Munich.

Beeckman, D. 1714. A voyage to and from the island of Borneo, in the East Indies. Reprinted 1973 by Dawson, Folkestone.

Bontius, J. 1658. Historiae naturalis et medicae Indiae Orientalis. *In* De India Utriusque re Naturali et Medica, ed. W. Piso, libri 14. Elsevirius, Amstelodamus.

Brandes, G. 1939. Buschi. Vom Orang-Säugling zum Backenwülster. Quelle & Meyer, Leipzig.

Buffon, G. L. L. Compte de. 1750–1779. Allgemeine Historie der Natur. Molle & Heinsius, Hamburg/Leipzig (in German; original published in 1745).

Camper, P. 1782. Natuurkundige verhandelingen over den Orang-Outang; en eenige andere Aapsoorten. Over den Rhinoceros met den dubbelen horen; en over het Rendier. Meijer & Warnars, Amesterdam.

Chasen, F. N. 1939–1940. The mammals of the Netherlands Indian Mt. Leuser Expedition 1937 to North Sumatra. *Treubia, 17:*479–502.

Cuvier, G. 1829. Le règne animal distribué d'après son organisation (2d ed.). D'Éterville & Crochard, Paris.

Dumortier, B. C. 1838. Notes sur les métamorphoses du crâne de l'orang-outang. *Bulletin de l'Academie Royale des Sciences et Belles-Lettres, 5:*756–766.

Edwards, G. 1758. Gleanings of natural history exhibiting figures of quadrupeds, birds, insects, plants etc., vol. 1. Royal College of Physicians, London.

Fick, R. 1895. Vergleichend anatomische Studien an einem erwachsenen Orang-Utan. *Archiv für Anatomie und Physiologie, Anatomische Abteilung, 1895:*1–100.

Galdikas, B. 1985. Subadult male orangutan sociology and reproductive behavior at Tanjung Puting. *American Journal of Primatology, 8:*87–99.

Gmelin, J. F. (ed.). 1788. C. Linnaeus. Systema Naturae per Regna Tria Naturae Secundum Classes, Ordines, Genera, Synonymis, Locis. Tomus I, Editio Decima Tertia, Aucta, Reformata Cura Jo. Fridr. Gmelin. Beer, Lipsia.

Gray, J. E. 1870. Catalogue of monkeys, lemurs and fruit-eating bats in the collection of the British Museum. British Museum (Natural History), London.

Harrisson, B. 1963. The education of young orangutans to living in the wild. *Oryx, 7:*108–127.

Hoppius, C. E. 1763. Anthropomorpha. Que Praeside D. D. Car. Linnaeo Proposuit Christianus Emmanuel Hoppius, Petropolitanus. Upsaliae 1760, Sept. 6. *In* Amoenitates Academicae; seu Dissertationes Variae, antehac Seorsim Editae, nunc Collectae et Auctae cum Tabulis Aeneis, Volumen Sextum, CV, ed. Linnaeus, C. Salvius, Holmiae: 63–76.

Hrdlička, A. 1907. Anatomical observations on a collection of orang skulls from western Borneo, with a bibliography. *Proceedings of the United States National Museum, 31:*539–568.

Huber, E. 1931. Evolution of Facial Musculature and Facial Expression. Johns Hopkins University Press, Baltimore.

Hunter, J. 1775. Disputatio Inauguralis Quaedam de Hominibus Varietatibus, et Harum Causis, Exponens etc. Thesaurus Medicus Universitatis, Edinburgh.

Kant, I. 1775. Von den verschiedenen Rassen der Menschen. IV: 11–30. Reproduced in 1964 *In* Werke in 6 Bänden, ed. W. Weischedel. Insel, Frankfurt.

Kant, I. 1785. Bestimmung des Begriffs einer Menschenrasse. VI: 65–82. Reproduced in 1964 *In* Werke in 6 Bänden, op. cit.

Kant, I. 1786. Mutmasslicher Anfang der Menschengeschichte. V: 85–102. Reproduced in 1964 *In* Werke in 6 Bänden, op. cit.

Kant, I. 1798. Anthropologie in pragmatischer Hinsicht. VI: 393–690. Reproduced in 1964 *In* Werke in 6 Bänden, op. cit.

Kleinschmidt, O. 1933. Über Stirnbeinhöhlen und Siebbeinzellen beim Orang. *Zeitschrift für Säugetierkunde, 8:*70–72.

Lethmate, J. 1976. Versuche zur Doppelstockhandlung mit einem jungen Orang-Utan. *Zoologischer Anzeiger, 197:*264–271.

Linke, K. 1973. Einige Beobachtungen zur Händigkeit von Anthromorphen beim Komfortverhalten. *Zoologische Garten, 43:*199–203.

Linnaeus, C. 1758. Systema Naturae. Regnum Animale etc., Editio Decima. Salvius, Holmiae (reproduced in 1894, Engelmann, Lipsia).

Linnaeus, C. 1766. Systema Naturae per etc, Tomus I, Editio Duodecima, Reformata. Salvius, Holmiae.

Linnaeus, C. 1767. Systema Naturae, per Regna etc., Tomus I, Editio Decima Tertia ed Editionem Duodecimam Reformatam Holmiensem. Thoma de Trattnern, Vindobona.

MacKinnon, J. 1974. The behaviour and ecology of wild orang-utans *(Pongo pygmaeus). Animal Behaviour, 22:*3–74.

Matschie, P. 1904. Einige Bermerkungen über die Schimpanzen. *Sitzungsberichte der Gesellschaft naturforschender Freunde zu Berlin, 1904:*55–69.

Mayer, A. F. J. C. 1849. Bemerkungen über den Bau des Orang-Outang-Schädels. *(Wiegmanns) Archiv für Naturgeschichte, 15:*352–357.

Müller, S. 1845. Ueber die auf den Sunda-Inseln lebenden ungeschwänzten Affen-Arten. *(Wiegmanns) Archiv für Naturgeschichte, 11:*72–90.

Opinion 114. 1929. Opinions rendered by the International Commission on Zoological Nomenclature. *Smithsonian Miscellaneous Collection, 73:*25–26.

Owen, R. 1830–1831. On the anatomy of the orang-utan (*Simia satyrus,* L.). *Proceedings of the Committee of Science and Correspondence of the Zoological Society of London,1:*4–5, 9–10, 28–29, 67–72.

Owen, R. 1839. Note sur les différences entre le Simia Morio (Owen), et le Simia Wurmbii dans la période d'adolescence, décrit par M. Dumortier, par M. Owen. *Annals des Sciences Naturelles, 11:*122–125.

Palmer, T. S. 1904. Index Generum Mammalium. Washington, D.C. (reprinted in 1968 by Cramer, Lehre/ New York).

Pilbeam, D. 1984. The descent of hominoids and hominids. *Scientific American, 250:*84–96.

Portelje, A. F. J. 1939. Triebleben bzw. intelligente Äusserungen beim Orang-Utan, *Pongo pygmaeus* Hoppius. Beobachtungen und Versuche an einem alteingefangenen und sofort verschifften "Riesen-Orang," nebst Bemerkungen über andere Orang-Utans im Garten der Königlichen Zoologischen Gessellschaft "Natura Artis Magistra," Amersterdam. *Bijdragen tot de Dierkunde, 18:*61–114.

Prey, U. 1950. Beobachtungen an den Orangs des Zirkus Krone (München). *Zoologische Garten, 17:*52–60.

Purchas, S. 1613. Hakluytes Posthumus, or Purchas His Pilgrimes, Contayning a History by Englishmen and Others. London (reproduced 1905–1907 by Machehouse & Sons, Glasgow).

Rensch, B. and Dücker, G. 1966. Manipulierfähigkeit eines jungen Orang-Utans und eines jungen Gorillas. Mit Anmerkungen über das Spielverhalten. *Zeitschrift für Tierpsychologie, 23:*874–892.

Rijksen, H. D. 1978. A Field Study on Sumatran Orang Utans (*Pongo pygmaeus abelii,* Lesson 1827): Ecology, Behaviour and Conservation. H. Veenman & Zonen, Wageningen.

Rodman, P. S. 1979. Individual activity patterns and the solitary nature of orangutans. *In* The Great Apes, pp. 234–255, ed. D. A. Hamburg and E. R. McCown. Benjamin/Cummings, Menlo Park.

Röhrer-Ertl, O. 1982. Über Subspecies bei *Pongo pygmaeus* Linnaeus, 1760. *Spixiana, 5:*317–321.

Röhrer-Ertl, O. 1983a. Zur Erforschungsgeschichte und Namengebung beim Orang-Utan, *Pongo satyrus* (Linnaeus, 1758); Synon. *Pongo pygmaeus* (Hoppius, 1763) (mit Kurzbibliographie). *Spixiana, 6:*301–332.

Röhrer-Ertl, O. 1983b. Hat Krieg eine biologische Wurzel? *Unitarische Blätter, 34:*293–296.

Röhrer-Ertl, O. 1984a. Orang-Utan Studien. Hieronymus, Neuried.

Röhrer-Ertl, O. 1984b. Zur Wanderungsaktivität beim Orang-utan, *Pongo satyrus* (Linnaeus, 1758) aufgrund durch Abschuβlisten gegebener Gruppierungen. *Verhandlungen der Deutschen Zoologischen Gesellschaft, 77:*208.

Röhrer-Ertl, O. and Frey, K. W. 1987. On secondary hyperparathyroidism in orang-utans, *Pongo satyrus* (L. 1758.). *Gegenbaurs Morphologisches Jahrbuch, Leipzig, 133:*361-383.

Rothschild, W. 1904. Notes on anthropoid apes. *Proceedings of the Zoological Society of London, 1904:*413–440.

Rudolphi, K. A. 1826. Ueber den Orang-Utan, und Beweis, dass derselbe ein junger *Pongo* sei. *Abhandlungen der physikalischen Klasse der Königlichen Akademie der Wissenschaften zu Berlin aus dem Jahre, 1826:*131–135.

Scheidegger, S. and Wendnagel, W. 1944–1950. Eine besondere Erkrankung des Skelettsystems, Osteodystrophia deformans beim Orang. *Zoologische Garten, 16:*66–74.

Schlegel, H. and Müller, S. 1839–1844. Bijdragen tot de Natuurlijke Historie von de Orang-oetan (*Simia satyrus*). Verhandelingen over de Natuurlijke geschiedenis van de Nederlandsche Overzeesche Bezittingen. *In* Folia A, Zoologie, pp. 1–28, ed. J. C. Temminck. Van den Hoek, Leiden.

Schultz, A. H. 1916. Der canalis cranio-pharyngeus persistens beim Menschen und bei den Affen. *Gegenbaurs Morphologisches Jahrbuch, 50:*417–426.

Schwartz, J. H. 1983. Palatine fenestrae, the orangutan and hominoid evolution. *Primates, 24:*231–240.

Schwartz., J. H. 1984. The evolutionary relationships of man and orang-utans. *Nature, 308:*501–505.

Schwartze, H. W. R. 1839. Descriptio Osteologia Capitis Simiae Parum adhuc Notae. Dissertatio Inauguralis quam Consensu Gratiosi Mediorum Heidelbergensium Ordinis. Scripsit H. W. R. Schwartze, Med. Chir. et Obst. Doctor, Hamburgensis, cum II Tab. lith, Nietackianus, Berolini.

Selenka, E. 1896. Die Rassen und der Zahnwechsel des Orang-Utan. *Sitzungsberichte der königl. preuss. Akademie der Wissenschaften, Phys.-Math. Classe (of 19.03.1896):*381–392.

Selenka, E. 1898. Rassen, Schädel und Bezahnung des Orangutan. *In* Menschenaffen (Anthropomorphae). Studien über Entwicklungsgeschichte der Tiere, 6 (1), pp. 1–99, ed. E. Selenka. Kreidel, Wiesbaden.

Selenka, E. 1899. Schädel des Gorilla und Schimpanse. *In* Menschenaffen (Anthropomorphae). Studien über Entwicklungsgeschichte der Tiere, 6 (2), pp. 95–160, ed. E. Selenka. Kreidel, Wiesbaden.

Sherborn, C. D. 1902. Index Animalium. Sive Index Nominum quae ab A. D. MDCCLVIII Generibus et Specibus Animalium Imposita Sunt. Sectio Prima a Kalendis Januariis MDCCLVIII usque ad Finem Decembris MDCCC. Academia, Cantabrigae.

Simonetta, A. 1957. Catalogo e sinonima annotata degli ominoidi fossili ed attuali (1758–1955). *Atti della Societa Toscana di Scienze Naturali Residente in Pisa, 64, ser. B:*53–112.

Soemmerring, S. T. 1785. Über die körperliche Verschiedenheit des Negers vom Europäer. Varrentrap, Sohn & Wenner, Frankfurt/Mainz.

Stiles, C. W. and Orleman, M. B. 1927. The nomenclature for man, the chimpanzee, the orang-utan, and the Barbary ape. *Hygienic Laboratory Bulletin, 145:*1–66.

Tiedemann, F. 1821. Icones Cerebri Simiarum. Morh & Winter, Heidelberg.

Tilesius, W. G. T. 1813. Naturhistorische Früchte der ersten kaiserlich-russischen, unter dem Kommando des Herrn v. Krusenstern glücklich vollbrachten Erdumseglung. Kaiserlich-Russischen Academie der Wissenschaften, St. Petersburg.

Tulpius, N. 1641. Observationes Medicae. Amstelodamus 1641 (reprinted in 1672 by Elsevirius, Amstelodamus).

Tyson, E. 1699. Orang-Outan, sive *Homo sylvestris* or, the Anatomie of a Pygmie Compared with that of a Monkey, an Ape, and a Man, etc. Bennent, Brown & Hun, London. Vesalius, A. 1542. De human; corpsoe fabrica. Padua. (Reprinted 1617, Ianssonius, Ammstelodamus.)

Vesalius, A. 1542. De humani corpore fabrica. Padua. (Reprinted 1617, Ianssonius, Amstelodamus.)

Vosmaer, A. 1778. Beschrijving de Zoo Zeltzaame als Zonderlinghe Aap-soort, Genaamd Orang-outang, van het Eiland Borneo. Meijer, Amsterdam.

Wegner, R. N. 1966. Variationen am Schädel von Gorilla und Orang. *Monatsbericht der Akademie der Wissenschaft zu Berlin, 8:*460–467.

Wheatley, B. R. 1884. On the question of authorship in "Academical dissertations." *Library Association of the United Kingdom, Transactions and Proceedings of the Fourth and Fifth Annual Meetings, 1884:*37–42.

Wurmb, F. Baron von. 1784. Beschryving van de groote Borneoosche Orang Outan of de Oost Indische *Pongo. Verhandelingen van het Bataviaasche Genootschap van Kunsten en Wetenschappen, 1784:*245–261.

Zimmerman, E. A. W. 1777. Specimen Zoologiae Geographicae, Quadrupedum. Domicilia et Migrationes Sistens Dedit, Tabulamque Mundi Zoographicum Adjunxit. Haat & Soc., Leiden.

2

Inter- or Intra-Island Variation? An Assessment of the Differences Between Bornean and Sumatran Orang-utans

JACKIE COURTENAY, COLIN GROVES, AND PETER ANDREWS

As is true of many species with geographically isolated populations, the taxonomy of the orang-utan has been the subject of much debate, especially in recent years. Such populations are problematic because the most important criterion for recognizing species, that of reproductive isolation (due to an intrinsic isolating mechanism, and not to mere geography), cannot be applied. Thus, degree of difference becomes the main consideration in determining the level of taxonomic relatedness (Mayr, 1969), as it is all the time for paleospecies.

The present range of the orang-utan is divided between the islands of Borneo and Sumatra (Figs. 2–1 to 2–3). The two populations are completely isolated geographically and hence reproductively. Although hybrids have been produced in captivity, their fertility relative to the pure types is uncertain. Preliminary data suggest that hybrid fertility is not especially reduced, at least compared to that of purebred Borneans (Markham, 1985). The taxonomy of the orang-utan therefore rests on the somewhat subjective concept of "degree of difference." Two of us (Andrews, 1984; Groves, 1986) have recently, and independently, put forward the suggestion that, because the degree of difference appears to be considerable, the two putative subspecies of orang-utan might better be considered to be separate species.

In this chapter we will discuss the numerous characters that have been listed as differences between Bornean and Sumatran orang-utans. We will examine first whether the differences are indeed as described; second, whether there are differences within either Borneo or Sumatra which affect the integrity of differences between the two populations; and third, what the differences might mean and how they can be interpreted. The characters are considered under four general headings: general appearance, behavior, cranial characters, and chromosomal and biochemical differences.

Throughout this chapter the two geographic isolates will be referred to as subspecies. This is partly for convenience and to follow traditional usage, but also because, to anticipate, this does seem to be, after all, most appropriate in the current state of our knowledge.

GENERAL APPEARANCE

A number of authors (for example, van Bemmel, 1968; Jones, 1969; MacKinnon, 1973; Mallinson, 1978) have listed characters of general external appearance which distinguish between Bornean and Sumatran orang-utans. Such characters include color, length, and structure of body hair, distribution of facial hair, growth and shape of the male's cheek flanges, body build, the size and shape of the throat pouch, and the presence or absence of a nail on the great toe.

Although some of these characters may indeed be useful in distinguishing the island of origin of various individuals, closer examination of the characters with reference to intra-island variability suggests that their taxonomic value may be limited.

Hair Color

Most authors describe Bornean orang-utans as ranging in color from red to deep maroon or blackish brown, while Sumatran orang-utans are lighter-colored—rusty red or light cinnamon. Rijksen (1978), however, notes considerable variability within a restricted region of Sumatra: he describes two extreme forms, a dark-haired type which is apparently similar in color to the Bornean, and a light-haired type which is the "typical" Sumatran reddish cinnamon to rusty red. The two extremes are linked by a range of intermediates, and the fact that all forms occur within

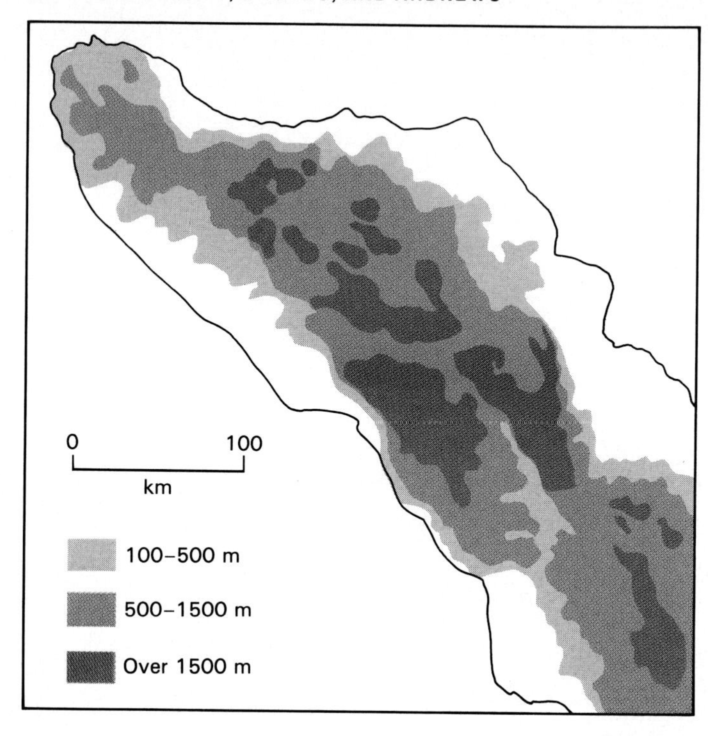

FIG. 2–1. Relief map of the Aceh (Atjeh) region of Sumatra.

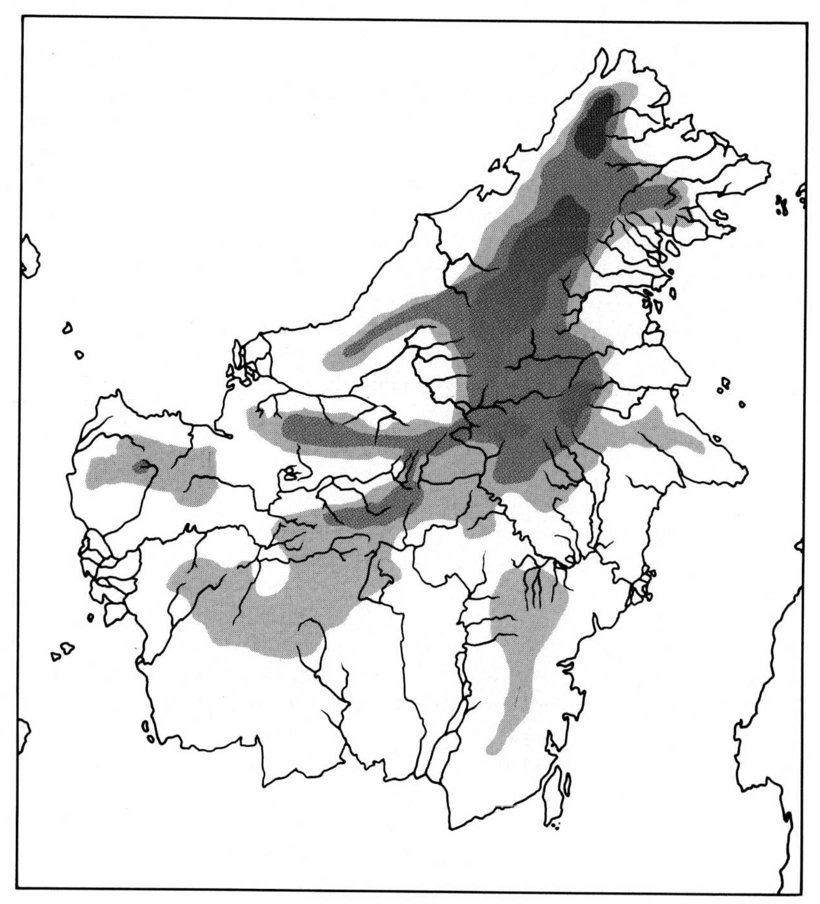

FIG. 2–2. Relief map of Borneo. Attitudes a.s.l. as in Fig. 2–1.

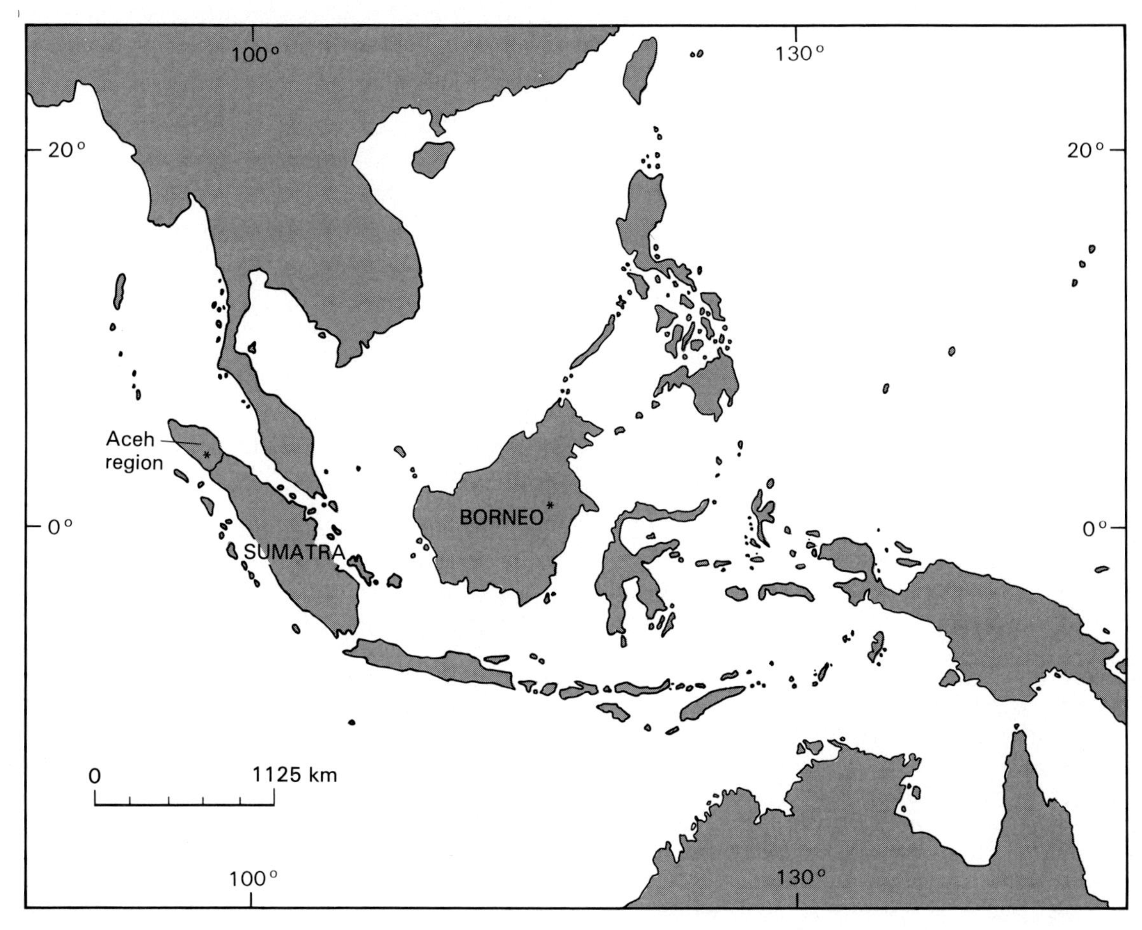

FIG. 2–3. Map of approximate distribution (*) of orang-utan.

the same population suggests that the difference between Sumatran and Bornean orangs might be more a matter of average color. Jones (1969) recognizes the considerable variation to be found in the Sumatran subspecies but claims that most Sumatran animals are indeed lighter than Bornean individuals. This is borne out by Rijksen (1978), who noted that the majority (60–75%) of Sumatrans were of the light type. Although in general terms it seems fair to use coloration as a distinguishing characteristic (particularly when combined with other features), the variability to be found at least in the Sumatran population dictates caution in its use as a taxonomic marker.

Hair Form and Structure

Hair is described as being longer, denser, and fleecier in Sumatran than in Bornean orang-utans. In cross section, the hairs of Bornean orang-utans are said to be flattish with a thick black pigment column; those of Sumatrans are thinner and rounder with a fine, often broken dark central pigment column (MacKinnon, 1973). Such a difference would be expected to be related to the overall difference in coloration between the two subspecies and, as such, variable; no figures or sample sizes were given by MacKinnon for the character.

The difference in hair density and texture is usually, in our experience, a good taxonomic differentiator, but may also be environmentally related. As will be seen from Fig. 2–3, the range of the Sumatran population is almost entirely restricted to Aceh (Atjeh), most of which is over 1,000 m in altitude, and, indeed, the orang-utan population apparently is limited to the higher altitudes. In contrast, Borneo is in general more low-lying (Fig. 2–2), and orang-utans there are definitely lowland animals (Groves, 1971). The denser, fleecier hair of the Sumatran form may relate to its predominantly higher-altitude habitat. Characters relating to pelage length must in any case be taken cautiously by taxonomists, as there is always the possibility that such features may be environmentally modifiable.

Facial Hair

Jones (1969) reports that a beard is always present in the Sumatran male, and it is lush and more pronounced than in the Bornean male. Other facial hair in the Sumatran male is also more noticeable: hair growth frequently extends to the upper lip, giving a moustached appearance, while the cheek pads are thickly covered with light silky hair. Long beards are also found in the Sumatran females. In contrast, Bornean males have less hair about the face and neck, and their beard is "short, scruffy and less pronounced" than in the Sumatran (Mallinson, 1978:71). The cheek flanges in the Bornean male are either naked or sparsely covered with short, shiny, bristly hairs, and the skin of the flanges is clearly visible.

These differences may well be of taxonomic importance, although they could also be a simple reflection of the denser, more fleecy hair covering of the Sumatran subspecies. It is impossible to say without further research whether these differences in hair are truly genetic or a developmental response to environmental conditions.

Presence of Nail on Hallux

Jones (1969:219), in describing the differences between the two subspecies, states that "in the Bornean animal . . . nails may be missing from the great toes." Rijksen (1978), however, observes that nails are often missing in Sumatran specimens, as well; in the extreme light-haired type, they are always missing on the great toes and frequently also on the thumbs. Since the light type is the more common in Sumatra, it would seem that the implied presence of nails in the Sumatran form must be based on observation by Jones of captive individuals which just happened, by chance, to possess nails; thus, no taxonomic importance should be attached to this character.

Body Build

Good morphometric data on relative size and shape of the two subspecies appear to be lacking. The often-quoted figures of Lyon (1911) are incorrect, as pointed out by Eckhardt (1975); when corrected, the two subspecies turn out to be much the same weight. Chasen (1940), however, records his subjective impression that Borneans are smaller than Sumatrans, while several authors (for example, Jones, 1969) note that Borneans are prone to obesity in captivity, whereas Sumatrans, more muscular and linear in build, rarely become overweight.

These differences, while necessarily somewhat vaguely expressed, could be seen as reflecting Bergmann's rule, i.e., the Sumatran form, reputedly the larger of the two, occupies the cooler (high-altitude) environment (Newman, 1975). In any event, quantitative data are required. It should be noted, too, that Lyon's weights are from gutted specimens, and actual weights of living wild orang-utans are completely unknown—even weights of captive individuals are scarce. Hick (1985) gives weights for a colony of apparently healthy (not overweight?) Bornean orang-utans in the Cologne zoo: two males 18 and 21 years old each weighed around 130 kg; a male 13 years old weighed 90 kg; and two females 14 and 18 years old weighed 55 and 60 kg, respectively.

Cheek Flanges

This is one character in which a very clear difference does appear to exist between the two subspecies. Jones (1969:223) describes the development and appearance of the cheek flanges as the "most easily recognised difference in adult and semi-adult males." In Bornean males, the flanges begin to develop at about 8 years and are usually completely formed by about 15 years; they tend to curve outward from the face, giving the impression of huge blinkers. In the Sumatran male, development of cheek flanges begins later, at about 10 years, and is rarely completed before age 20 (Jones, 1969) and never before 15 (Mallinson, 1978). The shape of the Sumatran male's flanges also differs: they tend to lie flat against the face, giving it a very wide appearance, especially in the midfacial region. Nonetheless, Galdikas (1980) illustrated two fully mature males, Throat Pouch and Nick, in which the flanges were considerably less developed than in the average Bornean male; and Kingsley (1982) found, in a captive study, that the presence of a more dominant male could inhibit flange development in a subordinate, maturing male, though not indefinitely. Individual and hormonal, as well as regional, differences are implied, as well as the distinct possibility that captive conditions may influence their development irrespective of inhibition effects.

It should be noted that von Koenigswald (1983) comments on the variability in the presence of cheek flanges within both Borneo and Sumatra, identifying (after Selenka, 1898) "races" in both areas in which they are said to be absent. This claim cannot, however, be accepted at face value, as the question of full maturity is not clear; it is only now, after three-quarters of a century, that the Selenka collection is being properly studied (Röhrer-Ertl, 1984, and this volume).

No author suggests any possible function, beyond the purely social one of a marker of full maturity, for these flanges, so possible environmental correlates cannot be discussed.

Throat Pouch

The throat pouch of the Bornean male is "very large and pendulous," while that of the Sumatran male, although noticeable, "does not reach the same proportions" (Jones, 1969:223). This is doubtless related to differences in the male's long call.

BEHAVIOR

Male's Long Call

The long call of the Bornean male is described by MacKinnon (1973) as long and drawn-out, whereas that of the Sumatran male is shorter and faster in tempo. The larger throat pouch in the former, acting as a resonating chamber, would seem to be related to this difference. The reasons for this difference are unclear; but the faster tempo of the Sumatran call may be more effective in a rugged, mountainous terrain, and MacKinnon suggests that the elongation of the call in Borneo may be a response to the wider dispersion of this race. In both subspecies, the call acts as a spacing mechanism between males and to advertise the location of the highest-ranking male to mature females (Rijksen, 1978).

Reproductive Behavior

The duration of the male-female bond has been listed by a number of authors (MacKinnon, 1974; Mallinson, 1978) as an important behavioral difference between the two subspecies. Both MacKinnon (1979) and Rijksen (1978) describe Sumatran orang-utan consortships as lasting between several days and several months, the former author stating that the male tends to remain with the female until after the birth of the infant. In contrast, Bornean males are reported to indulge in brief partnerships (no more than a few days) and then leave the female to rear the young alone. MacKinnon (1974, 1979) suggests that the longer-lasting bond observed in Sumatra may be a response to the presence of large predators (tigers) and aggressive competitors (siamangs), threats which are absent in Borneo. The prolonged presence of the male may thus serve to protect the infant from predation.

A further difference in reproductive behavior cited by MacKinnon (1979) is the performance of a courtship display by the male Sumatran form. In captivity, the display involves a bipedal strutting walk with the chest pushed forward and limbs extended; this has only once been observed in the wild, where a tree-swinging display, in which the long hair on the arms is swung around like a curtain, is more common. Courtship displays of any type are absent, or have not been described, in the Bornean orang-utan.

While intriguing, these reports of behavioral differences must be regarded as unconfirmed until quantitative research is undertaken. The modifiability of the behaviors under different environmental conditions, such as captivity, must also be made clear.

Sociability

The Sumatran orang-utan is considered by zoo personnel to be a more sociable animal than the Bornean. For example, in the Zurich zoo, a large social group of Sumatran orang-utans, including a fully mature male, is kept together (C. Schmidt, personal communication to C.P.G.), whereas in Cologne, an equally large group of Bornean orang-utans has to be split up (Hick, 1985). Markham (1980) made a survey of captive orang-utans in an attempt to determine whether there were consistent differences in social behavior between the two subspecies. She found that, on average, male Borneans are more aggressive and less inclined to share food than male Sumatrans; they are more aggressive even towards females and infants. On the other hand, there is a great deal of individual variability: "Each orang-utan had a distinct personality and in dealing with such highly intelligent animals in captivity, the keeper's knowledge of the individual was probably more important than knowledge of overall behaviour patterns" (Markham, 1980:40). The greater sociability of the Sumatran orang-utan, if confirmed by future observations, would perhaps be related to the need for the male to remain with the mother and infant.

CRANIAL MORPHOLOGY

Comparisons of cranial morphology are perhaps more useful than the more subjective observations so far discussed. Cranial characters can be measured and the measurements compared statistically to determine whether any significant differences exist between populations being examined. Most such analyses have, however, been somewhat limited in value up until now, owing to small sample sizes and, in particular, lack of precise information concerning the origin of specimens.

The notable exception to this is the recent work of Röhrer-Ertl (1984 and this volume). Based primarily on the enormous Selenka collection in Munich, and with full attention to precise locality, this author has for the first time succeeded in demonstrating that the differences between Bornean and Sumatran orang-utans really are greater than those between populations within either Borneo or Sumatra. In his cluster analysis (Röhrer-Ertl, 1984, Abb. 3), the primary split is between Sumatran and Bornean; within the Bornean sample, there is next a three-way split between (1) Sabah, (2) Kutai (eastern Kalimantan), and (3) western Kalimantan (with southwestern Sarawak). The skulls of Bornean and Sumatran subspecies can be sorted visually, as well, with over 95% accuracy: the average Bornean skull has a concave, dished face and a clearly expressed suborbital fossa, a "trumpet-shaped" piriform aperture, and an S-shaped zygomaticomaxillary suture; the average Sumatran skull has a straight, diagonal profile, no suborbital fossa, a "tubular" piriform aperture, and a diagonally straight or angular zygomaticomaxillary suture. It is interesting, parenthetically, that van Bemmel (1968:14) was evidenty able to recognize these differences in living animals, contrasting the "elongated, rather flat face, more or less O-shaped" of the Sumatran form with the "figure-8-shaped" faces of Bornean ones (specifically, from Sarawak).

One of us (C.P.G.) also measured adult skulls in the Selenka collection, and we compare them here with a few measured by each of us from Sarawak and northeastern Sabah in the British Museum (Natural History) and the Berlin Zoological Museum. The object is not to duplicate Röhrer-Ertl's study, but to obtain a different perspective: would some commonly utilized skull measurements differentiate between Bornean and Sumatran orang-utans, or are differences within Borneo (our Sumatran sample is too small to split up) liable to overshadow those between the two subspecies?

Analysis of variance and Student Newman-Keul tests were performed on various samples of the data to determine whether significant differences exist between the all-Borneo and Sumatra samples as well as between skulls from Sumatra, northwest Kalimantan, southwest Kalimantan, and Sabah. The northwest and southwest Kalimantan samples are separated by the Kapuas River; they correspond to Röhrer-Ertl's samples I–III and IV–V, respectively. The Sabah sample includes specimens from the Baram River District of northeastern Sarawak. All measurements of significance were at the $p < 0.05$ level. Means and ranges for each character are summarized in Table 2–1.

The all-Borneo vs. Sumatra comparisons revealed significant differences in nuchal length and bicondylar breadth in males, and in biorbital breadth in females; no other significant differences emerged. In all except two characters, the mean for Sumatran males falls between those for northwest and southwest Kalimantan, the significantly different Bornean means averaging to produce no overall difference from Sumatra. The averaging effect is particularly clear in the case of nuchal length in females: northwest Kalimantan, Sumatra, and southwest Kalimantan samples are all significantly different from each other, but there is no significant difference between the all-Borneo and Sumatra samples.

Comparisons between the samples have yielded a number of interesting results.

Males

No significant differences were noted in any feature between Sabah and northwest Kalimantan: the means are in every measurement extremely close. Between northwest and southwest Kalimantan, significant differences were found in face height, biorbital breadth, and basal length, while Sumatra was intermediate in every one of these measurements and not significantly different from either of the two Kalimantan samples. Bicondylar breadth was the only measurement

TABLE 2–1. Skull Measurements of Orang-utans

	NW Kalimantan			SW Kalimantan			Sabah			Sumatra		
	N	Mean	Range	*N*	Mean	Range	*N*	Mean	Range	*N*	Mean	Range
Males												
Face height	25	115.6	99–128	8	129.8	119–142	9	112.6	99–122	13	122.1	107–150
Cranial length	32	134.2	118–152	8	138.6	121–145	9	129.3	110–140	12	134.3	125–143
Bigonial breadth	26	127.4	110–144	6	132.8	125–139	6	125.6	108–142	9	126.1	111–146
Biorbital breadth	31	112.7	99–122	8	120.9	109–126	9	115.4	104–128	12	115.3	107–123
Basal length	25	173.1	152–188	7	181.0	173–191	9	174.7	161–187	13	174.8	164–184
Bicondylar breadth	24	136.8	117–147	6	140.3	126–147	9	135.8	124–144	9	134.0	129–142
Nuchal length	30	60.6	43–93	9	74.8	69–78	9	59.2	49–68	11	65.8	45–85
Females												
Face height	54	91.8	76–108	12	101.8	88–110				11	100.5	78–116
Cranial length	61	123.2	115–131	13	120.3	114–127				11	118.2	109–129
Bigonial breadth	51	100.5	83–112	12	100.7	87–106				9	100.0	91–114
Biorbital breadth	58	97.4	86–126	13	95.2	89–104				11	98.8	91–105
Basal length	54	141.3	125–155	9	142.0	134–160				11	140.4	130–150
Bicondylar breadth	50	116.5	103–130	12	113.9	105–126				9	116.7	105–123
Nuchal length	59	53.5	41–76	9	74.8	69–78				11	65.8	45–85

in which a significant difference was noted between an area of Borneo (southwest Kalimantan) and Sumatra, but no difference was revealed within Borneo. Significant differences were not forthcoming in any comparisons for cranial length or bigonial breadth. Nuchal length was significantly greater in southwest Kalimantan than in either northwest Kalimantan or Sumatra, but no difference was revealed between the latter two areas. On average, specimens from southwest Kalimantan are larger in all measurements utilized than those from the other three areas. Differences were not always statistically significant, although in five out of seven cases, southwest Kalimantan was found to be significantly different from the area with the smallest mean (usually Sabah).

Specimens from Sumatra were, on average, of intermediate size between those from northwest and southwest Kalimantan. They were, however, smallest for bicondylar breadth (significant compared to southwest Kalimantan only) and smaller than all except Sabah for bigonial breadth (not significant). This finding accords well with the general impression from the literature that Sumatran males have narrower jaws than Bornean, although the statistical significance is hardly overwhelming.

Northwest Kalimantan specimens are smallest in biorbital breadth, basal length, bicondylar breadth, and nuchal length; Sabah skulls in face height, cranial length, and bigonial breadth. In no character was either of these two samples significantly different from Sumatran samples. In general, therefore, it would seem that differences among Bornean orang-utans are as great as (or probably greater than) those observed between Bornean and Sumatran animals. The Sumatran measurements can be described in general terms as being intermediate in size between larger and smaller Borneans, but a little closer to those from the northern parts of Borneo.

Females

The data from females are less conclusive. Unlike the comparison for males, no clear-cut sequence of size appears in the female comparisons. In four of the seven characters (basal length, bigonial breadth, bicondylar breadth, and biorbital breadth), no significant differences are recorded for any comparisons. In all the remaining characters, differences within Borneo are as great as or greater than those between areas of Borneo and Sumatra. For nuchal length, all the measurements are significantly different. Comparisons of face height revealed significant differences between the two populations of Borneo, and between northwest Kalimantan and Sumatran populations (those from Sumatra being larger than northwest Kalimantan but smaller than southwest Kalimantan populations). For cranial length, the position was reversed, and the northwest Kalimantan population was significantly larger than those from both Sumatra and southwest Kalimantan.

An interesting result, which does not accord

well with descriptions of general appearance, concerns basal length, cranial length, and face height. Bornean animals are generally described as being more prognathous than Sumatrans, and examination of various photographs does suggest a difference. Yet basal length is not significantly different in either sex, and face height and cranial length are different only between northwest Kalimantan and Sumatra in females. Cranial length in females is the only one of these characters in which the Sumatran measurements are not intermediate. In this case, statistical analysis reveals that it is the northwest Kalimantan population which is significantly different from the other two, rather than that there is a general difference between Borneo and Sumatra. The reasons for this finding are unclear: it may be that the apparent prognathism of Bornean orang-utans is spurious, an effect of the dished face, as contrasted by Röhrer-Ertl (1984) with the Sumatran's straight, but in actuality equally prognathous, face.

Overview of Cranial Morphology

A superficial examination of the paleogeographic data (for example, see Haile, 1973) suggests that southwest Kalimantan may have been in contact with southern Sumatra for a longer period of time than was either northwest Kalimantan or northern Sumatra. The northern Sumatran population may have been restricted by rugged terrain, while the northwest Kalimantan/Sabah population was almost certainly isolated by the Kapuas River to the south and the North Sunda Valley to the west. If this is the case, the generally larger size of the southwest Kalimantan population could perhaps be explained by a more prolonged period of gene exchange with the larger and now extinct *Pongo pygmaeus palaeosumatrensis* (Hooijer, 1948). This is, of course, pure speculation, and more detailed information, including cranial data, on the extinct south Sumatran form is required before any meaningful assessment can be made. Analysis of dental measurements, to determine if the teeth of southwest Kalimantan orang-utans are larger than those of examples from elsewhere—as they would be were the hypothesis of a south Sumatra connection to be taken seriously—is also needed.

The cranial data, though inconclusive in many respects, do lead to a number of conclusions. The most obvious of these is that one cannot take for granted the homogeneity of any but the most geographically restricted samples. In the present analysis, even the relatively restricted Sumatran sample is, in a way, an invented population; a breakdown of it might lead to the discovery of yet further complexities. Subspecies must, therefore, be built from the ground upwards and tested at each stage. But although Röhrer-Ertl (1984), in just such a ground-upward analysis, has validated the two-subspecies concept in the orang-utan, the present univariate analysis has failed to do so and thus gives due warning that there may be a perfectly real and legitimate pattern which is nonetheless discoverable only by the simultaneous examination of many characters, i.e., multivariate analysis. Finally, a finding that "there are subspecies" is never likely to be the last word on interpopulation relationships.

BIOCHEMICAL AND CHROMOSOMAL EVIDENCE

The most convincing evidence of major differences between Bornean and Sumatran orang-utans has been provided by chromosomal and biochemical studies. In several cases, the differences found have been completely diagnostic, while in others, differences in gene frequency and degree of polymorphism have been observed. Yet it is still necessary to reiterate that locality data used in these studies are poorly controlled: all that is known of a given specimen is that it is Bornean or Sumatran, never the area within Borneo or Sumatra from which it came. Thus there remains the possibility, slight though it may be in this case, that regional rather than interisland differences are being sampled.

De Boer and Seuanez (1982) examined the karyotypes of 72 individuals (35 Bornean, 25 Sumatran, and 12 hybrids) and discovered intraspecific differences in at least one and possibly two chromosome pairs:

(1) Chromosome 2: Two forms of chromosome 2 were observed in the study population, and the evidence suggests a clear-cut difference between the subspecies in this character. All Bornean animals examined were found to be homozygous for type 1, and all Sumatrans homozygous for type 2; captive-bred hybrids were heterozygous. The two chromosome types could be derived from each other by a simple pericentric inversion which covered about a quarter of the chromosome length. The authors warn that "large karyotypic differences may exist between closely related species while very slight or apparently no differences are seen between very dis-

tantly related forms" (De Boer and Seuanez, 1982:136); they feel, however, that this difference among orang-utans in chromosome 2 clearly indicates a taxonomic distinction between the two subspecies.

(2) Chromosome 14: The same authors also observed a rare variant of chromosome 14 which was found only in Sumatran orang-utans and in one hybrid. Borneans were always homozygous for the normal form.

This difference in the structure of chromosome 2 raises some interesting questions concerning the fertility of hybrids. De Boer and Seuanez (1982) report that at the time nine hybrids had reproduced successfully, but it was as yet impossible to assess their total fertility, frequency of abortion and stillbirth, etc., relative to purebred animals. Markham (1985) lists the percentage of purebred and hybrid stock that have successfully reproduced (i.e., their offspring have survived infancy) in captivity. Of the captive-bred stock, the hybrid females are more successful than the hybrid males. She does not mention, however, whether the hybrids were breeding with each other or with animals of pure stock; and thus, even with this information, the actual fertility of hybrids is still difficult to assess.

A number of biochemical studies have also discovered apparently fully diagnostic differences between orang-utans from Borneo and Sumatra. Bruce and Ayala (1979) compared 23 blood proteins using electrophoretic techniques. Their study sample consisted of 69 individuals belonging to eight hominoid species, including eight Sumatran and three Bornean orang-utans. Different variants of the 17 red cell proteins and seven plasma proteins were studied. The two orang-utan subspecies appeared to differ in their alleles of adenosine deaminase (Ada): the "pongid allele," also carried by *Pan troglodytes* and *Pan paniscus,* was homozygous in the Bornean orang-utan, whereas the "human allele" was homogyzous in the Sumatran form. The very small sample size, of course, cannot rule out either interregional variation or polymorphism.

Differences in gene frequency and degree of polymorphism were also found in several other red cell proteins. Sumatran orang-utans emerged as being polymorphic for malate dehydrogenase (Mdh-S) and phosphoglucomutase (Pgm-2): in both cases the so-called "human" type was the most common, but the variant was not exceptionally rare, and all Borneans studied were homozygous for the human type. The gene frequency for the human-type variant of hemoglobin was low in Borneans, high in Sumatrans, but polymorphic in both. A more detailed study of hemoglobin by De Boer and Khan (1982) revealed that these differences between Bornean and Sumatran orang-utans were due to the frequency of two variants of the α chain. In their study the α_a variant was found to be high in Borneo, α_b high in Sumatra. They also report that a variant, *c*, of the β chain, which occurs, though infrequently, in Sumatrans, is absent from Bornean orang-utans. Perhaps the most important result of Bruce and Ayala's study was their calculation of genetic distance between Bornean and Sumatran subspecies. Based on 21 comparisons, the calculated genetic distance was 0.130, a figure comparable to that found between other subspecies.

Finally, there is the report by Wijnen et al. (1982) that the structure of glucose-6-phosphate dehydrogenase (G6PD) is slightly, but distinctly, different in the two subspecies. The activity of G6PD was tested in two different gels, and the variants, 1 and 2, were compared for pH optima and temperature sensitivity. Although similar in most respects and largely indistinguishable from the most common human type, the two variants were found to be clearly different in temperature sensitivity. The authors claimed that the alleles 1 and 2 are peculiar to Bornean and Sumatran orang-utans, respectively. Bruce and Ayala (1979) had found no such difference in their study, recording orang-utan G6PD to be of the "human" type in both subspecies; but it is quite likely that they would not have been distinguishable electrophoretically.

Ferris et al. (1981) and Wilson et al. (1985) found that differences in mitochondrial DNA between the two subspecies of orang-utan were much greater than within any other hominoid species (gibbons were not studied), and comparable to those between the species *Pan troglodytes* and *Pan paniscus* (Wilson et al., 1985, Fig. 8). If further study should find these differences to be consistent, a long period of separation for the two subspecies will be implied; but as the study is based on only three Sumatran and two Bornean specimens, there is a clear need for study of more specimens before the differences can be put down to taxonomic rather than simple polymorphic differentiation.

The biochemical and chromosomal data discussed above suggest two general findings: (1) broadly diagnostic differences do exist between orang-utans from the two different islands, namely in chromosome 2, Ada, and G6PD; (2), the Sumatran subspecies is more polymorphic than the Bornean. The second finding could be an artifact of the larger sample sizes for the Sumatran form in the case of the biochemical data, but the polymorphism of chromosome 14 would seem to be a real datum as the sample of karotypes was smaller in the case of the Sumatran.

CONCLUSIONS

The aim of this chapter has been to examine the differences between orang-utans from Borneo and Sumatra in an attempt to assess their degree of relatedness. The major conclusion seems to be that the differences are less impressive than might appear at first glance. This is for a number of reasons. Some of the differences may well be ontogenetic responses to different (high- vs. low-altitude) environments. Others are blurred by the wide range of variability to be observed on one or both islands, such that differences between samples from different parts of Borneo are greater than those between Borneo as a whole and Sumatra. Finally, many of the biochemical differences cannot be shown to differentiate Bornean from Sumatran taxa rather than one population within Borneo from one population within Sumatra.

We have therefore been forced to take a conservative position. Differences there certainly are, but until the long list of proposed differences can be verified as indicated above, we prefer to follow the recommendation of Mayr (1969:197) that allopatric populations of doubtful rank are best treated as subspecies rather than as a full species.

Groves and Holthuis (1985) have shown that the correct nomenclature of the subspecies of orang-utan is as follows:

Bornean orang-utan: *Pongo pygmaeus pygmaeus* Linnaeus, 1760.
Sumatran orang-utan: *Pongo pygmaeus abelii* Lesson, 1827

REFERENCES

Andrews, P. 1984. Review of L. E. M. de Boer, ed., The Orang Utan. Its Biology and Conservation. *Primate Eye, 23:*34–37.

Bemmel, A. C. V. van. 1968. Contribution to the knowledge of the geographical races of *Pongo pygmaeus* (Hoppius). *Bijdragen tot de Dierkunde, 38:*13–15.

Bruce, E. and Ayala, F. 1979. Phylogenetic relationships between man and the apes: electrophoretic evidence. *Evolution, 33:*1040–1056.

Chasen, F. N. 1940. Handlist of Malaysian Mammals. *Bulletin of Raffles Museum, 15:*i–xx, 1–209.

De Boer, L. E. M. and Khan, P. M. 1982. Haemoglobin polymorphisms in Bornean and Sumatran orangutans. *In* The Orang Utan: Its Biology and Conservation, pp. 125–134, ed. L. E. M. de Boer. W. Junk, The Hague.

De Boer, L. E. M. and Seuanez, H. 1982. The chromosomes of the orang utan and their relevance to the conservation of the species. *In* The Orang Utan: Its Biology and Conservation, pp. 135–170, ed. L. E. M. de Boer. W. Junk, The Hague.

Eckhardt, R. B. 1975. The relative body weights of Bornean and Sumatran orangutans. *American Journal of Physical Anthropology, 42:*349–350.

Ferris, S. D., Brown, W. M., Davidson, W. S. and Wilson, A. C. 1981. Extensive polymorphism in the mitochondrial DNA of apes. *Proceedings of the National Academy of Sciences of the USA, 78:*6319–6323.

Galdikas, B. M. F. 1980. Living with the great orange apes. *National Geographic, 157:*830–853.

Groves, C. P. 1971. *Pongo pygmaeus. Mammalian Species,4:*1–6.

Groves, C. P. 1986. Systematics of the Great Apes. *In* Comparative Primate Biology, vol. 1, pp. 187-217, ed. D. R. Swindler. Alan R. Liss, New York.

Groves, C. P. and Holthuis, L. B. 1985. The nomenclature of the orang utan. *Zoologische Mededelingen Leiden, 59:*411–417.

Haile, N. S. 1973. The geomorphology and geology of the northern part of the Sunda Shelf and its place in the Sunda Mountain system. *Pacific Geology, 6:*73–88.

Hick, U. 1985. Haltung, Zucht und kunstliche Aufzucht von Borneo-Orang-Utans *(Pongo pygmaeus pygmaeus)* im Kolner Zoo. *Zeitschrift des Kolner Zoo, 28:*113–132.

Hooijer, D. A. 1948. The prehistoric teeth of man and orang utan from Central Sumatra with notes on the fossil orang utan from Southern China. *Zoologische Mededelingen Leiden, 29:*175–301.

Jones, M. L. 1969. The geographical races of the orang utan. *In* Recent Advances in Primatology, pp. 217–223, ed. H. O. Hofer. S. Karger, Basel.

Kingsley, S. 1982. Causes of non-breeding and the development of the secondary sexual characteristics in the male orang utan: a hormonal study. *In* The Orang Utan: Its Biology and Conservation, pp. 215–229, ed. L. E. M. de Boer. W. Junk, The Hague.

Koenigswald, G. H. R. von. 1982. Distribution and evolution of the orang utan *Pongo pygmaeus* (Hoppius). *In* The Orang Utan: Its Biology and Conservation, pp. 1–15, ed. L. E. M. de Boer. W. Junk, The Hague.

Lyon, M. W. 1911. Mammals collected by Dr W. L. Abbott on Borneo and some of the small adjacent islands. *Proceedings of the U.S. National Museum, 40:*53–146.

MacKinnon, J. 1973. Orang Utans in Sumatra. *Oryx, 12:*234–242.

MacKinnon, J. 1974. The behaviour and ecology of wild orang utans *(Pongo pygmaeus). Animal Behaviour, 22:*3–74.

MacKinnon, J. 1979. Reproductive behaviour in wild orang utan populations. *In* The Great Apes, pp. 257–273, ed. D. A. Hamburg and E. R. McCown. Benjamin/Cummings, Menlo Park.

Mallinson, J. J. C. 1978. 'Cocktail' orang utans and the need to preserve pure-bred stock. *Dodo, 15:*69–77.

Markham, R. J. 1980. An investigation into the differences in social behaviour between the Sumatran *(Pongo pygmaeus abelii)* and Bornean *(Pongo pygmaeus pygmaeus)* sub-species of orang-utan in captivity. Unpublished honours thesis, University of East Anglia.

Markham, R. J. 1985. Captive orang utans: present status and future prospects. *Bulletin of Zoo Management, 23:*31–36.

Mayr, E. 1969. Principles of Systematic Zoology. McGraw-Hill, New York.

Newman, R. W. 1975. Human adaptation to heat. *In* Physiological Anthropology, pp. 80–92, ed. A. Damon. Oxford University Press, New York/London.

Rijksen, H. D. 1978. A field study on Sumatran orang utans (*Pongo pygmaeus abelii* Lesson 1827). H. Veenman & Zonen, Wageningen.

Röhrer-Ertl, O. 1984. Orang-Utan-Studien. Hieronymus, Neuried.

Selenka, E. 1898. Menschenaffen: Rassen. Schädel und Bezahnung des Orang-utan. Kreidel, Weisbaden.

Wijnen, J. T., Rijksen, H., De Boer, L. E. M., and Khan, P. M. 1982. Glucose-6-phosphate-dehydrogenase (G6PD) variation in the orang utan. *In* The Orang Utan: Its Biology and Conservation, pp. 109–118, ed. L. E. M. de Boer. W. Junk, The Hague.

Wilson, A. C., Cann, R. L., Carr, S. M., George, M., Gyllensten, U., Helm-Bychowski, K. M., Higuchi, R. G., Palumbi, S. R., Prager, E. M., Sage, R. D., and Stoneking, M. 1985. Mitochondrial DNA and two perspectives on evolutionary genetics. *Biological Journal of the Linnaean Society, 26:*375–400.

3

Diversity and Consistency in Ecology and Behavior

PETER S. RODMAN

This chapter reviews current data on the ecology and behavior of orang-utans. Sources of data are summarized in Table 3–1, with locations of field studies shown in Fig. 3–1. I conclude with a brief discussion of possible clues to the nature of *Sivapithecus* to be found in the adaptations of living orang-utans.

Geographical Dispersion

Orang-utans today are found in a "relict" geographical pattern (Fig. 3–1). Modern groups of the single species *Pongo pygmaeus* occupy forests of northern Sumatra *(P. p. abelii)* and of various parts of Borneo *(P. p. pygmaeus),* but fossils occur in Pleistocene deposits of Java as well as of mainland Asia (Hooijer, 1948; Koenigswald, 1982). The dispersion of orang-utans has therefore been severely restricted since the Pleistocene; it continues to be reduced today as forests of Sumatra and Borneo are cut both for timber and to accommodate the expanding human population.

Environment and Habitat Variation

The forests of southeast Asia are called "dipterocarp forests" because of the dominance of the family Dipterocarpaceae in the taxonomic composition of the forests. Generally the forest consists of a three-tiered structure: a lower, horizontally discontinuous stratum of shrubs, palms, and saplings of 1.0–5.0 m height; a middle stratum of small trees with crowns intermingled to form a horizontally continuous canopy between 6 and 25 m above ground; and an emergent, horizontally discontinuous stratum of the tallest trees, often dipterocarps, whose crowns rise above the middle canopy. On the ground there may be herbaceous cover. The vertical organization of the forest is populated by a set of epiphytes, such as "bird's nest" ferns, and is interlaced with diverse vines and lianas. Large emergent trees fall regularly, opening gaps in the middle canopy by dragging other trees down via interlacing lianas and by crushing smaller trees in their paths. Opportunistic plant species such as trees of the genus *Macaranga* colonize these gaps rapidly.

Despite a superficial appearance of uniformity of the southeast Asian forest, there is considerable variation between locations and within smaller areas. MacKinnon (1974) examined forests in the valleys and hills of the Ulu Segama (Borneo) and Ranun River (Sumatra) and found contrasts between the areas and between valleys and hills in each area. By way of comparison, he found that the "ten commonest species account for 27.5 per cent of the total of trees sampled in Segama valleys, 38.0 per cent in Segama hills, 45.5 per cent in Ranun valleys and 60.0 per cent in Ranun hills" (p. 9). In other words, diversity of tree species is higher in valleys than on hills, and higher in the Segama forest than in the Ranun forest. MacKinnon attributed vegetational differences to differences in rainfall and runoff between the valleys and hills and between the two study areas.

At Tanjung Puting, Galdikas (1978) identified two major forest types: peat swamp forest and lowland dipterocarp forest. Galdikas counted trees in three 0.2-hectare sample plots, of which two were in "lowland dipterocarp forest" (LDF) and the other in "peat swamp forest" (PSF). She found that the forest canopy is low, the proportion of dipterocarps is low, and the girths of trees are small compared with those of the dipterocarp forests of Malaya and Sabah reviewed by Whitmore (1975). Galdikas attributed these characteristics of the Tanjung Puting forest to the generally poor soils in central Kalimantan. Two samples of LDF shared approximately 40% of tree species, while each of these shared 10% or less of tree species with a sample of PSF. PSF forest, unlike LDF, was constantly inundated, and there was sparse growth of lower stratum trees and shrubs in this swampy environment.

The Mentoko forest contains a large number of mature, emergent ironwood trees (Lauraceae, *Eusideroxylon zwageri*) and large emergent dip-

TABLE 3–1. Field Studies of Orang-utans, 1967 to the Present

Investigator	Duration	Location	Site	Publications
Horr, D. A.	9/67–11/69	Sabah North Borneo	Lokan River	Horr, 1972, 1975, 1977; Horr and Ester, 1976
MacKinnon, J. R.	6/68–11/68 10/69–11/70 4/71–11/71	Sabah North Sumatra	Upper Segama River Ranun River	MacKinnon, 1969, 1971, 1973, 1974, 1977, 1979
Rodman, P. S.	4/70–8/71	E. Kalimantan East Borneo	Mentoko River Kutai Reserve	Rodman, 1973a,b, 1977, 1979, 1984a, 1984b; Rodman and Mitani, 1987
Galdikas, B. M. F.	11/71–1/75	Cn. Kalimantan Central Borneo	Sekonyer River Tanjung Puting Reserve	Galdikas, 1978, 1979, 1981, 1982, 1983, 1984, 1985a,b; Galdikas and Teleki, 1981
Rijksen, H. D.	6/71–8/74	North Sumatra	Ketambe River Gunung Loeser Reserve	Rijksen 1978; Rijksen and Rijksen-Graatsma, 1975
Wheatley, B. P.	2/75–9/76	E. Kalimantan East Borneo	Sengatta River	Wheatley, 1982
Schürmann, C.	6/75–4/79	North Sumatra	Ketambe River Gunung Loeser Reserve	Shürmann, 1981, 1982
Sugardjito, J.	4/80–6/80	North Sumatra	Ketambe River Gunung Loeser Reserve	Sugardjito, 1982
Mitani, J. C.	7/81–11/82	E. Kalimantan East Borneo	Mentoko River Kutai Reserve	Mitani, 1985a,b
Leighton, M. and Leighton, D. R.	8/77–9/79	E. Kalimantan East Borneo	Mentoko River Kutai Reserve	Leighton and Leighton, 1983

terocarps, which are hosts to climbing and strangling figs (Leighton and Leighton, 1983), as well as a high density of two species of the family Anacardiaceae (*Dracontomelum* spp. and *Koordersiodendron pinnatum*). At the Mentoko there are strong contrasts in structure of the forest between the alluvial area along the Sengatta River, where the middle canopy is continuous and the ground cover is thick, and the hills away from the river, where the canopy is filled with more and larger gaps and the ground cover is thin (Rodman, 1984a). Trees of the genus *Dracontomelum,* which are very important food trees for orang-utans, occur in the alluvial area along the rivers, while the related and equally important trees of the species *Koordersiodendron pinnatum* occur on higher ground away from the rivers. It is likely that these variations in habitat at the Mentoko are a consequence of variation in moisture in the soil.

The Ketambe forest appears to be similar to that of other sites in numbers of trees per hectare. This forest, however, contains an unusually high number of large, emergent, freestanding figs and a low proportion of dipterocarps (Rijksen, 1978). Since the large canopies of emergent figs are rich sources of food for orang-utans, this difference has a profound effect on utilization of the Ketambe forest by orang-utans. A particularly striking effect is the attraction of large numbers of individuals to single fruiting trees.

INDIVIDUAL BEHAVIOR AND ECOLOGY

Daily Activity Pattern

Orang-utans of all studies spend more than 95% of their waking hours feeding, resting, and moving between feeding and resting sites (Table 3–2). The balance of time is occupied with nest building and very limited social behavior. Day length varies with the time between sunrise and sunset through the year. The mean day length for habituated animals at the Mentoko in 1970–71 was 11 hr 24 min (N = 42 days) and varied from 11 hr 28 min for one adult male (N = 7 days) to

12 hr 8 min for a juvenile female (N = 3 days) (Rodman, 1979). Day lengths at Tanjung Puting varied from a mean of 9 hr 46 min for an adult male (N = 5 days) to a mean of 11 hr 29 min for an adult female (N = 59 days) (Galdikas, 1978:264). At the Mentoko site, the day consisted of heavy early morning feeding near the previous night's sleeping site, followed by several hours of reduced activity, with travel in the afternoon followed by increased feeding prior to moving to the sleeping site (Rodman, 1979). The morning and evening feeding peaks were usually in fruiting trees. The animals spent a greater proportion of the time feeding on leaves and other vegetation during afternoon travel than in the morning (Rodman, 1977). Overall, four habituated subjects at the Mentoko built new sleeping platforms on 75.2% of days when the end of the day was observed (N = 105), but two adult females constructed platforms essentially every night. A juvenile female often used an old nest and an adult male frequently slept in the crotch of a large tree without a sleeping platform (Rodman, 1979).

There is interesting variation among the studies for which there are reports on diurnal activity profiles (Table 3–2). Results of two studies 11 years apart at the Mentoko produced quite similar results when all observations are lumped together (Mitani, unpublished; Rodman, 1973a,b, 1979). Orang-utans spent less time moving at the Mentoko in both studies than at any of the other sites. At Tanjung Puting, feeding and travel time were both prolonged compared with other sites and resting time was short. Part of this difference among results of different studies may be due to different recording methods. Galdikas (1978) used waking hours for a base, while, for example, Rodman (1979) used time between 0530 and 1830 as a base. Thus Rodman included prewaking rest and postsleeping rest at the beginning and end of the day. Another source of variation is the differences among the populations of orang-utans themselves: orang-utans at Tanjung

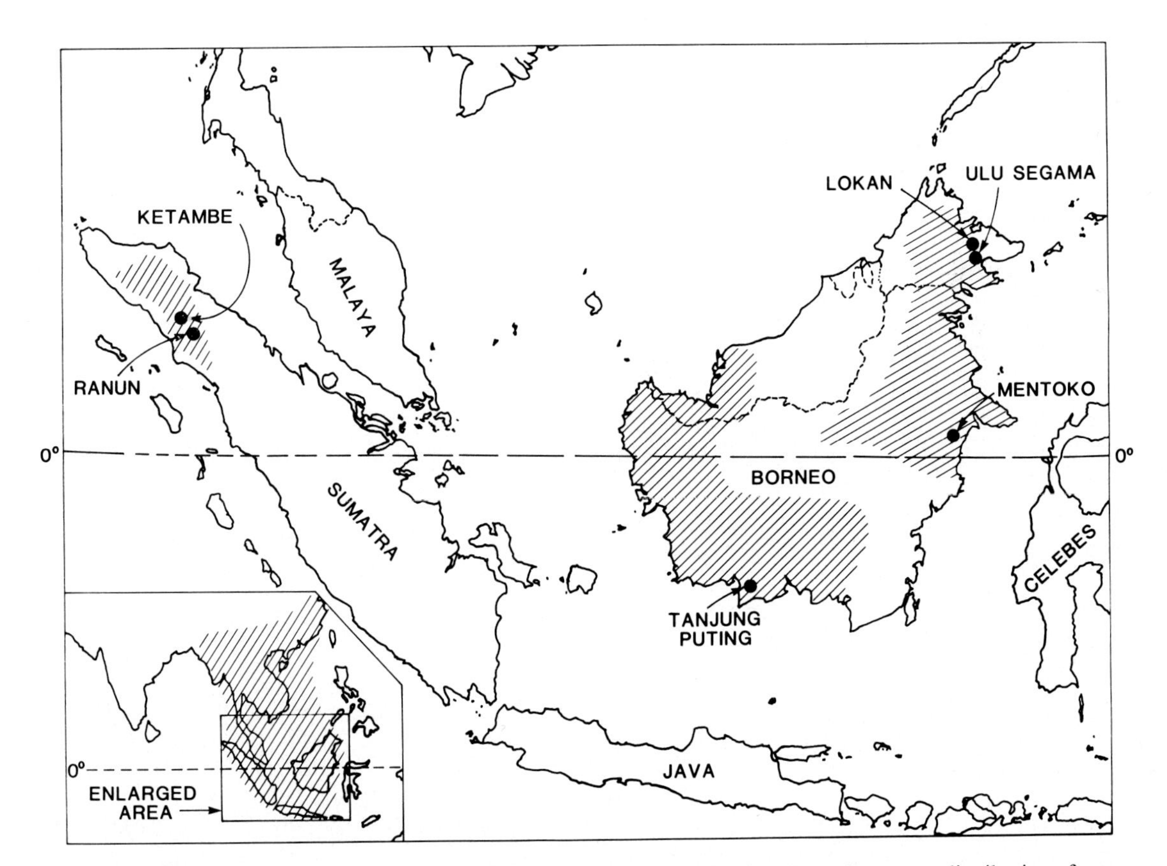

FIG. 3–1. Study sites and geographical dispersion of orang-utans. Cross-hatching shows the current distribution of orang-utans limited to parts of the islands of Sumatra and Borneo. The insert shows a probable distribution of orang-utans during the Pleistocene epoch (fossils assigned to *Pongo* occur in Pleistocene strata of China, Indochina, and Java where orang-utans no longer exist). Black dots indicate the six principal study sites described in the text.

TABLE 3–2. Activity Profiles of Orang-utans by Study Site and Age/Sex Class

	All Animals			Adult Females			Adult Males			Adoles. and Juv. Females			Subadult Males		
	Feed	Rest	Move	Feed	Rest	Move	Feed	Rest	Move	Feed	Rest	Move	Feed	Rest	Move
Mentoko 1970–71	44.5	41.6	11.4	31.4	55.7	10.6	57.0	32.6	9.3	42.1	42.0	12.6			
				47.5	36.3	13.1									
				39.5	46.0	11.8									
1982–84	46.9	43.0	9.9	60.5	22.3	17.0	32.7	59.6	6.7				60.0	28.2	11.7
				48.3	39.1	12.5	25.0	71.0	4.0				54.5	32.8	12.8
				44.1	46.1	9.6	49.7	45.2	5.0						
				51.0	35.8	13.0	35.8	58.6	5.2				57.2	30.5	12.2
Tanjung Puting	61.2	17.4	18.4	53.9	26.5	17.9	41.4	21.0	25.6	66.3	14.8	17.3	65.1	13.6	19.6
				66.3	13.5	18.6	60.4	17.1	20.8	72.3	7.1	19.2	64.0	16.6	16.9
				61.1	18.0	17.5	64.8	17.9	15.1						
				67.7	14.9	16.4	52.1	28.6	17.0						
				62.2	18.2	17.6	54.7	21.1	19.6	69.3	10.9	18.2	64.5	15.1	18.2
Ulu Segama	32.3	51.4	16.4	31.9	50.6	17.6	32.7	52.2	15.1						
Ranun	44.1	40.0	16.0												
Ketambe	42.5	43.5	14.0	43.0	44.0	13.0	42.0	43.0	15.0						

Sources: Mentoko 1970–71, Rodman (1973a,b, 1979); Mentoko 1982–84, Mitani, unpublished data; Tanjung Puting, Galdikas (1978); Ulu Segama and Ranun, MacKinnon (1974); Ketambe, Rijksen (1978).

Puting travel farther each day, feeding for shorter periods of time at more feeding sites in larger home ranges than those reported for the Mentoko (see Ranging Behavior, below). Thus the differences in rest and travel time each day make sense.

MacKinnon's (1974) observations at the Ulu Segama site suggest that orang-utans spend less time feeding and much more time resting than at other sites. Part of this difference might be due to MacKinnon's observation of unhabituated animals, but he used the same methods at Ranun where the activity profiles were more similar to those at other sites. His higher figure for time traveling is consistent with his report that orang-utans of the Ulu Segama ranged nomadically over large areas.

One adult male and one adult female observed by Rijksen (1978) at Ketambe were remarkably similar to each other in use of time, given contrasts between males and females found in studies at Tanjung Puting and the Mentoko. In the latter sites there is some important contrast in reported male-female differences. At Tanjung Puting, Galdikas (1978) found that males fed less and traveled slightly more than adult females. Mitani (unpublished) found a similar difference between males and females in feeding, although Rodman's (1979) figures showed an opposite difference between one male and two adult females at the Mentoko. Both Rodman and Mitani found that adult males moved for less time each day than adult females. MacKinnon (1974) lumped observations of subadult and adult males of the Ulu Segama area, which limits some critical comparisons between that population and others.

Time on Different Foods

There is little variation among several studies in the results of observation and analysis of dietary proportions by major food types (Table 3–3). Orang-utans at all sites feed predominantly on fruit (including seeds), leaves and shoots, bark, other vegetational parts including flowers and stems, and insects (termites and ants). The proportion of time (or of total feeding observations) spent feeding on fruit has varied between sites, but is always the major part of feeding, whereas feeding on leaves and shoots has been the second major part of feeding. Time feeding on "bark" (which includes rare feeding on wood, and which is normally feeding on the cambium of the bark) was third in four of five studies. Rijksen (1978:53) reported that 14.0% of feeding minutes were spent "in search for insects." In my own observations, feeding on insects included only feeding on termites from nests, so that the process was similar to feeding on other clumped food; searching for these termites was not a time-consuming activity (Rodman, 1977). Rijksen (1978), however, reported a wider variety of insects as food than have other field workers.

As noted above, the forest at Ketambe includes large numbers of very large freestanding fig trees, and Rijksen (1978:54) has reported that 54% of all observations of feeding on fruit were of feeding on figs (*Ficus* spp.). The fruiting pat-

TABLE 3–3. Diets of Orang-utans

	Dietary Proportions of Major Food Classes				
Site	Fruit	Leaves and Shoots	Bark and Wood	Other Vegetation	Insects
Mentoko (1970–71)[1]	53.8	29.0	14.2	2.2	0.8
Range for 12 months:	25.7–89.0	5.3–55.6	0.0–66.6	0.0–2.5	0.0–11.1
Tanjung Puting (1971–75)[2]	60.9	14.7	11.4	3.9	4.3
Range for 48 months:	16.4–96.1	0.0–39.6	0.0–47.2	0.0–41.1	0.0–27.2
Ulu Segama (1969–70)[3]	62.0	23.5	10.5	3.0	1.0
Range for 12 months:	7.7–90.8	7.5–76.9	0.0–37.0	—	0.0–9.1
Ranun (1971)[3]	84.7	10.2	4.5	—	0.6
Range for 5 months:	64.0–96.4	2.6–20.0	0.0–16.0	—	0.0–8.3
Ketambe (1971–75)[4]	58.0	25.0	3.0	—	14.0
Range for 14 months:	22.5–66.5	7.0–32.5	0.0–9.0	—	0.0–19.5

[1]Percentages of total minutes of feeding (Rodman, 1977).

[2]Percentages of total minutes of feeding (Galdikas, 1978).

[3]Percentages of total feeding observations, estimated from Fig. 19 in MacKinnon (1974).

[4]Percentages of total minutes of feeding (Rijksen, 1978). Ranges estimated from Fig. 30 in Rijksen (1978).

tern of figs as a genus is different from that of other fruiting trees and lianas at the Mentoko in that it is more uniform through the year, whereas other fruit sources tend to be temporally clumped into one or two seasons (Leighton and Leighton, 1983). Rijksen's observations of feeding by orang-utans suggest a similar uniform availability of figs at Ketambe (Figs. 28 and 29 in Rijksen, 1978). It seems likely that this abundance and uniformity of available figs would reduce variability in reliance on foods other than fruit at that site.

The term "fruit" has been applied to all stages of the development of the fruit in most studies (Galdikas, 1978; MacKinnon, 1974; Rijksen, 1978; Rodman, 1977), but there are significantly different physical and chemical properties of immature seeds, green fruits, and ripe (mature) fruits. Leighton (unpublished) examined the frugivorous community of birds and primates at the Mentoko and noted the stage of development of the "fruit" taken for all species, including orang-utans. Table 3–4 shows that orang-utans fed on the "mature mesocarp" (ripe fruit pulp), with or without the seed, of 64 of 95 species of fruit taken, and the immature mesocarp, with or without the seed, of 28 species. The mature seeds were normally swallowed and later defecated or discarded during "preparation." No mature seeds were actually "eaten" (i.e., crushed with the teeth as though to be digested) either with or without the mesocarp. Immature seeds were taken from 31 species, notably including five species of durians (Bombacaceae: *Durio* spp.). Many species of this genus of fruit are highly prized by humans for their ripe flesh, but orang-utans are seed predators of the majority of species they use.

Taxonomic Composition of the Diet

Table 3–5, which summarizes data from all studies, shows the composition of the diet by family of plant used and in descending order of number of species per plant family. In Table 3–5, species that have been identified in more than one site are counted once for each site. Table 3–6 shows the overlap of plant genera and species within sites. Overall, 74 plant families representing 325 genera, which, in turn, subsume 502 species, have been identified as food sources at the five study sites. Given the large number of species, it is surprising that there is little overlap between studies (Table 3–6), there being only 32 species appearing in food lists of more than one site.

The family of figs (Moraceae) is most diverse overall and is most diverse in each of the five study sites presented. As noted above, figs from large emergent freestanding fig trees were a major component of the diet of orang-utans at Ketambe. At the Mentoko in 1970–71, several species of *Ficus* provided large quantities of bark as food (Rodman, 1979), whereas at other sites fig trees primarily provided fruit.

Dipterocarps (species of the genus *Shorea*) (Rodman, unpublished) appeared unusually frequently in the sample from the Mentoko during 1970–71. Observations from the same site made in 1977–79 by Leighton and Leighton (unpublished) cannot be compared because these workers sampled only sources of fruit and seeds, while Rodman included observations of all plant foods. The bark of dipterocarps was an important component of the diets of orang-utans in Rodman's (1977) study, but dipterocarps were rare food sources in studies at other sites. This is one of the clearest differences among sites in di-

TABLE 3–4. Plant Reproductive Parts Consumed by Orang-utans at the Mentoko, 1977–79[1]

			This Part and One Other Part:				
	N	Only This Part	2	3	4	5	6
1. Mature mesocarp, swallowed seed	55	39					
2. Immature seed	31	26	2				
3. Immature seed and immature mesocarp	15	6	5	2			
4. Immature mesocarp	13	5	6	0	0		
5. Mature mesocarp, discard seed	8	6	1	1	0	0	
6. Mature seed with mature mesocarp	1	0	0	0	1	0	0

[1]Numbers in this table are numbers of plant species from which a particular reproductive part was taken. Numbers below the diagonal to the right are of species from which more than one reproductive part or combination was taken.

TABLE 3–5. Plant Families, Numbers of Genera and Numbers of Species[1]

	Study 1		Study 2		Study 3		Study 4		Study 5		Total	Total
Family	Gen	Spp	Gen	Spp	Gen	Spp	Gen	Spp	Gen	Spp	Genera	Species
Moraceae	3	5	1	21	5	19	2	11	3	17	14	73
Annonaceae	3	3	5	8	5	6	5	9	7	10	25	36
Euphorbiaceae	2	2	5	11	4	8	2	2	7	13	20	36
Leguminoseae	1	1			3	3	8	9	9	14	21	27
Sapindaceae	1	1	3	3	4	4	3	8	3	9	14	25
Anacardiaceae	2	2	4	5	1	1	2	3	6	10	15	21
Guttiferae	1	1	1	1	1	3	1	4	2	11	6	20
Meliaceae			2	3	3	5	5	7	3	4	13	19
Burseraceae	1	1	3	4			2	3	4	9	10	17
Fagaceae	1	1	2	4	2	2	2	4	2	3	9	14
Myrtaceae	1	1	1	1	1	1	1	2	2	9	6	14
Bombacaceae			1	5	2	3	1	3	1	2	5	13
Dipterocarpaceae	1	7	1	1			2	4	1	1	5	13
Ebenaceae			1	2	1	1	1	2	1	6	4	11
Asclepiadaceae					3	4	1	1	2	4	6	9
Dilleniaceae	1	2	1	2	1	1	1	2	1	2	5	9
Myristicaceae	1	1	1	1	1	1	2	3	2	3	7	9
Vitaceae			1	3	1	3	2	2	1	1	5	9
Lauraceae	2	3	1	1			1	1	3	3	7	8
Tiliaceae	1	1	1	2	1	1	1	1	2	3	6	8
Apocynaceae			1	1					4	6	5	7
Gnetaceae	1	1	1	1			1	3	1	2	4	7
Caesalpinoidea	1	1	5	5							6	6
Orchidaceae					3	3	2	2	1	1	6	6
Flacourtiaceae	1	1	1	1			1	3			3	5
Loganiaceae			1	1	1	1	1	2	1	1	4	5
Rosaceae							1	1	4	4	5	5
Sapotaceae	1	1	1	1	1	1	2	2			5	5
Araceae					3	3			1	1	4	4
Celestraceae					3	3			1	1	4	4
Graminae					3	3	1	1			4	4
Palmae	1	1			1	1	1	1	1	1	4	4
Rutaceae							2	2	2	2	4	4
Sterculiaceae					2	2			2	2	4	4
Combretaceae			1	1	1	1	1	1			3	3
Connaraceae				1	1	1			1	2	2	3
Convolvulaceae			1	1					1	2	2	3
Cornaceae					2	2			1	1	3	3
Icacinaceae			1	1	1	1			1	1	3	3
Magnoliaceae							2	2	1	1	3	3
Melastomataceae									2	3	2	3
Polygalaceae									1	3	1	3
Rubiaceae	1	1							2	2	3	3
Staphyleaceae					2	3					2	3
Urticaceae					2	3					2	3
Alangiaceae			1	2							1	2
Araliaceae					1	1			1	1	2	2
Cucurbitaceae					2	2					2	2
Filices					1	2					1	2
Menispermaceae					2	2					2	2
Oxalidaceae							1	1	1	1	2	2
Pteridophyta									2	2	2	2
Rhamnaceae							1	2			1	2
Thymelaeaceae									1	2	1	2

TABLE 3–5. (*Continued*)

Family	Study 1 Gen	Study 1 Spp	Study 2 Gen	Study 2 Spp	Study 3 Gen	Study 3 Spp	Study 4 Gen	Study 4 Spp	Study 5 Gen	Study 5 Spp	Total Genera	Total Species
Ulmaceae					1	1			1	1	2	2
Unknown	2	2									2	2
Actinidiaceae					1	1					1	1
Chrysobalinaceae			1	1							1	1
Flagellariaceae									1	1	1	1
Lecythidaceae							1	1			1	1
Linaceae									1	1	1	1
Malpighiaceae					1	1					1	1
Myrsinaceae									1	1	1	1
Pandanaceae					1	1					1	1
Papilionaceae			1	1							1	1
Piperaceae					1	1					1	1
Proteaceae					1	1					1	1
Rhizophoraceae					1	1					1	1
Saxifragaceae					1	1					1	1
Scitaminaceae									1	1	1	1
Simaroubaceae									1	1	1	1
Symplocaceae									1	1	1	1
Tetrameristicaceae									1	1	1	1
Zingiberaceae					1	1					1	1
Sums:	30	40	51	95	80	110	63	105	103	184	327	534

[1]Numbers in this table are of plant species and genera identified in each of five studies. The studies are (1) Mentoko 1970–71 (Rodman, unpublished); (2) Mentoko 1977–79 (Leighton and Leighton, unpublished); (3) Ketambe (Rijksen, 1978); (4) Ulu Segama (MacKinnon, 1974); and (5) Tanjung Puting (Galdikas, 1978). The author of this paper will be pleased to provide a complete list of species to anyone interested in the database from which this table was compiled.

etary composition, and the number of species identified as sources of the various plant parts in the diet shows that bark figured more heavily in the diets of orang-utans at the Mentoko than elsewhere (Table 3–7). Not surprisingly, Table 3–7 demonstrates that most food species were sources of fruit or seeds in all studies. Rodman's sample from the Mentoko in 1970–71 shows a high proportion of species used as sources of leaves (and shoots) and of bark compared with the three other samples in which all food types were recorded (i.e., excluding Leighton and Leighton's sample of fruits and seeds used). Compared with the Ulu Segama and with Ketambe, the proportions of species used for leaves and bark were high at Tanjung Puting as well, although not as high as at the Mentoko. This difference may reflect differing biases in collection of samples, or it may reflect differences in the feeding ecology of orang-utans at the four sites. If the latter is the real source of variation, orang-utans of the Mentoko appear to rely on a broader selection of food types, which may account for smaller range sizes than at other sites.

TABLE 3–6. Species and Genera of Food Plants Shared Between Sites

	1	2	3	4	5	Total Species	Total Genera
	Shared Genera \ Species						
1. Mentoko 1970–71	—	4	1	2	2	40	29
2. Mentoko 1977–79	11	—	2	4	5	95	51
3. Ketambe	8	18	—	1	8	110	79
4. Ulu Segama	10	25	22	—	9	105	62
5. Tanjung Puting	10	26	23	26	—	184	104
						502	325

[1]Numbers of genera shared between sites are entered below the diagonal. Numbers of species shared between sites are entered above the diagonal.

TABLE 3–7. Proportions of Identified Species from which Various Food Types Were Taken[1]

	Fruit and Seeds	Leaves and Shoots	Bark and Wood	Other Vegetation	Total Species
1. Mentoko 1970–71	17	16	15	1	40
	42.5%	40.0%	37.5%	2.5%	
2. Mentoko 1977–79	95	0	0	0	95
	100.0%	0.0%	0.0%	0.0%	
3. Ketambe	86	25	5	1	110
	78.2%	22.7%	4.5%	.9%	
4. Ulu Segama	97	4	2	6	105
	92.4%	3.8%	1.9%	5.7%	
5. Tanjung Puting	144	47	45	14	184
	78.3%	25.5%	24.5%	7.6%	
1 + 3 + 4 + 5	344	92	67	22	439
	78.4%	21.0%	15.3%	5.0%	

[1]The percentages sum to more than 100 because more than one plant part could be taken from one species.

Comparisons between sites with respect to taxonomic composition of the diets must be made with several sources of variation in mind. First, the number of plant taxa identified as sources of food increases with increasing study time (see Table 3–7 to compare results of Galdikas's 4-year study with Rodman's 15-month study) and with increasing attention to identification. Second, identifications are made in different herbaria by different botanists with varying familiarity with the complex flora of the area. Ideally each investigator would use the same approach to plant identification. There has been no such uniformity among the five principal studies. Finally, there may be differences in the diversity of the flora at the sites, which would determine the upper limits on numbers of food species. There have been no consistent sampling approaches among the studies to allow adequate comparison of these upper limits nor of the diversity of flora at the sites. There is much work to be done.

Feeding Postures

Cant (1987) followed two habituated adult females during 4 months at the Mentoko and recorded postures of these animals while they were feeding and traveling. In the majority of feeding bouts the animals either "sat" (above the branch, 49% of bouts) or used the "hand-foot hang I" posture, in which the animal hangs from a branch (or from clusters of small branches) (Rodman, 1977) by the hand and foot of the same side so that the body is suspended in more-or-less horizontal orientation with the hand or the hand and foot of the opposite side free (30% of feeding bouts). Cant found that in 44% of "hand-foot hang I" bouts, the "extra" foot was anchored, and in 45% of bouts it was used in feeding. Overall, the two females used suspensory postures of various sorts in about 49% of feeding bouts. Cant (1987) did not describe an additional posture noted by Rodman (1977) in which the animal hangs vertically upright by the feet, which extend upwards to grasp one or more branches or clusters of foliage at the animal's sides while one hand stabilizes the animal overhead. This posture and the "hand-foot hang I" described by Cant both demand extreme flexibility in the hip, but provide these large animals with unusually stable support in the canopy.

Techniques of Food Preparation

Fruit

The fruits of many species taken by orang-utans are small relative to the orang-utan's "gape," and consist of a hard seed surrounded by soft mesocarp and a "leathery" skin (e.g., fruits of the Anacardiaceae, such as *Dracontomelum* and *Koordersiodendron*). The fruits are taken with the lips or incisors directly from the branch, which may be bent towards the animal with a free hand or foot, or broken free and discarded when stripped of fruit. The animals normally process several fruits at a time with the lips and teeth, discard the skins from the mouth, and either spit out the seeds or swallow them. Leighton's (unpublished) observations show that swallowing the seed is the most common treatment

(Table 3–4). These seeds, which are swallowed in huge numbers, are subsequently eliminated with the mesocarp digested away. Durians are large and covered with sharp spines (for which they are named: *duri* in Indonesian means "thorn"). Rijksen (1978) found that orang-utans held the fruit carefully in both hands, plucked away spines with their incisors, and then bit through the exposed tough skin to the flesh and seeds within.

Leaves, Shoots, and Flowers

These plant parts normally require little special technique of acquisition except in posture. Cant (1987) noted that leaves taken tended to be from small, lower-canopy trees, so that "sitting" to feed on leaves was less likely than some form of suspension. Immature leaves from the base of the palm *Borassodendron borneensis,* which occurs on higher ground more than 0.5 km from the Sengatta River at the Mentoko, were taken by orang-utans (Rodman, 1977). The growing leaves are pulled from the plant's center, and the bases of the leaves are stripped with the incisors. This food is accessible only to orang-utans and humans, who eat the palm's "heart" as a vegetable.

Bark

Orang-utans feed on the bark of several species of the Moraceae (e.g., *Ficus* spp. and *Artocarpus* spp.) either by gnawing directly at the thin skin of the tree on its trunk and limbs or by stripping the bark from the tips of twigs. Bark of some species of the Dipterocarpaceae and other families requires more careful technique (personal observations): the thick skin is split with the upper canine; when the split in the bark is large enough, the animal reorients its jaws in order to grasp and pull the edge of the split bark with the incisors; when the opening is sufficiently larger, the animal pulls the bark away with a hand; finally, the growth layer, or cambium, is gnawed away from the inside of the piece of bark with the incisors, in a "corn-on-the-cob" fashion. The result is a wad of flesh and fiber that the animal chews and eventually discards from its mouth.

Insects

Rijksen (1978:89) described the clever manipulation of some ant and termite nests by orang-utans to feed on these insects. For example, orang-utans opened leaves that were rolled up, or crinkled or funnel-shaped, using their lips and hands, and then catching any emerging insects in the mouth. When feeding on ants of the genus *Camponotus,* orang-utans held the back of a hand against the tree, thus allowing the ants to crawl onto the hair, and subsequently picked the ants from the hairs with their lips. Their feeding techniques do not include using "manufactured" tools such as those of the chimpanzee for termiting (Goodall, 1968), although Rijksen (1978:92) has published a photograph of a young orang-utan using a small stick to disturb *Camponotus* ants in their nest. Rodman (1977) described one case in which a large male orang-utan bit chunks from a termite nest to reach the insects within, and Rijksen has described similar handling of termite and ant nests.

Sex Differences in Feeding

Rodman (1977) reported some sex difference in the use of major food types, based on his observations of one male and three females during a period of 2 months. The male spent less time feeding on fruit (58.6% of 144 feeding hours vs. 67.1% of 227 feeding hours) and termites (0.8% vs. 1.9%) and considerably more time feeding on bark (16.5% vs. 4.9% of feeding hours) than did the females. To the contrary, Galdikas (1978) reported that four adult males at Tanjung Puting differed from four adult females by feeding more on fruit, much more on termites, and less on bark. Galdikas's results were gathered during a period of 4 years compared with Rodman's observations during only 2 months and thus may be considered more representative of normal patterns. On the other hand, results of Rodman's more limited observations are strengthened by the uniformity of the general feeding regime of the male and females during these months (Rodman, 1977). Rijksen (1978) found that of 96 samples of orang-utans' feces at Ketambe, 76 contained insect remains, and that of the 20 samples lacking insect remains, 70% were from males. From this, Rijksen inferred that male orang-utans at Ketambe may feed less frequently on insects than females.

Ranging

Dispersion of Food

Rodman (1984b) reported a detailed analysis of time spent at food sources and distances between successive food sources of habituated subjects at the Mentoko in 1970–71. Orang-utans at the Mentoko fed for more than 90 min on nearly 25% of food sources and for 10 min or less on 11% of food sources (N = 355 sources on 72 whole days of observation of three habituated adults from August 1970 through July 1971). Thus, at the Mentoko, feeding was concentrated in large sources most of the time. Typically the long feeding bouts were in single trees where orang-utans fed on ripe fruit, flowers, or the bark of fig trees (Rodman, 1977). Considering only food sources at which orang-utans fed for more than 10 min, the median distance between successive food sources was less than 100 m, which indicates that food trees at the Mentoko were closely spaced. Similar analyses have not been done for other study areas.

Availability of Food

Galdikas (1978) counted the trees that were potential food sources for orang-utans in three sample plots. Of 177 species found in the three plots, 96 (54%) were identified as food species at Tanjung Puting. Within each plot, the number of individuals of food species was approximately 65% of all trees 10 cm or more in girth, and the number of edible species was 56% to 66% of all identified species. The majority of trees were immature, however, and the number of species with mature trees (able to produce flowers and fruits) varied from 8% to 17% of identified trees 10 cm or more in girth (N = 89/690, 117/707, and 50/703 in three plots). Galdikas (1978) reported no independent measure of seasonal abundance of fruit, but noted that the minimum number of fruit species eaten in a month was two (May 1973) and the highest was 36 (January 1973). Time spent feeding on fruit and flowers was directly related to the number of fruiting species eaten (Galdikas, 1978), suggesting strongly that fruit is preferred when it is available.

MacKinnon (1974) observed that none of the fruit species eaten by orang-utans in the Segama area was abundant. Of 28 "chief fruit species," 18 were "rare" (less than two per hectare), six were "occasional" (two to four per hectare), and four were "frequent" (four to eight per hectare). No species that were fruit sources were "abundant" (more than 10 per hectare). MacKinnon sampled 775 trees in the valleys and 640 trees in the hills of the Ulu Segama. In the valleys, six species were abundant, 22 frequent, approximately 22 occasional, and approximately 45 were rare (approximately 95 species; I have estimated the latter three totals from Fig. 9 in MacKinnon, 1974). In the hills, six species were abundant, eight frequent, 13 occasonal, and approximately 62 were rare (approximately 75 species; the latter two totals estimated from Fig. 9 in MacKinnon, 1974). Comparing the numbers of food species in each frequency class with the numbers of sampled species in the same classes shows no difference in the proportions (χ^2 = 2.10, 3 df). We might conclude that food species are no more rare than any species, but MacKinnon did not sort species into categories of mature and immature trees as Galdikas did. If we assume that the proportions of mature trees are similar to the proportions of all trees, we can conclude that in the Ulu Segama, fruit species fed on by orang-utans are no less common than the average tree species.

In the Ulu Segama area, fruiting was quite seasonal and was reflected in a very seasonal variation in frequency of feeding on fruit. The number of species of fruit eaten per month varied from three in May to 37 in July. During months of few fruit sources, orang-utans traveled farther per day (MacKinnon, 1974).

At Ketambe Rijksen (1978) censused 2,137 trees along transects and identified them with local "vernacular" species names. The number of vernacular species probably underestimates the number of true species. Of 52 "important" species of food plants, only four were "common" (nine or more trees per hectare), and 29 species were "rare" (few or single trees). Rijksen's observations and comments suggest that he found food sources to be rarer than the average species in the Ketambe forest.

The most precise analysis of seasonal variation in availability of fruits and seeds was presented by Leighton and Leighton (1983) as a result of their study of the frugivorous community of birds and mammals at the Mentoko site. They censused fruit patches in 30 half-hectare plots (first year) and 25 quarter-hectare plots (second year) each month and monitored all fruiting trees along the 27-km trail system in the 3-km^2 study area. At the same time they kept records of numbers of sightings of independent groups of all animal species. Results showed that fruits taken by orang-utans varied markedly in time, with an annual peak in supply (February through May) followed by a long period of scarcity. The rate of sightings of orang-utans varied with the peaks in fruiting, with striking peaks in February 1977 and April 1978. The variation in sightings of orang-utans demonstrates the existence of a nomadic component of the population that moves in relation to food availability.

Home Range Sizes

MacKinnon (1971) initially concluded that orang-utans of the Ulu Segama were highly nomadic, but he later modified this view somewhat to suggest that some of the members of the Ulu Segama population may have been "residents" (MacKinnon, 1974). Even the residents, however, apparently had ranges that were large relative to MacKinnon's primary 5-km^2 study area, and many individuals seemed to move long distances. Consequently, MacKinnon could not determine home range sizes.

Horr (1975) estimated that adult females occupied stable ranges of "¼ square mile or less" (ca. 0.64 km^2) in the Lokan River population, but that males used much larger ranges that he estimated to be 2 mi^2 (ca. 5.1 km^2). In an interesting modeling approach, Horr and Ester (1976) examined the effect on population growth of varying the size of male home ranges. By also taking into account rates of travel of males and females as well as some other parameters derived from observations, varying male range size had the effect of modifying the probability that estrous females would be contacted by males and thus conceive efficiently. The model showed that a range size of 5.76 km^2 maintained a stable population. A smaller range size led to decline and a larger range size had little effect on population size. This interesting result might provide important information about orang-utans if males stayed in stable ranges, but they do not. Instead, breeding males apparently move away when females in the local area are not nearing conception (see below). Thus, the "residency" of males during a period of observation and the sizes of their ranges when they are resident in the area probably depend primarily on the general reproductive condition of the local females (Rodman and Mitani, 1987).

Rodman (1973a,b) found that four adult females at the Mentoko site occupied ranges of 0.4 to 0.6 km^2, and that two adult males used ranges of approximately twice the size as the females during the 15 months of observation in 1970–71. Mitani found that the ranges of females at the Mentoko in 1981–82 were larger than 1.5 km^2 (Rodman and Mitani, 1987). Despite the difference in the two figures, what is most important is that the range sizes of females in both studies at the Mentoko are smaller than the range sizes at Tanjung Puting and Ketambe.

Galdikas (1978) reported female ranges of 5 to 6 km^2 at Tanjung Puting. The two "most observed" females used core areas of 2 to 3 km^2 where they spent most of their time. The ranges of adult males were complex. One "resident" male used a range of approximately the same size as the females' ranges from February 1972 to January 1973, when he left the region. He returned in March 1975 to occupy the same range until the end of the reported period in November 1975. This range was used by a second adult male from July 1973 through November 1975. A third adult male used a much larger range including the other males' ranges from July 1973 to November 1975. Thus, in the latter part of the study, there were three males occupying overlapping ranges, and their ranges also overlapped home ranges of several females.

At Ketambe one resident adult female occupied a range of 1 km^2, while another used a range of approximately 2 km^2 (Rijksen, 1978). The latter female shifted her range during the study to a different core area. One adult male was persistently resident, occupying an area of 2 km^2 throughout the study. A "nonresident" male remained in the study area for 2 months in 1971, left, and returned in 1973 for 4 months when his range overlapped the range of the resident male, whom he appeared to displace. The nonresident departed, but was resighted in December 1975, at the end of the study. A second nonresident moved in and out of the study area, apparently in relation to the fruiting of favored figs and durians. He also overlapped the resident male's range when present, and dominated the resident male in agonistic encounters.

Day Ranges

The path lengths of daily travel in each population studied are shown in Table 3–8. Day ranges appear to vary among populations in relation to the sizes of home ranges. The shortest day ranges were at the Mentoko (Rodman, 1977), where home range sizes were the smallest, and the longest day ranges were at Tanjung Puting, where home ranges are also the largest (although, again, MacKinnon's [1974] observations of ranging are difficult to fit into this interpretation). This relationship suggests that, although some of the variation in home range sizes reported results from differences in study methods as well as durations of study (e.g., contrast Rodman's and Galdikas's studies), variations in range sizes are also related to differences in the dispersion of food among the study areas. We may thus ten-

TABLE 3–8. Day Ranges (Lengths of the Daily Travel Paths)

Site and Study	Mean	
Mentoko 1970–71 (Rodman, 1977) Observations of one adult male, two adult females, and one juvenile female	305 m	$N = 76$
Ketambe (Rijksen, 1978)		
One adult male	480 m	(range of 180–1,250 m)
One adult female	550 m	(range of 100–1,800 m)
One subadult male	890 m	(range of 300–2,000 m)
Ulu Segama (MacKinnon, 1974)	500 m	
Tanjung Puting (Galdikas, 1978)		
Adult males	850 m	$N = 4$ males
Adult females	710 m	$N = 4$ females
Subadult males	642 and 999 m	Means of 2 subadult males
Adolescent females	698 and 854 m	Means of 2 adolescent females

tatively hypothesize that resources are most dispersed at Tanjung Puting, less so at Ketambe, and least dispersed at the Mentoko.

Locomotion

Sugardjito (1982), working at Ketambe with one adult male and one adult female, classified locomotion (travel between major food sources or between sleeping or resting places and food sources) into quadrumanous scrambling, brachiation, tree swaying, quadrupedal walking, and climbing. Cant (1987), who studied two adult females at the Mentoko, observed locomotion during travel and while the animals fed in trees. He recorded canopy level, type (tree or liana), size (five classes), and angle of the substrate for each locomotor bout as well as the mode of locomotion, to which he added some finer classes than the five categories used by Sugardjito. Table 3–9 shows the results of these two studies, which are alike in identifying the importance of "clambering" or "scrambling" in travel. This mode of locomotion is a kind of "catch-as-catch-can" use of the hands and feet to move forward more or less horizontally through the forest canopy. Cant found less brachiation than did Sugardjito. It may be important that Sugardjito observed that his female subject brachiated less than his male subject, which suggests that Cant's results are in part affected by observing only females. Tree swaying, in which the animal swings the tree it occupies back and forth unitl the arc brings it close enough to the next tree to allow crossing, is less common in Cant's than in Sugardjito's results. Sugardjito reported that tree swaying accounted for more than 20% of travel bouts of his male subject and only 9% of travel bouts of his female subject.

Cant's analysis showed that his subjects traveled in the lower forest strata, although they fed in all strata, and that two-thirds of travel was on parts of trees rather than lianas. The majority of travel (68%) was on branches and lianas less than 10 cm in diameter, and most of the supports were from 4 cm to less than 10 cm in diameter.

THE SOCIAL SYSTEM

Populations

All populations of orang-utans described so far consist of a relatively stable set of "resident" adult females who may have dependent offspring and may be accompanied by older immature individuals who are probably their offspring. These adult females occupy overlapping but nonidentical ranges. At two sites there has been observation of long periods in the lives of males

TABLE 3–9. Locomotor Modes during Travel and Feeding

	Brachiate	Quadrupedal	Inverted quadrupedal	Climb				Clamber horiz.	Tree sway	Quadrumanous scramble
				up	down	up diag.	down diag.			
Cant (1987)										
Travel (N = 3,496 m)	11%	13%	1%	12%	6%	5%	5%	41%	7%	
Feeding (N = 864 m)	9%	12%	2%	19%	14%	5%	6%	33%		
					Climb					
Sugardjito (1982)	21%	13%			10%				15%	41%

(Galdikas, 1978; Rijksen, 1978), and it is apparent that the lives of males are complex compared with those of females. One consequence of this complexity is variation from time to time and place to place in the male component of the local population. In all studies, however, there are a few adult males who are "residents," and there are transient males of several ages. The transients may be "subadults," meaning that they have not developed the secondary sexual characteristics of the fully mature male, including large body size, cheek flanges, and a throat sac. Some transients may also be physically mature males.

There is important variation in the size and composition of the populations at the four principal study sites (Table 3–10). MacKinnon (1974) reported observing approximately 70 different individuals during 12 months in his principal study area, Segama area A. Seventeen of these animals visited the area more than twice. Although these 17 individuals apparently ranged over large areas (as described above), they might be called residents.

At the Mentoko in 1970–71, Rodman (1973a,b) observed a smaller number of more sedentary orang-utans. Rodman reported that the resident males occupied nonoverlapping ranges and suggested that these males might be "territorial." Two subadult males passed "rapidly" through the area during the 15-month study. At the same location in 1981–82, Mitani (1985a,b) found a population that included two subadult males who were resident throughout his study as well as a set of transients, including eight males (six subadult and two adult), one solitary adult female, and an unknown number of additional unidentified subadult males. The resident adult male was dominant to all other males.

Rijksen (1978) recognized 22 individuals who appeared regularly during his 3-year study. One of the adult males was consistently a resident within the study area, while two moved in and out (see Ranging, above). The resident male persisted through Schürmann's subsequent study in the same area and apparently fathered at least one offspring born there (Schürmann, 1981, 1982).

Galdikas (1978) described a population of 54 known individuals occupying the large 20-km^2 study area, but she concentrated on a smaller set in a smaller part of the total area for most of her observations. She described a higher rate of social interaction among orang-utans than other field workers, largely because she studied a population in which more juvenile and adolescent animals were present.

Stochastic variation in the modal reproductive state of adult females apparently causes differences in patterns of sociality among the study populations (Rodman and Mitani, 1987). For example, the large number of juvenile and adolescent animals at Tanjung Puting (Table 3–10)

TABLE 3–10. Composition by Age and Sex of the Units of the Study Populations

	Mentoko[1] (1970–71)		Mentoko[2] (1981–82)		Ketambe[3]	Ulu Segama[4]	Tanjung Puting[5]
Adult Female	Res	Trans	Res	Trans			
Alone				1		1	
+ Infant	2				2		2
+ Juvenile or Adolescent	1		3		3	4	4
+ Infant and Adolescent or Juvenile	1				1	1	5
+ Adolescent and Juvenile							3
Adult Male	2		1	2	1	3	12
Subadult Male		2	2	6	4	2	9
Adolescent Female					1		1
Adolescent Male					1		

[1]From Rodman (1973a,b).

[2]From Mitani (1985a,b); Mitani does not give the composition of female-offspring units.

[3]From Rijksen (1978).

[4]From MacKinnon (1974).

[5]From Galdikas (1978).

resulted in very different frequencies of social aggregation in Galdikas's (1978) study than in Rodman's (1973) study at the Mentoko, and the persistence in some areas of "subadult" males, who may have been attracted by females who were near conception, provided Mitani (1985b) and Schürmann (1982) with many observations of sexual interaction compared with studies of populations in which females were generally sexually "quiescent" (i.e., pregnant or lactating). The important conclusion to be drawn is that apparent differences in these patterns result from constant characteristics of orang-utans in populations of different composition rather than from differences in the animals themselves or differences in methods or abilities of the observers.

The Life Cycle

Early development of orang-utans has been described in detail by Horr (1977) and Rijksen (1978). Early social experience is primarily with the mother because of the females' solitary habits. Other social experience depends, first, on whether there is an older sibling, and second, on whether there are other young animals with the mother's adult female neighbors. Complete dependence on the mother lasts several years. Galdikas (1978) estimated 4 years, and Rijksen (1978) estimated 5 years as the end of any transport by the mother. As young animals become more active and independent of the mother, they become more "playful" in encounters with neighboring peers. At Ketambe one juvenile male consistently initiated play with a juvenile siamang (Hylobatidae: *Hylobates syndactylus*), with little reciprocation, when the two fed in the same fruit tree (personal observation, 1985). Observations in 1970–71 at the Mentoko of a juvenile female revealed no such "playfulness" in her encounters with a pair of similar-aged "orphans" who followed the juvenile's putative mother (personal observation). Play among young orang-utans of Borneo has been reported by MacKinnon (1974) and Galdikas (1978), however, so that it would be incorrect to infer that Sumatran orang-utans differ from Bornean orang-utans in this regard.

Adolescence

As juvenile orang-utans mature, they reach "adolescence," and during this time they become increasingly independent in daily movements from the mother (Galdikas, 1978; Rijksen, 1978; Rodman, 1973a,b). A young female observed by Rijksen (1978) and by Schürmann (1982) began sexual cycles at approximately 7 to 8 years of age. Once reaching adolescence, she was quite proceptive sexually, seeking out the resident adult male and subadult males for copulation. She did not conceive until 5 years later, however, when the resident adult male apparently fathered her first offspring. Galdikas (1978, 1981) has described a very similar pattern of development in sexuality in another adolescent female.

Young females remain near the place of birth. This has been observed at Ketambe (Rijksen, 1978; Schürmann, 1982) and at Tanjung Puting (Galdikas, 1978, 1981). This inference from a small sample of direct observations is supported by the fact that strange females have rarely appeared in any study. The sedentary nature of females is significant socially because it means that neighboring females must be related rather closely.

The picture for adolescent males is not as clear. One adolescent male at Ketambe became independent of his mother and then was seen alone or in the company of "subadult" males in the study area (Rijksen, 1978). He was contacted briefly outside his mother's range and then was not seen at all for several years thereafter (Rijksen, 1978, citing 1976 personal communication from Schürmann). This single documented emigration is not sufficient basis for inferring that young males emigrate from their birthplaces, but indirect evidence supports the inference. In all populations studied, for example, there have been several transient males that appear briefly and may or may not appear again. The logical conclusion is that these males have emigrated from elsewhere—initially from their birthplaces. This conclusion is complicated, however, by the fact that "subadult" males have been residents during two studies (Mitani, 1985a,b; Rijksen, 1978). These males were sexually active, although most of their copulations were forced copulations with apparently nonovulating females (Mitani, 1985b; Schürmann, 1982). The best interpretation is that there is a floating population of males who have moved away from their natal areas and who settle for shorter or longer times in areas of potential reproductive opportunity. When reproductive opportunity persists, these males remain present, but they leave when reproductive opportunity decreases. It is important to keep in mind that some of the supposed "subadult" males may be older males who have never developed physically (see below).

Adulthood

First pregnancy has been observed in several females and comes at approximately 12 years of age (Galdikas, 1981; Schürmann, 1982). Once they become pregnant and from then on, young females become considerably less sociable, showing markedly reduced proceptivity and reduced nonsexual gregariousness (Galdikas, 1978). The interbirth interval has been estimated to be 7 years for one female, and indirect evidence indicates that it is generally more than 4 years (Galdikas, 1981). As adults, females are predominantly solitary, although they do not avoid other females when they meet in their overlapping ranges. Adult females are harassed by subadult males and resist the subadults' attempts at forced copulation (Galdikas, 1978, 1981, 1985a; MacKinnon, 1974; Mitani, 1985b; Schürmann, 1982). When they are sexually receptive (after many years of reproductive quiescence), they consort with and copulate cooperatively with some adult males (Galdikas, 1978, 1984; Mitani, 1985b; Rijksen, 1978; Schürmann, 1981, 1982). Although some evidence indicates female preference for "dominant adult" males (Galdikas, 1978, 1984; Rijksen, 1978; Schürmann, 1981), there is contradictory evidence that subadults are successful at impregnating females in spite of their social subordination to mature males (Mitani, 1985b).

Physical adulthood in males consists of the near-doubling in weight and the development of cheek flanges and of a large laryngeal sac (Rodman, 1984b). Study of captive animals has shown that the development of these secondary sexual characters in an immature male may be suppressed if the young male is housed with a physically mature male (Kingsley, 1982). There is a small amount of evidence that such suppression, or at least the lack of development, of these physical attributes may occur in the wild. Wallace (1869) reported collecting two males with fully adult dentitions and closed cranial sutures, but whose body size was similar to females and whose cheek pads were undeveloped. In addition, one "subadult" at Ketambe has been called a subadult for 15 years (van Noordwick, personal communication, 1986). Thus, some supposedly immature animals may be somatically undeveloped adults.

Many males observed in the wild do, however, have mature secondary sexual characteristics. In captivity these features develop between ages 6 and 12 (Kingsley, 1982). The pattern of range use by adult males (meaning those with developed secondary sexual characteristics) has been described above and it is complex. At Ketambe, one adult male was a persistent resident for many years (1971–1979) (Rijksen, 1978; Schürmann, 1982) and fathered at least one offspring in that time (Schürmann, 1982). Two other adult males moved in and out of the area, apparently dominating the resident in encounters (Rijksen, 1978). There was a similar pattern at Tanjung Puting during 1971–75 (Galdikas, 1978), when the dominant resident male departed for a long time and then returned, while other resident males remained in the area during his absence. The interpretation of male behavior is hampered by lack of information on what these adults do when they leave the study areas, but one hypothesis is that they depart when most females are pregnant or lactating and either seek female populations offering more reproductive opportunities (Rodman and Mitani, 1986) or seek good feeding grounds (Galdikas, 1978).

When in the presence of sexually receptive females, adult males compete aggressively for them (Galdikas, 1978, 1981, 1985a). They easily displace subadults from females (Galdikas, 1985a; Mitani, 1985b), although subadults may have contacted the receptive females first (Galdikas, 1985b). Adult males fight violently in the presence of females, and indirect evidence such as wounds, scars, and broken fingers suggests more aggression than has been observed directly (Galdikas, 1985a).

Adult males interact with other orang-utans indirectly through use of the "long call" (MacKinnon, 1974). This is a loud call similar in quality to the roar of a male lion given by mature males only, both spontaneously and in response to sudden stimuli such as falling trees. The throat sac is normally inflated when the call is given. Rodman (1973a,b), Galdikas (1978, 1983), and Schürmann (1981) proposed that the call is used for spacing between males as well as to attract receptive females. Mitani (1985a), however, has shown experimentally that the call displaces other males and that adult females avoid the caller rather than approaching him. Schürmann (1982) described in some detail interactions between a young female and an adult male in which the female was attracted to the male's long call. There is at least some sexual arousal in conjunction with the long call. The penis of at least one male was erect and ejaculation occurred during a long call given when the male was alone (Schürmann, 1982). Parts of the long call are given prior to copulation, and Schürmann suggested that a young female used startling stimuli (she broke off a dead "snag") to arouse the male to call, after which he copulated with her. Evidence about the function of the long call in male-female relations is thus contradictory.

The Social System

The social system may be simply summarized as one in which adults are solitary and males compete for access to ovulating females, resulting in effective polygyny and great sexual dimorphism. Females are sedentary, remaining near their natal ranges. Males emigrate from their natal areas. Young males pursue a reproductive strategy of "sneaky rape" (Galdikas, 1985b), and some males may persist in this pattern long past sexual maturity while they remain somatically immature. Other males mature physically and pursue a pattern of "conflict and consort" (Galdikas, 1985a) that leads them to remain for months or more in a smaller area and compete for local ovulating females. If these males are successful at their strategy, females consort with them around the time of ovulation.

MacKinnon (1974) asserted that there was a larger "community" of orang-utans in the Ulu Segama, but there is little evidence of larger community structure in three of the four populations studied. Patterns of social interactions among the solitary "units" of the population are omitted from discussion here because there are no systematic conclusions about the social system to be drawn that are not evident in describing the behavior of the individuals themselves.

Young animals, particularly adolescent and subadult males and females, are the most sociable component of the population, but there is no evidence that early social contact affects later adult social interactions. Adult females meet and occasionally travel together for short times. These are probably related females. Subadult males persistently follow adult females and forcibly copulate with them. Finally, adult males avoid each other and seldom contact animals of other age/sex classes except for ovulating females.

CONCLUSIONS: IMPLICATIONS FOR UNDERSTANDING FOSSIL HOMINOIDS

The fossil hominoid *Sivapithecus indicus* is closely allied with the living genus *Pongo* and shares some striking morphological characteristics with *Pongo*. Among these is the anterior maxillary dental complex, including small lateral incisors, broad medial incisors, and relatively short canines (Andrews and Cronin, 1982). Does review of the feeding ecology and behavior of *Pongo* illuminate the function of this dental complex? Bark is an unusual and persistent component of the diet of living orang-utans, and it is sought when favored mature fruit is not available (Rodman, 1977). Although chimpanzees *(Pan troglodytes schweinfurthii),* whose diet is quite similar to the diet of orang-utans in general characteristics (Galdikas and Teleki, 1981; Rodman, 1984b), consume bark when fruit is very scarce (Nishida, 1976), bark is not a consistent part of their diets and was very rarely eaten during one major study (Wrangham, 1977). I have described the technique of feeding on bark above (see Techniques of Food Preparation) and propose that the broad central maxillary incisors, small lateral maxillary incisors, and short maxillary canine of the orang-utan and *Sivapithecus* are related to that technique of opening and stripping bark. I suggest specifically that the use of the canine to penetrate the bark leads to unusual wear on the lateral maxillary incisors, and that this technique could be the source of selection favoring quite small lateral maxillary incisors. The method of handling bark could account for the remarkable wear evident on the lateral maxillary incisors of the fossil specimen GSP15000 (personal observation of casts) on which the right I^2 appears to have been sheared off by wear.

Walker (1973) has described the morphology and wear patterns of the anterior dentition of museum specimens of the great apes in detail, and his observations are highly germane to the hypothetical significance of bark-eating to the dentitions of orang-utans and *Sivapithecus*. He noted that, regarding specimens of *Pongo*, "In severely worn dentitions, particularly those of males, a considerable amount of the distal portion of the maxillary lateral incisors may be worn away by contact with the lower canine" (Walker, 1973: 39). I suggest that it is not so much the contact between maxillary incisors and the lower canine as it is the consistent interdental wear produced by penetration and opening of bark that contributes to this attrition of the lateral maxillary incisors. The bark-stripping technique would first cause wear as the upper canine penetrates the bark and pulls back to open the split in the bark. As the animal reorients the jaws and teeth to grasp the edge of the split, the edge of the bark would be forced between the distal border of the lateral maxillary incisor and the mesial border of the maxillary canine. As the orang-utan draws the head back to open the split further, the bark would wear against the distal border of the incisor. Any lateral movement (or slippage) during this use of the anterior dentition would produce laterally directed striations on the wear surfaces of the anterior dentition and ultimately heavy wear of the distal border of the maxillary lateral incisor. We might also expect wear on the mesial border of the mandibular canine.

Walker's (1973) description of wear striations on the incisors of museum specimens of the great apes provides some very weak support for this hypothesis. He found that wear striations were present on the maxillary incisors of 50% of specimens of *Gorilla*, but on only 7% and 11% of maxillary incisors of *Pan* and *Pongo*, respectively. Approximately 30% of specimens of all three apes had wear striations on the mandibular incisors. The majority of wear striations in all apes tended to fall along a sagittal plane, but Walker found that "there is a tendency for wear striations on lateral incisors to be oriented perpendicular to the sagittal plane (60–90 degrees). . . . *This was particularly true for the lateral mandibular incisors of* Pongo" (Walker, 1973: 54; my emphasis).

This hypothesis suggests a possible behavioral basis for the unusual morphological pattern of the anterior dentition of orang-utans. Testing the hypothesis will require several pieces of work in the field. It will be necessary to examine in detail the frequency and technique of handling bark by each of the great apes and at the same time to measure the physical properties and mechanical demands of bark as a food. If we find that bark handling is likely to be related to the anterior dental pattern in *Pongo,* then an important inference is that *Sivapithecus* also used bark as a food. If so, bark could have been a staple supporting *Sivapithecus* in forests that from modern perspective seem to have been inhospitable to hominoids.

Thickened molar enamel is a second character shared by *Pongo* and *Sivapithecus* (Andrews and Cronin, 1982), and this character may be related to "hard object feeding" as it seems to be in the New World monkey *Cebus apella* (Kay, 1981; Teaford, 1985). Thickened molar enamel may be a shared primitive character (Martin, 1985) that is less important than the shared derived anterior dental pattern to understanding the feeding adaptation of the *Sivapithecus,* however, and there is no convincing evidence in the feeding behavior of living *Pongo* that hard objects are even rarely of any importance to the diet. While it is not possible to rule out "extinct" patterns in extinct animals, such patterns should only be proposed when alternatives that take into account known behavior in relevant living animals fail as functional bases for observed fossil patterns. The first principle must be that the fossil hominoids must have been most like living hominoids, particularly in critical respects such as feeding and dental morphology.

The analysis of "The natural history of *Sivapithecus*" presented by Andrews (1983) must be criticized in light of this principle. Although Andrew's taphonomic analysis of likely habitats of *Sivapithecus* is a logical approach, and although global analyses of morphology of primates relative to feeding and other behavior (e.g., reference to analyses of incisor size and diet and to body size and diet) may be useful, Andrews nearly omits any reference to the behavior and ecology of living orang-utans (or other living hominoids). The resulting review presents a picture of a terrestrial, group-living herbivore that has no analog among the living primates. The habitat of *Sivapithecus* may have been quite different from that of living orang-utans, in which case Andrews's striking departure from relevant living models may be justified, but the evidence of habitat structure is not convincing enough to exclude the possibility that *Sivapithecus* occupied a forested habitat—though in a forest quite different taxonomically, if not structurally, from most that exist today. Understanding how the hominoid *Sivapithecus* occupied such habitat deserves a careful analysis with attention to the primary principle asserted here: we must first exclude the closest living morphological analogs as models for the natural history of fossils before proposing purely speculative alternative adaptations.

The orang-utan is a large, arboreal frugivore with an intriguing dependence on bark as an important food source. Cranial, dental, and postcranial configurations suggest an ancient adaptation to this mode of life, and any reconstruction of the ecology and behavior of fossils similar in morphology to *Pongo* should carefully reject all possibility of a similar existence before proposing alternative modes currently unknown among the apes.

ACKNOWLEDGMENTS

I thank M. Leighton, D. Leighton, and J. C. Mitani for generously allowing me to use their unpublished observations. I thank R. Palombit and K. Palombit for hospitality during a recent visit to Ketambe that refreshed my enthusiasm for understanding the orang-utan and other apes. I take this opportunity to thank C. L. Darsono for his continued generous assistance and hospitality in Jakarta, where he has consistently facilitated work by me and by my students. I have been supported by USPHS Grant RR00169 during preparation of this chapter.

REFERENCES

Andrews, P. J. 1983. The natural history of *Sivapithecus*. *In* New Interpretations of Ape and Human Ancestry, pp. 441–463, ed. R. L. Ciochon and R. S. Corruccini. Plenum Press, New York.

Andrews, P. J. and Cronin, J. E. 1982. The relationships of *Sivapithecus* and *Ramapithecus* and the evolution of the orangutan. *Nature, 297:*541–546.

Cant, J. G. H. 1987. Positional behavior of the Bornean orangutan *(Pongo pygmaeus). American Journal of Primatology, 12:*71–90.

Charles-Dominique, P. 1972. Ecologie et vie sociale de *Galago demidovii* (Fischer 1808; Prosimii). *Zeitschrift für Tierpsychologie, Suppl. 9:*7–42.

Galdikas, B. M. F. 1978. Orangutan Adaptation at Tanjung Puting Reserve, Central Borneo. Doctoral dissertation, University of California, Los Angeles.

Galdikas, B. M. F. 1979. Orangutan adaptation at Tanjung Puting Reserve: mating and ecology. *In* The Great Apes, pp. 194–233, ed. D. A. Hamburg and E. R. McCown. Benjamin/Cummings, Menlo Park.

Galdikas, B. M. F. 1981. Orangutan reproduction in the wild. *In* The Reproductive Biology of the Great Apes, pp. 281–300, ed. C. E. Graham. Academic Press, New York.

Galdikas, B. M. F. 1982. Orangutans as seed dispersers at Tanjung Puting, Central Kalimantan: implications for conservation. *In* The Orang Utan: Its Biology and Conservation, pp. 285–298, ed. L. E. M. de Boer. W. Junk, The Hague.

Galdikas, B. M. F. 1983. The orangutan long call and snag crashing at Tanjung Puting Reserve. *Primates, 24:*371–384.

Galdikas, B. M. F. 1984. Adult female sociality among wild orangutans at Tanjung Puting Reserve. *In* Female Primates: Studies by Women Primatologists, pp. 217–235, ed. M. F. Small. Alan R. Liss, New York.

Galdikas, B. M. F. 1985a. Subadult male orangutan sociality and reproductive behavior at Tanjung Puting. *American Journal of Primatology, 8:*87–99.

Galdikas, B. M. F. 1985b. Adult male sociality and reproductive tactics among orangutans at Tanjung Puting. *Folia Primatologica, 45:*9–24.

Galdikas, B. M. F. and Teleki, G. 1981. Variations in subsistence activities of female and male pongids: new perspectives on the origins of hominid labor division. *Current Anthropology, 22:*241–256.

Goodall, J. 1968. The behaviour of free-living chimpanzees in the Gombe Stream Reserve. *Animal Behaviour Monographs, 1:*165–311.

Hooijer, D. A. 1948. Prehistoric teeth of man and orang-utan from Central Sumatra, with notes on the fossil orang-utan from Java and Southern China. *Zool. Mededeel. Mus. Leiden, 29:*175–301.

Horr, D. A. 1972. The Borneo orang-utan. *Borneo Research Bulletin, 4:*46–50.

Horr, D. A. 1975. The Borneo orang-utan: population structure and dynamics in relationship to ecology and reproductive strategy. *In* Primate Behavior, vol. 4, pp. 307–323, ed. L. A. Rosenblum. Academic Press, New York.

Horr, D. A. 1977. Orang-utan maturation: growing up in a female world. *In* Primate Biosocial Development, pp. 289–321, ed. S. Chevalier-Skolnikoff and F. E. Poirier. Garland Publishing, New York.

Horr, D. A. and Ester, M. 1976. Orang-utan social structure: a computer simulation. *In* The Measures of Man: Methodologies in Biological Anthropology, pp. 3–53, ed. E. Giles and J. S. Friedlander. Peabody Museum Press, Cambridge, MA.

Kay, R. F. 1981. The nut-crackers—A new theory of the adaptations of the Ramapithecinae. *American Journal of Physical Anthropology, 55:*141–151.

Kingsley, S. 1982. Causes of non-breeding and the development of the secondary sexual characteristics in the male orang utan: a hormonal study. *In* The Orangutan: Its Biology and Conservation, pp. 215–229, ed. L. E. M. de Boer. W. Junk, The Hague.

Koenigswald, G. H. R. von. 1982. Distribution and evolution of the orang utan, *Pongo pygmaeus* (Hoppius). *In* The Orang Utan: Its Biology and Conservation, pp. 1–15, ed. L. E. M. de Boer. W. Junk, The Hague.

Leighton, M. and Leighton, D. R. 1983. Vertebrate responses to fruiting seasonality within a Bornean rain forest. *In* Tropical Rainforest: Ecology and Conservation, pp. 181–196, ed. S. L. Sutton, T. C. Whitmore, and A. C. Chadwick. Blackwell, Oxford.

MacKinnon, J. R. 1969. The Oxford University Expedition to Sabah 1968. *Oxford University Explorers Bulletin, 17:*53–70.

MacKinnon, J. R. 1971. The orang-utan in Sabah today. *Oryx, 11:*141–191.

MacKinnon, J. R. 1973. The orang-utan in Sumatra. *Oryx, 12:*234–242.

MacKinnon, J. R. 1974. The behaviour and ecology of wild orang-utans *(Pongo pygmaeus). Animal Behaviour, 22:*3–74.

MacKinnon, J. R. 1977. A comparative ecology of Asian apes. *Primates, 18:*747–772.

MacKinnon, J. R. 1979. Reproductive behavior in wild orang-utan populations. *In* The Great Apes, pp. 256–273, ed. D. A. Hamburg and E. R. McCown. Benjamin/Cummings, Menlo Park.

Martin, L. 1985. Significance of enamel thickness in hominoid evolution. *Nature, 314:*260–263.

Mitani, J. C. 1985a. Sexual selection and adult male orangutan long calls. *Animal Behaviour, 33:* 272–283.

Mitani, J. C. 1985b. Mating behaviour of male orangutans in the Kutai Game Reserve, Indonesia. *Animal Behaviour, 33:*391–402.

Nishida, T. 1976. The bark-eating habits in primates, with special reference to their status in the diet of wild champanzees. *Folia Primatologica, 25:*277–287.

Rijksen, H. D. 1978. A Field Study on Sumatran Orangutans (*Pongo pygmaeus abelii* Lesson, 1827): Ecology, Behaviour and Conservation. H. Veenman & Zonen, Wageningen.

Rijksen, H. D. and Rijksen-Graatsma, A. G. 1975. Orang utan rescue work in North Sumatra. *Oryx, 13:*63–73.

Rodman, P. S. 1973a. Population composition and adaptive organisation among orangutans of the Kutai Reserve. *In* Comparative Ecology and Behaviour of Primates, pp. 171–209, ed. R. P. Michael and J. H. Crook. Academic Press, London.

Rodman, P. S. 1973b. Synecology of Bornean Primates. Doctoral dissertation, Harvard University, Cambridge, MA.

Rodman, P. S. 1977. Feeding behaviour of orangutans in the Kutai Reserve, East Kalimantan. *In* Primate Ecology, pp. 383–413, ed. T. H. Clutton-Brock. Academic Press, London.

Rodman, P. S. 1979. Individual activity patterns and the solitary nature of orangutans. *In* The Great Apes, pp. 234–255, ed. D. A. Hamburg and E. R. McCown. Benjamin/Cummings, Menlo Park.

Rodman, P. S. 1984a. Structural differences in habitats of *Macaca fascicularis* and *Macaca nemestrina. American Journal of Physical Anthropology, 63:*209–210.

Rodman, P. S. 1984b. Foraging and social systems of orangutans and chimpanzees. *In* Adaptations for Foraging in Nonhuman Primates, pp. 134–160, ed. P. S. Rodman and J. G. H. Cant. Columbia University Press, New York.

Rodman, P. S. and Mitani, J. C. 1987. Orangutans: sexual dimorphism in a solitary species. *In* Primate Societies, pp. 146–154, ed. B. Smuts, D. L. Cheney, R. M. Seyfarth, T. Struhsaker, and R. W. Wrangham. University of Chicago Press, Chicago.

Schürmann, C. L. 1981. Courtship and mating behavior of wild orangutans in Sumatra. *In* Primate Behavior and Sociobiology, pp. 130–135, ed. A. B. Chiarelli and R. S. Corruccini. Springer-Verlag, Berlin.

Schürmann, C. L. 1982. Mating behaviour of wild orangutans. *In* The Orang Utan: Its Biology and Conservation, pp. 269–284, ed. L. E. M. de Boer. W. Junk, The Hague.

Sugardjito, J. 1982. Locomotor behaviour of the Sumatran orang utan *(Pongo pygmaeus abelii)* at Ketambe, Gunung Leuser National Park. *Malayan Nature Journal, 35:*57–64.

Teaford, M. F. 1985. Molar microwear and diet in the genus *Cebus. American Journal of Physical Anthropology, 66:*363–370.

Walker, P. L. 1973. Great Ape Feeding Behavior and Incisor Morphology. Doctoral dissertation, University of Chicago.

Wallace, A. R. 1869. The Malay Archipelago. Macmillan and Company, London.

Wheatley, B. P. 1982. Energetics and foraging in *Macaca fascicularis* and *Pongo pygmaeus* and a selective advantage of large body size in the orang-utan. *Primates, 23:*348–363.

Whitmore, T. C. 1975. Tropical Rain Forests of the Far East. Clarendon Press, Oxford.

Wrangham, R. W. 1977. Feeding behaviour of chimpanzees in Gombe National Park, Tanzania. *In* Primate Ecology, pp. 503–538, ed. T. H. Clutton-Brock. Academic Press, London.

4

The Phylogenetic Status of Orang-utans from a Genetic Perspective

JON MARKS

Evolution is, simply stated, a genetic transformation through time. While the locus, pace, and kinds of such change may be open to dispute, it was nevertheless recognized as early as the turn of the century that evolutionary differences among taxa could be approached as problems of heredity (Bateson, 1904; Morgan, 1903). When in the 1960s it became practical to study some aspects of the genetic proximity of taxa to one another, one of the earliest and most profound conclusions was that *Pongo* is genetically distinct from a cluster consisting of *Pan, Gorilla,* and *Homo* (Goodman, 1962). It was therefore inferred that the orang-utan lineage diverged somewhat earlier in time than the *Pan, Gorilla,* and *Homo* lineages separated from one another (Goodman, 1962, 1963; Simpson, 1963).

It is now generally believed that adaptive anatomical evolution does not keep pace well with molecular evolution and that the latter exhibits a long-term regularity in rate (Wilson et al., 1977). This obviously does not deny the probability of short-term fluctuations in rate and is consistent with the likelihood that much molecular change is not of adaptive significance (Kimura, 1983).

In this essay, I review the phenetic and cladistic data in support of the phylogenetic inference described above, particularly in light of recent alternative phylogenetic hypotheses proposed by Schwartz (1984a,b) and Kluge (1983). The works to be discussed fall broadly into three categories: studies at the level of proteins, chromosomes, and DNA.

PROTEIN DATA

In the 1950s, it was discovered that DNA contains the hereditary instructions which ultimately determine the structure of proteins. In the absence of technological capabilities for directly comparing hereditary materials, it was thought that, when it did become possible to compare proteins across taxa, such comparisons would yield a reflection of the genetic similarity among taxa. Nuttall had used immune reactions to vertebrate bloods as a gross measure of phylogenetic proximity as early as 1904; research in this area continued for the next half-century (Boyden, 1952), but this did not appreciably affect the analysis of hominoid relations until the work of Goodman (1962, 1963).

Three major approaches have been used in the study of protein similarities as indicators of phylogenetic divergence: (1) immunology, which depends upon the antibody reaction of serum drawn from an animal immunized against a particular antigen from a given species; (2) electrophoresis, which uses an electric field to separate allelic gene products and, thus, to determine whether proteins with similar migratory properties are shared among taxa or among individuals of the same taxon; and (3) amino acid sequencing, a direct analysis of protein structure. The first two methods yield distances and are therefore phenetic; the last yields potential cladistic data, as shared amino acid substitutions among taxa.

Phenetic Analyses

The major immunological research bearing on the placement of the orang-utan comes from three sources: whole serum reactions or specific protein reactions, agar gel diffusion (Goodman, 1962, 1963) and albumin and transferrin reactions using microcomplement fixation (reviewed in Sarich and Cronin, 1976). In all cases, the intensity of immunological reaction is the raw datum, which is then converted into arbitrary units of distance. The earliest gel diffusion data were not quantified, but an arbitrary scale allowing quantitative treatment was later introduced (Goodman and Moore, 1971).

Using microcomplement fixation, Sarich and Wilson (1967a) calculated a given distance among the apes and humans as an immunological distance (ID) or index of dissimilarity, defined as "the factor by which the antiserum concentration must be raised to give a reaction with a heterologous albumin equal to that given by the homologous one" (Sarich and Wilson, 1967b:143). Thus they presented values of, for example, 1.09 for *Homo*/anti-*Pan* and 1.22 for *Pongo*/anti-*Homo*. In subsequent presentations, the ID was transformed logarithmically to give an integer, for example, 4 for *Homo*/anti-*Pan* and 12 for *Pongo*/anti-*Homo*.

The Goodman and Moore (1971) study was based upon subjective assessments on an arbitrary 0–5 scale of the intensity and size of precipitin bands observed on Ouchterlony plates. A computer algorithm then calculated antigenic distances from these raw data. The distances were multiplied by 10 in subsequent presentations (Dene et al., 1976). In none of the seven analyses involving anti-human, anti-chimpanzee, or anti-gorilla sera (gorilla plasma, chimpanzee plasma, chimpanzee thyroglobulin, human γ-globulin, human ceruloplasmin, human prealbumin, human thyroglobulin, and human α_2-macroglobulin) did the orang-utan value fall within the range of those obtained for human, chimpanzee, and gorilla. The few anomalous results showed other taxa (the gibbon in two instances and colobines in one) as more similar to human than the orang-utan.

Since the immunological data are reviewed in detail elsewhere (Cronin, 1977, 1983; Dene et al., 1976; Goodman, 1975; Sarich and Cronin, 1976), it is sufficient to note here that these data are, in general, highly consistent (Andrews and Cronin, 1982). Values for humans, chimpanzees, and gorillas compared one to another rarely exceed 10, while values for orang-utans compared with any of these three large hominoids rarely fall below 11. Overall, a two-fold difference can be observed, with little, if any, overlap between *Pongo* vs. *Homo–Pan–Gorilla* and *Homo* vs. *Pan* vs. *Gorilla*.

What, however, are these measurements actually measuring? Obviously they reflect some aspect of genetic similarity: that is incontestable. What I wish to point out is that these values reflect the amount of difference accumulated along unique phylogenetic branches: that is, autapomorphic genetic characters, however nebulous they may be. Thus, these distances translate into phylogenetic branching series only insofar as the accumulation of uniquely derived differences occurs at a roughly equal rate or intensity in all lineages. If one taxon, say *Pongo*, were to accumulate differences in its genetic determinants of antigenic specificity twice as rapidly as *Homo, Pan,* and *Gorilla,* then a calculated immunological distance would overestimate the former taxon's phylogenetic divergence. Consequently, *Pongo* could be the sister taxon of any of the other three genera, but simply be highly divergent in its antigenic genes. In a comparison of all four genera to an outgroup, the relative rate test of Sarich and Wilson (1967b) might not be sensitive enough to detect a fluctuation of this magnitude. Such is the difficulty with phenetic data such as these. It should be emphasized, however, that there is no evidence whatsoever in support of such a dramatic fluctuation in immunological evolutionary rate for orang-utans.

A different phenetic approach involves a comparison of the allele frequencies of different proteins as judged by their migrations in an electric field, a technique known as electrophoresis. Bruce and Ayala (1978, 1979) examined more than 20 genetic loci from 15 orang-utans (eight Sumatran, three Bornean, four unidentified) and other anthropoids and calculated the genetic distance between taxa, which was "interpreted as the number of electrophoretically detectable allelic substitutions per locus that have accumulated in the two species being compared since they last shared a common ancestor" (Bruce and Ayala, 1978:1041–1042). Again, these are measurements of uniquely derived events, and radical differences in evolutionary rate would bias them heavily. Nevertheless, these measures were unable to distinguish with statistical significance any sister group among the orang-utan, chimpanzee, gorilla, and human. Bruce (1977:56) concluded: "The molecular data relevant to *Pongo*'s divergence from the greater apes and man is ambiguous. The general conclusion I reach is that *Pongo* is equally distant from all other hominoids." It must be noted that these genetic distances are sensitive to the small number of animals as well as of loci tested. Thus, the genetic distance between human and chimpanzee given as 0.39 is not considered to be statistically different from that calculated in a similar fashion by King and Wilson (1975) as 0.62 (Bruce and Ayala, 1978). In another immunological analysis, Wang et al. (1968) were also unable to discriminate the position of orang-utans within the human-pongid cluster.

Bruce and Ayala (1979) calculated a genetic distance between the two subspecies of *Pongo* of 0.13, which is consistent with the electrophoretic genetic distance encountered between subspecies of other higher taxa. Different alleles of adenosine deaminase (Ada) were also found to differentiate the Bornean and Sumatran subspecies.

Cladistic Analyses

A cladistic analysis is possible for amino acid sequences, but, surprisingly, very few proteins have in fact been sequenced. Published sequences are available for five orang-utan proteins (Goodman et al., 1985; Miyamoto and Goodman, 1986): fibrinopeptides A and B (Doolittle et al., 1971), hemoglobin α and β (Maita et al., 1978), and myoglobin (Romero-Herrera et al., 1976). A partial, but not phylogenetically informative sequence is also available for carbonic anhydrase I (Tashian et al., 1976).

To assess the phylogenetic position of the orang-utan with protein sequence data actually requires a fairly simple cladistic test, which takes advantage of the fact that evolutionary change generally consists of the transformation of characters from an ancestral (= primitive or plesiomorphic) state to a derived or apomorphic state. If *Pongo* is the sister of a *Homo–Pan–Gorilla* clade, we would expect there to be certain characters distributed in such a way that only the derived state will be found in *Homo–Pan–Gorilla* while the ancestral state will be possessed by *Pongo* and an outgroup. Alternatively, if *Pongo* is the sister of *Homo,* we would expect to find at least some derived characters unique to *Pongo* and *Homo,* with the ancestral states retained in *Pan, Gorilla,* and the outgroup. Figure 4–1 illustrates the way in which contrasting phylogenetic hypotheses can be tested cladistically.

Since humans and orang-utans are very similar genetically, there are few amino acid sites that are phylogenetically relevant. In general, the amino acid sequences attest to the historical reality of a human–chimpanzee–gorilla clade; some sites, however, appear to have experienced back-mutations, or reversions. None can be adduced to support an intimate human–orang-utan association, or an intimate great ape association.

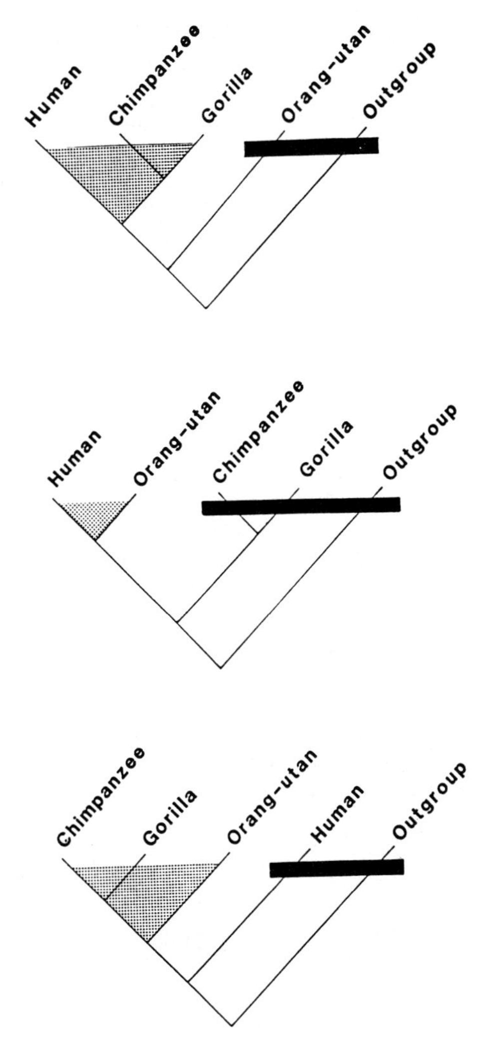

FIG. 4–1. Cladistic tests of phylogenetic hypotheses. Stippled areas designate presumptive clades, whose members share a novel (apomorph) character inherited from a unique common ancestor; dark boxes denote the distribution of the alternative character state (ancestral or plesiomorph) in a paraphyletic assemblage including the outgroup. *Top:* Expected distribution of characters on the hypothesis of an African ape–human linkage. *Middle:* Expected distribution of characters on the hypothesis of an orang-utan–human linkage. *Bottom:* Expected distribution of characters on the hypothesis of a great ape linkage.

For fibrinopeptide A, no differences exist among *Homo, Pan, Gorilla,* and *Pongo,* For fibrinopeptide B, human, chimpanzee, and gorilla are identical, but orang-utan differs at two positions (3 and 5). The character state possessed by *Pongo* is also shared by *Hylobates* and some cercopithecines. For α-globin, human, chimpanzee, and gorilla are again identical, and orang-utan differs at three positions (12, 23, and 57); however, the orang-utan appears to be unique (autapomorphic) relative to the other large hominoids because the rhesus macaque also shares the human character state (Dickerson and Geis, 1983).[1] For β-globin, human and chimpanzee are identical, and gorilla differs at position 104. Interestingly, where human, chimpanzee, and orang-utan share an arginine at this position, gorilla and rhesus macaque possess lysine. Considering either state to be a simple derived character yields an unsupportable phylogeny; however, the two amino acids have similar properties, and convergent evolution or a revertant mutation is likely the cause of this distribution. At position 125, human, chimpanzee, and gorilla share proline; orang-utan and rhesus macaque share glutamine. At position 87, human, chimpanzee, and gorilla share threonine, the orang-utan has lysine, and the rhesus macaque has glutamine; since the "outgroup" (i.e., the rhesus) possesses yet a different amino acid, no phylogenetic conclusion can be drawn. Finally, for myoglobin, gorilla has an autapomorphic state at position 22 while the chimpanzee has an autapomorphic state at 116. At position 23, human, chimpanzee, and gorilla share glycine; orang-utan, gibbon, and two cercopithecines share serine. At position 110, a reversion must be invoked to explain the sharing by orang-utan and the two cercopithecines of serine and by humans, the chimpanzee, gorilla, and gibbon of cysteine.

Thus, we find four amino acid substitutions linking humans to the African apes, and none supporting a great ape clade, or a human–orang-utan linkage. However, for two other substitutions the most parsimonious phylogenetic reconstruction is inaccurate (i.e., yields a phylogeny inconsistent with any under serious consideration). Reversion mutations and "multiple hits" at the same site are an occupational hazard in studies of molecular evolution and must occasionally be invoked.

CHROMOSOMAL DATA

The chromosomes of orang-utans have little to say of a definitive nature concerning the phylogenetic position of orang-utans. Orang-utans retain the diploid number of $2n = 48$ (Chiarelli, 1961), which is primitive for the great apes and humans. The separate chromosomes homologous to human arms $2p$ and $2q$ are unambiguously visible in the G-banded karyotype of *Papio* and *Macaca,* which refutes the contention of Kluge (1983) that a hypothetical fission is a synapomorphic character uniting the great apes: the derived state is a fusion unique to the human lineage which forms the human second chromosome.[2]

Orang-utans have more variability among their karyotypes than appears to be the case for the other great apes. Seuanez et al. (1979) found an inversion of orang-utan chromosome 2 which distinguishes *P. pygmaeus pygmaeus* from *P. p. abelii* (see Fig. 4–2). Additionally, there is a segregating polymorphism in the species that cuts across the subspecific distinction—an inversion of chromosome 9 in both taxa (Seuanez et al., 1976b). Interestingly, this polymorphism can be used to document the most serious problem inherent in the primate cytogenetics literature—that of interobserver replication. According to Seuanez et al. (1976b), in 15 out of 46 chromosome-9s analyzed, the same variant (i.e., inverted) chromosome 9 had been previously interpreted as a variant chromosome 7 by Lucas et al. (1973), an insertion by Dutrillaux et al. (1975), a different insertion by Turleau et al. (1975), and two nested inversions by Seuanez et al. (1976a). Other, less well-characterized chro-

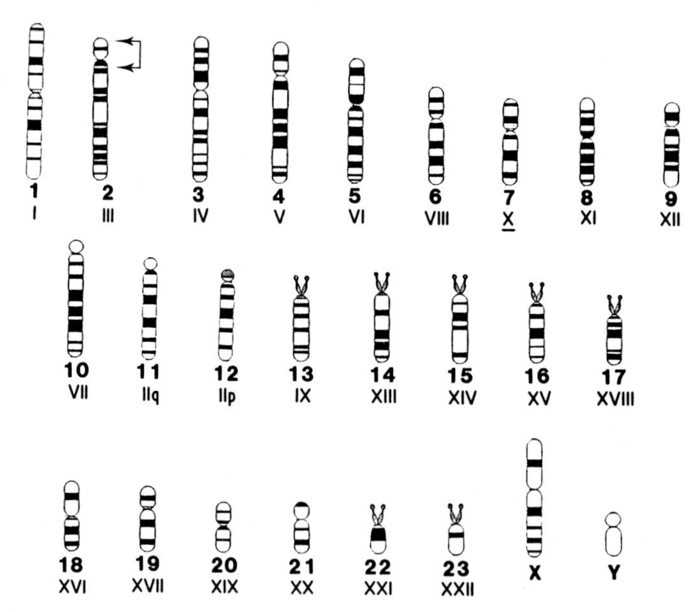

FIG. 4–2. Karyotype of the Sumatran orang-utan, after the Stockholm Conference (1978). Arabic numbers designate specific chromosomes of the orang-utan karyotype; Roman numerals indicate the human homolog corresponding to the particular orang-utan chromosome. The standard karyotype of the Borneo orang-utan is generated by an inversion of chromosome 2, shown.

mosomal polymorphisms have now been reported through the use of restriction endonuclease digestion to generate chromosome bands (De Stefano et al., 1986).

When comparing the G-banded karyotypes of orang-utans to those of humans, we are therefore obliged to take such comparisions with a degree of circumspection, as they are often contradictory in detail. With this is mind, certain generalizations can, however, be made from all the studies of the great ape karyotypes (Bianchi et al., 1985; Dutrillaux, 1979; de Grouchy et al., 1978; Stanyon, 1982; Yunis and Prakash, 1982). First, the number of chromosomes that differ is approximately the same for *Homo* vs. *Pan,* or *Gorilla,* or *Pongo.* Second, the magnitude of differences observed is greater for *Homo* vs. *Pongo* than for *Homo* vs. *Pan* or *Gorilla.* Chromosomes 3, 7, and 17 from the human karyotype, for example, bear striking similarities to chromosomes in each of the great ape species, but the *Pongo* homolog has a strikingly different form from that of *Homo, Pan,* and *Gorilla* (Paris Conference, 1971: Supplement, 1975).

Phenetically, therefore, it is warranted to state that *Pongo* is chromosomally more distant from *Homo* than are *Pan* and *Gorilla.* But, given the variability which exists in the rates of chromosomal evolution among, for example, the hylobatids and papionins (Marks, 1982, 1983a), it is not possible to translate cytogenetic distance into phylogenetic relationship.

What sort of phylogenetic (i.e., cladistic) argument can be made for the placement of *Pongo* on the basis of cytogenetic characters? Again, we are faced with the problem that the cytogenetic data are not as reliable as we would like them to be—indeed, the most comprehensive reviews on the subject contain major flaws. Miller (1977) used four cytogenetic characters to link *Homo* to *Gorilla:* distribution of satellite DNA, presence of the minor base 5-methylcytosine, C-band on chromosome XVI, and size of Y chromosome. The first two characters are correlated and primitive for the great apes and humans, but secondarily lost in *Pan.* The third is not clearly a homology, as the C-bands in *Homo* and *Gorilla* are on opposite arms. The fourth may be a valid homology, but it involves the most labile chromosome. Yunis and Prakash (1982) used three cytogenetic characters to link *Homo* and *Pan:* inversions on chromosomes IIp, VII, and IX. But the first two features are imperceptible even on the photomicrographs published by Yunis and Prakash, while the third is most plausibly the result of homoplasy—parallel but different inversions in humans and the chimpanzee (Dutrillaux, 1977).[3]

The proper outgroup for a cytogenetic study of orang-utans in this case is the chromosomally conservative Papionini, not the chromosomally derived Hylobatidae (Marks, 1982). Ambiguous though the chromosomal data may be, I know of no cytogenetic characters distributed in such a way as to link orang-utans with either humans or chimpanzees/gorillas. On the other hand, "Q-brilliance"—the property of intense fluorescence after quinacrine treatment—is shared among mammals only by *Homo, Pan,* and *Gorilla* (Pearson et al., 1971). Interpretations of chromosome morphology, such as inversions or translocations in the karyotype of one species relative to that of another, have yet to be represented by an explicit phylogenetic comparison using the papionin karyotype as an outgroup. The closest approach to such a study is that of Stanyon (1982), which encompassed a comparison among papionin karyotypes, and also among the great apes and humans. Stanyon then compared an interpretive "hominoid ancestral karyotype" (though gibbons were not considered) to that of *Macaca.* While Stanyon (1982; Stanyon and Chiarelli, 1982) invokes derived inversions in chromosomes III and VII to link *Homo, Pan,* and *Gorilla,* his closer descriptions reveal less certainty. Thus, for chromosome III,

> . . . (T)he changes may be complex. A translocation from another chromosome is possible. The homolog in chimpanzee, gorilla, and man is altered further by at least one pericentric inversion. Two forms of the chromosome exist in the orangutan. Apparently, considerable change is allowed in this chromosome (Stanyon, 1982:95).

And for chromosome VII,

> No good homolog was found [between *Macaca* and the "hominoid ancestral karyotype"]. This chromosome has undergone many rearrangements among the hominoids where all species differ from each other. Chromosome 21 in [the "hominoid ancestral karyotype"] may represent a fissioned product of the chromosome in *Macaca.* Coupled with a pericentric inversion a good approximation of the ["hominoid ancestral karyotype"] is derived (Stanyon, 1982:95).

By contrast, Dutrillaux (1979), comparing almost haphazardly the karyotypes of many primate species, considered four rearrangements as linking humans and the African apes: a pericentric inversion (i.e., involving both arms) of chromosome II, a pericentric inversion of chromosome VII, a paracentric inversion (i.e., involving one arm) of chromosome X, and a pericentric inversion (or "other intrachromosomal" rearrangement—this is ambiguous in the presentation) of chromosome XI. The poor quality of the published preparations makes it difficult to independently assess these judgments; it is certainly worth noting, however, that Dutrillaux's and Stanyon's analyses converge on chromosome VII as representing a valid synapomorphy of *Homo, Pan,* and *Gorilla.*

Nucleolar organizers (NORs) are chromosomal localizations which contain the tandemly duplicated genes coding for the major RNA components of the ribosome. The great apes and humans all have genus-specific distributions of these sites. Most hylobatids and all papionins have but a single NOR per haploid genome; this is a plausible ancestral condition for the Hominoidea. All large apes and *Homo* have multiple sites, but not equal numbers of them. Therefore, both expansions and contractions must be invoked to account for the distribution of NORs and remain faithful to any defensible cladogram of hominoid relationships (Fig. 4–3). To reconstruct with plausibility the evolution of hominoid NOR sites requires the observation that the set of NOR-bearing chromosomes of *Pan, Gorilla,* and *Homo* each form different subsets of the NOR-bearing chromosomes of *Pongo* (Henderson et al., 1974, 1976, 1977, 1979). I think this makes the orang-utan's distribution the most likely candidate for the ancestral NOR distribution among the great apes and *Homo* (Marks, 1983b). *Pan, Gorilla,* and *Homo* therefore represent secondary cutbacks of NOR sites, with *Gorilla* bearing the most derived state with but two NORs per haploid genome. This is, of course, a plausibility and consistency argument, not a strict cladistic test.

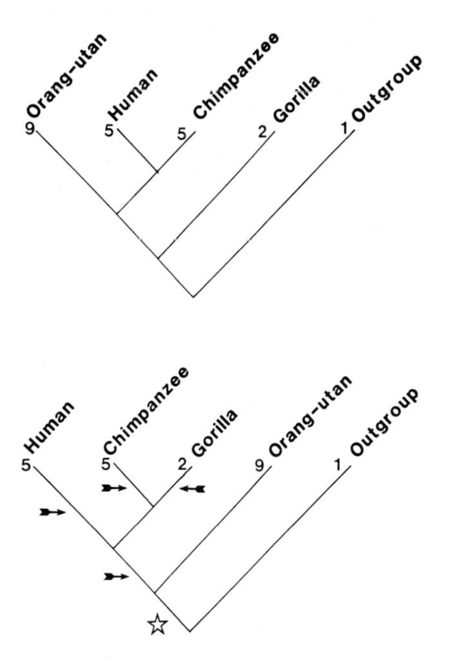

FIG. 4–3. Number of nucleolar organizers per haploid genome. *Top:* Relationships of hominoid taxa on the assumption that species with similar numbers of NORs are closely related. *Bottom:* Phylogenetic relationships of the hominoid taxa, with inferred changes in the number of NORs superimposed. The star indicates a proliferation of NOR sites; arrows indicate lineages in which NORs have independently been cut back (see Marks, 1983a for a detailed discussion).

DNA DATA

Direct studies of the genome, with their consequent bearing on the phylogenetic placement of the orang-utan, fall into three categories, each of which comprises a lamentably small number of studies, but which together present a highly synoptic picture. These studies are (1) DNA hybridization, (2) mitochondrial genome analysis, and (3) nuclear genome analysis.

Sturtevant (1913) is credited with formalizing the theory that genes are arranged linearly along the chromosome. This theory, while true as stated, carried with it a nonsequitur: the implicit derivation from the suggestion "genes are linearly arranged on chromosomes" of the notion that "chromosomes are linear arrangements of genes." However, it has become clear since the 1960s that the chromosome is an indescribable mass (or mess) of genetic material, of which in humans and their close relatives approximately 10% consists of genes—in the loosest sense of that term. Of the genic DNA, it now appears that but a tenth is actually expressed phenotypically. Information is thus a rare commodity on the chromosome; noninformation is apparently more common by two orders of magnitude. How can we say, then, that a chromosome is a linear arrangement of genes? If anything, the chromosome is a linear arrangement of nongenes! A chromosome, in a metaphor suggested by Ohno (1982), is a one-dimensional genetic desert, with genes representing rare oases.

The human genome can be analytically subdivided into three components: highly repetitive (including "satellite" DNA; Jones, 1977), middle repetitive (including *Alu* repeats; Doolittle, 1985; Schmid and Shen, 1985), and unique sequence (including genes). Repetitive and unique-sequence DNA appear to account for roughly equal proportions of the genome. Yet even among genes there is widespread duplication: genes come in clusters, the products of tandem duplications of ancestral genes, which may subsequently diverge in structure and function. These duplicated genes and their nonfunctional brethren pseudogenes are therefore serially homologous.

As the great bulk of the genome is not expressed phenotypically, it is difficult to conceive of natural selection playing a major role in the evolution of the genome, though it may certainly be involved in the evolution of genes (Gillespie, 1986). The mutational forces affecting the evolution of the genome are poorly understood. Repetitive DNA can spread horizontally through the genome of a species, or can be lost in an evolutionary instant. In unique-sequence DNA we recognize three broad mutational processes: a single base change, or point mutation; homologous recombination, in which two similar sequences cross over, which may lead to duplication, deletion, or correction of one of the sequences; and nonhomologous (or illegitimate) recombination, which involves the joining of dissimilar sequences.

With these cryptic processes now known to be at work, it becomes increasingly difficult to gauge the phylogenetic significance of isolated genomic data. When limited to a specific evolutionary mode (e.g., single base changes) over a specific region (e.g., the β-hemoglobin gene cluster), molecular data are excellent tools—perhaps the ultimate tools—for basing phylogenetic judgments. But does the absence of DNA satellite II in the chimpanzee genome unite *Homo, Gorilla,* and *Pongo* as a clade (Jones, 1977), or does it simply reflect the secondary loss of this DNA fraction from the chimpanzee genome? Similarly, does the presence of a truncated immunoglobulin ϵ pseudogene in the genomes of *Homo* and *Gorilla* unite them as a clade (Ueda et al., 1985), or reflect instead the presence a labile gene cluster?

In my opinion it is the latter, in both cases. One can speculate on why the urge to rewrite hominoid phylogeny on the basis of a single genomic character exists. But given the fact that the genome is still *terra incognita* and the complexities of its evolutionary dynamics are only beginning to be grasped, it is surely unwise for traditional systematics to yield absolute hegemony to molecular genetics. Genomic data constitute a valuable set of tools, but are not yet the ultimate tool-set; and when they conflict strongly with other lines of evidence, the possibility must be entertained that the fault lies with the genomic data, not with the phylogeny of the organisms.

DNA Hybridization

As a double-stranded molecule, the DNA polymer is held together by hydrogen bonds connecting the A-T and G-C base pairs. These bonds are easily broken by heating the DNA; the result is single-stranded or denatured DNA. Slow cooling allows the DNA to renature by pairing once again with its complementary strand. If the DNA is unable to pair with a perfect complement, but is compelled to make an imperfect match with DNA from a different species—perhaps because the latter is in great excess—the result is "heteroduplex" DNA. Heteroduplex DNA has fewer bonds, and therefore requires less energy to become denatured. If the temperature is monitored carefully, a measure of the difference between heteroduplex and homoduplex DNA can be obtained, as the difference between the temperatures at which the two DNAs are 50% denatured. This temperature difference is taken to indicate percentage of DNA base-pairs mismatched, based on the empirical relationship of approximately 1° C difference per mismatch.

This technique affords a gross measure of genomic similarity but may conceal some sources of error that would undermine the conclusions for very closely related species. For example, the fact that genes are now known to occur in clusters of multiple copies of varying degrees of divergence allows the possibility that the heteroduplex DNA is not paired as perfectly as it could be. That is, gene *A1* (species *a*) may not be paired with gene *A1* (species *b*), but with gene *A2* (species *a* or *b*). With respect to the orang-utan's phylogenetic placement, however, these data are very highly consistent with both the protein and chromosomal data, in making the orang-utan the sister group of a phenetic cluster consisting of *Homo, Pan,* and *Gorilla.*

The first published data using this technique come from Hoyer et al. (1972): orang-utan–gorilla hybrids were found to differ by 2.3%, orang-utan–chimpanzee by 1.8%, human–chimpanzee by 0.7%, and human–gorilla by 1.4%. However, there was a problem with complementarity of the hybridization results—while orang-utan–human hybrids differed by 1.9%, human–orang-utan were found to differ by 2.9%, requring the tempering of phylogenetic conclusions by the recognition of a significant magnitude of experimental error. Benveniste and Todaro (1976) obtained values of 2.3–2.6% for human–chimpanzee–gorilla *inter se* (and reciprocals), and a single value of 4.5% for orang-utan–gorilla. Nearly a decade later, Sibley and Ahlquist (1984) published the results of an immense battery of DNA hybridization experiments on the unique-sequence portion of primate genomes. They concluded that the unique-sequence DNAs of human, chimpanzee, and gorilla all differ by about 3.8% from that of orang-utan. This contrasts with values of 1.8–2.4% obtained for human–gorilla–chimpanzee *inter se.* O'Brien et al. (1985) replicated this study in part, obtaining values of 1.9% for human–chimpanzee–gorilla and 3.5% for human–orang-utan and gorilla–orang-utan (chimpanzee–orang-utan was not performed).

These more recent phenetic data afford convincing evidence that the unique-sequence DNA of orang-utans is more different from human, chimpanzee, and gorilla than any of these three hominoids is from any other. For any of these three hominoids to be a sister of the orang-utan would require that the orang-utan genome on the whole is evolving nearly twice as fast as any of the other genomes. Although that is not impossible to conceive of, it seems highly unlikely for the following reason: the unique-sequence genome, as previously discussed, is for the most part nongenic, not transcribed, and if not exactly nonfunctional, then at least of difficult-to-imagine function. The rate of change of such DNA would be a function solely of the mutation rate, or if weak selection were considered to operate, the rate would also very inversely with population size (Kimura, 1983). While it would be rash to think that all noncoding DNA is equally absolutely neutral to natural selection, it is likely that the DNA hybridization technique, sampling a huge genomic fraction, swamps the effects of deviations in evolutionary rate of specific DNA sequences which may accrue from directional selection or conservative stabilizing selection.

Another interesting result obtained using this method comes from the study of Benveniste and Todaro (1976), who hybridized type C virus DNA to the nuclear genomes of various catarrhine primates. They found that the genomes of African genera (e.g., *Pan, Gorilla, Colobus, Papio, Cercopithecus*) hybridized strongly to the viral DNA derived from *Papio cynocephalus,* while Asian genera (*Hylobates, Pongo,* and *Presbytis*) did not. *Homo* fell clearly in the latter category, thus implying that modern humans have an Asian ancestry. But these data do not suggest divorcing *Homo* from *Pan–Gorilla* any more than they suggest divorcing *Presbytis* from *Colobus,* and can be accounted for quite readily by the hypothesis that the baboon type C virus is capable of horizontal transmission (Temin, 1976), which is well within the orthodoxy of modern genetics. An Asian origin for "man," can, of course, be applied at a variety of taxonomic levels: the family Hominidae, the genus *Homo,* species *H. sapiens,* or subspecies *H. s. sapiens.*

Mitrochondrial DNA

Several classes of cytoplasmic subcellular organelles, of which the mitochondrion is one, are known to exist outside the nucleus of the cell. The function of mitochondria is to generate ATP (adenosine triphosphate), whose energy (stored in the bond between the third phosophate and the rest of the molecule) is required in most cellular activities. Mitochondria not only perform the chemical reactions that produce this energetic bond, but also produce the proteins that carry out these reactions. Mitochondria contain their own DNA, as well. The 16,500 nucleotides of the human mitochondrial genome incorporate little wasted space (unlike nuclear genomes) and contain not only genes for the enzymes catalyzing the reactions, but genes for the transfer RNAs and ribosome components, as well (Grivell, 1983). An entire human mitochondrial DNA sequence is known (Anderson et al., 1981), and information is accumulating on intraspecific variation in the mitochondria of *Homo sapiens* (Cann et al., 1984; Nei, 1985).

A comparison among representative mitochondrial genomes of the great apes may be expected to yield information independently of other evolutionary data, since there is no clear evidence of a relationship between the mitochondrial genome and either outward phenotype or nuclear genotype, which are the sources of all other evolutionary data. Ferris et al. (1981) used restriction enzyme sites, which are specific sequences of four to six nucleotides, to sample the similarity of mitochondrial genomes of the great apes, human, and an outgroup, the lar gibbon. The phenetic conclusions of this study yielded a group consisting of *Homo–Pan–Gorilla* as an unresolved trichotomy, with *Pongo* further removed. Subsequent statistical treatments (Nei and Tajima, 1985; Templeton, 1983) have not challenged the placement of the orang-utan.

Brown et al. (1982) presented the first homologous nucleotide sequences (a stretch of 896 base pairs) from these five hominoid genera, thus enabling a direct cladistic analysis at the level of nucleotide substitutions. Although a human–chimpanzee–gorilla group is unresolvable phenetically or cladistically, the orang-utan is clearly outside this assemblage. The orang-utan sequence differs from the human sequence at 144 positions, while chimpanzee–gorilla differs at 93, chimpanzee–human at 79, and gorilla–human at 92. If we perform on these data the cladistic tests discussed earlier, we find 29 sites at which human, chimpanzee, and gorilla are identical and at which orang-utan shares a different nucleotide with the outgroup; there is one site at which humans and orang-utans are identical and the chimpanzee–gorilla–outgroup possesses a different nucleotide. There are also but three sites at which chimpanzee, gorilla, and orang-utan share the identical nucleotide and the human–outgroup has a different one—this tests and rejects the phylogeny of Kluge (1983).

Mitochondrial DNA studies are not, however, ultimate solutions to evolutionary problems. Mitochondrial genomes, for example, are inherited only through the maternal line, and are therefore not subject to the neo-Darwinian consequences of biparental nuclear Mendelian inheritance (Wilson et al., 1985). If, for example, speciation occurs by hybridization, the hybrid will be equidistant from two species in its nuclear genome, but may possess the mitochondrial genome of only one. This may cause a disparity between phylogenies derived from nuclear vs. those derived from mitochondrial genomes (DeSalle and Giddings, 1986). Further, the general evolutionary dynamics of mitochondrial DNA evolution and their consequences are still not clearly known (Avise and Lansman, 1983; Brown, 1983; Latorre et al., 1986). Therefore, with regard to hominoid phylogeny, it would be unwise to accept these data in the absence of all others. Yet insofar as mitochondrial DNA studies provide an independent test of phylogenetic relationships, we are obliged to find the strong concordance between these studies and the others encouraging.

Nuclear DNA Sequencing

At the finest levels of genetic analysis, which have been accessible for only a few years, there exists but a single published study comparing homologous nuclear DNA sequences across the Hominoidea. This is the study of Koop et al. (1986) on η (eta) globin.

Blood globins certainly constitute the best-known genetic system from a structural, functional, populational, and evolutionary viewpoint. In humans (and probably all amniotes), the genes encoding these oxygen-transport proteins are localized into two clusters: α-like and β-like. These genes are all serially homologous or, in the jargon of molecular evolution, paralogous (Fig. 4–4). The human β-cluster comprises a gene encoding a β-like globin expressed in the embryo (ϵ), two fetal genes (γ), a dysfunctional pseudogene ($\psi\beta$), a minor adult gene (δ), and a major adult gene (β). The human $\psi\beta$ is homologous (orthologous) to a gene which is functional in other mammalian orders (Goodman et al., 1984); for this reason it has been renamed η by Harris et al. (1984).

Pseudogenes should be particularly useful as phylogenetic markers, since (being unexpressed) they are presumed to be unaffected by the constraining effects of stabilizing selection or by the possibly confusing effects of directional selection. Koop et al. (1986) compared a sequence of more than 2.2 kilobases of DNA containing the (ψ)η orthologs of human, chimpanzee, gorilla, orang-utan, and four outgroups. Excluding three short stretches where a complex series of deletions or insertions has occurred, each nucleotide position can be considered a phylogenetic character. There are 17 sites at which human, chimpanzee, and gorilla share one nucleotide, and orang-utan and the outgroups share another; there are no sites at which human and orang-utan share one nucleotide, and chimpanzee, gorilla, and the outgroups share another; and there are no sites at which chimpanzee, gorilla, and orang-utan share one nucleotide, and human and the outgroups share another. Again, an independent set of biomolecular data supports a *Homo–Pan–Gorilla* linkage, not only phenetically, but cladistically as well.

There are cladistically useful nuclear DNA data from only one other gene locus, the α1-glo-

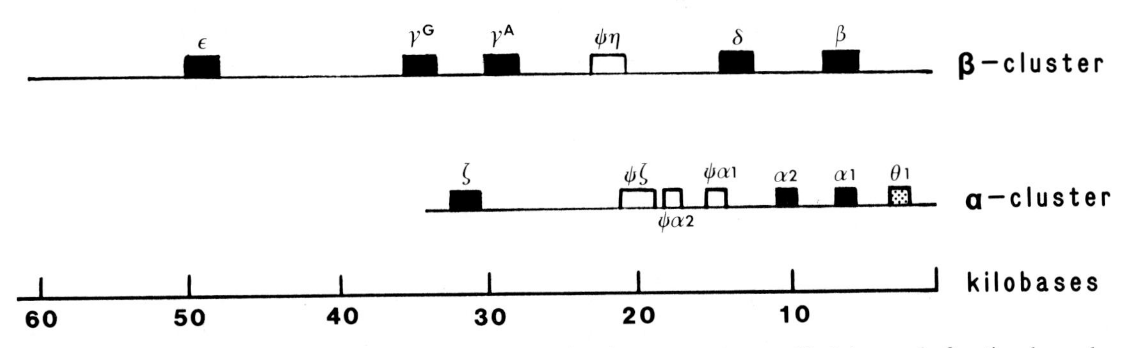

FIG. 4–4. Structure of the blood globin gene clusters. Functional genes are given as black boxes, dysfunctional pseudogenes are given as white boxes. θ1, of unknown functional status, is given as a stippled box.

bin gene, from the other blood globin cluster. The human DNA sequence has been determined (Michelson and Orkin, 1983); the orang-utan DNA sequence has also been determined (Marks et al., 1986). For the chimpanzee, however, only a complementary DNA sequence has been published (Liebhaber and Begley, 1983). This is actually a sequence derived from a mature messenger RNA, which means that the introns (intervening sequences) have been spliced out and are not part of the determined sequence. As an outgroup, we can use the α1-globin gene of an olive baboon (Shaw et al., in 1987). The sequences are given in Fig. 4–5, with the human sequence printed in full.

Of 31 informative (i.e., variable) nucleotide sites, 28 are unique to a single taxon—one is unique to human, one to chimpanzee, seven to orang-utan, and 19 to baboon. Another site is ambiguous in that it has more than one substitution. The remaining two sites can be used to test a phylogenetic hypothesis: of a proposed human–chimpanzee–orang-utan clade, which taxon shares its character state with the outgroup (therefore presumably retaining the primitive state) while the other two taxa are united as a clade (by their sharing a derived state)?

In the two remaining informative sites, human and chimpanzee share the identical nucleotide, while orang-utan and the outgroup (baboon) share a different nucleotide. These data support the view that human and this African

```
H  ATG.GTG.CTG.TCT.CCT.GCC.GAC.AAG.ACC.AAC.GTC.AAG.GCC.GCC.TGG.GGT.AAG.GTC.GGC.GCG
Ch                       C          CC A           G             T               C
O                        C          CC A           A             G               C
B                        A          AA C           G             T               A

H  CAC.GCT.GGC.GAG.TAT.GGT.GCG.GAG.GCC.CTG.GAG.AGG.ATG.TTC.CTG.TCC.TTC.CCC.ACC.ACC
Ch       T       G
O        C       C
B        T       G

H  AAG.ACC.TAC.TTC.CCG.CAC.TTC.GAC.CTG.AGC.CAC.GGC.TCT.GCC.CAG.GTT.AAG.GGC.CAC.GGC
Ch                   C                       C          C            G GTC
O                    C                       T          C            G GAC
B                    C                       C          A            C AAA

H  AAG.AAG.GTG.GCC.GAC.GCC.CTG.ACC.AAC.GCC.GTG.GCG.CAC.GTG.GAC.GAC.ATG.CCC.AAC.GCG
Ch                       C         AA            C                         A C
O                        G         AA            C                         A C
B                        G         CT            G                         C G

H  CTG.TCC.GCC.CTG.AGC.GAC.CTG.CAC.GCG.CAC.AAG.CTT.CGG.GTG.GAC.CCG.GTC.AAC.TTC.AAG
Ch         GCC       T               G
O          GCC       C               T
B          AAG       C               G

H  CTC.CTA.AGC.CAC.TGC.CTG.CTG.GTG.ACC.CTG.GCC.GCC.CAC.CTC.CCC.GCC.GAG.TTC.ACC.CCT
Ch       A                           C           C           C
O        G                           C           C           T
B        G                           T           T           C

H  GCG.GTG.CAC.GCC.TCC.CTG.GAC.AAG.TTC.CTG.GCT.TCT.GTG.AGC.ACC.GTG.CTG.ACC.TCC.AAA
Ch
O
B

H  TAC.CGT.TAA
Ch
O
B
```

FIG. 4–5. DNA sequence of the α1-globin gene of four catarrhine primates: H, human; Ch, chimpanzee; O, orang-utan; B, baboon. Only the coding sequence is given; intron sequences are not available for the chimpanzee. Human sequence is from Michelson and Orkin (1983); chimpanzee sequence is from Liebhaber and Begley (1983); orang-utan sequence is from Marks et al. (1986); baboon sequence is from J.-P. Shaw, J. Marks, and C.-K. J. Shen (1987).

ape share a unique ancestry, and reject the phylogenetic hypothesis of Schwartz (1984a,b), which would predict a chimpanzee–outgroup similarity, and that of Kluge (1984), which would predict a human–outgroup similarity. These data are admittedly skimpy, but they constitute an independent phylogenetic test, and are wholly consistent with the (unlinked) $\psi\eta$ data.[4]

CONCLUSIONS

Early on, molecular data suggested that the orang-utan is the extant sister of a clade consisting of human, chimpanzee, and gorilla (Goodman, 1962). This scheme had been proposed by morphologists as early as 1916 (Gregory, 1916 [but not 1927]; Miller, 1920; Morton, 1927; Sonntag, 1924), but the anatomical relations of the great apes were sufficiently ambiguous four decades later to give the molecular data bearing on this problem disproportionately great weight. Subsequent to the pioneering immunological studies, phenetic and cladistic analyses of proteins, chromosomes, nuclear DNA, and organelle DNA have together strongly supported the Morton–Sonntag–Goodman phylogeny. Four general conclusions can be drawn from the genetic data: (1) human, chimpanzee, and gorilla are more similar genetically to one another than any is to the orang-utan; (2) insofar as genetic similarity may reflect recency of common ancestry (since rates of genetic evolution appear not to vary widely across closely related taxa), human, chimpanzee, and gorilla are more closely related to one another than any of them is to the orang-utan; (3) where cladistic analyses are possible, the phenetic inferences are supported; and (4) in general, among human, chimpanzee, and gorilla, the genetic relations are ambiguous.

Each of these statements, it must be recognized, is occasionally contradicted. Genetic characters can be adduced in which chimpanzee, gorilla, and orang-utan are more similar (e.g., chromosome number); gorilla, orang-utan, and human are more similar (e.g., genomic satellite II); etc. There is a 2-kilobase repeat from the human X chromosome which finds a close homolog in the gorilla, a divergent homolog in the chimpanzee, and no homolog in the orang-utan (B. Hamkalo, personal communication). Rates of genetic evolution have been shown to fluctuate in the Catarrhini at the level of proteins (Hewett-Emmett et al., 1976), chromosomes (Marks, 1983a), and DNA (Britten, 1986; Li et al., 1985). Further, genetic similarity is a function of many processes, only some of which are known, and sometimes phenetic similarity may well conceal complex or poorly known genomic processes. And cladistic analyses have also been ambiguous—exasperatingly so—with respect to the genetic relations within the human–chimpanzee–gorilla triad (Ferris et al., 1981; Hixson and Brown, 1986, Koop et al., 1986).

The study of evolution is usually approached via one of two paths. One is a historical, particularizing path, whose subject is the reconstruction of unique events; this is the study of phylogeny. The other is a generalizing path, whose subject involves relationships among classes of phenomena. This latter path leads to studies of adaptation, evolutionary mechanisms, and, at the risk of sounding whiggish, of evolutionary advancement. While the dichotomy as I have just presented it is taken from the work of anthropologist Leslie White (1960), and was recognized at least implicitly by an earlier generation of biologists (e.g., Dobzhansky, 1937), it has more recently become a signal distinction for modern biologists under the phrase "pattern and process" (Eldrege and Cracraft, 1980; Rieppel, 1986; Stanley, 1979).

Although the focus of this chapter has been on pattern (i.e., historical reconstruction), it should be clear that our interpretation of data, and especially genomic data, is strongly constrained by our knowledge of the evolutionary processes that generated the particular phylogenetic pattern we are attempting to interpret. Our knowledge of the evolutionary processes in the genome is quite sketchy at the present, and surely far from complete—that much should also be clear. However, with respect to the phylogenetic patterns detectable among the hominoid primates, I believe that the genetic data point to a single most likely pattern: that orang-utans constitute the sister group of a clade consisting of *Homo, Pan,* and *Gorilla.*

We no longer need to debate the general problem of hominoid phylogeny as if the status of the problem were the same in 1987 as it was in the time of Hrdlička and Keith. It is not: considerable progress has indeed been made in the study of human origins since the 1920s. But far be it from any modern scientist to maintain that a problem of science is "closed," and is therefore no longer accessible to the skeptic. I wish to suggest, rather, that as a result of the advancements in the field over the last decades, the interesting problems of hominoid evolution are now and will be for some time in the evolutionary *processes* inferrable from the deciphered phylogenetic pattern. Why, for example, are there suites of anatomical similarities between human and orang-utan, as Schwartz (1984a,b) has demonstrated? What is the significance of the ambiguous genetic relations within the African clade? How much of the difference among these taxa is simply noise in the system? Are there specific genetic changes that can be associated with the adaptive shifts of the hominid lineage, or of adaptive divergence in general? If so, are they single-base changes, repetitive sequences, transposable elements or perhaps some feature of genomic change not yet recognized? How and why do rates of evolution of chromosomes, genes, anatomies, and behaviors vary?

These questions can only be asked sensibly, however, when we recognize the predication of these "process" questions on a prior establishment of the underlying phylogenetic pattern. "[T]he most important reason for relating organismal and molecular evolution to each other," wrote Simpson at the very dawn of molecular evolutionary studies, "is not simply the testing of hypotheses or the validation of methods. It is the balancing of points of view and the achievement of more complete explanations" (Simpson, 1964:1535).

REFERENCES

Anderson, S., Bankier, A., Barrell, B., de Bruijn, M., Coulson, A., Drouin, J., Eperon, I., Nierlich, D., Roe, B., Sanger, F., Schrier, P., Smith, A., Staden, R. and Young, I. 1981. Sequence and organization of the human mitochondrial genome. *Nature, 290:*457–465.

Andrews, P. and Cronin, J. 1982. The relationships of *Sivapithecus* and *Ramapithecus* and the evolution of the orang-utan. *Nature, 297:*541–546.

Avise, J. and Lansman, R. A. 1983. Polymorphism of mitochondrial DNA in populations of higher animals. *In* Evolution of Genes and Proteins, pp. 147–164, ed. M. Nei and R. K. Koehn. Sinauer Associates, Sunderland.

Bateson, W. 1904. Opening address, Section D. Zoology. *Nature, 70:*406–413.

Benveniste, R. E. and Todaro, G. J. 1976. Evolution of type C viral genes: evidence for an Asian origin of man. *Nature, 261:*101–107.

Bianchi, N. O., Bianchi, M. S., Cleaver, J. E. and Wolff, S. 1985. The pattern of restriction enzyme-induced banding in the chromosomes of chimpanzee, gorilla, and orangutan and its evolutionary significance. *Journal of Molecular Evolution, 22:*323–333.

Boyden, A. 1952. Fifty years of systematic serology. *Systematic Zoology, 1:*19–30.

Boyer, S. H., Noyes, A. N., Timmons, C. F. and Young, R. A. 1972. Primate hemoglobins: polymorphisms and evolutionary patterns. *Journal of Human Evolution, 1:*515–543.

Britten, R. J. 1986. Rates of DNA sequence evolution differ between taxonomic groups. *Science, 231:*1393–1398.

Brown, W. M. 1983. Evolution of animal mitochondrial DNA. *In* Evolution of Genes and Proteins, pp. 62–88, ed. M. Nei and R. K. Koehn. Sinauer Associates, Sunderland.

Brown, W. M., Prager, E. M., Wang, A. and Wilson, A. C. 1982. Mitochondrial DNA sequences of Primates: tempo and mode of evolution. *Journal of Molecular Evolution, 18:*225–239.

Bruce, E. 1977. A Study of the Molecular Evolution of Primates Using the Techniques of Amino Acid Sequencing and Electrophoresis. Ph.D. dissertation, University of California, Davis.

Bruce, E. and Ayala, F. 1978. Humans and apes are genetically very similar. *Nature, 276:*264–265.

Bruce, E. and Ayala, F. 1979. Phylogenetic relationships between man and the apes: electrophoretic evidence. *Evolution, 33:*1040–1056.

Cann, R., Brown, W. and Wilson, A. C. 1984. Polymorphic sites and the mechanism of evolution in human mitochondrial DNA. *Genetics, 106:*479–499.

Chiarelli, B. 1961. Chromosomes of the orangutan *(Pongo pygmaeus). Nature, 192:*285.

Cronin, J. 1977. Anthropoid evolution: the molecular evidence. *Kroeber Anthropological Papers, 50:*75–84.

Cronin, J. 1983. Apes, humans, and molecular clocks: a reappraisal. *In* New Interpretations of Ape and Human Ancestry, pp. 115–150, ed. R. L. Ciochon and R. S. Corruccini. Plenum Press, New York.

Dene, H. T., Goodman, M. and Prychodko, W. 1976. Immunodiffusion evidence on the phylogeny of the primates. *In* Molecular Anthropology, pp. 171–196, ed. M. Goodman and R. Tashian. Plenum Press, New York.

DeSalle, R. and Giddings, L. V. 1986. Discordance of nuclear and mitochondrial DNA phylogenies in Hawaiian *Drosophila. Proceedings of the National Academy of Sciences, USA, 83:*6902–6906.

De Stefano, G. F., Romano, E. and Ferrucci, L. 1986. The Alu I-induced bands in metaphase chromosomes of organgutan. *Human Genetics, 72:*268–271.

Dickerson, R. and Geis, I. 1983. Hemoglobin. Benjamin/Cummings, Menlo Park.

Dobzhansky, T. 1937. Genetics and the Origin of Species. Columbia University Press, New York.

Doolittle, R. F., Wooding, G. L., Lin, Y. and Riley, M. 1971. Hominoid evolution as judged by fibrinopeptide structures. *Journal of Molecular Evolution, 1:*74–83.

Doolittle, W. F. 1985. The evolutionary significance of middle-repetitive DNAs. *In* The Evolution of Genome Size, pp. 443–487, ed. T. Cavalier-Smith. John Wiley, New York.

Dutrillaux, B. 1977. New chromosome techniques. *In* Molecular Structure of Human Chromosomes, pp. 77–94, ed. J. Yunis. Academic Press, New York.

Dutrillaux, B. 1979. Chromosomal evolution in primates: tentative phylogeny from *Microcebus murinus* (Prosimian) to man. *Human Genetics, 48:*251–314.

Dutrillaux, B., Rethoré, M.-O. and Lejeune, J. 1975. Comparaison du caryotype de l'orang-outang (*Pongo pygmaeus)* a celui de l'homme, du chimpanzeé, et du gorille, *Annales di génétique, 18*:153–161.

Eldredge, N. and Cracraft, J. 1980. Phylogenetic Patterns and the Evolutionary Process. Columbia University Press, New York.

Ferris, S., Brown, W., Davidson, W. and Wilson, A. C. 1981. Extensive polymorphism in the mitochondrial DNA of apes. *Proceedings of the National Academy of Sciences, USA, 78:*6319–6323.

Ferris, S. D., Wilson, A. C. and Brown, W. M. 1981. Evolutionary tree for apes and humans based on cleavage maps of mitochondrial DNA. *Proceedings of the National Academy of Sciences, USA, 78:*2432–2436.

Gillespie, J. H. 1986. Natural selection and the molecular clock. *Molecular Biology and Evolution, 3:*138–155.

Goodman, M. 1962. Evolution of the immunologic species specificity of human serum proteins. *Human Biology, 34:*104–150.

Goodman, M. 1963. Serological analysis of the systematics of recent hominoids. *Human Biology, 35:*377–436.

Goodman, M. 1975. Protein sequence and immunological specificity: their role in phylogenetic studies in Primates. *In* Phylogeny of the Primates, pp. 219–248, ed. W. P. Luckett and F. S. Szalay. Plenum Press, New York.

Goodman, M. and Cronin, J. 1982. Molecular anthropology: its development and current directions. *In* A History of American Physical Anthropology, pp. 105–146, ed. F. Spencer. Academic Press, New York.

Goodman, M., Czelusniak, J. and Beeber, J. E. 1985. Phylogeny of Primates and other eutherian orders: a cladistic analysis using amino acid and nucleotide sequence data. *Cladistics, 1:*171–185.

Goodman, M., Koop, B., Czelusniak, J., Weiss, M. and Slightom, J. 1984. The η-globin gene: its long history in the β-globin gene family of mammals. *Journal of Moelcular Biology, 180:*803–823.

Goodman, M. and Moore, G. W. 1971. Immunodiffusion systematics of the Primates. I. The Catarrhini. *Systematic Zoology, 20:*19–62.

Gregory, W. K. 1916. Studies on the evolution of the Primates. *Bulletin of the American Museum of Natural History, 35:*239–355.

Gregory, W. K. 1927. How near is the relationship of man to the chimpanzee and gorilla stock? *Quarterly Review of Biology, 2:*549–560.

Grivell, L. A. 1983. Mitochondrial DNA. *Scientific American, 248(3):*78–89.

Harris, S., Barrie, P., Weiss, M. and Jeffries, A. 1984. The primate $\psi\beta 1$ gene: an ancient β-globin pseudogene. *Journal of Molecular Biology, 180:*785–801.

Henderson, A. S., Atwood, K. and Warburton, D. 1976. Chromosomal distribution of rDNA in *Pan paniscus, Gorilla gorilla beringei,* and *Symphalangus syndactylus:* comparison to related primates. *Chromosoma, 59:*47–55.

Henderson, A. S., Warburton, D. and Atwood, K. C. 1974. Localization of rDNA in the chimpanzee *(Pan troglodytes)* chromosome complement. *Chromosoma, 46:*435–441.

Henderson. A. S., Warburton, D., Megraw-Ripley, S. and Atwood, K. C. 1977. The chromosomal localization of rDNA in selected lower primates. *Cytogenetics and Cell Genetics, 19:*281–302.

Henderson, A. S., Warburton, D., Megraw-Ripley, S. and Atwood, K. C. 1979. The chromosomal location of rDNA in the Sumatran orangutan, *Pongo pygmaeus albei* [sic]. *Cytogenetics and Cell Genetics, 23:*213–216.

Hewett-Emmett, D., Cook, C. and Barnicot, N. 1976. Old World monkey hemoglobins: deciphering phylogeny from complex patterns of molecular evolution. *In* Molecular Anthropology, pp. 257–276, ed. M. Goodman and R. Tashian. Plenum Press, New York.

Hixson, J. E. and Brown, W. M. 1986. A comparison of the small ribosomal RNA genes from the mitochondrial DNA of the great apes and humans: sequence, structure, evolution, and phylogenetic implications. *Molecular Biology and Evolution, 3:*1–18.

Hoyer, B., van de Velde, M., Goodman, M. and Roberts, R. 1972. Examination of hominid evolution by DNA sequence homology. *Journal of Human Evolution, 1:*645–649.

Jones, K. W. 1977. Repetitive DNA and primate evolution. *In* Molecular Structure of Human Chromosomes, pp. 295–326, ed. J. Yunis. Academic Press, New York.

Kimura, M. 1983. The Neutral Theory of Molecular Evolution. Cambridge University Press, New York.

King, M.-C. and Wilson, A. C. 1975. Evolution at two levels in humans and chimpanzees. *Science, 188:*107–116.

Kluge, A. G. 1983. Cladistics and the classification of the great apes. *In* New Interpretations of Ape and Human Ancestry, pp. 151–177, ed. R. L. Ciochon and R. S. Corruccini, Plenum Press, New York.

Koop, B. F., Goodman, M., Xu, P., Chan, K. and Slightom, J. 1986. Primate η-globin DNA sequences and man's place among the great apes. *Nature, 319:*234–238.

Latorre, A., Moya, A. and Ayala, F. J. 1986. Evolution of mitochondrial DNA in *Drosophila subobscura. Proceedings of the National Academy of Sciences, USA, 83*:8649–8653.

Li, W.-H., Luo, C.-C. and Wu, C.-I. 1985. Evolution of DNA sequences. *In* Molecular Evolutionary Genetics, pp. 1–94, ed. R. J. MacIntyre. Plenum Press, New York.

Liebhaber, S. A. and Begley, K. 1983. Structural and evolutionary analysis of the two chimpanzee α-globin mRNAs. *Nucleic Acids Research, 11:*8915–8928.

Lucas, M., Page, C. and Tanmer, M. 1973. Chromosomes of the orangutan. *Jersey Wildlife Preservation Trust, Annual Report,* pp. 57–58.

Maita, T., Araya, A., Goodman, M. and Matsuda, G. 1978. The amino acid sequences of the two main components of adult hemoglobin from orangutan *(Pongo pygmaeus). Hoppe-Seylers Zeitschrift für Physiologische Chemie, 359:*129–132.

Marks, J. 1982. Evolutionary tempo and phylogenetic inference based on primate karyotypes. *Cytogenetics and Cell Genetics, 34:*261–264.

Marks, J. 1983a. Rates of karyotype evolution. *Systematic Zoology, 38*:207–209.

Marks, J. 1983b. Hominoid cytogenetics and evolution. *Yearbook of Physical Anthropology, 25:*125–153.

Marks, J., Shaw, J.-P. and Shen, C.-K. J. 1986. The orangutan adult α-globin gene locus: duplicated functional genes and a newly detected member of the primate α-globin gene family. *Proceedings of the National Academy of Sciences, USA, 83:*1413–1417.

Michelson, A. M. and Orkin, S. H. 1983. Boundaries of gene conversion within the duplicated human α-globin genes. *Journal of Biological Chemistry, 258:*15245–15254.

Miller, D. A. 1977. Evolution of primate chromosomes. *Science, 198:*1116–1124.

Miller, G. S. 1920. Conflicting views on the problem of man's ancestry. *American Journal of Physical Anthropology, 3:*213–245.

Miyamoto, M. M. and Goodman, M. 1986. Biomolecular systematics of eutherian mammals: phylogenetic patterns and classification. *Systematic Zoology, 35:*230–240.

Morgan, T. H. 1903. Evolution and Adaptation. Macmillan, New York.

Morton, D. J. 1927. Human origin. *American Journal of Physical Anthropology,10:*173–203.

Nei, M. 1985. Human evolution at the molecular level. *In* Population Genetics and Molecular Evolution, pp. 41–64, ed. T. Ohta and K. Aoki. Springer-Verlag, Berlin.

Nei, M. and Tajima, F. 1985. Evolutionary change of restriction cleavage sites and phylogenetic inferences for man and apes. *Molecular Biology and Evolution, 2:*189–205.

Nuttall, G. H. F. 1904. Blood Immunity and Blood Relationship. Cambridge University Press, London.

O'Brien, S., Nash, W., Wildt, D., Bush, M. and Benveniste, R. 1985. A molecular solution to the riddle of the giant panda's phylogeny. *Nature, 317:*140–144.

Ohno, S. 1982. The common ancestry of genes and spacers in the euchromatic region: *Omnis ordinis hereditarium a ordinis priscum minutum. Cytogenetics and Cell Genetics, 34:*102–111.

Paris Conference (1971): Supplement. 1975. Standardization in human cytogenetics. *Birth Defects: Original Article Series, 11(9).* The National Foundation–March of Dimes, New York.

Pearson, P. L., Bobrow, M., Vosa, C. G. and Barlow, P. W. 1971. Quinacrine fluorescence in mammalian chromosomes. *Nature, 231:*326–329.

Rieppel, O. 1986. Species are individuals: a revew and critique of the argument. *In* Evolutionary Biology, vol. 20, pp. 283–317, ed. M. K. Hecht, B. Wallace, and G. Prance. Plenum Press, New York.

Romero-Herrera, A. E., Lehmann, H., Castillo, O., Joysey, K. A. and Friday, A. E. 1976. Myoglobin of the orangutan as a phylogenetic enigma. *Nature, 261:*162–164.

Sarich, V. M. Cronin, J. E. 1976. Molecular systematics of the Primates. *In* Molecular Anthropology, pp. 141–170, ed. M. Goodman and R. Tashian. Plenum Press, New York.

Sarich, V. M. and Wilson, A. C. 1967a. Immunological time scale for hominid evolution. *Science, 158:*1200–1202.

Sarich, V. M. and Wilson, A. C. 1967b. Rates of albumin evolution in primates. *Proceedings of the National Academy of Sciences, USA, 58:*142–148.

Schmid, C. W. and Shen, C.-K. J. 1985. The evolution of interspersed repetitive DNA sequences in mammals and other vertebrates. *In* Molecular Evolutionary Genetics, pp. 323–358, ed. R. J. MacIntyre. Plenum Press, New York.

Schwartz, J. H. 1984a. The evolutionary relationships of man and orangutans. *Nature, 308:*501–505.

Schwartz, J. H. 1984b. Hominoid evolution: a review and a reassessment. *Current Anthropology, 25:*655–672.

Seuanez, H. N., Fletcher, J., Evans, H. J. and Martin, D. E. 1976a. A chromosome rearrangement in an orangutan studied with Q-, C-, and G-banding techniques. *Cytogenetics and Cell Genetics, 17:*26–34.

Seuanez, H. N., Fletcher, J., Evans, H. J. and Martin, D. E. 1976b. A polymorphic structural rearrangement in the chromosomes of two populations of orangutans. *Cytogenetics and Cell Genetics, 17:*327–337.

Shaw, J.-P., Marks, J. and Shen, C.-K. 1978. Evidence that the recently discovered θ1-globin is functional in higher primates. *Nature, 326*:717–720.

Sibley, C. G. and Ahlquist, J. 1984. The phylogeny of the hominoid primates, as indicated by DNA-DNA hybridization. *Journal of Molecular Evolution, 20:*2–15.

Simpson, G. G. 1964. The meaning of taxonomic statements. *In* Classification and Human Evolution, pp. 1–31, ed. S. L. Washburn. Wenner-Gren, New York.

Simpson, G. G. 1964. Organisms and molecules in evolution. *Science, 146:*1535–1538.

Sonntag, C. F. 1924. The Morphology and Evolution of the Apes and Man. John Bale, London.

Stanley, S. M. 1979. Macroevolution: pattern and process. Freeman, San Francisco.

Stanyon, R. 1982. Chromosome Evolution in the Primates: *Macaca* to *Homo.* Ph.D. dissertation, Pennsylvania State University. University Microfilms, Ann Arbor.

Stanyon, R. and Chiarelli, B. 1982. Phylogeny of the Hominoidea: the chromosome evidence. *Journal of Human Evolution, 11:*493–504.

Stockholm Conference. 1978. An international system for human cytogenetic nomenclature. *Birth Defects: Original Article Series, 14(8).* S. Karger, Basel, for the National Foundation–March of Dimes.

Sturtevant, A. H. 1913. The linear arrangement of six sex-linked factors in *Drosophila,* as shown by their mode of association. *Journal of Experimental Zoology, 14:*43–59.

Tashian, R., Goodman, M., Ferrell, R. E. and Tanis, R. J. 1976. Evolution of carbonic anhydrase in primates and other mammals. *In* Molecular Anthropology, pp. 301–320, ed. M. Goodman and R. Tashian. Plenum Press, New York.

Temin, H. 1976. The DNA provirus hypothesis. *Science, 192:*1075–1080.

Templeton, A. R. 1983. Phylogenetic inference from restriction endonuclease cleavage site maps with particular reference to the evolution of humans and the apes. *Evolution, 37:*221–244.

Turleau, C., de Grouchy, J., Chavin-Colin, F., Mortelmans, J. and van den Berghe, W. 1975. Inversion pericentrique du 3, homozygote et heterozygote, et translocation centromerique du 12 dans un famille d'orangs-outangs: implications evolutives. *Annales de Genetique, 18:*227–233.

Ueda, S., Takenaka, O. and Honjo, T. 1985. A truncated immunoglobulin ϵ pseudogene is found in gorilla and man but not in chimpanzee. *Proceedings of the National Academy of Sciences, USA, 82:*3712–3715.

Wang, A. C., Shuster, J., Epstein, A. and Fudenberg, H. H. 1968. Evolution of antigenic determinants of transferrin and other serum proteins in primates. *Biochemical Genetics, 1:*347–358.

White, L. A. 1960. Panel five: social and cultural evolution. *In* Evolution After Darwin: vol. 3, Issues in Evolution, p. 236, ed. S. Tax and C. Callender. University of Chicago Press, Chicago.

Wilson. A. C., Cann, R., Carr, S., George, M., Gyllenstein, U., Helm-Bychowski, K., Higuchi, R., Palumbi, S., Prager, E., Sage, R. and Stoneking, M. 1985. Mitochondrial DNA and two perspectives on evolutionary genetics. *Biological Journal of the Linnean Society, 26:*375–400.

Wilson, A. C., Carlson, S. S. and White, T. J. 1977. Biochemical evolution. *Annual Review of Biochemistry, 46:*573–639.

Yunis, J. and Prakash, O. 1982. The origin of man: a chromosomal pictorial legacy. *Science, 215:*1525–1530.

NOTES

1. Position 23 of α-globin is somewhat ambiguous. Humans have glutamic acid, as do chimpanzees; gorillas and orang-utans have aspartic acid. Goodman and Cronin (1982) use this as a shared derived substitution linking humans and chimpanzees on the basis of a partial gibbon α-globin sequence (Boyer et al., 1972) indicating that gibbons also have aspartic acid at this position. However, the rhesus macaque and olive baboon have glutamic acid, as do humans and chimpanzees. Multiple substitutions of amino acids with similar properties have occurred at this site, making determinations of evolutionary polarity difficult.

2. According to the conventions adopted by the Paris Conference (1971; Supplement, 1975) and the Stockholm Conference (1978), chromosomes of the great apes may be designated in either of two ways. An Arabic numeral designates a particular chromosome of the species karyotype, while a Roman numeral designates the chromosome of the human karyotype to which it is homologous. Thus, chromosome 12 of the chimpanzee karyotype may also be designated as IIp; chromosome 23 of the orang-utan karyotype as XXII, etc. Both notations are given in Figure 2, and are used throughout this chapter.

3. The Yunis and Prakash (1982) study is particularly unfortunate, as it is the least reliable interpretively while at the same time being the most technically advanced. Using "high-resolution" G-banding, they presented photomicrographs and interpretive drawings of the chromosomes of human, chimpanzee, gorilla, and orang-utan. Although the inversions which are claimed to link human and chimpanzee are clear enough in the diagram, they are highly ambiguous in the photomicrographs. Yunis and Prakash report a nucleolar organizer (NOR) on *Gorilla* XIII, XXI, and XXII, and on *Pongo* IIp, IIq, VII, IX, XIII, XIV, XV, and XXII. No evidence or citation is given, and this conflicts with previous studies finding NORs only on *Gorilla* XXI and XXII and on *Pongo* IIp, IIq, IX, XIII, XIV, XV, XVIII, XXI, and XXII (Henderson et al., 1976, 1979). Yunis and Prakash claim to have performed C-banding, but do not discuss it, and to have examined five orang-utans, five gorillas, and nine chimpanzees, but do not discuss any variability among individuals. Further, in their Figure 3, Yunis and Prakash indicate that they also karyotyped two cercopithecines, but do not present or discuss these data. No subsequent report has appeared.

4. Additional nucleotide sequence data have now been collected for the orang-utan (fetal) γ-globin genes. Slightom (personal communication) and colleagues find at least 20 shared derived sites linking human, chimpanzee, and gorilla.

ACKNOWLEDGMENTS

I wish to express my gratitude to Francisco Ayala and Jeffrey Schwartz for their encouraging comments and to Barbara Hamkalo, Jerry Slightom, Jeng-Pyng Shaw, and Ben Koop for access to unpublished results.

5

History, Morphology, Paleontology, and Evolution

JEFFREY H. SCHWARTZ

Although it might seem to most systematists that issues surrounding the phylogenetic relationships of and among the large hominoids are essentially resolved—indeed, that there really are no issues of major significance—it has not always been this way, nor is there reason to believe even now that this is truly the case. There has not been a history of general agreement on the relationships of the extant taxa (e.g., see reviews by Schwartz, 1987, and Schultz, 1936), and the relevant fossils, as well, have had their share of taxonomic about-faces (e.g., see Andrews and Tekkaya, 1980; Kay and Simons, 1983; Pilbeam, 1982; Simons and Pilbeam, 1965, 1972).

HISTORICAL OVERVIEW

Perhaps not now widely appreciated is the fact that, in pre-evolutionary and pre-Darwinian views of the ordering of the organic world, the orang-utan figured prominently as the intermediary form or "link" in the Great Chain of Being between an ascending hierarchy of human "races" and the "lower" primates (see reviews by Röhrer-Ertl, this volume, and Schwartz, 1987). Indeed, the favorable comparisons that could be, or were at least thought to have been made between humans and orang-utans provoked many scholars to conclude that only the acquisition of language served to distinguish the "lowest" of humans from the red ape (see Schwartz, 1987, and references therein).

It was Darwin who turned attention away from southeast Asia, with regard to the orang-utan and eventually to human origins, as well. Although he had made only passing reference to primate evolution in *The Origin of Species* (1859)—and that was restricted to shedding light on human origins—Darwin did forcefully argue 12 years later, in volume I of *The Descent of Man*, not only that Africa was the seat of human evolution but that the closest living relatives of modern humans were the knuckle-walking African apes. Although he did depend on the comparative anatomical work of Huxley (e.g., 1863) and others for demonstrating that *Homo sapiens* is as much a mammal as are those conventionally regarded as primates, and is actually closely related to a small group within the order Primates, Darwin's arguments were not morphological ones. Rather, Darwin relied on selectionist, transformationist, and other sorts of scenarios to arrive at his conclusion.

Since Darwin was largely concerned in volume I of *The Descent* with the development of civilized humans from a more primitive, savage state (the better-known volume II deals with sexual selection), he sought to demonstrate "that man has risen, though by slow and uninterrupted steps, from a lowly condition to the highest standard as yet attained by him in knowledge, morals and religion" (Darwin, 1871:141). By identifying fleetness and bipedalism as major attributes of modern humans, Darwin apparently thought he had found the requisite prehuman transitional evolutionary stage in the knuckle-walking African apes:

> If the gorilla and a few allied forms had become extinct, it might have been argued, with great force and apparent truth, that an animal could not have been gradually converted from a quadruped into a biped, as all the individuals in an intermediate condition would have been miserably ill-fitted for progression. But we know (and this is well worthy of reflection) that the anthropomorphous apes are now actually in an intermediate condition; and no one doubts that they are on the whole well adapted for their conditions of life (Darwin, 1871:50).

Darwin's emphasis on the relatedness of humans to the African apes in particular differs from the conclusion reached by Huxley, who (contrary to popular opinion) regarded the "Man-like apes" as a group to which "Man" is related "inasmuch as he differs less from them than they do from other families of the same order" (Huxley, 1863:145). In addition to citing St. George Mivart's reference to the orang-utan as a peculiar and aberrant form (Darwin, 1871: 149), Darwin also argued that southeast Asia could not have been the "birthplace of man" because potential human ancestors "would not have been exposed to any special danger, even if far more helpless and defenseless than any existing savages, had they inhabited some warm continent or large island, such as Australia, New Guinea, or Borneo, which is now the home of the orang" (Darwin, 1871:62). Indeed, as Dart (1925) would later reiterate in defending *Australopithecus* as the ancestor of humans and the link with the African apes, Darwin saw Africa, especially southern Africa, with its "greater degree [of] dangerous beasts . . . [and] . . . more fearful physical hardships," as a more likely environment in which the earliest hominids could have evolved (Darwin, 1871:61).

In his discussion of "Affinities and Genealogy," Darwin went further in developing his argument:

> In each great region of the world the living mammals are closely related to the extinct species of the same region. It is therefore probable that Africa was formerly inhabited by extinct apes closely allied to the gorilla and chimpanzee; and as these two species are now man's nearest allies, it is somewhat more probable that our early progenitors lived on the African continent than elsewhere. (Darwin, 1871:151)

But although it may be extremely reasonable (especially from our present vantage point) to conclude that some part of human evolution, if not aspects of the diversification of hominoids in general, is indeed preserved in the fossil record of Africa, this does not in and of itself lead inexorably to the conclusion that humans and African apes are closely related. And nowhere in *The Descent* is there a morphological argument in support of Darwin's declaration that the "two species [of African ape] are now man's nearest allies."

Darwin bemoaned the absence of a fossil record to fill in "the great break in the organic chain between man and his nearest allies." He also felt that "at some future period . . . the civilized races of man will almost certainly exterminate, and replace, the savage races throughout the world." If the apes were to be exterminated as well, this would widen "the break between man and his nearest allies . . . for it will intervene between man in a more civilized state, as we may hope, even than the Caucasian, and some ape as low as a baboon, instead of as now between the negro or Australian and the gorilla" (Darwin, 1871:152). And this is the essence of the argument for uniting humans with the African apes.

Although the latter part of the 19th and the first half of the 20th centuries witnessed attempts at mustering anatomical evidence in support of the relatedness of humans to one or more of the great apes, there remained tendencies to perceive the demonstration of phylogenetic affinity through less objective means. For example, Gregory (e.g., 1921:98), a sometime proponent of the relatedness of humans and the African apes, invoked as strong evidence in favor of this hypothesis the "proof" that "the gorilla nose, although repulsive in human eyes, has in it all the 'makings' of a human nose, and acquires chiefly a forward and downward growth of its tip to be transformed into a subhuman type . . . the lowest existing types of human nose (fig. 302) [a Tasmanian] indeed retain[ing] much of the gorilloid heritage." Gregory followed this remark with the additional observation that "the fundus of the eye of a chimpanzee . . . exhibits the most detailed resemblance to that of a negro" and (p. 95) found "convincingly human the appearance of the pregnant female chimpanzee," as well as noting that "the pendent breasts of old female chimpanzees and gorillas . . . [are] suggestive of human relationship." Subsequently, he (1927) added that humans and chimpanzees were similar in their "psychology."

From the late 1800s onward there were alternative theories on who was related to whom among the hominoids (see Schultz, 1936, for review), but the most successful was the one suggested by Huxley (1863) and defended with great conviction by Schultz (e.g., 1936, 1968, and references therein): the great apes themselves constituted a group which in turn shared a common ancestor with humans (see review by Schwartz, 1987). Although ultimately internally inconsistent, the argument begins by recognizing the gorilla as the most similar overall to humans, then the chimpanzee, the orang-utan, and, last, the gibbon, but by also perceiving there to be a gap between these taxa and other primates that is greater than the dissimilarities that exist between any of them. Among the large hominoids, Huxley and, especially, Schultz tried to demonstrate that the great apes are more similar to one another than any of them is to a human. Gregory, for example, used a version of this approach in associating humans with the African apes, although, as is evident from the discussion above, he, as did others, relied on perceptions of transformation—how one thing could be converted into another—to argue the case.

Regardless of the ensuing groupings among the hominoids, the legacy which Huxley left was that the orang-utan is less like humans than the gorilla or even the chimpanzee. The more popular schemes of hominoid relationships reflect this general impression, with overall similarity being translated directly into a scheme of phylogenetic closeness—which represents an interpretive leap that Huxley and Schultz specifically addressed, pointing out that it was not necessarily warranted. Only Haeckel (1896) ventured to suggest that the orang-utan may in fact be closer evolutionarily to humans than any other hominoid because of apparent uniquenesses that these two hominoids share in the morphology of the brain.

Neither the great ape–human nor the orang-utan–[African ape–human] theory of relatedness was based firmly on features particularly unique to each nested clade. However, the distinctive musculoskeletal anatomy of the knuckle-walking African apes had been known, at least in part, as early as the late 1800s (cf. Schultz, 1936) and continues to warrant serious consideration of a sister relationship between the chimpanzee and gorilla (Andrews, 1987; Martin, 1986; Tuttle, 1974). But while the acceptance of a great ape clade relied heavily on the perception of the great apes' being in general more similar to one another than any was to humans, the suggestion that humans and the African apes might be united within Hominoidea did have the support of Keith's (1916) observation that these three taxa alone develop true frontal sinuses; while this remains a loosely valid interpretation, the precise account is that these hominoids are distinguished by the development of ethmoidal air cells (Cave and Haines, 1940).

Since not until recently have there been attempts to delineate other potential synapomorphies reflecting the unity of a human–African ape clade (e.g., Andrews, 1985, 1987; Delson and Andrews, 1975; Groves, 1986; Martin, 1986; Schwartz, 1984a,b), and since the Huxley-Schultz notion of a human–great ape assemblage persisted in force well beyond the endeavors of Keith, Gregory, and others, it seems, historically, that the impetus for dropping the former theory of hominoid relationships and endorsing an arrangement in which the orang-utan was seen as the sister of a human–African ape clade came from an increasing acceptance during the 1960s of the results of biomolecular approaches to hominoid phylogeny (see review by Goodman and Cronin, 1982). These immunologic and electrophoretic studies tended to show that humans appear more similar to one or another or both of the African apes than to either the orang-utan or the gibbon, although virtually all combinations and permutations had some support at one time or another. But, with the acceptance of Zuckerkandl and Pauling's (1962) a priori assumption that overall genomic similarity could be equated with evolutionary closeness because change is supposed to be essentially unidirectional and additive, there emerged a general consensus that, indeed, Darwin had been right all along: the chimpanzee and gorilla "are now man's nearest allies."

PRACTICAL AND THEORETICAL CONSIDERATIONS FOR RECONSTRUCTING HOMINOID PHYLOGENY

Shifting from a human–great ape to a human–African ape theory of relatedness made little difference in the interpretation of apparent fossil hominoids and their presumed relationships to extant taxa. A generalized dryopithecine "stock" could still have given rise, on the one hand, to the specific dryopithecines which, in turn, may have been ancestral to the extant apes, and, on another, to *Ramapithecus,* the ancestral hominid (see review by Fleagle and Jungers, 1982). The association of *Sivapithecus, Gigantopithecus,* and other apparently thick molar-enameled fossil hominoids with *Ramapithecus* and, thus, hominids (e.g., Simons and Pilbeam, 1972; Pilbeam, 1979) did nothing to alter the primary structure of hominoid relationships. Even when *Sivapithecus,* with which *Ramapithecus* may be synonymous, was found to have dentofacial features which are otherwise specifically characteristic of the orang-utan (Andrews and Cronin, 1982; Andrews and Tekkaya, 1980; Pilbeam, 1982; Ward and Kimbel, 1983), the general tendency has been for primate systematists to relocate the fossils to an orang-utan clade and then try to explain why *Sivapithecus*, for example, can be hominid-like in its teeth and jaws and yet otherwise so orang-utan-like. In the early 1900s, when *Sivapithecus* was known from just a portion of a mandible with teeth, a similar argument ensued: Pilgrim (1915), who had discovered the specimen, thought it was more human-like than anything else, but Gregory (1915) quickly and adamantly responded that *Sivapithecus* compared better morphologically with the orang-utan.

Without a fossil record predating *Australopithecus,* the ancientness of the hominid "lineage" has been rethought, especially by paleoanthropologists, and brought in line with biomolecularly based dates (cf. Andrews and Cronin, 1982; Pilbeam, 1984, 1986; Sarich, 1983; Sibley and Ahlquist, 1984). But, at the same time, biomolecular systematists have been refining their levels of resolution to the point of finding that humans and chimpanzees appear to differ less overall in their nuclear and mitochondrial DNA, hemoglobin, and amino acid sequences than either does when compared to any other hominoid (e.g., Brown et al., 1982; Goodman et al., 1982, 1983; Hasegawa et al., 1985; Koop et al., 1986; Miyamoto et al., 1987; and references therein). When rephrased as a phylogenetic conclusion, these results have been interpreted as demonstrating that the chimpanzee is more closely related to humans than it is to the gorilla, its obvious morphological sister taxon.

Although a human–chimpanzee (or perhaps it is a chimpanzee–human) theory of relationship may emerge as the most robust of alternative biomolecularly based theories, it is important to point out that there is little morphological support for it (Andrews, 1987, and personal communication; Schwartz, 1984a,b, 1987; see below for discussion of Groves, 1986). Indeed, if one were to accept as valid a human–chimpanzee sister group in the face of what appears to be an exceedingly robust morphologically based theory of relationship—the sister relationship of the chimpanzee and the gorilla—one would have to reject morphology as being in any way reflective of the relatedness of taxa, any taxa. Thus, neontologists could not sort out with credibility the relationships of extant taxa and fossils could not be "fit" into a biomolecularly generated scheme. With regard to the latter point, there could not exist a paleontological datum from which one could calibrate any molecular clock. However, and especially with regard to hominoids, there is more than one biomolecularly based phylogeny available for consideration (see reviews by Andrews, 1987, and Marks, this volume).

But it is of historical interest to note how much debate on the relationships of the hominoids has taken place when, in reality, so little of it has been concerned with the theoretical, philosophical, and methodological underpinnings of the phylogenetic assumptions. Clearly, as is obvious with the vision hindsight provides, morphologically based (as much as they were so) theories of relationships among the hominoids rarely resulted from the delineation first of apparent synapomorphies. Although one can argue that overall genomic similarity does indeed reflect synapomorphy, one can do so only by accepting first the notion that genomic change is unidirectional and additive and that parallelism is unlikely to or cannot occur. Thus the conclusion that humans and one or both of the African apes must constitute a closely related group within Hominoidea because these taxa appear to be the most similar overall, in whatever the biological system being studied, will not necessarily be replicated by an approach based on the dissection of apparently homologous features into their respective hierarchical levels of primitiveness and derivedness. Thus, when a "cladistic" approach is applied to biomolecular data, alternative theories of relatedness among humans and the African apes emerge in addition to that in support of a human–chimpanzee sister group (Andrews, 1987), or the data are considered insufficient to resolve a trichotomy among these three hominoids (Marks, this volume). In any case, with the paucity of such data across a numerous and diverse array of taxa, the "principle of commonality" cannot be used to assess character polarity, and assumptions of relatedness must be accepted in order to pursue character polarity analysis via the "outgroup" approach.

RELATIONSHIPS OF AND AMONG THE EXTANT LARGE HOMINOIDS

Since so much of the more than 100 years following Huxley's (1863) publication of "On the relations of man to the lower animals" has been devoted to the supposed morphological reasons for adhering to one rather than another of various theories of relationship among the hominoids, it is perhaps reasonable to focus the remainder of this chapter on morphology and the unravelling of phylogenetic relationshps among extant and fossil taxa.

In this regard one finds, when attempting to define clades solely on the basis of features unique to and distinctive of the taxa under scrutiny, that the longstanding notion of a great ape assemblage gains meager support from morphology other than in the form of various limb proportions. For example, in the context of a broader comparison among anthropoids, the great apes are distinguished by having a vaginal vestibule that is deep and bears irregular margins and into which the urethra and vagina open via a short anteroposterior cleft (e.g., Hill, 1951). Contrary to earlier studies, Dahl (this volume) demonstrates that the labia majora of all great apes do not atrophy during development and, thus, atrophy of the labia majora is not, as once thought, another synapomorphy of the group. In other anatomical regions, Schultz (1961) found that the vertebral spines (C1-T1) of the great apes are distinctively robust and elongate, and Anderton (this volume) reminds us, but with caution, that these hominoids stand out in aspects of lower limb myology (e.g., variability in the flexor pollicis longus).

As noted above, there are few features that are suggestive of a sister group relationship between humans and the chimpanzee. I state this in spite of the long list (25) of characters presented by Groves (1986) as being derived states common to *Pan* and *Homo* because these features are not synapomorphies in the sense that they are unique to these two taxa either alone among hominoids or in the broader context vis-à-vis the outgroup of monkeys. Thus, for example, it may be true that humans and chimpanzees develop bilobate livers, but because orang-utans and gibbons do also, this shared feature cannot be synapomorphic for humans and chimpanzees alone. Similarly, although the cranial end of the heart in humans and chimpanzees may indeed lie opposite the second rib, it also does so in gorillas and gibbons and therefore cannot be synapomorphic for the former two taxa. In some instances, distinctions between proposed characters states are blurred by overlapping ranges (e.g., in the onset of puberty and in the position of the caudal end of the heart), and in yet others, the data have since been updated. Indeed, Andrews (1987) used the very same table of characters among hominoids in his endeavor to delineate synapomorphies of humans and chimpanzees, and, by using the criteria outlined above, could not find solid justification for more than five features that might reflect the unity of these two hominoids (Andrews, personal communication). Thus, of the more than 200 features provided by Groves (1986), a list of potential morphological synapomorphies of humans and chimpanzees consists of the following: coincident fusion of ankle epiphyses and elbow and hip, reduction of body hair dorsally, development of three right lung lobes, parotid gland free from the sternocleidomastoid muscle, and axis of ear bones over 90°. I would concur with Andrews's conclusions.

The historically most obvious possibility—the relatedness of humans and gorillas (see review by Schwartz, 1987)—can be argued on the basis of large body size, similarity in various relative body proportions (ibid.), and the most frequent development of a peroneus tertius muscle (Jungers, personal communication). In addition, Andrews (personal communication) and I have reviewed Groves's (1986) tables and agree on the following as being potentially synapomorphic of humans and gorillas: fluorescent bodies in the sperm, brachial index less than 80, a sparsity of chest hair, a flexed rectum, ear breadth less than 70% of ear length, and the development of ovaries greater than 40 mm in length. Although it is true that humans and African apes develop uteri that are broader than the orang-utan's (suggested by Andrews [1987] as a synapomorphy of humans and the African apes), it appears that if there is a synapomorphy to be found in this part of the anatomy, it is between humans and gorillas (with uteri of 52–55 and 52 mm, respectively); *Pan* (breadth of uterus = 37–40 mm) would be the sister of such a human–gorilla group, with the orang-utan even more primitive in this regard (28 mm). If these figures were corrected for female body size, however, the interpretation might very well be different. Nevertheless, it is interesting to note that morphology is actually slightly more suggestive of the unity of humans and gorillas than of humans and chimpanzees.

Although there is some morphology suggestive of alternative theories of relatedness of each African ape, there is considerable musculoskeletal evidence in support of the unity of the chimpanzee and the gorilla. These features are in large part associated with the African apes' knuckle-walking and include the development of friction pads over the middle phalanges of the hand, distension and modification of the distal articular surface of the radius, and shortened flexor tendons of the hand, which prevent simultaneous extension of the fingers and wrist (e.g., Schultz, 1936; Tuttle, 1974). Andrews (1987) also points to the African apes' having on average six sacral vertebrae, in contrast to five or fewer in other hominoids and anthropoids (cf. Schultz, 1968), as well as these apes' tendencies toward developing short ethmolacrimal and frontomaxillary contacts. Groves's (1986) tables reveal that the African apes are apparently synapomorphic in lacking a deep head of the flexor pollicis brevis and in having a small parotid gland. And Martin (1985) has demonstrated that, among the large hominoids, the chimpanzee and gorilla are set apart by their extreme emphasis on the deposition of pattern 1 enamel, which comes to comprise 40% of the enamel capping their molars. These and other features that emerge as potentially synapomorphic of the African apes (as a result of their restricted distribution among catarrhines) are listed in Table 5–1, which is modified from Andrews (1987) and Groves (1986).

At this point, and by accepting the hypothesis that the large hominoids do indeed constitute a monophyletic group (e.g., see Andrews, 1985, 1987; Delson and Andrews, 1975; Groves, 1986; Martin, 1986; Schultz, 1936, 1968), the question that becomes relevant is, "What are the relationships of the orang-utan?" If the great apes do not constitute a monophyletic group, then two possibilities arise: either orang-utans are the sister

TABLE 5–1. Possible Synapormorphies of the African Apes, in Comparison to *Homo* and *Pongo*

	Pan	*Gorilla*	*Homo*	*Pongo*
Sacral vertebrae	6	6	≦5	≦5
Ethmolacrimal contact (%)	63–93	49	100	100
Frontomaxillary contact (%)	30–50	30–50	0	0
Enamel	Thin	Thin	Thick	Intermed. thick
Pattern 3	~60%	~60%	~90–95%	~80%
Pattern 1	~40%	~40%	5–10%	~20%
Rate of deposition	~300 μm @ 1.5 μm/day	~400 μm @ 1.5 μm/day	20–30 μm @ 2.5 μm/day	200 μm @ 2.5 μm/day + 50 μm @ 1.8 μm/day
Metacarpals				
Dorsal transverse ridges on leads	X	X	0	0
Dorsal extension of articular surface	X	X	0	0
Dorsal ridges on distal radius & scaphoid	X	X	0	0
Volar & ulnar inclination of radius distally	X	X	0	0
Humerus				
Trochlear ridge	Prom.	Prom.	Mod.	Mod.
Olecranon fossa depth	Marked	Marked	Mod.	Mod.
Flexor pollicis longus	Obs.	Obs.	X	Atten./ (2 acc. fl.)
Flexor tendons of hand	Short	Short	Normal	Normal
Friction pads on middle digits of hand	X	X	0	0
Parotid gland	Small	Small	Large	Large

Abbreviations: 0 = not present; X = present; intermed. = intermediate; mod. = moderate; prom. = prominent; obs. = obsolescent; atten. = attenuated; acc. = accessory; fl. = flexor. In this and the other tables, as well as in the text, shared derivedness (= potential synapomorphy) was determined in the context of the broader comparisons provided in Groves (1986), Schwartz (1984a,b; 1986), and the contributions to this volume.

taxon of a human–African ape assemblage, or orang-utans and humans are sister taxa and this group is the sister of the African ape group. Each of these competing theories of relatedness can claim the support of synapomorphy.

The Relatedness of Humans and African Apes

As discussed above, humans can be united with the African apes on the basis of the development of ethmoidal air cells, which give rise, but typically from different sites of the ethmoid among the three taxa, to true frontal sinuses (Cave and Haines, 1940). This is distinctive of these hominoids among primates, not just among Hominoidea. Shea (this volume) suggests that frontal sinus development in humans and the African apes may be due to their development of klinorhynchy, which may be synapomorphic of these three hominoids, but Brown and Ward (this volume) disagree with this interpretation; in this regard, the discussion by Winkler et al. (this volume) of frontal sinus development as a functional correlate of different stresses on ecto- and endocranial surfaces of the frontal bone is relevant and permits a different interpretation for the lack of these structures in the orang-utan. In light of an increasing number of fossil taxa of potential hominoid affinity which have frontal sinuses but of unknown origin, Andrews (personal communication) is inclined to reject frontal sinus/ethmoid air cell development altogether as significant in delineating relationships among the large hominoids.

Weinert (1933), and most recently Andrews (1987) and Groves (1986), have suggested that early fusion of the os centrale and scaphoid in the wrist, as found in humans and the African apes, unites these taxa to the exclusion of the orang-utan, which also fuses these two carpals but typically does so later in development. In responding to Weinert, Schultz (1936) pointed out that the gibbon is similar to humans and the African apes in fusing the os centrale and scaphoid early, and thus the broader comparison diminishes the significance of the similarity between humans and the African apes. These data thus lend themselves to the conclusion that carpal fusion would have characterized the last common ancestor of all hominoids, and that such fusion would most likely have been early since this is the prevalent condition. In this regard, the orang-utan emerges as unique among the hominoids. But, whatever the ultimate resolution of the matter, fusion of the os centrale and scaphoid does not necessarily provide primary support for uniting humans and the African apes.

Andrews (1987) proposed as being synapomorphic of hominids, chimpanzees, and gorillas a particular configuration of the supraorbital region: namely, that the supraorbital tori are developed and continuous and that they are delineated posteriorly by a noticeable postorbital sulcus. However, in an earlier paper, Andrews and Cronin (1982) did note that this description does not pertain to modern humans, and Kimbel (1986) has recently demonstrated that the species of *Australopithecus*, as well, do not conform to this description in that these hominids lack a postorbital sulcus and possess a toral configuration markedly different from that typical of the African apes. Thus, at best, a presumably favorable comparison in supraorbital morphology exists between the chimpanzee and gorilla and various fossil *Homo*.

In his review of the relationships of and between the hominoids, Martin (1986) pointed to the shape of the second upper incisor as having significance in uniting humans and the African apes. But while he delineated this apparent synapomorphy as the development of a spatulate I^2, he also noted that this does not characterize the gorilla, which has a more conical tooth. This feature could, however, be cited as a synapomorphy of humans and the chimpanzee, but, as Martin extensively documents, the evidence in favor of accepting the African apes as sister taxa is overwhelming. I^2 is therefore seemingly uninformative about hominoid relationships.

Ward and Kimbel (1983) and Ward and Pilbeam (1983) have argued that aspects of subnasal morphology are indicative of the relatedness of hominids and the African apes. They based this conclusion primarily on comparisons between *Australopithecus* and the extant apes. The pertinent configuration, which they called the "African pattern," is defined on the basis of "stepping down" of the floor of the nasal cavity (caused by a steep declination of the posteriorly extended pole of the nasoalveolar clivus) and the division by the vomer of a transversely broad incisive fossa into two chambers or fossae (i.e., there is a fossa on either side of the vomer). In modern *Homo* in particular, the floor of the nasal cavity is typically flat, the nasoalveolar clivus posteriorly does not extensively override the maxilla, the incisive "fossa" is not broad, and thus this hominid does not develop the African pattern.

Delson (1985) and Martin (1986) have questioned the human–African ape synapomorphy of the African pattern, suggesting instead that such a configuration would have had significance at the level of the last common ancestor of all extant large hominoids. In support of this interpretation is the fact that hominids and African apes are not the only hominoids in which the vomer intervenes between two fossae in the floor of the nasal cavity. This configuration is also found in the orang-utan (e.g., Schwartz, 1983). But this configuration is not, in any of these hominoids, due to the vomer's bisecting a single incisive fossa into two chambers. These fossae—one on each side of the vomer—exist as receptacles for right and left nasopalatine nerves (if not also arteries) and are the openings of canals that extend through the bone of the palate. Since all other primates otherwise possess two anterior palatine fenestrae which perforate the palate and thus communicate between the oral and nasal cavities, it is reasonable to conclude that the last common ancestor of at least the extant great apes and all hominids would have been derived in the truncation of these fenestrae into the incisive canals which retain their openings (=incisive fossae) upon the floor of the nasal cavity (Schwartz, 1983).

Another reason for concluding that the African pattern as defined above should be considered significant only at the level of the last common ancestor of the extant large hominoids is that hominids—fossil and extant and, as far as is known, juvenile and adult—are characterized by the development of a single incisive foramen that perforates the palate on its oral side (e.g., Schwartz, 1983). This single incisive foramen is the confluence of the two incisive canals coursing through the palate. In contrast, juvenile and adult gorillas retain the separateness of the two incisive canals; this distinctiveness is also especially noticeable in juvenile chimpanzees, although, in some adults, the canals' distinctiveness is obscured or even obliterated orally (Kay and Simons, 1983; Schwartz, 1983). The early development and its persistence into the adult of a single incisive foramen in hominids would thus serve to unite these taxa apart from the African apes and add further support to the conclusion that an African pattern of subnasal morphology is present in *Australopithecus* only as a result of primitive retention.

Of the anatomical attributes that Andrews

TABLE 5–2. Possible Synapomorphies of *Homo* and the African Apes, in Comparison to *Pongo*

	Homo	*Pan*	*Gorilla*	*Pongo*
Ethmoidal air cells[1]	X	X	X	0
Dental eruption relative to epiphyseal fusion	Late	Medium	Medium	Early
Middle ear depth (mm)	8.5–9.2	8.8–9.0	9.9–12.5	6–8
Axillary organ	Large	Large	Large	Small
Apocrine gland distribution	None	Sparse	Sparse	Abund.
Eccrine gland distribution	Very abund.	Abund.	Abund.	Mod. abund.
Single larynx tuberculum cuneiforme	Small	Small	Small	Large
Type I aorta	3–11%	21%	11%	63–100%
Pattern intensity: palmar relative to plantar	50	100	100	170
Heart: anterior papillary muscle	Single	Single	Single	Multiple
Recurrens ulnaris artery: split	Ant.+post.	Ant.+post.	Ant.+post.	Communis/interossea
Adenohypophysis: embryonic pars intermedia	0	0	0	Large[2]
Restriction of palatine ridges	$\leq M^1$	$\leq M^1$	$\leq M^1$	$\leq M^2$
Progesterone level during ovulatory cycle	High	High	High	Low

Abbreviations: 0 = not present; X = present; mod. = moderately; abund. = abundant; ant. = anterior; post. = posterior.

[1]Given the number of fossil taxa now known to have frontal sinuses, this feature and/or ethmoidal air cell development may emerge as being characteristic of hominoids in general.

[2]It may be that *Pongo* is autapomorphic since this structure is rare in monkeys and but weakly developed in gibbons.

(1987) and Groves (1986) list as reflective of the relatedness within Hominoidea of humans and the African apes, those features that Andrews (personal communication) and I agree are potential synapomorphies—because of their restriction to these taxa alone—are presented in Table 5–2, as are Schultz's (1958) observation that the palatine ridges of these hominoids do not extend posteriorly beyond the level of M^1 and the determination by Czekala et al. (this volume) that progesterone levels during ovulation are higher in these three taxa than in the orang-utan.

The Relatedness of Humans and Orang-utans

Recent inquiries into hominoid relationships have uncovered an unexpected number of possible synapomorphies shared by humans and the orang-utan (Schwartz, 1983, 1984a,b, 1987). One of these apparently derived characters—thick molar enamel—had been acknowledged in the literature but had not been considered as being revealing phylogenetically because of an adherence to a human–African ape theory of relatedness and African apes were known to have thin molar enamel. Thick molar enamel was, however, considered significant in the delineation of a clade of hominoids that included the hominids as well as various Miocene taxa (e.g., *Ramapithecus, Sivapithecus, Ouranopithecus, Gigantopithecus*) (Pilbeam, 1979; Simons and Pilbeam, 1972). With the discovery of specimens of *Sivapithecus* that demonstrated quite convincingly via details of the facial skeleton and teeth that this Miocene hominoid was related to the orang-utan (Andrews and Cronin, 1982; Andrews and Tekkaya, 1980; Pilbeam, 1982), all thick-enameled Miocene hominoids were grouped with the orang-utan, and thick molar enamel was abandoned as a feature whose significance was other than at the level of the last common ancestor of all large hominoids (e.g., Andrews, 1985; Pilbeam, 1986; Ward and Pilbeam, 1983).

The apparent lack of significance of thick molar enamel within the large hominoid clade also gained support from the microanatomical work of L. Martin (1985), who pointed out that there were differences beyond those of simply thin vs. thick enamel. For example, whereas the gibbon has thin enamel because it deposits a layer of fast-forming pattern 3 enamel for a brief period of time, the African apes have thin enamel because they shift from pattern 3 to pattern 1 enamel, which is laid down over a longer period of time, but at a much slower pace. *Sivapithecus* and modern *Homo* have relatively very thick molar enamel as a result of depositing fast-forming pattern 3 enamel over a long developmental period, whereas the orang-utan only has "intermediately" thick molar enamel because it shifts after a time from laying down pattern 3 to laying down slow-forming pattern 1 enamel.

Martin concluded that the data are best interpreted developmentally as indicating that the last common ancestor of the large hominoids had thick pattern 3 molar enamel, as in modern *Homo*, and that all configurations and thicknesses among the large hominoids were derived or retained from that ancestral condition; this conclusion was also based on the interpretation of the data in the context of a human–African ape theory of relatedness (cf. Martin, 1985 [caption of Fig. 3]; but see Martin, 1986, for revised interpretation). However, since in at least *Pan* (of the great apes) the period of dental growth is approximately two-thirds as long as in modern humans (cf. Bromage, 1985; Bromage and Dean, 1985; and references therein), it would seem unlikely that the last common ancestor of the large hominoids would have had the exact molar enamel features now found in any of the extant taxa. Thus it may be developmentally more reasonable to hypothesize that the last common ancestor of the extant large hominoids attained, at most, a moderately thick layer of pattern 3 enamel, which would also be synapomorphic for the clade since it is derived relative to the condition in gibbons.

Certainly, the configurations seen, respectively, in the African apes and in modern humans and *Sivapithecus* can be derived from this suggested ancestral state and sister groups hypothesized. As Martin (1985, 1986) emphasizes, the most obvious association is of the chimpanzee with the gorilla, because in both hominoids pattern 3 enamel secretion is truncated, converting to pattern 1 enamel, which comes to constitute 40% of the total enamel deposited. In the opposite direction are *Sivapithecus* and modern *Homo*, which are equally distinctive (and thus potentially synapomorphic relative to the ancestral condition hypothesized here) in developing a markedly thick layer of pattern 3 enamel. Thus it would appear that the enamel data yield at least two possible sister groups: chimpanzee + gorilla and *Homo* + *Sivapithecus*, respectively.

Undiscussed so far is the orang-utan, whose "intermediately" thick enamel and distinctive composition of pattern 1 enamel do not lend themselves easily to the interpretation of the phylogenetic relationships of this hominoid. Although Martin (1985) argues that the orang-utan's emphasis on pattern 1 enamel cannot be homologous with that in the African apes, that conclusion would seem to be valid only if enamel deposition was indeed a series of discrete and discontinuous steps. However, the delineation of patterns 3 vs. 1 is artificial to the extent that amelogenesis (from the dentine-enamel juncture to the surface of the crown) is a continuous process, the distinction between decussating "pattern 3" enamel and nondecussating "pattern 1" enamel being the result of alterations in path and pace of a single ameloblast rather than the replacement of one cell-based deposit by another (see review by Osborn and Ten Cate, 1976). Thus, the process of truncating pattern 3 deposition and converting to a longer-than-usual period of prism pattern 1 secretion could be homologous in the great apes, with the African apes being most derived within the clade by virtue of their depositing a thick layer of pattern 1 enamel at a consistently slower daily rate than the orang-utan. Certainly, the enamel prism data can be seen as presenting a stepwise pattern of pattern 1 secretion: i.e., from 2.5 μm/day to 1.8 μm/day (as seen in the orang-utan) to 1.5 μm/day (as in the African apes).

Although Martin (1985) argues that the "intermediately" thick enamel of the orang-utan is not homologous with the thick enamel of modern humans, he allows for a morphocline between the thick pattern 3 enamel of *Sivapithecus* and the "intermediately" thick pattern 3+1 enamel of the orang-utan because other features of dental as well as facial anatomy appear synapomorphic for these two hominoids. However, if the enamel configuration in the orang-utan can be derived from that seen in *Sivapithecus*, it should also be derivable from that found in *Homo sapiens*, which is the same as in *Sivapithecus*. Thus, while Martin's detailed analyses have indeed pointed out that our previous use and understanding of thick vs. thin molar enamel was naive, his data do not contradict the association of the orang-utan both with *Sivapithecus* (as a close relative) and *Homo* (as a member of a larger *Sivapithecus-Homo* clade that excludes the African apes).

R. D. Martin (personal communication) pointed out that there were differences of potential interest among the extant large hominoids in the lengths of their gestation periods. As Cross and Martin (1981) had calculated, the gestation period of the orang-utan appeared to be approximately 270 days, which is nearly equivalent to that of humans (~267 days), and thus these two hominoids would be distinguished by having the longest gestation periods among primates. In response to the suggestion that such long, and essentially equally long, gestation periods in humans and the orang-utan might reflect the relatedness of these two hominoids (Schwartz, 1984a,b), Andrews (1987) countered by arguing that this similarity is of "doubtful significance" when "gestation time is corrected for body size." This particular disagreement is now moot since there are more recent and reliable data on the orang-utan. According to Martin and MacLarnon (personal communications), gestation length in the orang-utan is approximately 250 days; the values on gestation length in the gorilla and chimpanzee remain approximately 260 and 235 days, respectively. However, these figures should be corrected for the scaling effects of body size.

According to Martin and MacLarnon (1985), gestation period (G) typically scales to body weight (W) as follows in mammals:

$$G = kW^{0.1}.$$

Hence, relative gestation lengths can be calculated from the ratio $G/W^{0.1}$. With mean female *Gorilla* body weight being 82475 gm, *Homo* being 55000 gm, *Pan* being 40300 gm, and *Pongo* being 37078 gm, values for the ratio $G/W^{0.1}$ indicate a decreasing order: *Homo* (89.6) > *Pongo* (87.3) > *Gorilla* (83.8) > *Pan* (81.4). Interestingly, as do humans, the orang-utan has a gestation period that is much longer than would be predicted for an animal of its body size. Thus, while absolute gestation period length would distinguish humans and the gorilla from the other large hominoids, the more significant and potentially phylogenetically revealing similarity is between humans and the orang-utan.

Since human females are essentially continuously sexually receptive during the ovulatory cycle and orang-utans, especially in captive situations, appear to be similarly disposed, this might represent another potential synapomorphy of these two hominoids (see Schwartz, 1984a,b, and references therein). But, while acknowledging that copulation in the gorilla is restricted to the peak of the ovulatory cycle and is most frequent in the chimpanzee during that brief period, Andrews (1987) objected to the proposed synapomorphy because "in no species does [continuous receptivity of females] reach human levels." This may indeed be sufficient reason to rule out this attribute as a potential synapomorphy of humans and orang-utans, but one should consider the broader implications of the argument: in order to be synapomorphic, shared derived features must always be of the same level of derivedness and cannot represent derivations upon the original (syn)apomorphy.

Quite rightly, Andrews (1987) took to task the suggestion that reduced lower third molars united humans and orang-utans (cf. Schwartz, 1984a,b), pointing out that one can find similar configurations in the chimpanzee and gibbon. Andrews also rejected the lack of upper molar cingula as a viable synapomorphy of humans and the orang-utan. Although I based this suggestion on my own observations as well as the data tables in Delson and Andrews (1975), I yield to the most recent survey of Swindler and Olshan (this volume), who characterize the orang-utan and hominids as having the least amount of upper molar cingulum development among the hominoids and conclude that the orang-utan and hominids are derived among hominoids in this regard. I would go further and suggest that these hominoids are synapomorphic in their degree of cingular diminution.

Of the potential synapomorphies outlined in Schwartz (1984a) as suggesting the unity of humans and orang-utans, Andrews (1987) accepted only the presence of a foramen lacerum in the basicranium, delay in ossification of proximal humeral and distal radial epiphyses, and the lack of keratinized ischial callosities as being possibly viable. These and the potential synapomorphies delineated in Schwartz (1984a) but not addressed by Andrews (1987), as well as others discussed elsewhere (Schwartz, 1984b, 1985, 1987) and derivable from Groves's (1986) tables, are listed in Table 5–3. In addition, an oval-shaped occlusal outline of the upper molars, various features of the upper and lower posterior deciduous molars, the frequent absence of the deep head of the flexor digitorum brevis of the lower limb, and the development of a large fetal adrenal gland appear to be distinctive of humans and orang-utans (see, respectively, Swindler and Olshan, Swarts, Anderton, and Graham, this volume) and are thus included in Table 5–3, as well.

FOSSIL EVIDENCE RELEVANT TO THE EVOLUTION OF THE ORANG-UTAN

As mentioned throughout the preceding sections, there is considerable morphological support for the suggestion that the Miocene *Sivapithecus* (= ?*Ramapithecus*) is related to the orang-utan (Andrews and Cronin, 1982; Andrews and Tekkaya, 1980; Pilbeam, 1982; Ward and Kimbel, 1983; Ward and Pilbeam, 1983; as well as Brown and Ward, Shea, and Swindler and Olshan, this volume). In addition to a compressed, elongate, slit-like incisive foramen on the oral side of the palate and narrow incisive fossae in the floor of the nasal cavity, other apparent synapomorphies of these two hominoids include narrow interorbital breadth, high and rounded orbits, long and narrow nasal bones, a flattened zygoma bearing relatively large foramina located superior to the lower rim of the orbit, posterior thinning of a maxilla that is markedly overridden anteriorly by the posterior pole of the premaxilla, and an I^1 that is noticeably larger than the I^2.

Enlargement of the P_3 metaconid region has also been cited as a synapomorphy of *Sivapithecus* and the orang-utan, but this feature more broadly characterizes the other thick-molar-enameled Miocene hominoids which appear to be in some way associated with a larger orang-utan clade, e.g., *Gigantopithecus, Ouranopithecus, "Sivapithecus"* from Lufeng, China, and possibly *"Bodvapithecus"* (at least in part) from Hungary. Similarly, the development of orang-utan-like pseudofrontal sinuses (which result from expansion of the maxillary sinuses through the nasal and into the supraorbital region [Cave and Haines, 1940]) in *Sivapithecus indicus* (Ward, 1986) and apparently in the Lufeng hominoid, if not *"Bodvapithecus"* as well (Brown and Ward, this volume), may emerge as yet another feature representative of a larger orang-utan clade. Within this large clade there would appear to be a hierarchical arrangement of taxa based on subnasal morphology: i.e., since the Lufeng hominoid (Brown and Ward, this volume) and *Ouranopithecus* (Bonis and Melentis, 1985) possess and thus, like *Australopithecus*, would have retained the so-called African pattern, a subclade consisting of *Sivapithecus* and the orang-utan is distinguished by modifications of subnasal, premaxillary, and maxillary morphology (see discussion above).

Further and more detailed analyses of the *Kenyapithecus* (Ishida et al., 1984; Pickford, 1985), Buluk (Leakey and Walker, 1985), and Kalodirr (= *Afropithecus*) (Leakey and Leakey, 1986) hominoid material will certainly have bearing on the proposition of there being a larger orang-utan clade and of its possible relatedness to hominids, as well as on the more popular alternative, the relatedness of hominids and the African apes. These fossil taxa appear to retain primitive (i.e., *Proconsul*-like) contours of the floor of the nasal cavity, but it is unknown at present what their incisive fossae and canals are like and whether they possessed one or two incisive foramina on the oral cavity side of the palate. There have been preliminary suggestions that molar enamel in these hominoids is thick, but more definitive analyses are still lacking. *Afropithecus*, at least, had a frontal sinus, but whether this structure is derived from an ethmoid that is subdivided by air cells or is an extension of the maxillary sinus has yet to be determined. Both genera are described as having upper canine rotation (although Andrews [personal communication] would disagree in the case of *Afropithecus*), which would initially suggest an association of these hominoids with hominids and/or the orang-utan clade (cf. Ward, 1983), but in the absence of other corroborative synapomorphies it would be premature to align *Kenyapithecus* and *Afropithecus* with specific hominoids.

In short, and while at present there seem to be grounds for establishing an orang-utan clade comprising the extant taxon and various Miocene forms, fossil data shed little light on the affinities of the orang-utan to other extant hominoids.

CONCLUSIONS

Regardless of what scheme of phylogenetic relationships of the orang-utan is agreed upon, it is most critical that the underlying reasons be fully understood. Obviously, history and the weight of received wisdom provide no such authority or, for that matter, significant corroboration of any theory of hominoid phylogeny. Studies of especially the past few years, however, have attempted to generate phylogenetic hypotheses of relatedness among the hominoids based on the delineation of the hierarchical levels of significance of the similarities which are all too often—but unjustifiably—taken as reason enough to claim relationships between taxa; these studies have at least begun to add substance to the morphological side of the issue.

Although apparent morphologic synapomorphy in support of human–great ape, human–African ape, and human–orang-utan clades can be garnered, robustness (in terms of being highly corroborated, not the subjective weighting of characters) is not equivalent among these alternative hypotheses. It would appear that the orang-utan could be closely related to humans, in spite of what we might otherwise expect to learn through study of history, morphology, and even the fossil record. If we are true to the logic and philosophy of evolutionary science, this must be the conclusion at present, regardless of the apparent demonstration of potential morphological synapomorphy between humans and

TABLE 5–3. Possible Synapomorphies of *Homo* and *Pongo*, in Comparison to the African Apes

	Homo	*Pongo*	*Pan*	*Gorilla*
Delayed ossification of epiphyses at birth:				
Proximal humerus	X	X	0	0
Distal radius	X	X	0	0
Scapula				
General shape	Short/deep	Short/deep	Mod. short/mod. deep	Mod. short/deep
Supraspinous fossa	Reduced	Reduced	Mod. red.	Mod. red.
Spine	Most horiz.	Most horiz.	Mod. horiz.	Mod. horiz.
Coracoid process deflected	X	X	0	0
Talar tubercle	Short	Short	Long	Long
Deep head of flexor digitorum brevis, lower limb	Rare	Rare	X	X
Sole: thenar relative to hypothenar pattern	>>	>>	=	=
Ethmo-sphenoid contact (%)	97	99	77	50
Subnasal floor	Smooth	Variably stepped/smooth	Stepped	Stepped
Foramen lacerum	X	X	0	0
Incisive foramen/foramina				
Adult	Single	Single	Weakly double	Double
Juvenile	Single	Single	Double	Mrkd double
Upper canine rotation[1]	X	X	0	0
Upper molars				
Lingual shape	Oval	Oval	Squarer	Squarer
Lingual cingulum	0	0	Rare	Occ.

Upper anterior deciduous molar				
Protocone relative to paracone height	>	>	<	<
Lower anterior deciduous molar				
Protoconid	Anterior	Anterior	Central	Central
Paracristid	Angled	Angled	Straight	Straight
Talonid basin	Closed	Closed	Open	Open
Lower posterior deciduous molar trigonid	Short	Short	Mod.	Mod.
Cerebral asymmetries	Mrkd	Mrkd	Mod.	Mod.
Sylvian sulcus asymmetry	Mrkd	Mrkd	Mod.	Mod.
Estriol levels				
Menstrual cycle	High	High	Low	Low
Pregnancy	High	High	Low	Low
Relative size of fetal adrenal gland	Large	Large	Small	Small
Estrous cycle	0	0	X	X
Gestation period scaled to body weight	89.6	87.3	81.4	83.8
Anogenital tumescence during menstrual cycle	0	0	X	X
Position of mammary glands	Almost axillary	Almost axillary	Mod. sep.	Mod. sep.
Gall bladder	Sl. bend	Straight	Bend	Bend
Parotid gland accessory lobes	Slight	Small	None	None
Hair length	Mrkd	Mrkd	Mod.	Mod.
Ischial callosities	0	Unker.	Ker.	Ker.
Vallate papillae (#)	8–12	7–12	6–8	3–8

Abbreviations: 0 = not present; X = present; mod. = moderate(ly); red. = reduced; horiz. = horizontal; occ. = occasional; intermed. = intermediately; sl. = slight; mrkd = marked; (un)ker. = (un)keratinized.

[1]Disagreement on the identification and distribution of this feature may warrant its rejection as a synapomorphy here.

the African apes. The features that by themselves appear to be reflective of the relatedness of humans and the African apes, by dint of a more robust and highly corroborated theory of relationship, either must be seen as primitive retentions or parallelisms, or they must be rethought altogether. For example, instead of being the result of primitive retention, the "lack" of ethmoidal complexity and perhaps also frontal sinus development in the orang-utan might be due to the invasion of these regions by the maxillary sinuses, which thereby supplant and/or interfere with such developments; and, as Winkler et al. (this volume) argue, lack of frontal sinus development in the orang-utan appears to be correlated with its having less morphological disparity between the inner and outer tables of the frontal bone, compared, for instance, with the gorilla, which does develop frontal sinuses. The configuration seen in humans and the African apes would thus have characterized the last common ancestor of all large hominoids, not a proposed ancestor of these three hominoids alone.

There are, however, other sources of data that impinge upon our view of hominoid phylogeny, and these come from biomolecular and chromosomal studies. But, as reviewed above, the scheme of relationships increasingly favored by biomolecular systematists dissociates the chimpanzee from the gorilla and aligns it with humans. There have also been suggestions based on chromosomal data that humans are most closely related to the gorilla (Miller, 1977) or perhaps even to the orang-utan (Mai, 1983); in order to unite humans with the African apes, the most parsimonious interpretation of the myoglobin data demands the conclusion that the gibbon is more closely related to this triad than is the orang-utan (Romero-Herrera et al., 1976); and some electrophoretic analyses result in the chimpanzee being the sister of the orang-utan, a group to which humans and then the gorilla are related (Bruce and Ayala, 1979, see diagram in Fig. 1, p. 1048), while any number of possible arrangements of the hominoids have been suggested by other electrophoretic as well as immunologic reaction studies of blood serum proteins (Goodman, 1962). Although these latter studies have not gained popularity, they do represent alternative theories of relatedness among the hominoids which struck their authors as being viable interpretations of the data. But perhaps the most profitable approach to these data is a cladistic one, which although at present hampered by taxonomically incomplete data sets, has been attempted (Andrews, 1987; Marks, this volume). In this regard, however, Marks feels that the data are unable to resolve a trichomotous relationship among these hominoids, whereas Andrews concludes that an African ape sister group can be further distinguished. The level at which Marks prefers to accept lack of resolution is, however, precisely where most studies of late are proposing the uniting of chimpanzees with humans rather than with gorillas.

And herein lies the crux of the matter: if one accepts one particular hypothesis, the features in support of the others cannot be true synapomorphies. Thus, for example, if humans are most closely related to the African apes or the chimpanzee alone among the hominoids, the morphological features otherwise uniquely shared by humans and the orang-utan must be either parallelisms or primitive retentions. And if the apparent synapomorphies of humans and orang-utans are merely primitive retentions or parallelisms, this, as much as the dissociation of the chimpanzee from the gorilla, would seem to call into question the viability of any morphologically based theory of relatedness, regardless of whether the taxa involved are fossil or extant. Should an appeal to the case of the orang-utan seem unwarranted on the grounds that biomolecular systematists have not at present found support for it, one should not forget that, until the recent suite of refined analyses uniting chimpanzees and humans, there was essentially no biomolecular corroboration of the accepted African ape sister group either.

There is no a priori reason why a biomolecularly based phylogeny should necessarily falsify a competing morphologically based theory of relatedness. Like it or not, parallelism and primitive retention on molecular levels are real and cannot be coped with simply by approaching phylogenetic reconstruction phenetically and in terms of maximum parsimony (see criticism by Andrews, 1987). But although the technology necessary to provide the sequence data that are so crucial to undertaking a cladistic analysis exists, it is still, and may prehaps have to be, at least in part, an a priori assumption that sequence or serial similarity does indeed represent true homology.

The basic realization that must be achieved is that, regardless of the particular theory of hominoid relatedness one prefers, one is making a choice among competing, alternative theories. A human–chimpanzee sister group (to which the gorilla, the orang-utan, and then the gibbon are related) may be most consistent with the interpretation of certain types of biomolecular data, but it gains meager support from morphology.

The hypothesis that humans and orang-utans are sister taxa which share a common ancestor with an African ape clade is highly corroborated morphologically but at odds with current interpretations of biomolecular data. A human–African ape cluster, to which the orang-utan is related, is a compromise position with regard to both biomolecular and morphological analyses.

While it is becoming popular to ask how morphology might be interpreted given a particular biomolecularly based phylogenetic arrangement of taxa, it would seem to be equally viable to ask the question the other way round. If the chimpanzee is closely related to the gorilla, and if the orang-utan is closely related to humans, what are the possible consequences? At present, and of the three most robust of the available competing theories of hominoid relatedness, I am inclined toward the hypothesis that humans and orang-utans, and chimpanzees and gorillas, constitute respective sister groups which, in turn, shared a common ancestor. But even if neither of these two groupings should survive my or any other attempts at falsification (and Andrews, Groves, and I shall be collaborating on a review of the hominoids), it is obvious that the "case" of the orang-utan has provided us with a focus for airing methodological differences and a forum for refining our perceptions of evolutionary relatedness.

ACKNOWLEDGMENTS

Discussions and exchange of ideas, data, and unpublished manuscripts with many colleagues have greatly influenced this contribution. In particular, I thank Peter Andrews, Eric Delson, William Kimbel, Lawrence Martin, Robert D. Martin, and Steven Ward for their time, criticisms, and encouragement, and R. D. Martin and Ann MacLarnon for data and analysis pertinent to the understanding of gestation lengths in the large hominoids. I especially thank Peter Andrews for scrutinizing with me various tables in Groves (1986), and for reviewing the last two versions of this chapter and providing comments and suggestions which I have attempted to incorporate, making this version not only different but I hope better.

REFERENCES

Andrews, P. 1985. Family group systematics and evolution among catarrhine primates. *In* Ancestors: The Hard Evidence, pp. 14–22, ed. E. Delson. Alan R. Liss, New York.

Andrews, P. 1987. Aspects of hominoid phylogeny. *In* Molecules and Morphology in Evolution: Conflict and Compromise, pp. 21-53, ed. C. Patterson. Cambridge University Press, Cambridge.

Andrews, P. and Cronin, J. E. 1982. The relationships of *Sivapithecus* and *Ramapithecus* and the evolution of the orang-utan. *Nature, 297:*541–545.

Andrews, P. and Tekkaya, I. 1980. A revision of the Turkish Miocene hominoid *Sivapithecus meteai. Paleontology, 9:*85–95.

Bonis, L. de and Melentis, J. 1985. La place du genre *Ouranopithecus* dans l'évolution des hominidés. *Comptes Rendus de l'Académie des Sciences, Paris, 300:*429–432.

Bromage, T. G. 1985. Taung facial remodelling: a growth and development study. *In* Hominid Evolution: Past, Present and Future, pp. 239–245, ed. P. V. Tobias. Alan R. Liss, New York.

Bromage, T. G. and Dean, M. C. 1985. Re-evaluation of the age at death of immature fossil hominids. *Nature, 317:*525–527.

Brown, W. M., Prager, E. M., Wang, A. and Wilson, A. C. 1982. Mitochondrial DNA sequences of primates: tempo and mode of evolution. *Journal of Molecular Evolution, 18:*225–239.

Bruce, E. J. and Ayala, F. J. 1979. Phylogenetic relationships between man and the apes: electrophoretic evidence. *Evolution, 33:*1040–1056.

Cave, A. J. E. and Haines, R. W. 1940. The paranasal sinuses of the anthropoid apes. *Journal of Anatomy, 136:*493–523.

Cross, J. F. and Martin, R. D. 1981. Calculation of gestation period and other reproductive parameters for primates. *Dodo, Journal of the Jersey Wildlife Preservation Trust, 18:*30–43.

Dart, R. A. 1925. *Australopithecus africanus:* the man-ape of South Africa. *Nature, 115:*195–199.

Darwin, C. 1859. On the Origin of Species (Harvard University Press facsimile of 1st ed., 1964).

Darwin, C. 1871. The Descent of Man, vol. I. John Murray, London.

Delson, E. 1985. The earliest *Sivapithecus? Nature, 818:*107–108.

Delson, E. and Andrews, P. 1975. Evolution and interrelationships to the catarrhine primates. *In* Phylogeny of the Primates, pp. 405–446, ed. W. P. Luckett and F. S. Szalay. Plenum Press, New York.

Fleagle, J. G. and Jungers, W. L. 1982. Fifty years of higher primate phylogeny. *In* A History of American Physical Anthropology 1930–1980, pp. 187–230, ed. F. Spencer. Academic Press, New York.

Goodman, M. 1962. Immunochemistry of the primates and primate evolution. *Annals of the New York Academy of Sciences, 102:*219–234.

Goodman, M., Braunitzer, G., Stangl, A. and Schrank, B. 1983. Evidence on human origins from haemoglobins of African apes. *Nature, 303:*546–548.

Goodman, M. and Cronin, J. E. 1982. Molecular anthropology: its development and current directions. *In* A History of American Physical Anthropology 1930–1980, pp. 105–146, ed. F. Spencer. Academic Press, New York.

Goodman, M., Olson, C. B., Beeber, J. E. and Czelusniak, J. 1982. New perspectives in the molecular biological analysis of mammalian phylogeny. *Acta Zoologica Fennica, 169:*1–73.

Gregory, W. K. 1915. Is *Sivapithecus* Pilgrim an ancestor of man? *Science, 42:*341–342.

Gregory, W. K. 1921. The origin and evolution of the human dentition (Concluded). *Journal of Dental Research, 3:*87–223.

Gregory, W. K. 1927. Dawn-man or ape? Reprint from *Scientific American,* September.

Groves, C. 1986. Systematics of the great apes. *In* Comparative Primate Biology: vol. 1, Systematics, Evolution, and Anatomy, pp. 187–217, ed. D. R. Swindler. Alan R. Liss, New York.

Haeckel, E. 1896. The Evolution of Man, vols. 1 and 2 (translated from the German). D. Appleton and Company, New York.

Hasegawa, M., Kishino, H. and Yano, T. 1985. Dating of the human-ape splitting by a molecular clock of mitochondrial DNA. *Journal of Molecular Evolution, 22:*160–174.

Hill, W. C. O. 1951. The external genitalia of the female chimpanzee, with observations of the mammary apparatus. *Proceedings of the Zoological Society of London, 121:*131–145.

Huxley, T. H. 1863. Man's Place in Nature. D. Appleton and Company, New York.

Ishida, H., Pickford, M., Nakaya, H. and Nakano, Y. 1984. Fossil anthropoids from Nachola and Samburu Hills, Samburu District, Kenya. *African Study Monographs Supplementary Issue, 2:*73–85.

Kay, R. F. and Simons, E. L. 1983. A reassessment of the relationship between later Miocene and subsequent Hominoidea. *In* New Interpretations of Ape and Human Ancestry, pp. 577–624, ed. R. L. Ciochon and R. S. Corruccini. Plenum Press, New York.

Keith, A. 1916. Lo schema dell'origine umana. *Rivista di Anthropologia, 20:*1–20.

Kimbel, W. 1986. Calvarial morphology of *Australopithecus afarensis*: a comparative phylogenetic study. Ph.D. dissertation, Kent State University.

Koop, B. F., Goodman, M., Xu, P., Chan, K. and Slightom, J. L. 1986. Primate η-globin DNA sequences and man's place among the great apes. *Nature, 319:*234–237.

Leakey, R. E. and Leakey, M. G. 1986. A new Miocene hominoid from Kenya. *Nature, 324:*143–146.

Leakey, R. E. F. and Walker, A. 1985. New higher primates from the early Miocene of Buluk, Kenya. *Nature, 318:*173–175.

Mai, L. L. 1983. A model of chromosome evolution and its bearing on cladogenesis in the Hominoidea. *In* New Interpretations of Ape and Human Ancestry, pp. 87–114, ed. R. L. Ciochon and R. S. Corruccini. Plenum Press, New York.

Martin, L. 1985. Significance of enamel thickness in hominoid evolution. *Nature, 314:*260–263.

Martin, L. 1986. Relationships among extant and extinct great apes and humans. *In* Major Topics in Primate and Human Evolution, pp. 161–187, ed. B. Wood, L. Martin, and P. Andrews. Cambridge University Press, New York.

Martin, R. D. and MacLarnon, A. M. 1985. Gestation period, neonatal size and maternal investment in placental mammals. *Nature, 313:*220–223.

Miller, D. A. 1977. Evolution of primate chromosomes. *Science, 198:*1115–1124.

Miyamoto, M. M. and Goodman, M. 1986. Biomolecular systematics of eutherian mammals: phylogenetic patterns and classification. *Systematic Zoology, 35:*230–240.

Miyamoto, M. M., Slightom, J. L. and Goodman, M. 1987. Phylogenetic relations of humans and African Apes from DNA sequences in the $\psi\eta$-globin region. *Science, 238:*369–373.

Osborn, J. W. and Ten Cate, A. R. 1976. Advanced Dental Histology (3rd edition). John Wright & Sons, Bristol.

Pickford, M. 1985. A new look at *Kenyapithecus* based on recent discoveries in western Kenya. *Journal of Human Evolution, 14:*113–143.

Pilbeam, D. R. 1979. Recent finds and interpretations of Miocene hominoids. *Annual Review of Anthropology, 8:*333–352.

Pilbeam, D. R. 1982. New hominoid skull material from the Miocene of Pakistan. *Nature, 295:*232–234.

Pilbeam, D. R. 1984. The descent of hominoids and hominids. *Scientific American, 250* (March): 84–97.

Pilbeam, D. R. 1986. Hominoid evolution and hominoid origins. *American Anthropologist, 88:*295–312.

Pilgrim, G. E. 1915. New Siwalik primates and their bearing on the question of the evolution of man and the Anthropoidea. *Record of the Geological Survey of India, 45:*1–74.

Romero-Herrera, A. E., Lehmann, H., Castillo, D., Joysey, K. A. and Friday, A. E. 1976. Myoglobin of the orang-utan as a phylogenetic enigma. *Nature, 261:*162–164.

Sarich, V. M. 1983. Retrospective on hominoid macromolecular systematics (appendix). *In* New Interpretations of Ape and Human Ancestry, pp. 137–150, ed. R. L. Ciochon and R. S. Corruccini. Plenum Press, New York.

Schultz, A. H. 1936. Characters common to higher primates and characters specific for man. *Quarterly Review of Biology, 11:*259–283, 425–455.

Schultz, A. H. 1958. Palatine ridges. *Primatologia, 1:*127–138.

Schultz, A. H. 1961. Vertebral column and thorax. *Primatologia, 5:*1–65.

Schultz, A. H. 1968. The recent hominoid primates. *In* Perspectives on Human Evolution, vol. 1, pp. 122–195, ed. S. L. Washburn and P. C. Jay. Holt, Rinehart and Winston, New York.

Schwartz, J. H. 1983. Palatine fenestrae, the orangutan and hominoid evolution. *Primates, 24:*231–240.

Schwartz, J. H. 1984a. The evolutionary relationships of man and orang-utans. *Nature, 308:*501–505.

Schwartz, J. H. 1984b. Hominoid evolution: a review and a reassessment. *Current Anthropology, 25:*655–670.

Schwartz, J. H. 1985. Morphological, chromosomal, and macromolecular considerations for the interpretation of orang-utan phylogeny (abstract). *American Journal of Physical Anthropology, 66:*227–228.

Schwartz, J. H. 1987. The Red Ape: Orang-Utans and Human Origins. Houghton Mifflin, Boston.

Sibley, C. G. and Ahlquist, J. E. 1984. The phylogeny of the hominoid primates, as indicated by DNA-DNA hybridization. *Journal of Molecular Evolution, 20:*2–15.

Simons, E. L. and Pilbeam, D. R. 1965. Preliminary revision of the Dryopithecinae (Pongidae, Anthropoidea). *Folia Primatologica, 3:*81–152.

Simons, E. L. and Pilbeam, D. R. 1972. Hominoid paleoprimatology. *In* The Functional and Evolutionary Biology of the Primates, pp. 36–63, ed. R. H. Tuttle. Aldine, Chicago.

Tuttle, R. H. 1974. Darwin's apes, dental apes, and the descent of man: normal science in evolutionary anthropology. *Current Anthropology, 15:*389–398.

Ward, S. C. 1983. Canine implantation patterns in Miocene hominoids (abstract). *American Journal of Physical Anthropology, 60:*268.

Ward, S. C. 1986. Paranasal pneumatization of the frontal bone in early Miocene hominoids (abstract). *American Journal of Physical Anthropology, 69:*276.

Ward, S. C. and Kimbel, W. J. 1983. Subnasal alveolar morphology and the systematic position of *Sivapithecus. American Journal of Physical Anthropology, 61:*157–172.

Ward, S. C. and Pilbeam, D. R. 1983. Maxillofacial morphology of Miocene hominoids from African and Indo-Pakistan. *In* New Interpretations of Ape and Human Ancestry, pp. 211–238, ed. R. L. Ciochon and R. S. Corruccini. Plenum Press, New York.

Weinert, H. 1933. Ursprung der Menschheit: über den engeren Anschluss des Menschengeschlechts an die Menschenaffen. Ferdinand Enke, Stuttgart.

Zuckerkandl, E. and Pauling, L. 1962. Molecular disease, evolution, and genic heterogeneity. *In* Horizons in Biochemistry, pp. 189–225, ed. M. Kasha and B. Pullman. Academic Press, New York.

II

REPRODUCTIVE PHYSIOLOGY, BEHAVIOR, AND ANATOMY

To begin this section, Charles E. Graham presents an overview of orang-utan reproductive physiology. Here, he suggests that the rather "uncomplex" social structure in conjunction with the predominantly arboreal habit of the orang-utan has had an impact upon the evolution of its pattern of sexual behavior and, consequently, on the development of secondary sexual characteristics. Remarkable secondary sexual characteristics typify fully mature males, who can perhaps inhibit maturation of young sexual competitors in the same territory or by so far unknown mechanisms (see chapter by Susan Kingsley). In contrast, females exhibit no dramatic secondary sexual characteristics evident to human beings. The mating system seems to depend on choice exercised by receptive females, who can usually remain concealed in the forest canopy from unwelcome males. Although the internal anatomy of all the large hominoids is similar, there are differences in the orang-utan in reproductive physiology and metabolism. Most important, orang-utans exhibit a luteal peak of androsterone excretion (precursor unknown), unlike human females. The fetal adrenal gland is large in humans and the orang-utans, resulting in greater conversion of dehydroepiandrosterone sulfate to estriol during pregnancy in these hominoids compared to the African apes. Finally, orang-utans show some individuality in placental androgen metabolism, particularly in the presence of an active Δ-4-5α-reductase system.

In complementary fashion, Ronald D. Nadler reviews sexual and reproductive behavior in the orang-utan and presents his latest contributions to this fascinating area of study. After summarizing the earlier history of studies on captive orang-utans, from Yerkes and Yerkes on, Nadler brings us to the present state of these investigations, which are focusing on both the proximate and ultimate factors contributing to the regulation of sexual and reproductive behavior. Proximate factors include (1) phase of the menstrual cycle; (2) hormones of the female, especially of estradiol, testosterone, and progesterone associated with cycle phase; and (3) environmental conditions that differentially favor male or female sexual initiative. The mating system or, specifically, the number of males that typically compete for sexual access to estrous females, represents an ultimate factor or influence on mating (and genital anatomy). Under natural conditions and in restricted-access tests with female choice, mating occurs as a result of the female's sexual initiative, primarily during the midcycle periovulatory phase, when estradiol and testosterone levels are elevated and progesterone levels are low. In the typical laboratory free-access pair test, in which male dominance prevails, mating can occur on a daily basis, irrespective of cycle phase. Orang-utans thereby demonstrate the following: (1) cyclicity in sexual behavior associated with estrus in the female, and (2) some emancipation from hormonal control as, for example, is reflected in the sexual initiative of the male. These characteristics of the orang-utan bear some resemblance to comparable dimensions of human sexuality.

On a different level of study, Nancy Czekala, Susan Shideler, and Bill Lasley provide some of the basic data necessary to further our understanding of female reproductive hormone patterns among the hominoids, the gibbon included. Their data result from analysis of urinary hormone profiles of estrogen and progestin metabolites during the normal menstrual cycle, implantation, and pregnancy. All hominoids are similar during the follicular phase of the ovarian cycle in the pattern of estrogen excretion, but, in the luteal phase, estrogen levels are much lower in the gibbon (as in the macaque), as is progesterone excretion as well. In the latter case, the large hominoids excrete large amounts of pregnanediol-3-glucuronide, with humans excreting only slightly more than the gorilla, but up to 100 times as much as orang-utans. However, during pregnancy, the orang-utan is most like humans in its total estrogen profile, which reflects the relative size of the fetal adrenal component, and both are similarly different from the African apes. The latter features of pregnancy may perhaps provide the stronger structure-function argument, and thus more characteristics, for associating humans with the orang-utan than would ovulatory progesterone profiles.

Susan Kingsley has focused on physiological development in the less well-known male orang-utan and here presents these data in comparison with those she also collected on the male gorilla. After determining the testosterone concentrations in urine samples from a large number of male orang-utans, ranging from 1 to 16 years of age, and an almost equal number of male gorillas, aged from 1 to 31 years, Kingsley found that changes through life in male orang-utans and male gorillas are similar to those already documented for human and chimpanzee males. Thus, adolescent gorillas had urinary testosterone concentrations which were 4 times those of juveniles, while levels in adult gorillas were 2.4 times those of adolescents. In orang-utans, subadult males (that is, males with no cheek flange development) had urinary testosterone levels that were 2.3 times greater than those of juveniles. Fully adult male orang-utans (that is, those with cheek flanges) had urinary testosterone concentrations that were 2.4 times greater than those of nonflanged males. By observing cheek flange development in two male orang-utans over a 2-year period, Kingsley found in only one of the two animals that cheek flange development was preceded by increases in testosterone and estrogen levels. Increases in urinary testosterone and estrogen concentrations were not correlated with cheek flange development in the second male, which may be because this male's flanges had already started to grow by the beginning of the study. This, in turn, suggests that the actual process of cheek flange development in the orang-utan may be associated with such hormonal excretions only during the early stages.

The study of reproductive behavior and physiology of the orang-utan would not be complete without an assessment of the anatomy of the external genitalia, particularly in light of varous generalizations pertaining to both the male and female which have continued to be repeated in the literature. In his investigation of the anatomy of the external genitalia of both sexes of the orang-utan, Jeremy Dahl reminds us that these structures are actually poorly known and are thus deserving of more detailed attention. His study of seven male and six female orang-utans (most of which were Sumatran), of varying ages, thus provides new, and different, data. For example, (1) the labia majora of female *Pongo* develop into conspicuous structures in the adult and do not, as previously thought, atrophy; (2) although the labia minora and associated prepuce tend to lose their identity in adult individuals, these structures are conspicuous in juveniles and, together with the labia majora, resemble the arrangement in adult human females; (3) the structure of the orang-utan clitoris is quite variable; (4) in males, the scrotum and testes are inconspicuous and small, as earlier described, but there is subspecific variability; (5) the penis is quite similar to that of the chimpanzee in basic organization, but highly specialized ridges occur on the distal portion; and (6) in contradiction to common dogma, the male orang-utan's external penis is relatively long, and some individuals even develop a bulbous glans penis. With regard to the similarity in penile structure between the orang-utan and chimpanzee, Dahl suggests that this is actually the result of independent evolution in the two hominoids. Dahl also emphasizes that the external genitalia of both sexes of the orang-utan, while being similar to humans or other hominoids in various ways, are also distinctive among extant hominoids.

This section concludes with a presentation by Hiroaki Soma of his latest studies on the morphology of the orang-utan placenta. Because he is one of the few scholars to pursue such analyses, Soma's data are important for assessing the normal morphology, pathology, and, eventually, the phylogenetic significance of the placenta in hominoids. To date, Soma has been able to study the placentas of nine orang-utans and compare them to humans and chimpanzees. He has found that the orang-utan placenta is usually larger than that of the chimpanzee, although the insertion of the umbilical cord tends to be eccentric in the placentas of both hominoids. In comparison to humans, orang-utans and chimpanzees have a higher frequency of placenta extrachorialis, which, in several cases, was found in association with placental infarcts and marginal hemorrhages. In the chorionic plate of the two apes, Soma was able to identify massive subchorionic fibrin. In microscopic and ultrastructural features of the chorionic villi, the placentas of humans, orang-utans, and chimpanzees are similar. Given the degree of similarity among these three hominoids in placental morphology, Soma concludes that further study of ape placentas will shed light on the delineation of the causes of high perinatal mortality rates in humans.

6

Reproductive Physiology

CHARLES E. GRAHAM

The reproductive physiology of the orang-utan has not been systematically studied, and indeed until recent years, has been little studied at all. Knowledge is still meager because of the rarity of this species and the difficulty of observing orang-utans in their natural environment. However, since the last major review of the reproductive biology of the great apes (Graham, 1981a), several useful contributions on the orang-utan have appeared, including a review devoted to the biology of the species (de Boer, 1982). Despite the difficulty of field studies, data are gradually accumulating that, combined with some important laboratory studies, have radically changed our ideas of the sexual biology of the species.

Opinions differ on the degree of phylogenetic relationship of the orang-utan to humans, compared with other great apes. The reproductive physiology of the orang-utan is best evaluated from the perspective of the species' overall biology. Primarily solitary dwellers of the canopy of tropical jungles, orang-utan females have an unusual opportunity to use concealment to control males' access, a theme developed by other contributors to this volume. This characteristic sets orang-utans apart from other great apes in their sociosexual biology.

Early accounts of sexual interaction based on captive animals suggested a "rape," or "forced copulation," pattern (Nadler, this volume). Both laboratory and field studies have shown that this is partly an artifact of males' unrestricted access during captivity (Nadler, this volume).

In the natural environment, the normally solitary female has a greater opportunity to determine when mating occurs, based on her own cyclic receptivity and proceptivity. In this respect orang-utans are markedly different from other great apes, which has important implications for theories of sexual selection.

A comparative review of the reproductive biology of orang-utans will provide the necessary factual background to appreciate orang-utan reproduction in a more holistic way. This chapter is designed to complement the chapter by Nadler on sexual behavior and sexual selection, by providing an overview of the reproductive physiology and morphology of the species (in a comparative perspective when possible). It attempts to be comprehensive in providing a reference path to all key papers on orang-utan reproduction. In some instances this is done by citing secondary sources, since some primary sources are quite old, inaccessible, or deal with single cases.

MORPHOLOGY

Before reviewing the reproductive physiology of the orang-utan, it will be helpful to describe the sexual morphology of the species, at least insofar as it differs from the other great apes and humans. I also wish to draw attention to the chapter by Dahl in this volume on the external genitalia of the orang-utan.

Sexual Dimorphism

There is a significant size difference between the sexes (females 35–50 kg; males 45–100 kg), with the sex/weight ratio being as high as 1:2 (Eckhardt, 1975; MacKinnon, 1974).

Size dimorphism begins at puberty (Short, 1979), in addition to which males develop massive lateral facial skin folds (flanges), changes in the laryngeal structure, a characteristic scent, and a muscle-covered throat pouch. These features do not develop at puberty in all animals, but are often delayed for a number of years to the second decade of life, as discussed below (Kingsley, 1982). Kingsley (1982) developed a means of measuring the extent of flange development by dividing the total face width by the horizontal breadth; thus photographs taken at various distances and angles could be used for this purpose.

Male Morphology

Based on limited data, the testes of the orang-utan are about the same absolute size as those of the gorilla; however, in comparison to body weight, they are larger (Short, 1979). Compared to other primates, the orang-utan has very small testes (Schultz, 1938). Related to the small testis size, the scrotum is also small, being a mere bulge of black skin.

Studies on the penis, summarized by Short (1979), indicate that the flaccid organ is concealed by the prepuce. When drawn out or erect the adult organ is at least 4 cm in length and clearly visible at a distance; Dahl (this volume) and Nadler (this volume) discuss penile size and display in more detail. Penile structure is illustrated by Kinzey (1971:102). The baculum, or os penis, present in all great apes, but not in humans, is longest in the orang-utan (11–15 mm).

The seminal vesicles are larger than those of humans or the gorilla (8.0–10.0 × 2.5 × 1.2 cm in two animals). Each is a single blind-ending convoluted tube, quite unlike the branched duct of human males (Short, 1979). The vas deferens does not possess terminal diverticula (Mijsberg, cited by Eckstein, 1958).

The prostate gland is not grossly divided into caudal and cranial portions as in other great apes and humans. Whether there is a distinction at the microscopic level has not been reported. Since the cranial portion has a role in semen coagulation in other species, and orang-utan semen does coagulate, it is suspected that microscopic counterparts of the cranial and caudal portions exist.

Although Eckstein (1958) stated that there appears to be no record of bulbourethral glands in the orang-utan, Kinzey (1971) published a redrawn illustration of a gross dissection by Poursargues which shows these glands; however, their large size leads one to wonder if this detail is exaggerated or imagined, especially as other authors have not noted them. Nevertheless, bulbourethral glands, although not visible grossly, were discovered in serial transverse sections in chimpanzee reproductive tracts by Graham and Bradley (1972). These glands may remain to be discovered by similar methodology in the other great apes.

Female Morphology

Adult Female

The female shows no obvious secondary sexual characteristics. A cyclic sexual and labial swelling is absent and the breasts remain undeveloped until the end of the first pregnancy. Towards the end of pregnancy, the nipples grow long and pendulous (Eckstein, 1958). The external genitalia are described in detail by Dahl (this volume), who reports that the labia majora are large and conspicuous in adults, in contrast to previous reports.

The ovaries are relatively small (Short, 1979). Eckstein (1958) cited evidence that there is practically no thecal cell development; this contrasts with chimpanzees, in which Graham and Bradley (1972) reported considerable thecal or interstitial cell development in many subjects. Because of the limited amount of material examined by Eckstein, one suspects that more orang-utan material must be examined before any definite conclusion can be drawn.

The oviducts were reported by Eckstein (1958) to be similar to those of women and chimpanzees, but more convoluted. The uterus is considerably smaller than that of the human, and its fundus is flattened in the sagittal plane.

The urethrovaginal septum is deficient, so that the urethra and vagina open into a vestibule, as in the other great apes, and unlike women, in whom the septum is thicker and more complete; the vestibule appears as a prolongation of the vagina. Consequently, the urethral meatus, which lacks a true papilla, is concealed within the vestibule, in contrast to humans.

Placental Morphology

The only complete study, including five orang-utan and 11 chimpanzee specimens, was conducted by Soma (1983), using light microscopy and scanning and transmission electron microscopy. The findings generally confirmed those of previous workers who noted that the placenta of great apes closely resembles that of humans (see also Soma, this volume).

The orang-utan placenta is discoid; the umbilicus contains two arteries and a single vein, and is eccentrically inserted. In most instances the vessels were arranged in the magistral pattern in which the two arteries extend to opposite margins of the placental disc; in two cases the arteries branched repeatedly to form a disperse pattern. Placenta extrachorialis (circumvallate or circummarginate placenta) occurred in 80% of cases, high by comparison to humans. Subchorionic fibrin, intervillous fibrin, and thrombi in the intervillous space were noted, but similar observations are made in human placentas unassociated with fetal abnormality.

In a detailed study of a single orang-utan placenta, Vacek (1974) noted a close similarity to the human placenta, with the exception that electron microscopy revealed differences in the pattern of distribution of certain cytoplasmic organelles in the trophoblast and decidual cells. Patterns of fibrin deposition similar to those reported by Soma (1983) were noted. The orang-utan placenta showed a lesser differentiation of the trophoblast into zones recognizable at the submicroscopic level, in contrast to the human placenta.

REPRODUCTIVE PHYSIOLOGY

In reviewing the reproductive physiology of the orang-utan, a developmental approach will provide a systematic basis for assessing our knowledge of the subject, and show where further research is needed.

Neonatal Period

The first weeks after birth are an endocrinologically active period, followed by a prolonged period of endocrine quiescence during preadolescent growth. The transition from hormonal activity to inactivity has been termed the infant endocrine transition (IET; Hobson et al., 1981). In infant chimpanzees, gonadotropins are suppressed at birth, probably due to inhibition by elevated maternal steroid hormone levels. However, after removal of maternal suppression, gonadotropins rise rapidly during the first week after delivery, reach a zenith at 3 months, and subsequently decline to a nadir at 1 year. Similar changes are anticipated in the orang-utan. The reason for the decline in gonadotropins at 1 year of age (i.e., the mechanism that induces the IET) is unknown.

At birth, another endocrine system, the fetal adrenal zone, is active in humans and great apes. This zone of the adrenal gland is so hypertrophied that the entire gland is enlarged. Regression of this zone after birth is associated with a decrease in the levels of serum dehydroepiandrosterone sulfate as well as in the 3β-hydroxy-Δ^5 steroids found in the urine of newborn humans, chimpanzees, and orang-utans (Hobson et al., 1981; Shackleton, 1974). The fetal adrenal zone provides the precursor (dehydroepiandrosterone sulfate, DHAS) of estriol, the dominant urinary estrogen of human pregnancy and of the third trimester in apes. Orang-utans produce relatively more total urinary estrogen per unit of creatinine and more estrone than gorillas but an equivalent amount to humans (Czekala et al., 1983), whereas chimpanzees and gorillas produce somewhat less. It is therefore interesting that the fetal adrenal zones of the orang-utan and human, in proportion to body weight, are larger than in gorillas and chimpanzees, and are closer to each other in size than either is to the fetal adrenal zone in gorillas or chimpanzees. Czekala et al. (1983) suggested that this may be related to the higher total estrogen levels, and specifically estriol levels, in humans and orang-utans during pregnancy, compared with gorillas and chimpanzees.

Infancy

After the infant adrenals and pituitary become quiescent, little sexual endocrine activity occurs until adolescence approaches (Hobson et al., 1981). This interval has not been studied in the orang-utan.

Puberty

The mechanism that initiates puberty in primates is not understood. For a discussion of this topic see Hobson et al. (1981).

Males as young as 8 years old have fathered infants in captivity (Graham-Jones and Hill, 1962); however, young adult males in the wild have not been observed to begin copulating until 10 years of age. Of course, field studies rarely provide precise data about the age of adolescents.

Although male orang-utans become sexually mature and sometimes capable of reproduction at 7–10 years of age, full development of secondary sexual characteristics is frequently delayed, at least in captive animals, for another 3–7 years (Kingsley, 1981; Ulmer, 1958, cited in Kingsley, 1982). These characteristics include prominent lateral, fatty facial skin folds (flanges), the large, muscle-covered throat pouch, laryngeal development, a final phase of body growth, lengthening and thickening of the hair, and a musty scent. Bornean orang-utans also often develop a fatty crown. Once this final stage of development is initiated, it may be completed within a year. The endocrine changes associated with these events are discussed below.

An early study (Selenka, 1898, cited in von Koenigswald, 1982) reported that some geographically isolated populations of Bornean orang-utans lack flanges altogether. Based on this observation, von Koenigswald suggested that the evolution of cheek flanges may be a recent event. If so, they may be a genetically labile feature that may not occur in 100% of individuals from other populations of the species. Alternatively, Selenka may have only observed young adults with delayed flange development.

Reports of captive females indicate the onset of menarche at about age 7, with copulation occurring a year later. However, mating patterns in captive orang-utans differ from those observed in their native habitat (Nadler, this volume). Galdikas (1981) noted juvenile females approaching males, although only adolescents actually copulated. Galdikas estimated that first parturition does not occur in the wild animals until age 12 at earliest. Observation of a year's cyclicity before conception in one individual suggested a period of adolescent sterility such as has been observed in chimpanzees and humans. However, based on other reports, Werff ten Bosch (1982) concluded that the period of orang-utan adolescent sterilty is rather short.

Adult Female

Menstrual Cycle

The duration of the menstrual cycle may be estimated from the intermenstrual interval, from the interval between successive periods of proceptivity, or from endocrine data. No labial or sexual swelling, such as is exhibited by gorillas or chimpanzees, is seen in the orang-utan.

Determinations of the intermenstrual interval have yielded variable results (Graham, 1981). However, the most complete data (Nadler, this volume) on six animals and 35 cycles indicate that in mature captive individuals (11–22 years old) the range is 23–33 days with a mean of 27.8 and SE of ± 1.1 (see also Inaba et al., 1983). More data are needed to determine possible age-related trends.

When the female controls her availability to the male, as in the experiments described by Nadler (this volume, FAT tests), copulation occurs for 2–8 days of the menstrual cycle. Under such conditions the interval between periods of mating could provide useful estimates of cycle length. However, no such data have been reported for this purpose.

Hormone profiles throughout the menstrual cycle have been studied from several points of view. Nadler et al. (1984) and Nadler (this volume) reported the serum profiles of estradiol, progesterone, testosterone, and luteinizing hormone (LH). The pattern and level of estradiol and progesterone are very similar to the other great apes and humans. Testosterone shows a midcycle peak on the same day as the LH peak, with a rather slow fall through the first 4 days of the luteal phase. The testosterone peak is well-defined and relatively short in duration, as in gorillas (Nadler et al., 1979, 1983); the testosterone peak in chimpanzees (Nadler et al., 1986) is about twice the magnitude and more prolonged, perhaps being related to the longer period of proceptivity in chimpanzees compared with orang-utans and gorillas. Nadler et al. (1984) noted that orang-utan LH cross-reacts with antibodies to human LH to a lesser extent than does that of gorillas and chimpanzees, indicating greater biochemical differences.

Urinary levels of estrone, estradiol, estriol, pregnanediol, and androsterone were studied by Collins et al. (1975) throughout three menstrual cycles of two orang-utans. Estrone was the predominant estrogen metabolite and estriol the minor one, as in chimpanzees, gorillas, and humans.

Pregnanediol showed the expected rise during the luteal phase, although levels were significantly less than in humans. Of interest was a corresponding luteal peak of androsterone. Although women have higher overall androsterone levels (Kirchner and Lipsett, 1964), they do not show the cyclic pattern and luteal rise observed in orang-utans. Although one might expect the orang-utan luteal androsterone to arise by metabolism of progesterone, as in the rhesus monkey, that explanation has not yet been validated. In chimpanzees administered ^{14}C-progesterone, ^{14}C-androsterone was not recovered. Androsterone excretion has not been determined during the menstrual cycle of gorillas and chimpanzees. Therefore, although we can assume that androsterone excretion in the luteal phase of the orang-utan reflects luteal activity, its precursor remains unknown.

Czekala et al. (this volume) have determined total estrone conjugates throughout the menstrual cycle of the orang-utan. They found that the pattern and quantity of estrone conjugates were similar for all the species of ape they studied as well as for humans. Pregnanediol-3-glucuronide also showed the expected luteal peak. The authors compared the estrogen conjugate profiles during the cycle of conception in orang-utans, other apes, and women, and noted an early, rapid rise in estrone after conception in each species.

Inaba et al. (1983) described urinary estrogen levels during five sequential menstrual cycles of one 11-year-old, nulliparous orang-utan. Estrone, estradiol, and estriol concentrations, indexed to creatinine concentration, were determined by radioimmunoassay in 16-hr nocturnal samples. The overall pattern was similar to that described by Collins et al. (1975), with the major difference that estriol was found to be the dominant estrogen, rising to a sharply defined follicular peak twice the amplitude of either estrone or estradiol, but without a clearly defined luteal peak in pooled data normalized to the day of the follicular peak.

Some of the summarized cycle data (mean ±SE) are of interest from this particular animal:

Length of menstrual cycle	28.8 ± 0.6
Day of estrogen peak	14.6 ± 1.0
Interval: estrogen peak-menses	14.0 ± 0.4

There is a discrepancy in the proportion of estriol vs. other metabolites obtained in this study as compared with the study by Collins et al. (1975) in which very careful metabolite identification using classical methods was performed. Inaba et al. (1983) did not use independent methods to validate the identity of the estrogens measured. However, if their findings were validated, it would mean that the orang-utan alone among nonpregnant apes has the ability to excrete estrogens primarily as estriol. This would require a reevaluation of the relationships between these species as well as the significance of estriol excretion during pregnancy discussed elsewhere in this chapter. More technical details about the methods for preserving urine during the 16-hr collection period and the specificity of the radioimmunoassays employed by Inaba et al. (1983) are needed.

Pregnancy

Estimates of pregnancy duration in the orang-utan have been reviewed by Martin (1981). The best data are probably Martin's, which yielded an estimated range of 227–275 days, with 245 days being close to the mean, based upon mating records and rectal palpations. It is unfortunate, however, that such techniques as the nonhuman primate pregnancy test, validated for use in orang-utans (Davis, 1977; Hobson, 1976; Hodgen et al., 1977; Woodard et al., 1976) have not been utilized effectively to define the duration of pregnancy more exactly. Czekala et al. (1981) used this method repeatedly during a single orang-utan pregnancy (see below).

Orang-utans exhibit a subtle labial swelling during pregnancy that is of some use in pregnancy diagnosis to experienced observers (Schultz, 1938). According to Lippert (1974) this swelling appears 2–4 weeks after conception and disappears after delivery. Some chimpanzees show an erratic sexual swelling during pregnancy. No swelling has been noted in gorillas.

Czekala et al. (1981, 1983, and this volume) have studied hormone excretion during orang-utan pregnancy. The profile of LH/CG (luteinizing hormone/chorionic gonadotropin) bioactivity (rat interstitial cell testosterone assay) throughout the single menstrual cycle studied exhibited a considerable elevation during the first 10–12 weeks of pregnancy, followed by a fall to low but detectable levels. These low levels could also be detected during most of pregnancy by the nonhuman primate pregnancy test. Total urinary immunoactive estrogens were also followed throughout pregnancy; they began to rise shortly before the fall in CG and reached their highest levels 2–3 weeks before parturition, when a decline occurred. The authors found total immunoreactive estrogens to be a useful measure for the study of endocrine patterns in primates, especially in nonexperimental environments, such as zoos. The authors also studied the ratio of estrone, estradiol, and estriol early and late in pregnancy in various primate species. The four human and orang-utan pregnancies studied each showed considerable increase in excretion of estriol relative to the other estrogens as pregnancy progressed. The total amount of immunoreactive estrogen excreted in orang-utans and humans was quite similar, but was considerably less in a number of gorillas that were also studied. This difference could be accounted for by the lower estriol excretion in gorillas.

Czekala and her colleagues (1983 and this volume) developed the concept that the high level of estriol in the pregnant orang-utan and human is due to the greater development of the fetal adrenal zone in these species compared to the gorilla and chimpanzee. The latter two species produce an equivalent amount of estrone to the human, while estriol levels are considerably lower. The role of the fetal adrenal gland in production of estriol has already been discussed.

Paradoxically, while the orang-utan is most like the woman in the excretion of estrogens during pregnancy, it is least like the woman in its excretion of progesterone metabolites during the menstrual cycle. It excretes much less pregnanediol than either the chimpanzee or gorilla; pregnanediol is the primary metabolite of progesterone in the woman.

Bonney and Kingsley (1982) studied the endocrinology of five pregnancies in four orang-utans. They measured urinary concentrations of estrone, estradiol-17β, estriol, pregnanediol-3α-

glucuronide, and CG. The partial 24-hr samples were expressed as a function of creatinine excretion, according to accepted practice. Estrone was initially the major estrogen metabolite, but was surpassed by estriol in the second and third trimesters. Estriol exceeded the estrone level 10-fold at term. Estradiol was a minor metabolite.

Maximum CG levels were excreted during the first trimester, but were detectable throughout pregnancy, as in woman and chimpanzee. An early peak of pregnanediol-3α-glucuronide coincided with the CG peak, declined, and thereafter levels rose steadily until term. Only a small quantity of pregnanediol is excreted during the luteal phase (Collins et al., 1975), and it is unknown if other important progesterone metabolites exist during pregnancy. Unfortunately, androsterone excretion was not examined in this study.

The orang-utan appears to resemble the other great apes and woman in that estriol is the major estrogen excreted during pregnancy. In this respect one must contrast the menstrual cycle, in which estriol is no more than a minor metabolite. Clearly the apes cannot convert estrogens efficiently to estriol without the assistance of the fetoplacental unit. Therefore one must carefully discriminate between the two physiological states when generalizing about estrogen metabolism in great apes. We can also contrast the apes with rhesus monkeys and baboons, where estrone, not estriol, is the major urinary estrogen metabolite during pregnancy.

It is important to note that orang-utan CG shows parallelism with human CG in its reaction with antibody to human CG, indicating similarity of protein structure between the two species. This suggests that human CG antibodies can be used to measure orang-utan CG, as they have been to measure LH (Nadler et al., 1984). However, orang-utan CG is less closely related to human CG than is gorilla and chimpanzee CG (Chen and Hodgen, 1976). Elsewhere I mentioned that orang-utan somatomammotropin is immunologically indistinguishable from the human analog, with similar implications. This knowledge has not been applied to the study of great apes so far.

Bonney and Kingsley (1982) drew three generalizations from their data and that of others to support the idea that the large hominoids resemble each other and differ from cercopithecoids in three principal respects:

1. Estriol rather than estrone is the major estrogen metabolite after the first trimester (presumably reflecting a different fetoplacental metabolic interaction).
2. Estrogen-metabolite excretion rises rapidly after the first trimester and is markedly elevated at term.
3. CG reaches a maximum at the end of the first trimester, decreases, and is maintained at low levels until term.

The other studies I have cited are consistent with these generalizations.

Evidence that orang-utan CG is less similar to human CG than is that of other great apes is cited by Bonney and Kingsley (1982) in support of the idea that orang-utans are less closely related to man than are the African apes. Contrary to the authors' assertion, however, androsterone has not been shown to be a progesterone metabolite in the orang-utan; thus the significance of the luteal-phase peak of androsterone is unclear. Because androsterone excretion has not been examined in the other great apes, no comparative conclusion can be drawn except to note the difference from women. Comparative study of this question may be helpful in elucidating ape affinities.

Immunological similarities between apes and humans not shared by lower primates often make apes of interest for particular types of study. In humans, chimpanzccs, and orang-utans, four specific proteins, in addition to CG, occur in the circulation during pregnancy. These proteins are immunologically similar in apes and humans: some monkeys possess related proteins, but they are immunologically dissimilar (Lin and Halbert, 1978). One of these proteins has been identified as chorionic somatotropin. The other three appear to be immunologically identical to the human pregnancy-associated plasma proteins PaPP-A, -B, and -C. The function of the other three proteins is unknown. Knowledge that chorionic somatotropin is immunologically similar in humans, orang-utans, and chimpanzees has potential practical applications, especially in radioimmunoassay. In addition, several studies cited by Lin and Halbert (1978) indicate that immunization against two of these proteins, PAPP-C (also known as pregnancy-specific B1-glycoprotein) and somatomammotropin, can decrease fertility or induce abortion in primates or rodents. This observation has potential significance for contraception development and pregnancy management.

Disturbances of the immune system of orang-utans during pregnancy have also been described. Van Foreest and Socha (1981) have described a case of erythroblastosis fetalis caused by maternal-fetal incompatibility in an orang-utan family. Extensive testing indicated that the transplacental immunization of the fetus was caused by maternal-fetal incompatibility related to an orang-utan red cell antigen that, in turn, is related to the Rh blood factors. This suggests that orang-utans share with humans and with the other great apes an Rh-like antigenic structure of red cells. The infant received repeated exchange transfusions and survived. Since the orang-utan is an endangered species, blood grouping of potential breeding partners to avoid incompatibility, screening sera of pregnant females for early detection of transplacental immunization, and conducting exchange transfusions to preserve the infant's life could be helpful management tools. These management strategies were discussed in a more general way in an earlier publication (Socha and Moor-Jankowski, 1979).

Parturition

Bo (1971) published a comparative review of parturition in primates, and Werff ten Bosch (1982) contributed additional information on the orang-utan. In one description of parturition, stage I of delivery was characterized by vaginal discharge, indications of pain, and an increased level of activity; the duration was 10 min. During stage II, the animal examined the vulva, exhibited evidence of pain, and assumed a right lateral or dorsal decubitus position. As contractions increased, the head appeared. The mother rested on the right forearm and knee, and grasped the head with the left hand, apparently assisting in expulsion, which took 13 min. During stage III, the placenta was expelled and the mother licked but did not eat it. Soma (1983) also noted that captive orang-utans typically do not consume the placenta. Other accounts of parturition (Werff ten Bosch, 1982) report a characteristic posture: the female places her head on the floor and leans on the lower arm and elbow, with her legs extended so that the pelvis reaches upward. In one instance the father unexpectedly physically assisted in the delivery with apparent skill.

In another instance described by Graham-Jones and Hill (1962), the orang-utan neonate began to breathe within 30 sec of birth, exhibited considerable alertness, and was suckling within 4 hr. The mother exhibited great care for the infant, licking and sucking it, breathing into its mouth, examining it, and carefully holding it. Other individuals may exhibit lesser degrees of attention to the infant.

Postpartum Physiology

Werff ten Bosch (1982) noted that menstrual cycles resume about 70 days after weaning. Based on more extensive chimpanzee data (Nadler et al., 1981), one would expect considerable variability to become apparent if a sufficient number of orang-utan pregnancies were examined.

Nadler et al. (1981) have published data on postpartum physiology in chimpanzees and orang-utans and have shown that their endocrine patterns are very similar. Four orang-utans that retained their infants for 66–77 weeks were studied by obtaining twice-monthly plasma samples. The elevation of follicle-stimulating hormone (FSH) and depression of LH characteristic of the postpartum period in higher primates were observed. Estradiol levels were generally low, with occasional random elevations. Progesterone levels remained below 0.4 ng/ml, with rare exceptions. Two animals showed elevations of progesterone levels above 5 ng/ml 4–16 weeks after infant removal, indicating resumption of cyclicity. Maintenance of postpartum amenorrhea and its characteristic endocrine profile is associated with the repeated suckling stimulus. In the four orang-utans studied, the percentage of time spent in ventroventral contact and in nipple contact was observed. Nipple contact occupied 19% of the time during the first 3 postpartum months, then declined to 7% by the sixth postpartum month. Ventroventral contact occupied 66–95% of the time during the first 4 months, then declined to about 17% after 8 months. These data suggest that the intense nipple contact and physical interaction observed early in the postpartum period need not be maintained at such a high level to prolong postpartum amenorrhea. It is possible, however, that an initially high level of stimulation is required to establish the maternal physiological profile. It is also likely that frequency of suckling, rather than duration of suckling, is more important in maintaining lactation and amenorrhea (Nadler et al., 1981).

Buss (1971) has reviewed lactation among primates and collected data from the literature on milk composition in orang-utans. There is about twice as much casein as residual whey proteins in milk from orang-utans and baboons, whereas milk from chimpanzees and humans has less casein than whey proteins. Unfortunately the number of animals, the number of samples, and the stage of lactation and status of the infant were not provided, so that the milk composition data are difficult to interpret. The stage of lactation and suckling status of the infant are major determinants of milk composition. For this reason the published milk composition data cannot be considered reliable and further investigation is needed.

Adult Male

Studies on sexual physiology of male orang-utans have been limited. A beginning has been made in male developmental endocrinology, sperm biology, and limited aspects of testicular steroidogenesis (see Steroid Hormone Metabolism, below).

Endocrinology

Circulating testosterone levels are slightly higher than in humans: 1,606 and 2,367 ng/ml in an 11-year-old subadult and a 14-year-old adult, respectively (Short, 1979). Obviously these few values are insufficient to permit any meaningful generalization.

Although orang-utans become capable of successful insemination in captivity at 7–10 years of age (MacKinnon, 1974; Rijksen, 1978), they often do not develop their striking secondary sexual characteristics until up to 7 years later (Ulmer, 1958, cited in Kingsley, 1982). Kingsley (1982) has examined some of the social and endocrine correlates of this phenomenon. Although the methodology used in these studies is not ideal, due to the constraints of the captive situation, the insights gained are important. Urine samples from captive orang-utans were collected from the cage floor or traps in the floor drain. The estrogen assay employed an antiserum to 17β-estradiol that significantly cross-reacted (27.9%) with estrone; consequently the assay is considered to measure "estrogens." The testosterone antibody also lacked specificity, cross-reacting 5.1% with dehydrotestosterone and 3.4% with dehydroepiandrosterone; Kingsley represented these data as "testosterone." The concentrations of both hormones in urine were expressed as a ratio of the hormone concentration and the urinary concentration of sodium chloride (the usual convention of expressing hormone concentrations as a function of urinary creatinine concentration could not be used because of technical problems). This relationship should be borne in mind when "hormone levels" are referred to below.

Among the 20 orang-utans Kingsley studied (1982), there was a gradual increase in the levels of both hormones with age, when the animals were classified according to the following categories: juveniles and adolescents, nonflanged adults, adults growing flanges, and flanged adults. Actively breeding and nonbreeding males with fully developed secondary sexual characteristics had estrogen levels at least double those of nonflanged adults. Testosterone in flanged males ranged from equal to twice the level of nonflanged males. Flange development was associated with an increase in the urinary level of both hormones and, in one male that was studied longitudinally during flange growth, flange development commenced a few months after the initial rise in hormone levels.

Kingsley (1982) noted anecdotally that the retardation of flange development, which does sometimes occur, is a function of the social environment, possibly being associated with the presence of a fully flanged male, or, according to Werff ten Bosch (1982), even a very dominant zookeeper. Males caged away from other males develop their flanges soon after sexual maturity and this development is then complete by about age 10. However, if caged with a flanged male, flange development in the younger animal is usually retarded by approximately 3 years, but sometimes for as long as 7 years.

Kingsley (1982) studied a flanged and an unflanged male housed together. Flange development eventually began in the younger animal during the study, being preceded by a rise in estrogen and testosterone. This suggests that if the presence of a flanged adult inhibits the final stages of sexual maturation in a cage mate, the suppression is only temporary. The proximate cause for flange development is likely an increase in testicular steroid secretion, but the means by which social environment influences steroidogenesis presents a fascinating subject for further study. It is of interest to note that testosterone and estrogen declined in both animals in this experiment after separation.

Endocrine concentrations were compared in breeding and nonbreeding flanged males. The males that had failed to breed showed strikingly higher urinary estrogen levels than the successful breeders, which were in close visual, olfactory, and auditory proximity. Behavioral observations suggested that the nonbreeding animals were the object of dominance displays from the breeders. The breeding flanged males did not direct aggressive behavior against one another. These observations suggest that the unsuccessful breeders had a socially subordinate status.

The social and endocrine correlates of sexual maturation and potency in orang-utans are extremely interesting and remind one of the effects of dominance and social structure changes on male macaques (Rose et al., 1971, 1975). A full understanding of these relationships awaits an analysis of the source of increased steroid secretion associated with flange development and analysis of the estrogen associated with inability to breed. Flange development and associated changes such as hair growth are reminiscent of the changes associated with adrenarche in humans and chimpanzees (Hobson et al., 1981), although the timing is different. Therefore an adrenal factor must be seriously considered. No measures of adrenal steroidogenic activity were carried out.

It is difficult to say what significance these findings have for wild orang-utans, where males enjoy considerable spatial separation. Clearly there are important implications for captive management. These socioendocrine relationships certainly deserve further investigation, especially using blood samples so that absolute levels of sex steroids can be determined and related to gonadotropin and sex steroid binding levels.

Semen

Methods have been developed for collection of semen from living great apes (Gould et al., 1978; Warner et al., 1974), including orang-utans. Electroejaculation of eight orang-utans yielded 0.2–3.2-ml semen samples. On average, approximately 80% of the coagulated ejaculate liquefied during incubation for 30 min at 37° C. Sperm concentration varied from 10 to 128 $\times$ 10^6/ml. Approximately half of the sperm were motile and about 60% were alive (stain exclusion test). The amount of semen was comparable to that obtained from chimpanzees by similar means and the extent of liquefaction was also similar; however, orang-utan sperm concentrations were roughly 10-fold less than in chimpanzees, whether the chimpanzee samples were obtained by electroejaculation or by natural means (Warner et al., 1974). Perhaps the lesser concentrations in orang-utans are related to relatively infrequent mating opportunities in the wild.

Electroejaculation is an important source of material for artificial insemination and for the study of sperm biology. Neither of these applications has been much developed, however. Gould et al. (1975) and Martin and Gould (1975) examined the morphology of orang-utan sperm by light and scanning electron microscopy. Orang-utan spermatozoan populations showed greater homogeneity than in humans or gorillas and less variation in their total dry weight, which is a measure of the amount of DNA in the sperm (Seuanez et al., 1977). The sperm have a characteristic outline: the anterior portion of the spermatozoan head (bearing the acrosome) is flattened or expanded laterally, with a resulting constriction at the posterior acrosomal margin. The posterior region of the head is somewhat thickened, but less than in other hominoids. An apical ridge is often present. The sperm lack a fluorescent "F" body (Seuanez et al., 1976). Martin and Gould (1975) provided comparative data for a number of species. Kelly et al. (1976) have reported the presence of high concentrations of 19-hydroxyprostaglandin E_1 in the semen of two orang-utans as well as many other primates. It was not found in nonprimates, and its biological significance is unknown.

STEROID HORMONE METABOLISM

Metabolism of steroid hormones by great apes has most recently been reviewed by Wright et al. (1981). In general, all the apes show closer similarities to humans than do monkeys. The orang-utan does show some unusual and interesting features, some of which were alluded to above.

Much of our information about orang-utan steroid hormone metabolism has been gained from determining urinary levels of steroids during the menstrual cycle and pregnancy (see above). *In vivo* studies using radiolabeled steroids have not been reported, except on the placenta.

In parallel studies, Graham et al. (1972) and Collins et al. (1975) performed comparative studies on steroid hormone excretion throughout the menstrual cycles of chimpanzees and orang-utans, respectively. In both taxa, the relative abundance of hydrolyzed estrogen metabolites in urine was first estrone, then estradiol, and last estriol. Midcycle and luteal estrone peaks were well-defined, but corresponding estradiol and estriol peaks were not as apparent. On the other hand, rhesus monkeys excrete small amounts of estriol (Hopper and Tullner, 1970) but do not usually show a luteal phase estrogen peak, thus suggesting that the corpus luteum in monkeys cannot secrete estrogens. Orang-utans excrete relatively more estriol than chimpanzees, the amount of estrone excreted by the latter hominoid only slightly exceeding the amount of estriol excreted by orang-utans.

The relative size of the midcycle and urinary estrogen peaks in chimpanzees and women is interesting. In women, the midcycle peak is usually greater than the luteal peak, whereas in chimpanzees they are usually about the same, and in the orang-utan the luteal peak is relatively much larger. However, the absolute amount of luteal estrogen excretion in apes is less than in humans.

When the estrogens in pooled urine from orang-utans and chimpanzees are compared, a rather similar profile emerges. In contrast with women, both apes excrete a high proportion of unconjugated steroid (although this could possibly be an artifact due to the method of collection of 24-hr ape urine samples). Orang-utans excrete only slightly more estrogen as the glucosiduronate than as the sulfate, in contrast to both chimpanzees and women, in which the glucosiduronate fraction is dominant.

Unconjugated pregnanediol, pregnanediol sulfate, and pregnanediol glucosiduronate have been identified in menstrual cycle orang-utan urine. The sulfated fraction is relatively larger than in the chimpanzee, which excretes relatively little pregnanediol sulfate. As with estrogens, the ratio of unconjugated progestins is relatively high in orang-utans.

The plasma progesterone concentrations of orang-utans and women are quite similar, yet the amount of pregnanediol excreted by the orang-utan is significantly lower. This may be due to a lower rate of metabolism of progesterone relative to pregnanediol in the orang-utan. Because androsterone is a known metabolite of progesterone in monkeys and baboons, it was measured throughout orang-utan menstrual cycles (Collins et al., 1975) and a dramatic luteal peak of androsterone was demonstrated. However, significant amounts were also excreted during the follicular phase, when progesterone secretion is low. Androsterone has not been identified as a metabolite of progesterone or 17-hydroxyprogesterone in chimpanzees, thus suggesting that the androsterone in orang-utans arose from an ovarian precursor other than progesterone.

Axelson et al. (1984) have done an elegant comparative study of the steroid sulfate fractions of plasma in pregnant rhesus monkeys, chimpanzees, and orang-utans, and compared the data to pregnant women. The two apes closely resemble women in the pattern of steroid sulfates, although the concentration was 4–5 times lower. In contrast, the steroid sulfate pattern in rhesus monkeys was completely different, yielding mostly C-19 steroids. The variety of steroid sulfates in the rhesus was also greatly reduced compared to the apes.

Metabolism of steroid hormones by specific tissues in orang-utans has so far been studied only by Preslock (1979) with reference to the testis and by Ainsworth and Ryan (1969) with regard to placental progesterone metabolism. An orang-utan placenta demonstrated extensive conversion of Δ-4-androstenedione-4-^{14}C to estrone and estradiol-17β and of pregnenolone-7α-^{3}H to progesterone. The former conversion resembles other primates while the latter compares closely with the human and thus differs from lower primates. An active Δ-4-5α-reductase system was also demonstrated in a term orang-utan placenta, whereas its activity is insignificant or absent in human placentae.

It is suggested that the low blood levels of progesterone found in late pregnancy sera of many species are the consequence of the reductase systems such as are found in the orang-utan placenta. It is not known why pregnenolone compounds are more efficiently converted than progesterone to 5α-pregnane compounds.

In summary, female orang-utans generally resemble humans and other apes rather closely in their steroid metabolism. However, the luteal excretion of androsterone remains an unexplained anomaly. Relative to humans, primitive features retained by the orang-utan include the low estriol excretion during the menstrual cycle, the active Δ-4 pathway in the placenta, the high proportion of unconjugated metabolites excreted, and the relatively low levels of steroid hormones excreted.

Incubation of testicular fragments with labeled pregnenolone or progesterone yielded principally testosterone (Preslock, 1979). However, a number of intermediates were found, indicating that both a Δ-4 and a Δ-5 pathway are active, with the Δ-5 predominant, as in humans. An unidentified metabolite was observed that did not match any previously isolated testosterone metabolite from other primates. This observation might conceivably be related to the unusual pregnanediol and androsterone secretion observed in the luteal phase (Collins et al., 1975). No other studies of metabolism are known in male orang-utans, nor have any other studies of testicular metabolism been reported in great apes.

CONCLUSIONS

The orang-utan is fundamentally similar to the other great apes in its reproductive morphology and physiology and shows about the same degree of difference as the other apes do from humans: in most respects, all the great apes resemble humans more closely than do monkeys and other mammals.

Orang-utans do exhibit some unique features, however. They are solitary in habit, with consequences for sexual behavior which consequently occurs in a nonsocial setting. All ape species differ in the pattern of secondary sexual characteristics that are most evident (to humans). All the great apes show sexual dimorphism in size, but gorillas have no additional striking physical secondary sex characteristics. Chimpanzees show the development of conspicuous sexual swelling in females, but no special features in the male, except for the relatively large penis. Orang-utans, on the other hand, show no special features in the female, but conspicuous lateral facial flanges in the male.

The suppression of facial flange development in male orang-utans in the presence of a dominant male under captive conditions is an interesting socioendocrine problem, but its relationship to the wild condition is unknown. Can dominant males within audible range of adult unflanged males influence flange development, perhaps by means of the long call? The high estrogen levels in flanged nonbreeders also requires explanation.

There are differences in steroid metabolism between orang-utans and gorillas. The larger fetal adrenal in orang-utans and humans seems to explain the greater amount of estriol excretion in these species. However, Inaba et al. (1983) reported that estriol was the dominant estrogen metabolite during the orang-utan menstrual cycle, in contrast to two other studies on the orang-utan menstrual cycle, and several studies on chimpanzees and gorillas. If the findings of Inaba et al. are confirmed, the pattern of estrogen excretion will have to be reassessed in the other great apes as well. In the absence of confirmation of Inaba and colleagues' findings, it appears that the primary difference in the menstrual cycle between great apes and humans is the low rate of estrogen production.

Orang-utans, unlike women, produce a luteal peak of urinary androsterone, the source of which is unknown. Although progesterone is a precursor of androsterone in rhesus monkeys, it appears not to be in chimpanzees. Thus, the origin of the luteal androsterone peak in orang-utans is of considerable comparative interest. Androsterone excretion has not been studied in the other great apes. The low concentration of pregnanediol excretion in orang-utans and chimpanzees vs. women has not been explained satisfactorily in terms of the comparable circulating levels of its precursor, progesterone, in these three species.

The orang-utan also shows some individuality in placental androgen metabolism, especially the presence of an active Δ^4-5α-reductase system.

Debate concerning the phylogenetic relationship of the Asian vs. the African apes to human progenitors continues. Reproductive studies provide limited information on this topic, but it does seem that the orang-utan has accumulated and retained some individualized features as a result of its geographical isolation. The solitary behavior and habit is perhaps the most striking feature. Since the geographical site of the divergence of apes and humans is still open to speculation, it is difficult to know how to interpret the orang-utan data.

ACKNOWLEDGMENT

The author is indebted to Delwood C. Collins, Ph.D., for a critical review of the manuscript.

REFERENCES

Ainsworth, L. and Ryan, K. J. 1969. Steroid hormone transformation by endocrine organs from pregnant mammals. 5. Biosynthesis and metabolism of progesterone and estrogens by orangutan placental tissue in vitro. *Steroids, 14:*301–314.

Axelson, M., Graham, C. E. and Sjovall, J. 1984. Identification and quantitation of steroids in sulfate fractions from plasma of pregnant chimpanzee, orang-utan, and rhesus monkey. *Endocrinology, 114:*337–344.

Bo, W. J. 1971. Parturition. *In* Comparative Reproduction of Nonhuman Primates, pp. 302–314, ed. E. S. E. Hafez. Charles C. Thomas, Springfield, IL.

Boer, L. E. M. de (ed.) 1982. The Orang Utan: Its Biology and Conservation. Dr. W. Junk Publishers, The Hague, The Netherlands.

Bonney, R. C. and Kingsley, S. 1982. Endocrinology of pregnancy in the orang-utan *(Pongo pygmaeus)* and its evolutionary significance. *International Journal of Primatology, 3:*431–444.

Buss, D. J. 1971. Mammary glands and lactation. *In* Comparative Reproduction of Nonhuman Primates, pp. 315–333, ed. E. S. E. Hafez. Charles C. Thomas, Springfield, IL.

Chen, H. C. and Hodges, G. D. 1976. Primate chorionic gonadotropins: antigenic similarities to the unique carboxyl-terminal peptide of the HCG subunit. *Journal of Clinical Endocrinology and Metabolism, 43:*1414–1417.

Collins, D. C., Graham, C. E. and Preedy, J. R. 1975. Identification and measurement of urinary estrone, estradiol-17S, estriol, pregnandiol and androsterone during the menstrual cycle of the orangutan. *Endocrinology, 96:*93–101.

Czekala, N. M., Benirschke, K., McClure, H. and Lasley, B. L. 1983. Urinary estrogen excretion during pregnancy in the gorilla *(Gorilla gorilla)*, orangutan *(Pongo pygmaeus)* and the human *(Homo sapiens). Biology of Reproduction, 28:*289–294.

Czekala, N. M., Hodges, J. K. and Lasley, B. L. 1981. Pregnancy monitoring in diverse primate species by estrogen and bioactive luteinizing hormone determinations in small volumes of urine. *Journal of Medical Primatology, 10:*1–15.

Davis, R. R. 1977. Pregnancy diagnosis in an orangutan using two prepared test kits. *Journal of Medical Primatology, 6:*315–318.

Eckhardt, R. B. 1975. The relative body weights of Bornean and Sumatran orangutans. *American Journal of Physical Anthropology, 42:*349–350.

Eckstein, P. 1958. II. Reproductive organs. *In* Handbook of Primatology, pp. 542–629, ed. H. Hofer, A. H. Schultz, and D. Starck. S. Karger, Basel.

Galdikas, B. M. F. 1981. Orangutan reproduction in the wild. *In* Reproductive Biology of the Great Apes, pp. 281–300, ed. C. E. Graham. Academic Press, New York.

Gould, K. G., Martin, D. E. and Hafez, E. S. E. 1975. Mammalian spermatozoa. *In* SEM Atlas of Mammalian Reproduction, pp. 42–57, ed. E. S. E. Hafez. Igaku Shoin, Tokyo.

Gould, K. G., Warner, H. and Martin, D. E. 1978. Rectal probe electroejaculation of primates. *Journal of Medical Primatology, 7:*213–222.

Graham, C. E. (ed.) 1981a. Reproductive Biology of the Great Apes. Academic Press, New York.

Graham, C. E. 1981b. Menstrual cycle of the great apes. *In* Reproductive Biology of the Great Apes, pp. 1–44, ed. C. E. Graham. Academic Press, New York.

Graham, C. E. and Bradley, C. F. 1972. Microanatomy of the chimpanzee genital system. *In* The Chimpanzee, vol. 5, pp. 77–126, ed. G. H. Bourne. S. Karger, Basel.

Graham, C. E., Collins, D. C., Robinson, H. and Preedy, J. R. K. 1972. Urinary levels of estrogens and pregnanediol and plasma levels of progesterone during the menstrual cycle of the chimpanzee: relationship to the sexual swelling. *Endocrinology, 91:*13–24.

Graham-Jones, O. and Hill, W. C. O. 1962. Pregnancy and parturition in a Bornean orang. *Proceedings of the Zoological Society of London, 139:*503–510.

Hobson, B. M. 1976. Evaluation of the subhuman primate tube test for pregnancy in primates. *Laboratory Animals, 10:*87–89.

Hobson, W. C., Fuller, G. B., Winter, J. S. D., Faiman, C. and Reyes, F. I. 1981. Reproductive endocrine development in the great apes. *In* Reproductive Biology of the Great Apes, pp. 83–104, ed. C. E. Graham. Academic Press, New York.

Hodgen, G. D., Turner, C. K., Smith, E. E. and Bush, R. M. 1977. Pregnancy diagnosis in the orangutan *(Pongo pygmaeus)* using the subhuman primate pregnancy test kit. *Laboratory Animal Science, 27:*99–101.

Hopper, B. and Tullner, W. W. 1970. Urinary estrone and plasma progesterone levels during the menstrual cycle of the rhesus monkey. *Endocrinology, 86:*1225–1230.

Inaba, T., Imozi, T. and Saburi, T. 1983. Urinary estrogen levels during the menstrual cycle of the orangutan. *Japanese Journal of Veterinary Science, 45:*857–859.

Kelly, R. W., Taylor, P. L., Hearn, J. P., Short, R. V., Martin, D. E. and Marston, J. H. 1976. 19-Hydroxyprostaglandin E1 as a major component of the semen of primates. *Nature, 260:*544–545.

Kingsley, S. R. 1981. The Reproductive Physiology and Behaviour of Captive Orangutans *(Pongo pygmaeus).* Ph.D. thesis, University of London.

Kingsley, S. R. 1982. Causes of non-breeding and the development of the secondary sexual characteristics in the male orang-utan: a hormonal study. *In* The Orang-utan: Its Biology and Conservation, pp. 215–229, ed. L. E. M. de Boer. W. Junk, The Hague.

Kinzey, W. G. 1971. Male reproductive system and spermatogenesis. *In* Comparative Reproduction of Nonhuman Primates, pp. 85–114, ed. E. S. E. Hafez. Charles C. Thomas, Springfield, IL.

Kirchner, M. A. and Lipsett, M. B. 1964. The analysis of urinary steroids using gas-liquid chromatography. *Steroids 3:*277–294.

Koenigswald, G. H. R. von. 1982. Distribution and evolution of the orang-utan, *Pongo pygmaeus* (Hoppius). *In* The Orang-Utan: Its Biology and Conservation, pp. 1–16, ed. L. E. M. de Boer. W. Junk, The Hague.

Lin, T. M. and Halbert, S. P. 1978. Immunological relationships of human and subhuman primate pregnancy-associated plasma proteins. *International Archives of Allergy and Applied Immunology, 56:*207–223.

Lippert, W. 1974. Beobachtungen zum Schwangerschafts- und Geburtsverhalten beim Orang-utan *(Pongo pygmaeus)* in Tierpark Berlin. *Folia Primatologica, 21:*108–134.

MacKinnon, J. 1974. The behavior and ecology of wild orang-utans *(Pongo pygmaeus). Animal Behaviour, 22:*3–74.

Martin, D. E. 1981. Breeding great apes in captivity. *In* Reproductive Biology of the Great Apes, pp. 343–375, ed. C. E. Graham. Academic Press, New York.

Martin, D. E. and Gould, K. G. 1975. Normal and abnormal hominoid spermatozoa. *Journal of Reproductive Medicine, 14:*204–209.

Nadler, R. D., Collins, D. C. and Blank, M. S. 1984. Luteinizing hormone and gonadal steroid levels during the menstrual cycle of orang-utans. *Journal of Medical Primatology, 13:*305–314.

Nadler, R. D., Collins, D. C., Miller, L. C. and Graham, C. D. 1982. Menstrual cycle patterns of hormones and sexual behavior in gorillas. *Hormones and Behavior, 17:*1–17.

Nadler, R. D., Graham, C. E., Collins, D. C. and Gould, K. G. 1979. Plasma gonadotropins, prolactin, gonadal steroids and genital swelling during the menstrual cycle of lowland gorillas. *Endocrinology, 105:*290–296.

Nadler, R. D., Graham, C. E., Collins, D. C. and Kling, O. R. 1981. Postpartum amenorrhea and behavior of great apes. *In* Reproductive Biology of the Great Apes, pp. 69–82, ed. C. E. Graham. Academic Press, New York.

Nadler, R. D., Herndon, J. G. and Wallis, J. 1986. Adult sexual behavior: hormones and reproduction. *In* Comparative Primate Biology, vol. 2A (Behavior and Ecology), pp. 363–407, ed. G. Mitchell and J. Erwin. Alan R. Liss, New York.

Preslock, J. P. 1979. Testicular steroidogenesis in the orang-utan *Pongo Pygaeus* (sic). Society for the Study of Reproduction, 12th Annual Meeting, August 21–24, Quebec. Abstract 95, p. 59A.

Rijksen, H. D. 1978. A field study on Sumatran orang-utans *(Pongo pygmaeus abelii* Lesson, 1827): Ecology, Behavior and Conservation. H. Veenman & Zonen, Wageningen, Nederland.

Rose, R. M., Bernstein, I. S. and Gordon, T. P. 1975. Consequences of social conflict on plasma testosterone levels in rhesus monkeys. *Psychosomatic Medicine, 37:*50–61.

Rose, R. M., Holaday, J. Q. and Bernstein, I. S. 1971. Plasma testosterone, dominance rank and aggressive behaviour in male rhesus monkeys. *Nature, 231:*366–368.

Schultz, A. H. 1938. Genital swelling in the female orang-utan. *Journal of Mammalogy, 19:*363–366.

Seuanez, H. N., Carothers, A. D., Martin, D. E. and Short, R. V. 1977. Morphological abnormalities in spermatozoa of man and great apes. *Nature, 270:*345–347.

Seuanez, H., Robinson, J., Martin, D. E. and Short, R. V. 1976. Fluorescent (F) bodies in the spermatozoa of man and the great apes. *Cytogenetics and Cell Genetics, 17:*317–326.

Shackleton, C. H. L. 1974. Steroid excretion in the neonatal period: a comparative study of the excretion of steroids by human, ape and rhesus monkey infants. *Journal of Steroid Biochemistry, 5:*113–118.

Short R. V. 1979. Sexual selection and its component parts, somatic and genital selection, as illustrated by man and the great apes. *Advances in the Study of Behavior, 9:*131–158.

Socha, W. W. and Moor-Jankowski, J. 1979. Serological materno-fetal incompatibility in nonhuman primates. *In* Nursery Care of Nonhuman Primates, pp. 35–42, ed. G. C. Ruppenthal and D. J. Reese. Plenum Publishing Corp., New York.

Soma, H. 1983. Notes on the morphology of the chimpanzee and orang-utan placenta. *Placenta, 4:*279–290.

Vacek, Z. 1974. The histochemistry and light and electron microscopy structure of the orangutan *(P. pongo pygmaeus)* placenta. *Folia Morphologica, 22:*165–174.

Van Foreest, A. W. and Socha, W. W. 1981. Transplacental immunization in the course of incompatible pregnancy in zoo orangutans. *American Association of Zoo Veterinarians Annual Proceedings,* pp. 57–69.

Warner, H., Martin, D. E. and Keeling, M. E. 1974. Electroejaculation of the great apes. *Annals of Biomedical Engineering, 2:*419–442.

Werff ten Bosch, J. J. van der. 1982. The physiology of reproduction of the orang-utan. *In* The OrangUtan: Its Biology and Conservation, pp. 201–214, ed. L. E. M. de Boer. W. Junk, The Hague.

Woodard, D. K., Graham, C. E. and McClure, H. M. 1976. Comparison of hemagglutination inhibition pregnancy tests in the chimpanzee and the orang-utan. *Laboratory Animal Science, 26:*922–927.

Wright, K., Collins, D. C., Musey, P. I. and Preedy, J. R. K. 1981. Comparative aspects of ape steroid hormone metabolism. *In* Reproductive Biology of the Great Apes, pp. 163–191, ed. C. E. Graham, Academic Press, New York.

7

Sexual and Reproductive Behavior

RONALD D. NADLER

When Yerkes and Yerkes (1929) published their exhaustive treatise on knowledge of the great apes at that time, little was known about the social behavior of wild orang-utans and virtually nothing was known of their sexual relations under natural conditions. Reports by early explorers and naturalists had suggested correctly that orang-utans were the least sociable and gregarious of the great apes, but as regards their reproductive behavior there was essentially no information: for example, "Except during the mating season the adult males live alone, and it is said that the females go off by themselves to bear their young" (Yerkes and Yerkes, 1929:136). This paucity of data on orang-utans in the wild persisted, in fact, for more than 40 years afterward. As a result, most information on sexual behavior of orang-utans up until the 1970s was derived from reports on animals living in zoos.

The impression conveyed by these combined reports on captive orang-utans was provocative in the suggestion that this species, more than any of the other apes or monkeys, bore superficial similarities to humans in several aspects of their sexual interactions. These similarities included (1) frequent copulation during the female cycle, irrespective of cycle phase (Asano, 1967; Coffey, 1972; Fox, 1929; Heinrichs and Dillingham, 1970) or the occurrence of pregnancy (Asano, 1967; Coffey, 1972, 1975), i.e., an apparent absence of estrus; (2) the use of varied copulatory positions, including the *more hominum* or ventro-ventral (V-V) copulatory position (Coffey, 1975; Fox, 1929); and (3) relatively prolonged periods of copulation, in comparison to most mammals other than humans (Coffey, 1971, 1975; Graham-Jones and Hill, 1962). It was recognized relatively early, moreover, that the orang-utan was the only extant hominoid, other than the human, in which females exhibited no genital swelling during the cycle (although it was also known that the female orang-utan's genitals do swell during pregnancy) (Graham-Jones and Hill, 1962; Lippert, 1974; Schultz, 1938). It should be noted that there was some conflicting evidence as well, suggesting cyclicity in sexual activity during the menstrual cycle, but few details were provided (Chaffee, 1967; Coffey, 1975).

In addition to the foregoing sexual attributes, and, perhaps, the characteristic of orang-utan sexuality to which reference is made most frequently, is the male's initiation of copulation (Coffey, 1972; Fox, 1929; Graham-Jones and Hill, 1962), commonly described as "rape" (MacKinnon, 1971, 1974; Rijksen, 1978). In recognition of the admonitions against using anthropomorphic terminology to describe the behavior of animals (Beach, 1978; Dewsbury, 1984; Estep and Bruce, 1981), the term "forcible copulation" (Nadler, 1981) is used herein to describe the male orang-utan's use of force to initiate copulation against the female's resistance. Since this behavior occurs in the wild as well as in captivity, its existence is clearly not an artifact of captivity.

MODERN RESEARCH: FIELD STUDIES

The modern era of research on orang-utans was initiated by several field studies that were relatively short, included few sightings, and served primarily to establish the feasibility of conducting such research on wild orang-utans (Davenport, 1967; Milton, 1964; Okano, 1965; Schaller, 1961; Yoshiba, 1964). Subsequent studies by MacKinnon (1971, 1974), Rodman (1973), and Horr (1975) were all conducted during the late 1960s and did much to clarify the social organization of orang-utans and to describe in general terms their limited social relations. These studies confirmed that orang-utans are the most arboreal of the great apes and that they live a relatively solitary life. A mother with one or two offspring is the most common social unit, whereas fully adult males are predominantly solitary. Adult males establish home ranges that may overlap to some degree and that may include the home ranges of one to several females.

When subadults leave their natal groups, females are believed to establish home ranges near their mothers, whereas the males initially become nomadic. The first data on sexual behavior in the wild were obtained by MacKinnon (1971, 1974), who reported that seven of eight copulations he observed in Bornean orang-utans were forcibly initiated by the males: "The females showed fear and tried to escape from the males, but were pursued, caught and sometimes struck and bitten" (MacKinnon, 1971:176). The remaining (eighth) copulation was between immature animals.

MacKinnon thought it was "unlikely that such uncooperative matings often lead to pregnancy" and proposed that "it is probably only in the context of pair-bonds that successful mating occurs in wild orang-utans." No coooperative matings were observed by him in Borneo, however, although some consortships were reported.

These field studies also suggested that the male orang-utan's loud vocalization, the "long call" (MacKinnon, 1971), served two functions: a spacing function with respect to other males and an attraction function with respect to estrous females. MacKinnon, who did observe three cooperative matings in Sumatra, proposed that this subspecies might be more sociable than the Bornean subspecies, but later studies have not supported this assessment.

Although the data of MacKinnon, Rodman, and Horr are in general agreement, somewhat different models were proposed to describe the reproductive strategies of orang-utans. They all agreed, for example, that the forcible copulations by subadult males represent a low-return strategy that is used until the male is sufficiently dominant to establish its own home range. MacKinnon (1979), in addition, proposed that males go through three stages in their development. As subadults, they are very active sexually and nonaggressive. As mature adults they become more aggressive and engage in sexual activity less frequently, but presumably, more productively. Finally, as old adults they become highly aggressive guardians of their earlier reproductive investments, but are impotent. The development of such a latter stage of impotency in male orang-utans is questionable since it has not been reported by others. These three studies of MacKinnon, Rodman, and Horr, therefore, provided considerable insight into the social relations of orang-utans, but few details about species-typical sexual behavior, other than its infrequent occurrence.

Four further field studies by Rijksen (1978), Galdikas (1979, 1981, 1984), Schürmann (1981, 1982) and Mitani (1985) did much to clarify the sexual and reproductive life of wild orang-utans. Rijksen (1978) proposed that forcible copulations, carried out primarily by subadult males, served a dominance, rather than a reproductive function. Although not specifically defined, this view suggests that forcible copulations establish the male's dominance over females and, presumably, make him attractive to females when he becomes a fully mature male with a home range. Galdikas (1984), who observed congenial as well as aggressive interactions between subadult males and females, emphasized the former. She proposed that subadult males attempt to consort with females in a nonaggressive manner, essentially an extended period of courtship. When this fails, forcible copulation may be pursued. There is the further possibility that forcible copulation reflects a stage in the social learning process of a male that has been reared in the absence of opportunities for early social experience with peers.

The most important results of these later field studies, however, were the first detailed observations in the wild of cooperative mating between fully adult males and females of both subspecies, and documentation of the major involvement of the female in initiating sexual interactions (Galdikas, 1979, 1981; Schürmann, 1981, 1982). These results supported the earlier proposal that reproduction in orang-utans occurs primarily, if not exclusively, within consortships between fully adult males and females (MacKinnon, 1971). The data suggested, furthermore, that these consortships are initiated by the female, which, at the time of estrus, seeks out a fully adult resident male with which to copulate. One or, perhaps, a few males may be available to the female as a result of the overlapping of male ranges. The location of the male(s) is presumably known to the female on the basis of the "long call," emitted daily by the male. Once the female has located the male and has approached him, either individual may initiate a cooperative act of copulation. Such copulations generally last for 10–15 min. The duration of estrus, estimated from the number of days of copulation while in consort, is 5–6 days.

Although in some instances the male may position the female for copulation, the female may also actually mount the male. In the latter instances, also observed under certain conditions in the laboratory (see below), the male remains relatively inactive while directing his erect penis toward the female. The female may first inspect the male's penis and then mount and even deliver the pelvic thrusts leading to ejaculation.

The field research clarified and expanded

upon the data gathered on orang-utans living in zoos. With respect to the frequent, sometimes daily copulation observed in captivity, the data on wild orang-utans suggested that this is a distortion of the species-typical pattern brought about by unnatural living conditions that exist in zoos. Whereas in the wild, adult male and female orang-utans live apart from each other except when a female is in estrus, in captivity males and females generally live together continuously. Under the latter conditions, and for reasons that are not completely clear, the fully adult male engages in a pattern of forcible sexual initiation which, under natural conditions, is performed primarily by subadult males and only rarely by fully adult ones.

If the male orang-utan in the wild ceases to forcibly initiate copulation in part because his large size renders him less able to pursue and catch a smaller and hence more mobile nonreceptive female, then his reliance on this behavior in captivity may be opportunistic and made possible by the female's inability to avoid and/or escape from the male while confined in the relatively small area of a zoo cage. Developmental factors could be involved as well, since most, if not all, captive male orang-utans in zoos have matured in captivity. As such, they may have never encountered conditions that in the wild would lead them to abandon the pattern of forcible copulation they developed as subadults. Further discussion of forcible copulation by male orang-utans and the conditions under which it occurs is presented below in the context of laboratory studies of sexual behavior.

EVOLUTIONARY INFLUENCES ON REPRODUCTIVE BEHAVIOR OF ORANG-UTANS

As a result of the increased data available on orang-utans (and the other great apes) in their natural habitat(s), several investigators developed hypotheses regarding the evolutionary influences on sexual and reproductive behavior of the great apes and the selection pressures that account for the interspecific differences among them (because of insufficient data on pygmy chimpanzees, this species was not included in these analyses). The description of these contributions which follows is necessarily abbreviated and does not include the supporting data available in the original sources.

Building on the theory of sexual selection (Darwin, 1871), Short (1977, 1981) proposed that certain anatomical differences among the great apes can be interpreted as adaptations, functionally appropriate to their species-typical mating systems, copulatory frequencies, and the performance of courtship displays, i.e., "form reflects function" (Short, 1981:321). Specifically, differences in the testis:body weight ratios of the great apes, presumably reflecting differences in testis function, were attributed to differences in copulatory frequencies, whereas differences in penile size and visibility were attributed to differences in the degree to which the penis was used in male displays. With respect to these characteristics, the chimpanzee was clearly ranked higher than the orang-utan and gorilla, whereas the differences between the latter two species were considered to be small.

Nadler (1977b), elaborating on a hypothesis proposed by R. M. Yerkes (1939), attributed differences in copulatory frequency (in the cycle) among the great apes to the differential initiation of sexual activity by males and females of the three species. It was proposed that the greater the role of the male in initiating copulation, the more frequently copulation would occur in the cycle and the less closely it would be restricted to the presumptive time of ovulation. Since the pattern of mating for wild orang-utans had not been described when this hypothesis was proposed, the mating pattern observed in captivity was used for comparison with those of the other apes. Since it was the subadult pattern of daily forcible initiation of copulation which was observed in the laboratory, use of these data resulted in an erroneous ranking of the male orang-utan as the more assertive in sexual initiative. Regardless of this error, the relationship between male sexual initiative and the frequency of mating in the cycle conformed to the predicted pattern.

Harcourt and Stewart (1977) and Harcourt (1981:313) attributed male sexual initiative (and display) to intermale competition at the time of estrus, i.e., "the number of males competing for estrous females." Chimpanzees were ranked highest on intermale competition, since essentially all the adult males in a community may court and copulate with the same estrous female. In gorilla groups, there is only one male, the silverbacked leader, that copulates with the parous females of the group. In the case of orang-utans, there may be more than a single male available to a female at the time of estrus because male home ranges overlap to some degree. The orang-utan, therefore, was ranked between the chimpanzee and gorilla with respect to intermale competition at estrus.

TABLE 7–1. Male Reproductive Behavior and Anatomy of the Great Apes in Relation to the Number of Males in Competition for Estrous Females, Ranked from Greatest (1) to Least (3)

	Chimpanzee	Orang-utan	Gorilla
Intermale competition at estrus	1	2	3
Male sexual initiative	1	2	3
Courtship display	1	2	3
Penis size and visibility	1	2	3
Periovulatory restriction of copulation	3	2	1
Preovulatory copulation	1	2	3
Testis:body weight ratio	1	2	3
Body weight	3	2	1

Consistent with their ranking on intermale competition, chimpanzees were also ranked highest on sexual initiative and the presence of a courtship display (Harcourt, 1981). Orang-utans were ranked intermediate on sexual initiative, but approximately equivalent to gorillas with respect to the absence of a conspicuous courtship display.

Table 7–1 presents an integration of the previous analyses of reproductive behavior and anatomy of the great apes, including some modifications, organized with respect to intermale competition for estrous females (Harcourt, 1981). It can be seen in this analysis that the orang-utan is positioned intermediate to the chimpanzee and gorilla on all of the categories considered.

One refinement over the previous analyses that accounts for the present rankings relates to penis size and visibility. Previously, the orang-utan was thought to be approximetley equivalent to the gorilla in this regard, with both these larger apes less conspicuously endowed than the chimpanzee. However, recent observations of the adult orang-utan's penis during "male presenting" (Schürmann, 1981, 1982) and a "penile display" (Nadler, 1982) suggest that the appropriate ranking of this species is, in fact, greater than the gorilla and, therefore, in the intermediate position shown in Table 7–1 (also see Dahl, this volume). Another difference between the present analysis and the earlier ones relates to the category of preovulatory copulation. In one of the earlier analyses, copulatory frequency was defined in terms of daily rates of copulation, with the orang-utan ranked lowest among the apes (Harcourt, 1981). Assessed by this measure, therefore, copulatory frequency is not directly related to the testis:body weight ratios of the three apes.

Daily rates of copulation (actually ejaculation) could be an appropriate index of testis function if sperm competition was at the level at which intermale competition was expressed (Harcourt et al., 1981). This is unlikely to be the case for gorillas and orang-utans, and it is probably not the only arena for intermale competition in chimpanzees. Although sperm competition may be operative in the opportunistic matings of chimpanzees, it would be a less significant factor in consortships, during which the greatest percentage of conceptions are thought to occur (Tutin and McGinnis, 1981).

It is possible, of course, that differences between gorillas and orang-utans, with respect to testis function and copulatory frequency, are not significant and that this accounts for the lack of a strong correlation among the three species. On the other hand, it may be that a different dimension of copulatory frequency better reflects the selection pressures on testis size. One such possibility is the number of days of copulation (in the cycle) that precede ovulation, i.e., preovulatory copulation. In the three great apes, it is the estrous female that primarily determines the duration of preovulatory copulation by any particular male: in gorillas and orang-utans by determining the time for establishing proximity with the male at the onset of estrus, and in chimpanzees by determining the time for joining a particular male in consortship. The duration of preovulatory copulation for any given male, therefore, can be approximately equivalent to the duration of estrus. Since ovulation occurs at the termination of estrus, conception requires that a male maintain adequate sperm/semen concentrations to impregnate a female following a number of days of copulation that could equal the duration of estrus. The duration of estrus is greatest in the chimpanzee (10 days), least in the gorilla (2 days), and intermediate in the orang-utan (5 days), thereby accounting for the rankings of preovulatory copulation in Table 7–1.

Although there appears to be a strong correlation among the great apes between this dimension of copulatory frequency and relative testis size, it is not at all clear that the relationship is a causal one, i.e, that the testis:body weight ratios of the apes evolved to accommodate different durations of preovulatory copulation. There are data on human males, however, that indicate that sperm concentrations of ejaculates in our species do decline over a number of days of ejaculation in excess of 3.5 times per week (Freund, 1962). These results suggest that effects of ejaculation can operate in an additive fashion over days, and thus support the use of preovulatory copulation in the great apes as a measure reflecting the selective pressure on testis function.

The discussion above suggests that several aspects of reproductive anatomy and behavior of the great apes developed as a result of sexual selection, and that differences among the apes are related to the number of males that typically compete for a female at the time of estrus (Harcourt, 1981). Since the ranking of body weight for the great apes is related inversely to the ranking of intermale competition, body weight is also related to the other categories in Table 1. Body weight, however, is most likely a function of natural selection and does not appear to account as well for the ranking of reproductive characteristics of the great apes as does intermale competition for estrous females.

MODERN RESEARCH: LABORATORY STUDIES

Laboratory studies of sexual and reproductive behavior of orang-utans have been directed toward three related issues: (1) cyclicity in sexual behavior, (2) the temporal association between sexual activity and the time of ovulation, and (3) the relationship between sexual behavior and the concentrations of endogenous hormones of the female (Nadler, 1982). These issues are related, in that cyclicity in sexual behavior is generally reflected in the occurrence or increased frequency of copulation during the midcycle periovulatory phase of the female cycle when certain hormones implicated in the regulation of sexual behavior, e.g., estradiol and testosterone in the female, are at elevated levels. This research also provided the opportunity to test Yerkes's (1939) hypothesis that among the higher primates, females tend to restrict mating to the time of ovulation whereas males primarily account for mating at other times in the cycle.

In an initial study, oppositely sexed orang-utans were observed daily in traditional laboratory pair tests conducted in a single cage, i.e, the free-access test (FAT) paradigm (Nadler, 1977a). The term "free-access" refers to the conditions when two individuals are confined together in a single cage and are freely accessible to each other for the entire duration of the test. The FAT is contrasted with the restricted-access test (RAT), described below, in which one of the individuals, generally the female, determines whether and when copulation may occur. These two test paradigms are useful for investigating different dimensions of sexual responsiveness (Beach, 1976). The FAT is especially useful for assessing female attractivity and receptivity, whereas the RAT with female choice is more appropriate for assessing female proceptivity (Nadler et al., 1985).

The results of the study using the FAT paradigm conformed to Yerkes's hypothesis to an extreme degree. During tests of 30 min duration, copulation occurred on a near-daily basis as a result of the male's attempting to initiate copulation forcibly every time the female was introduced into the test cage. The females attempted to avoid and escape from the males, but given their restriction to the test cage they were easily caught and subdued. Three of the four pairs tested copulated on every day of the cycle and the fourth pair copulated on over 40% of the days. Thus, mating occurred quite frequently at times in the cycle dissociated from ovulation and the males were responsible for its initiation.

In the same study, tests of longer duration, i.e., 5–6 hr, provided an additional perspective on the mating of orang-utans. Although the first copulations each day were, as in the shorter tests, forcibly initiated by the males on a near-daily basis, multiple copulations on a given day occurred most frequently during the midcycle phase. Because the recording of copulations beyond the first 30 min of these longer tests was accomplished by time-sampling on videotape, the gender of the initiator could not be determined reliably. Affiliative behavior, however, such as grooming and social contact of the males by the females, also occurred most frequently during the midcycle phase, as did all of the relatively few examples of proceptive and masturbatory behavior by the females. Other evidence of cyclicity in female proceptive behavior was obtained in a study of a pair of orang-utans living in a zoo (Maple et al., 1979). Although the male of that pair initiated copulation frequently throughout the entire study, the female's proceptive behavior was observed only during three periods of 4–6 days each, separated by intervals of

26 and 30 days. These data, therefore, provided some evidence of cyclicity in sexual behavior of orang-utans and suggested that a midcycle enhancement of female sexual responsiveness might be contributory to that cyclicity.

The results of the FATs suggested rather clearly that the female orang-utan's prerogatives regarding mating were severely compromised under these test conditions by the male's dominance and forcible initiation of copulation. Yerkes (1939), who drew the same conclusion about the mating of chimpanzees under similar conditions, stated that an accurate description of female sexual responsiveness (in captivity) could only be obtained if it were possible to reduce the inordinate influence of the male. A laboratory study was designed to accomplish this and thereby obtain a clearer reflection of female proceptivity during the cycle.

In contrast to the earlier FATs, in which the male and female were confined in a single compartment, in these later tests, termed RATs with female choice, the male and female were placed in separate compartments of the test cage and the size of the doorway between the compartments was adjusted so that only the female could pass through it (Fig. 7–1). This effectively gave the female the prerogative of determining whether and when in the cycle copulation could occur and, in that sense, simulated an aspect of female choice shown to be characteristic of females in the wild (Galdikas, 1979, 1981; Schürmann, 1981, 1982).

FIG. 7–1. Male orang-utan attempting to get through the reduced doorway during a restricted-access pair-test.

Eleven oppositely sexed pairs of orang-utans, consisting of various combinations of six females and four males, were tested daily in the RATs during a total of 24 menstrual cycles (Table 7–2). Daily urine samples were also collected during the menstrual cycles of three females for radioimmunoassay of estrone glucuronide (E_1G) and pregnanediol glucuronide (PG).

The intermenstrual intervals, calculated from a number of cycles for each female, suggested that the females were cycling normally. Mean and median cycle lengths were approximately 28 days (range: 23–33 days) (Table 7–3), comparable to earlier reports (Aulmann, 1932; Nadler, 1977a; Napier and Napier, 1967). Although differences in behavior were apparent among the different pairs, several common characteristics were revealed.

When first confronted by the reduced doorway, the males attempted to get through it (Fig. 7–1) and sometimes, during initial tests, did so successfully. Once an effective size to exclude the male was determined, however, the behavior of the pairs underwent a series of changes as the animals apparently adjusted to the modified test conditions. Initially, some of the females entered the male's compartment as they had done in the earlier FATs with similar results. The males chased and forcibly tried to copulate with them. The females then became more cautious, either remaining in their own compartment or, when they did enter the male's compartment, positioning themselves so they could escape through the doorway before being caught. Some of the males then began ignoring the females when the females first entered their compartment until the females either got close to them or far from the doorway. In these circumstances, the males quickly seized the females and forcibly tried to copulate with them. Even though the males were much larger and apparently stronger than the females, and despite the fact that the males had been capable of forcibly copulating with the females in the FATs, they were not usually successful in the RATs. Whereas in the FATs the females stopped struggling once the males had wrestled them to the cage floor, the females in the RATs generally struggled free and escaped

TABLE 7–2. Copulation and Female Sexual Initiation during the Menstrual Cycle in Restricted-Access Tests with Female Choice

Test Pair (female-male)[1]	Number of Cycles	Copulation Days per Cycle (Mean ± SE)	Percentage of Copulation Days per Cycle	Percentage of Female-initiated Copulations
G-L (S)[2]	2	4.0 ± 2.0	18	71
G-D (S)	5	2.4 ± 0.5	11	67
G-B (S)	2	3.5 ± 1.5	17	100
I-L (S)	1	8.0 ± 0.0	47	25
I-D (S)	3	2.7 ± 1.2	14	25
I-B (S)	1	2.0 ± 0.0	9	0
T-D (S)	2	6.0 ± 4.1	25	0
T-B (S)[2]	1	4.0 ± 0.0	17	75
L-B (S)	2	5.0 ± 1.0	41	0
P-D (B)[2]	3	3.6 ± 0.3	14	64
J-B (S)	2	6.5 ± 0.5	37	100
Total	24	4.3 ± 0.6	23	48

[1]S = Sumatran, B = Bornean.

[2]Conception occurred in last cycle of testing.

through the doorway. Within the first several tests, however, all the males stopped pursuing the females and became quite passive, to the point that they frequently failed to respond even later when the females slapped at them and pulled their hair.

Under these conditions, in which the females regulated sexual access, all the pairs copulated, but mating in the cycle was reduced to 4 days on average (23% of the days on which tests were conducted), i.e., the species-typical rate (Table 7-2). This contrasts markedly with the near-daily mating by several pairs in the FATs (mating during 40–100% of the tests). In general, a high percentage of female-initiated copulations within a pair in these RATs was associated with a relatively low percentage of days of copulation per cycle, but the correlation was not significant.

During the RATs, the females differed in the extent to which they exhibited proceptive behavior or sexual initiative, but all except female "L" were proceptive toward at least one male. Proceptive behavior consisted of the female approaching the male, combined with genital-presenting, rubbing the genitals on the male's body (including the head, back, and hand), manual or oral manipulation of the male's penis, or actually mounting the male, effecting intromission and performing pelvic thrusts, several times leading to ejaculation by the male. Those instances in which the female mounted the male were usually preceded by a male "penile display" characterized by the male reclining on his back and directing his erect penis toward the female (Fig 7-2). All four of the males performed penile displays with at least one female partner, and four of the six females (J, G, P, and T) mounted and performed pelvic thrusting in copulation with males (Fig. 7-3). In one instance the male displayed his penis in this manner for 45 min be-

TABLE 7–3. Length of the Menstrual Cycle in 35 Cycles from Six Female Orang-utans

Female[1]	Age (yrs)	Number of Cycles	Cycle Length: Mean ± SE (days)	Cycle Length: Median (days)	Cycle Length: Range (days)
Guchi (S)	11.5	7	24.4 ± 0.4	24.8	23–26
Ini (S)	19.5	10	25.6 ± 0.4	25.5	24–28
Tupa (S)	20.0	6	26.3 ± 0.3	26.5	25–27
Lada (S)	20.0	5	30.4 ± 0.9	29.5	29–33
Paddi (B)	21.0	4	31.3 ± 0.6	31.0	30–33
Jowata (S)	22.0	3	28.7 ± 0.9	29.0	27–30
Total	—	35	27.8 ± 1.1	27.7	23–33

[1]S = Sumatran, B = Bornean.

FIG. 7–2. Female orang-utan looking at "penile display" during a restricted-access pair-test.

FIG. 7–3. Female orang-utan exhibiting proceptive behavior by mounting the male during a restricted-access pair-test.

fore the female entered his compartment and mounted him. Although not previously reported at the time these observations were made in the laboratory, very similar examples of "male presenting" and female mountings were subsequently described for orang-utans living in the wild (Schürmann, 1981, 1982). It appears, therefore, that the female orang-utan plays a major role in initiating sexual activity under natural conditions and that display of this species-typical pattern of sexual interactions in captivity requires that the female control whether and when mating may occur.

Other reflections of sexual responsiveness consisted of masturbation, observed in all of the males and four of the females (J, G, I, and T). Masturbation by the males, sometimes leading to ejaculation, was performed by thrusting the penis against the bars, fencing, or walls of the cage or by manipulating the penis with hand or foot. The frequency of masturbation by the males varied by individual and male-female pair and ranged from once or twice a cycle to near-daily, irrespective of mating. Female masturbation consisted of rubbing the genitals against the bars, the cage floor, or the male's body or manipulating the genitals manually. Female masturbation tended to be associated with mounting of a male by that female. For four of six pairs in which the female masturbated, she also mounted the male, whereas for five pairs in which the female did not masturbate, there was only one case of the female mounting the male. The difference between these proportions, however, was not statistically significant ($\chi^2 = 2.54$, df = 1, n.s.).

Female masturbation, proceptivity, and days of copulation by the pairs, normalized to the day of maximal levels of estrone glucuronide (E_1G) in the cycle, are shown for 11 cycles of three females from which urine was collected for hormone assays (Fig. 7–4). Copulations extended over a major portion of the cycle; as early as 14 days before the E_1G peak and as late as 14 days afterward. There was a substantial increase in copulations, however, beginning about 7 days prior to the peak in E_1G and continuing to a maximum on the day of the peak. The number of copulations then declined in association with

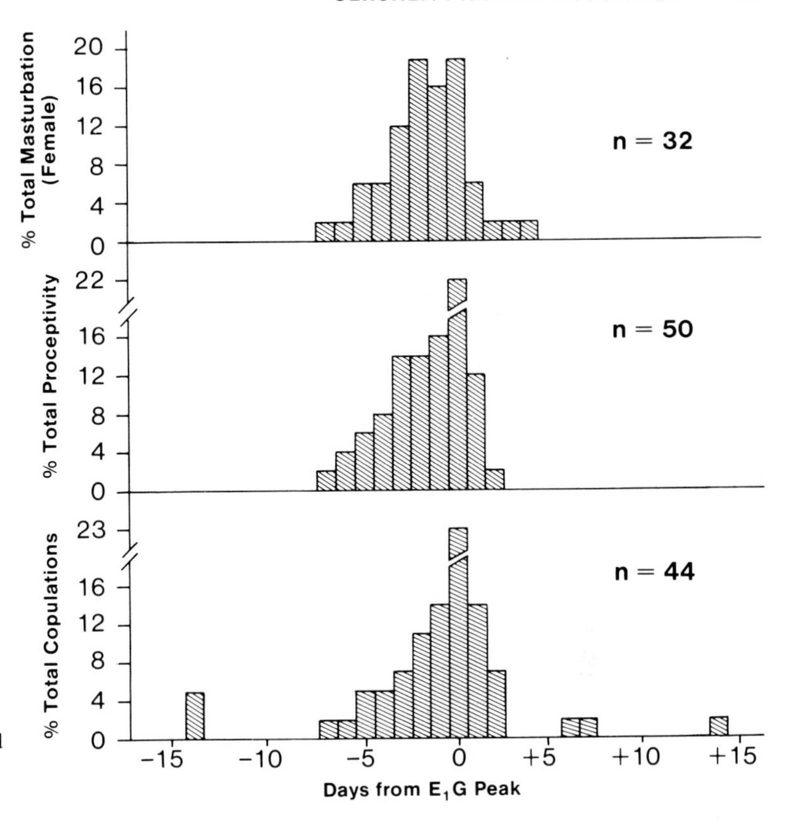

FIG. 7–4. Female orang-utan masturbation and proceptivity and total days of copulation, normalized to the day of peak levels of estrone glucuronide in the cycle.

decreased levels of E_1G and increased levels of pregnanediol glucuronide (PG). Female proceptive behavior and masturbation were more circumscribed in the cycle than were the days of copulation, spanning only 10-day and 12-day intervals, respectively. Moreover, 78% of all instances of proceptive behavior and 72% of all female masturbation occurred within the 5-day interval from day −3 to day +1, in comparison to 69% of all copulations for the same interval. Indications of heightened female sexual responsiveness, therefore, were more closely associated with the peak levels of E_1G, i.e., the periovulatory period, than were the indications of heightened male sexual responsiveness.

A study of circulating hormone levels in five of the six female orang-utans used in this behavioral study (Nadler et al., 1984) demonstrated that the midcycle peak in 17β-estradiol, the major precursor of E_1G in urine, was followed within 1 day by a peak in testosterone levels (and by peak LH levels) (Fig. 7–5). The orang-utan thereby resembles other primates with respect to the closely associated midcycle peaks in estrogen and testosterone. These are the two hormones, moreover, that are most closely identified with the facilitation of sexual behavior in primates, as exemplified by the extensive experimental studies on rhesus monkeys (Baum et al, 1977; Herbert and Trimble, 1967; Johnson and Phoenix, 1976, Keverne, 1976; Michael and Wellegalla, 1968; Wallen and Goy, 1977). Elevated levels of progesterone, on the other hand, are associated with an inhibition of sexual activity.

When considered in relation to these endocrine data, the results of the RATs with orang-utans suggest that sexual behavior in this species occurs (1) cyclically, as a result primarily of the female's role in regulating sexual interactions, and (2) in close association with the presumptive time of ovulation, when (3) levels of 17β-estradiol and testosterone in the female are elevated and prior to the time when maximal levels of progesterone are reached.

CONCLUSIONS

The research on orang-utans provided support for several hypotheses regarding factors that contribute to the regulation of sexual behavior. In relation to the other great apes, various aspects of the orang-utan's sexual behavior and anatomy conform to predictions based on the number of males competing for sexual access to females at

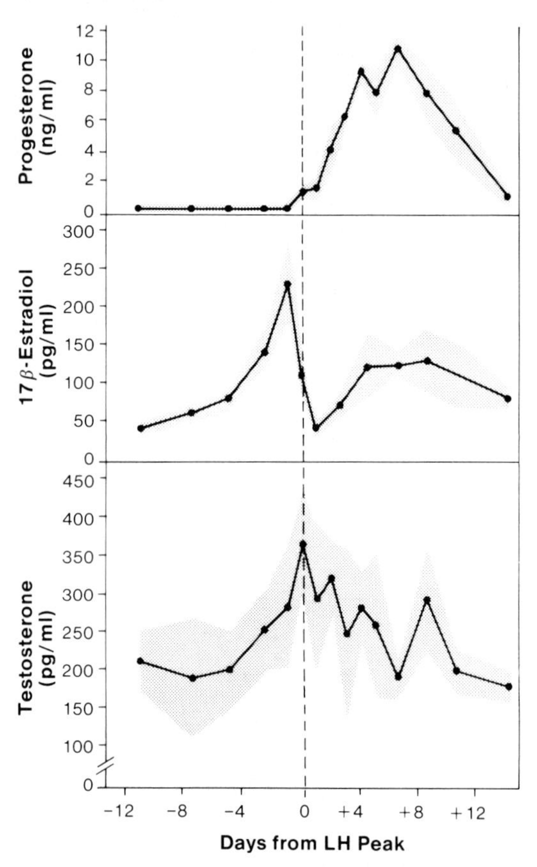

FIG. 7–5. Serum levels of progesterone, 17β-estradiol, and testosterone during the menstrual cycles of five female orang-utans, normalized to the day of the peak in LH levels. Shaded area represents the standard error of the mean.

the time of estrus (Harcourt, 1981). Consistent with Zuckerman's (1930) early proposal regarding the action of the "follicular hormone," mating (during the follicular phase) increased with increasing levels of estrone glucuronide. As proposed by several investigators (Beach, 1942; Yerkes, 1939; Young and Orbison, 1944), mating in this primate species was less rigidly restricted to the presumptive time of ovulation, i.e., less rigidly controlled by hormones of the female than it is in nonprimate mammals. Consistent with Yerkes's (1939) qualification of this hypothesis, it was the female orang-utan that restricted mating to the presumptive time of ovulation, whereas the male accounted for mating temporally dissociated from ovulation. Finally, environmental conditions that differentially favored male (FATs) or female (RATs) regulation of mating clearly influenced, if not determined, the frequency and distribution of mating in the cycle.

Although the question of cyclicity in human sexual behavior has been investigated extensively, results have been equivocal and much of the early research suffered from a variety of methodological problems (James, 1971). Some provocative evidence of a midcycle enhancement in human sexual responsiveness, though certainly not conclusive (e.g., Bancroft et al., 1983), has been presented, as well as a possible explanation for the absence of such evidence in many of the earlier studies (Adams et al., 1978). Adams et al. (1978:1145) reported that male-initiated sexual behavior occurred throughout the cycle, irrespective of cycle phase, but that female-initiated behavior, "both autosexual and female-initiated heterosexual behavior" did show a significant increase at midcycle. It was proposed that the previous failures to detect such cyclicity in human sexual responsiveness were probably due in part to the "use of measures of sexual behavior that are primarily determined by the male partner."

These data on human sexual behavior, therefore, are quite analogous to the data on sexual behavior of the orang-utan (and the other great apes). The finding that both human and great ape females restricted their sexual initiative to the midcycle, periovulatory phase, or exhibited increased sexual initiative at this phase of the cycle, suggests that this pattern of female sexual responsiveness may be characteristic of all the hominoids. The related finding in these species that it was the male that primarily accounted for sexual activity temporally dissociated from the periovulatory period suggests that Yerkes's (1939) hypothesis regarding the differential regulation of sexual behavior by males and females may also apply to human sexual relations. The apparent pervasiveness of this pattern of sexual interactions among the hominoids speaks to the value of the orang-utan (and the other great apes) for comparative research on the biological basis of human sexual behavior.

ACKNOWLEDGMENTS

Preparation of this manuscript was supported by USPHS NIH Grants RR-00165 to the Yerkes Regional Primate Research Center and HD-19060 to the author. The Yerkes Center is fully accredited by the American Association for Accreditation of Laboratory Animal Care. The research in which the restricted-access tests (RATs) were used was supported by NSF Grant BNS 79-23015 and a NIH Biomedical Research Support Grant to Emory University. The radioimmunoassays for estrone glucuronide and pregnanediol glucuronide were conducted in the laboratory of D. C. Collins at the VA Hospital in Decatur, GA. The following individuals contributed to the collection of behavioral data for the RATs: M. A. Codner, C. J. Juno, L. K. Maret, F. E. McKean, L. C. Miller, K. Riedy, R. S. Schonwetter, A. J. Volpi, and J. T. Wright.

REFERENCES

Adams, D. B., Gold, A. R. and Burt, A. D. 1978. Rise in female-initiated sexual activity at ovulation and its suppression by oral contraceptives. *New England Journal of Medicine, 299:*1145–1150.

Asano, M. 1967. A note on the birth and rearing of an orang-utan at Tama Zoo, Tokyo. *International Zoo Yearbook, 7:*95–96.

Aulmann, G. 1932. Gegluckte Wachzucht eines Orang-Utan im Dusseldorfer Zoo. *Zoologische Garten, 5:*81–90.

Bancroft, J., Sanders, D., Davidson, D. and Warner, P. 1983. Mood, sexuality, hormones, and the menstrual cycle. III. Sexuality and the role of androgens. *Psychosomatic Medicine, 45:*509–516.

Baum, M. J., Everitt, B. J., Herbert, J. and Keverne, E. B. 1977. Hormonal basis of proceptivity and receptivity in female primates. *Archives of Sexual Behavior, 6:*173–192.

Beach, F. A. 1942. Central nervous mechanisms involved in the reproductive behavior of vertebrates. *Psychological Bulletin, 39:*200–226.

Beach, F. A. 1976. Sexual attractivity, proceptivity, and receptivity in female mammals. *Hormones and Behavior, 7:*105–138.

Beach, F. A. 1978. Sociobiology and interspecific comparisons of behavior. *In* Sociobiology and Human Nature, pp. 116–135, ed. M. S. Gregory, A. Silvers and D. Sutch. Jossey-Bass Publishers, San Francisco.

Chaffee, P. S. 1967. A note on the breeding of orang-utans at Fresno Zoo. *International Zoo Yearbook, 7:*94–95.

Coffey, P. F. 1971. Breeding the Bornean orang utan *Pongo pygmaeus pygmaeus* (Linnaeus). *Annual Report of the Jersey Wildlife Preservation Trust,* pp. 32–36.

Coffey, P. F. 1972. Breeding Sumatran orang-utan *Pongo pygmaeus abelii* Lesson 1827. *Annual Report of the Jersey Wildlife Preservation Trust,* pp. 15–17.

Coffey, P. F. 1975. Sexual cyclicity in captive orang-utans *Pongo pygmaeus* with some notes on sexual behavior. *Annual Report of the Jersey Wildlife Preservation Trust,* pp. 54–55.

Darwin, C. 1871. *The Descent of Man, and Selection in Relation to Sex.* John Murray, London.

Davenport, R. K., Jr. 1967. The orang-utan in Sabah. *Folia Primatologica, 5:*247–263.

Dewsbury, D. A. 1984. Comparative Psychology in the 20th Century. Hutchinson Ross, Stroudsburg, PA.

Estep, D. Q. and Bruce, K. E. M. 1981. The concept of rape in non-humans: a critique. *Animal Behaviour, 29:*1272–1273.

Fox, H. 1929. The birth of two anthropoid apes. *Journal of Mammalogy, 10:*37–51.

Freund, M. 1962. Interrelationships among the characteristics of human semen and factors affecting semen specimen quality. *Journal of Reproduction and Fertility, 4:*143–159.

Galdikas, B. M. F. 1979. Orangutan adaptation at Tanjung Puting Reserve: mating and ecology. *In* The Great Apes: Perspectives on Human Evolution, pp. 195–233, ed. D. A. Hamburg and E. R. McCown. Benjamin/Cummings, Menlo Park.

Galdikas, B. M. F. 1981. Orangutan sexuality in the wild. *In* Reproductive Biology of the Great Apes: Comparative and Biomedical Perspectives, pp. 281–300, ed. C. E. Graham. Academic Press, New York.

Galdikas, B. M. F. 1984. Orangutan sociality at Tanjung Puting. *American Journal of Primatology, 9:*101–119.

Graham-Jones, O. and Hill, W. C. O. 1962. Pregnancy and parturition in a Bornean orang. *Proceedings of the Zoological Society of London, 139:*503–510.

Harcourt, A. H. 1981. Intermale competition and the reproductive behavior of the great apes. *In* Reproductive Biology of the Great Apes: Comparative and Biomedical Perspectives, pp. 301–318, ed. C. E. Graham. Academic Press, New York.

Harcourt, A. H., Harvey, P. H., Larson, S. G. and Short, R. V. 1981. Testis weight, body weight and breeding system in primates. *Nature, 293:*55–57.

Harcourt, A. H. and Stewart, K. J. 1977. Apes, sex, and societies. *New Scientist, 76:*160–162.

Heinrichs, W. L. and Dillingham, L. A. 1970. Bornean orang-utan twins born in captivity. *Folia Primatologica, 13:*150–154.

Herbert, J. and Trimble, M. R. 1967. Effect of oestradiol and testosterone on the sexual receptivity and attractiveness of the female rhesus monkey. *Nature, 216:*165–166.

Horr, D. A. 1975. The Borneo orang-utan: population structure and dynamics in relation to ecology and reproductive strategy. *In* Primate Behavior: Developments in Field and Laboratory Research, vol. 4, pp. 307–323, ed. L. A. Rosenblum. Academic Press, New York.

James, W. H. 1971. Coital rates and the pill. *Nature, 234:*555–556.

Johnson, D. F. and Phoenix, C. H. 1976. The hormonal control of female sexual attractiveness, proceptivity and receptivity in rhesus monkeys. *Journal of Comparative and Physiological Psychology, 90:*473–483.

Keverne, E. B. 1976. Sexual receptivity and attractiveness in the female rhesus monkey. *Advances in the Study of Behavior, 7:*155–200.

Lippert, W. 1974. Beobachtungen zum Schwangerschafts- und Geburtsverhalten beim Orang-utan *(Pongo pygmaeus)* im Tierpark Berlin. *Folia Primatologica, 21:*108–134.

MacKinnon, J. R. 1971. The orang-utan in Sabah today. *Oryx, 11:*141–191.

MacKinnon, J. R. 1974. The behaviour and ecology of wild orang-utans *(Pongo pygmaeus). Animal Behaviour, 22:*3–74.

MacKinnon, J. R. 1979. Reproductive behavior in wild orangutan populations. *In* The Great Apes: Perspectives on Human Evolution, pp. 257–273, ed. D. A. Hamburg and E. R. McCown. Benjamin/Cummings, Menlo Park.

Maple, T. L., Zucker, E. L. and Dennon, M. B. 1979. Cyclic proceptivity in a captive female orang-utan *(Pongo pygmaeus abelii). Behavioural Processes, 4:*53–59.

Michael, R. P. and Welegalla, J. 1968. Ovarian hormones and the sexual behavior of the female rhesus monkey *(Macaca mulatta)* under laboratory conditions. *Journal of Endocrinology, 41:*407–420.

Milton, O. 1964. The orang-utan and rhinoceros in North Sumatra. *Oryx, 7:*177–184.

Mitani, J. C. 1985. Mating behaviour of male orangutans in the Kutai Game Reserve, Indonesia. *Animal Behaviour, 33:*392–402.

Nadler, R. D. 1977a. Sexual behavior of captive orang-utans. *Archives of Sexual Behavior, 6:*457–475.

Nadler, R. D. 1977b. Sexual behavior of the chimpanzee in relation to the gorilla and orang-utan. *In* Progress in Ape Research, pp. 191–206, ed. G. H. Bourne. Academic Press, New York.

Nadler, R. D. 1981. Laboratory research on sexual behavior of the great apes. *In* Reproductive Biology of the Great Apes: Comparative and Biomedical Perspectives, pp. 191–238, ed. C. E. Graham. Academic Press, New York.

Nadler, R. D. 1982. Reproductive behavior and endocrinology of orang utans. *In* The Orang Utan: Its Biology and Conservation, pp. 231–248, ed. L. E. M. de Boer. W. Junk, The Hague.

Nadler, R. D., Collins, D. C. and Blank, M. S. 1984. Luteinizing hormone and gonadal steroid levels during the menstrual cycle of orangutans. *Journal of Medical Primatology, 13:*305–314.

Nadler R. D., Herndon, J. G. and Wallis, J. 1986. Adult sexual behavior: hormones and reproduction. *In* Comparative Primate Biology, vol. 2A (Behavior and Ecology), pp. 363–407, ed. G. Mitchell and J. Erwin. Alan R. Liss, New York.

Napier, J. R. and Napier, P. H. 1967. A Handbook of Living Primates. Academic Press, London.

Okano, T. 1965. Preliminary survey of the orang-utan in North Borneo. *Primates, 6:*123–128.

Rijksen, H. D. 1978. A Field Study on Sumatran Orang-utans (*Pongo pygmaeus abelli* Lesson 1827). Ecology, Behaviour and Conservation. H. Veenman and Zonen, Wageningen.

Rodman, P. S. 1973. Population composition and adaptive organization among orang-utans of the Kutai Reserve. *In* Comparative Ecology and Behaviour of Primates, pp. 171–209, ed. R. P. Michael and J. H. Crook. Academic Press, London.

Schaller, G. B. 1961. The orang-utan in Sarawak. *Zoologica, 46:*73–82.

Schultz, A. H. 1938. Genital swelling in the female orang-utan. *Journal of Mammalogy, 19:*363–366.

Schürmann, C. L. 1981. Courtship and mating behavior of wild orang-utans in Sumatra. *In* Primate Behavior and Sociobiology, pp. 130–135, ed. A. B. Chiarelli and R. S. Corruccini. Springer-Verlag, Berlin.

Schürmann, C. L. 1982. Mating behaviour of wild orang-utans. *In* The Orang Utan: Its Biology and Conservation,pp. 269–284, ed. L. E. M. de Boer. W. Junk, The Hague.

Short, R. V. 1977. Sexual selection and the descent of man. *In* Reproduction and Evolution, pp. 3–19, ed. J. H. Calaby and C. H. Tyndale-Biscoe. Griffin Press, Netley, South Australia.

Short, R. V. 1981. Sexual selection in man and the great apes. *In* Reproductive Biology of the Great Apes: Comparative and Biomedical Perspectives, pp. 319–341, ed. C. E. Graham. Academic Press, New York.

Tutin, C. E. G. and McGinnis, R. P. 1981. Sexuality of the chimpanzee in the wild. *In* Reproductive Biology of the Great Apes: Comparative and Biomedical Perspectives, pp. 239–264, ed. C. E. Graham. Academic Press, New York.

Wallen, K. and Goy, R. W. 1977. Effects of estradiol benzoate, estrone, and propionates of testosterone or dihydrotestosterone on sexual and related behaviors of ovariectomized rhesus monkey. *Hormones and Behavior, 9:*228–248.

Yerkes, R. M. 1939. Sexual behavior in the chimpanzee. *Human Biology, 11:*78–111.

Yerkes, R. M. and Yerkes, A. W. 1929. The Great Apes. Yale University Press, New Haven.

Yoshiba, K. 1964. Report of the preliminary survey of the orang utan in North Borneo. *Primates, 5:*11–26.

Young, W. C. and Orbison, W. D. 1944. Changes in selected features of behavior in pairs of oppositely-sexed chimpanzees during the sexual cycle and after ovariectomy. *Journal of Comparative Psychology, 37:*107–143.

Zuckerman, S. 1930. The menstrual cycle of the primates. Part I. General nature and homology. *Proceedings of the Zoological Society of London, 45:*691–754.

8

Comparisons of Female Reproductive Hormone Patterns in the Hominoids

N. M. CZEKALA, S. E. SHIDELER, AND B. L. LASLEY

Until recently there has been no singular, unified approach to the study of primate reproductive physiology that would allow accurate comparisons to be made between species. Historically, only blood samples or 24-hr urine collections were considered to be adequate to monitor hormone changes throughout the ovarian cycle, implantation, pregnancy, and parturition. Neither of these approaches is practical for large, longitudinal studies, although both have been used in smaller studies. Venipuncture is stressful to animals not trained to present an arm or leg for sample collection. Too few trained animal subjects of various species are available at any one time to make serum sampling a viable approach for broad comparative studies. Twenty-four-hour urine collections are feasible in all species, but the procedure requires isolating subjects, which not only obliterates sexual and other social behavior but is known to stress some individuals to the point where they are affected physiologically.

Over the past 10 years an alternative approach to the study of primate reproduction has been developed and applied to an array of primate species, including most hominoids (Lasley, 1985). This approach consists of a method of hormone measurement that utilizes daily random urine samples to determine the rate of ovarian steroid excretion, which, in turn, provides a complete record of major reproductive events. The overall approach has several advantages over blood sampling or 24-hr urine collections, and allows objective comparison of ovarian cycles within and between individual subjects and species.

Random daily urine samples are generally easy to collect on a routine basis in captivity. Animal subjects as small and flighty as marmosets or as large and intractable as gorillas and orang-utans rapidly adapt to the procedure (Czekala et al., 1983; Heger and Neubert, 1983). Collection and preservation of the samples require no specialized skill, equipment, or labor. So little is required in the sampling that it is efficient to collect continuously as part of the daily routine, and to analyze samples on a selective, retrospective basis. By doing this, infrequent events such as conceptions, abortions, and births can be evaluated as incidental events in a broader surveillance program. For the continuous recording of hormonal events, urine evaluations are unparalleled by other approaches. Daily hormone records spanning 2 years have been collected for a number of individual gorillas, gibbons, and other hominoid and monkey species. Questions relating to ovulatory patterns in young females as well as ovarian dysfunction in older females are now approached through longitudinal studies that were not possible with blood collection.

Methods for urine evaluation of ovarian hormones have progressed rapidly in the last few years. Labor-intensive and technically complex collection and evaluation procedures have been replaced by simplification and direct assaying methods. The assays themselves are less time-consuming and more cost-effective than previously existing technologies. The future adaptation of urine assays to enzyme systems provides the basis for urine analysis kits that can be used in nonlaboratory settings (Czekala et al., 1986).

The data reviewed in this report focus on the comparison of hominoid reproductive physiology. Ovarian cycles, conceptions, and pregnancies, and other known parameters of reproductive significance are described and compared between the human, gorilla, chimpanzee, orang-utan, gibbon, and macaque.

DISCUSSION

Figure 8–1 illustrates the ovarian cycles of six anthropoid primates: human, gorilla, chimpanzee, orang-utan, gibbon, and macaque. The urinary estrone conjugate profiles presented in this

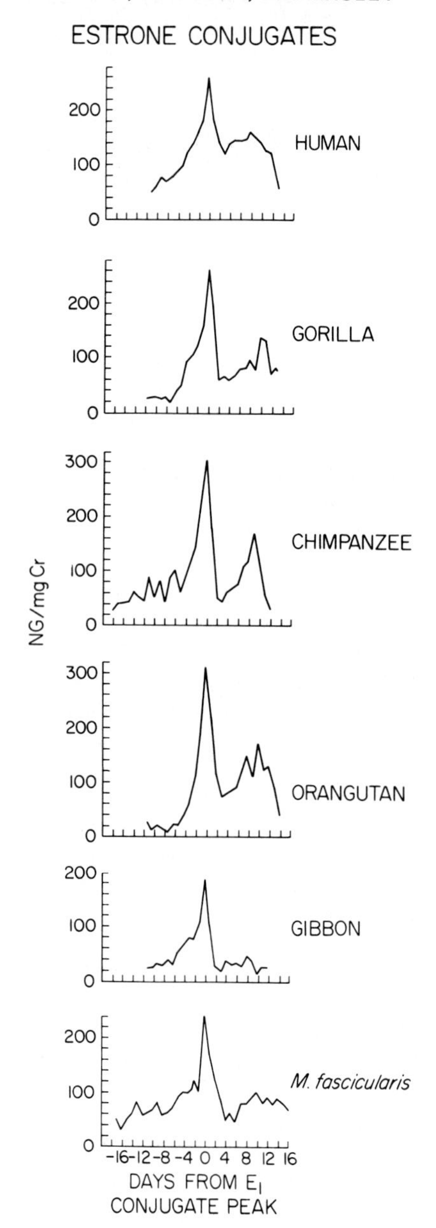

FIG. 8–1. Urinary estrone conjugate (ng/mg Cr) profiles from normal cycles of representative individuals of six anthropoid primates are aligned to days before and after the highest estrone conjugate level.

figure are representative examples taken from individuals of each of the six primates. The follicular phase (the first day of menstruation to estrone conjugate peak) is a particularly sensitive part of the primate ovarian cycle that is easily affected by poor health, stress, or other environmental variables. It is apparent that follicular phase length of the primates represented in Fig. 8–1 is variable. This variation may not reflect real differences between species and may be an artifact of the condition of the individual animal represented. The pattern of estrone conjugate excretion is, however, similar in all species, increasing from 20–50 ng/mg creatinine (Cr) during the early follicular phase to approximately 200–300 ng/mg Cr in the late follicular phase or at midcycle (day 0). The duration of the luteal phase (the day after estrone conjugate peak to menstruation) also is similar in most species with the exception of the gibbon, in which it is abbreviated. The pattern of the luteal phase increase in estrone conjugate excretion is more variable and may reflect phylogenetic differences. Comparing the least similar taxa (the human and the macaque), a large difference in the maximum level of luteal phase estrogen excretion is seen. The human, gorilla, chimpanzee,

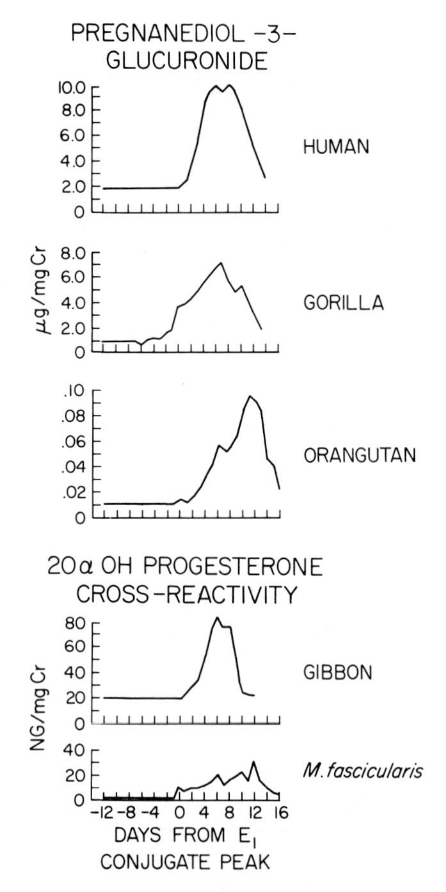

FIG. 8–2. In the top three panels, urinary pregnanediol-3-glucuronide (μg/mg Cr) values from representative individuals during a normal cycle are aligned to the peak estrone conjugate value. In the bottom two panels (gibbon and *M. fascicularis*), 20αOH progesterone cross-reactivity values are likewise exhibited.

and orang-utan have similar estrogen levels during this phase of the cycle, while the gibbon approximates the macaque in its low luteal estrogen levels.

The pattern of progesterone metabolite excretion during the ovarian cycles of five species is illustrated in Fig. 8–2. Again, the length of the luteal phase as reflected in progesterone metabolite profiles is consistent, agreeing with the estrone conjugate profiles. In general, progesterone metabolite excretion is qualitatively similar but quantitatively different between species. Pregnanediol-3-glucuronide is the primary progesterone metabolite excreted by the human and accounts for approximately 30% of the progesterone in circulation. In comparison to other primates, the human female is unique in her high levels of pregnanediol excretion, which reach 10 μg/mg Cr during the luteal phase. The gorilla exhibits similar pregnanediol trends, and excretes slightly less than the human, reaching luteal phase levels of 8.0 μg/mg Cr. The orang-utan excretes 100% less pregnanediol than the human, only reaching levels of 0.1 μg/mg Cr during the luteal phase. Both the gibbon and the macaque excrete very low levels of pregnanediol; the major progesterone metabolite in these primates is a compound immunoreactively similar to 20α-OH-progesterone (20αOHPo) which is believed to reflect the same ovarian activity that pregnanediol does in the other anthropoids. The temporal pattern seen in the 20αOHPo cross-reactivity profile of the gibbon is similar to that of the great apes and humans. The gibbon exhibits higher levels of 20αOHPo than the macaque: levels approach 80 ng/mg Cr in the gibbon in contrast to 40 ng/mg Cr in the macaque.

The difference between the gibbon and other hominoids in the amount and kind of progesterone metabolite excreted during the ovarian cycle is commensurate with other features that appear to phylogenetically distinguish the gibbon from the great apes. At the present time it is not known whether these differences reflect differences in hormone production rate, in metabolism, or both; it is also not known if siamangs exhibit the same differences as seen in gibbons and macaques.

Estrone conjugate profiles of ovulatory cycles resulting in conception and implantation in the human, gorilla, orang-utan, and macaque are depicted in Fig. 8–3. As discussed above, the ovulatory or midcycle estrogen peak is comparable among all anthropoids, whereas the secondary or luteal phase increase in estrogen varies quantitatively among taxa. By day +10 after conception, a rapid increase of estrogen ensues, such

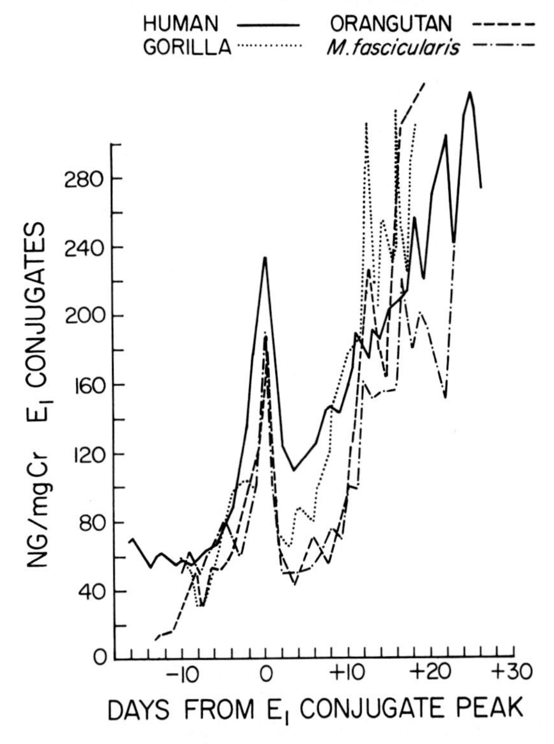

FIG. 8–3. Urines collected during ovulation and early implantaton were evaluated for estrone conjugates (ng/mg Cr) in four anthropoid primates. The values are aligned to the peak estrone conjugate value during ovulation in each individual.

that by day +12 estrogen levels are quantitatively and qualitatively very similar among taxa. This surge during early pregnancy in estrogen excretion is ovarian in origin and is an indirect effect of trophoblastic gonadotropin (Lasley et al., 1985).

The patterns of estrogen profiles during primate pregnancy indicate that, in general, pregnancies are less affected by environmental variables than are ovarian cycles. During pregnancy, estrogen excretion increases as gestation progresses. The estrogen profiles obtained during pregnancy are largely the result of fetal adrenal hormone output, they reflect fetal well-being, and provide another basis for interspecific comparisons. The pregnancy profiles throughout gestation of a human, orang-utan, gorilla, common chimpanzee, and pygmy chimpanzee are presented in Fig. 8–4. Comparison of total urinary estrogen profiles indicates that the human and orang-utan are more similar to each other during pregnancy that either is to the gorilla.

The implication of these observations is of some consequence: since estrogen excretion during pregnancy is predominantly of fetoplacental

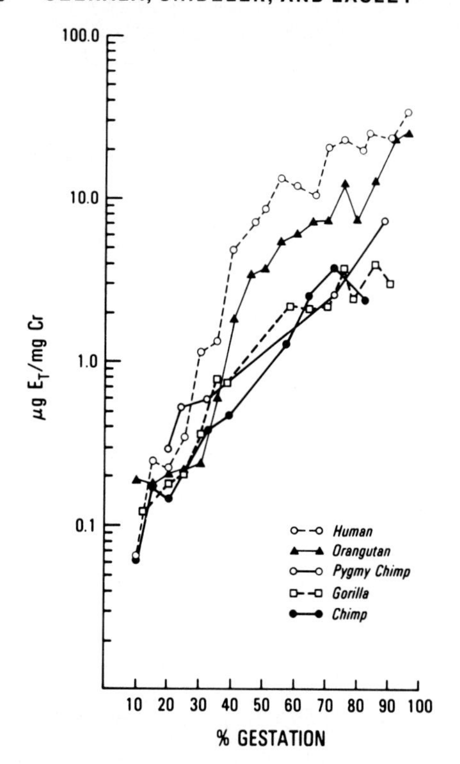

FIG. 8–4. Urinary total immunoreactive estrogens (E_T) during pregnancy in five hominoid species are compared to the stages of gestation. Gestational length used for calculation for each species: human, 270 days; orang-utan, 258 days; pygmy chimpanzee, 255 days; gorilla, 255 days; common chimpanzee, 225 days. Values are graphed on logarithmic scale.

origin, these profiles suggest that two types of fetoplacental units might exist within the hominoids. From these data, it appears that humans and the orang-utan develop one type of fetoplacental unit while the gorilla and chimpanzee possess another. The difference between the two fetoplacental units might be attributed to (1) a difference in hepatic steroid metabolism, (2) a difference in placental metabolism, or (3) a difference in fetal production of estrogen precursors. The third possibility is the most likely and it is also the most provocative, since it would indicate the existence of morphological differences among the hominoids in the fetal adrenals. Analysis of the types of estrogen present during gestation in each taxon supports the conclusion that a fetal rather than placental or maternal difference does exist among the large hominoids (Czekala et al., 1983).

Unlike most other mammals, primates possess a fetal adrenal composed of a definitive zone and a fetal zone. The fetal zone produces the precursor of estriol while the placenta converts this precursor to estriol. Hominoids are unique in their ability to produce large quantities of estriol during pregnancy. Since the fetal adrenal may be the source of the difference in hormone excretion observed in hominoid pregnancies, histological comparisons of prenatal stillbirths and neonatal fetal adrenals were made to see if morphological differences among taxa exist.

Comparisons of fetal adrenal size and morphology between the human, orang-utan, gorilla, and chimpanzee are shown in Table 8–1. Since the fetal zone portion of the primate fetal adrenal produces the precursor for estriol, the size of the fetal zone in each species will reflect the amount of precursor source and therefore the amount of estriol produced. When considered against the total estrogen profiles of pregnancy, the human and orang-utan fetal adrenals are similar, with larger fetal zones and thus lower ratios (Table 8–1); the chimpanzee and gorilla fetal adrenals are similar to each other, with smaller fetal zones and higher ratios. As the *p* values indicate, the fetal adrenals of the human and orang-utan are not significantly different from one another, whereas the fetal adrenals of both the chimpanzee and the gorilla are significantly different from that of the human fetus.

CONCLUSION

The approach used in this review to study primate reproductive physiology allows direct com-

TABLE 8–1. Comparison of the Ratios of the Size of the Definitive Zone:Fetal Zone in the Fetal Adrenal of Five Hominoid Species

Species	Definitive Zone / Fetal Zone		*p* Value
Human	0.110 ± 0.01	$N=9$	—
Orang-utan	0.095 ± 0.02	$N=5$	n.s.
Chimpanzee	0.240 ± 0.05	$N=10$	0.05
Gorilla	0.270 ± 0.05	$N=4$	0.002

The *p* values are computed as a comparison to human values.

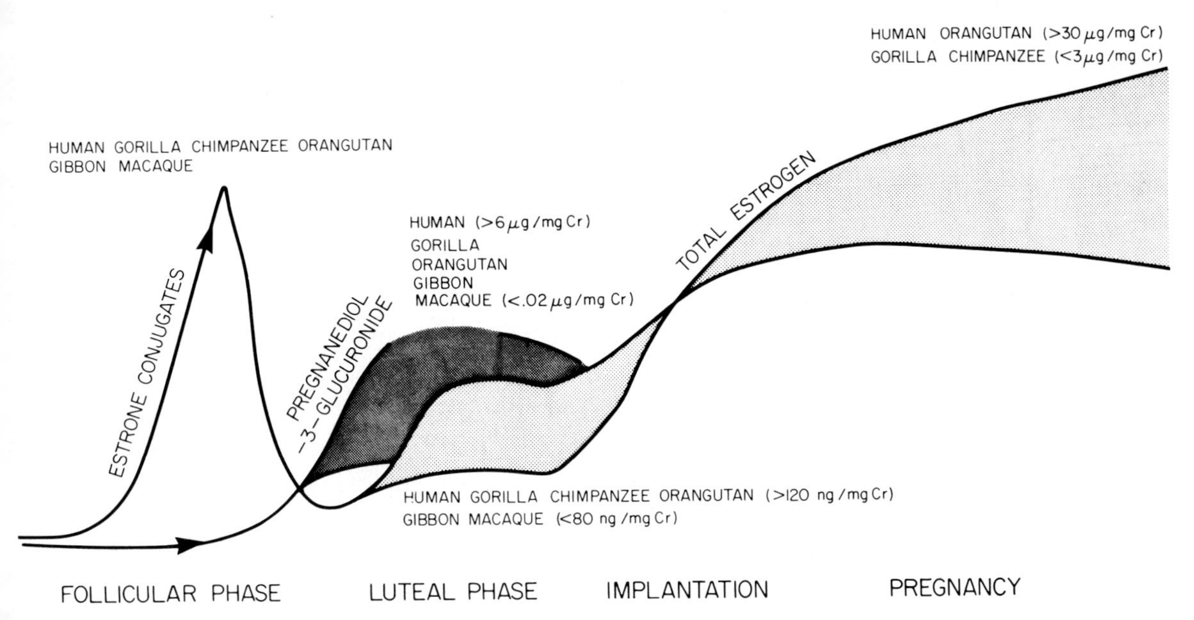

FIG. 8–5. This composite graph of hormonal aspects of comparative reproduction shows the relationship of estrone conjugate, pregnanediol-3-glucuronide (Pdg), and total estrogen excretion between several primate species. During the follicular phase, estrone conjugate values (unshaded area) are equivalent between the six species listed. During the luteal phase this same hormone is excreted in varying quantities by each species; humans and great apes excrete similar quantities while the gibbon and macaque excrete lower quantities. If conception occurs, the estrone conjugates increase uniformly but during gestation maximum levels of total estrogens vary with species: the human and orang-utan have higher estrogen levels than do the gorilla and chimpanzee. During the luteal phase the excretion of pregnanediol-3-glucuronide (darkest shading) varies markedly between taxa. The human excretes the most Pdg with the quantity decreasing as the species list above the Pdg curve descends.

parisons between species and, thus, provides insight into phylogenetic differences in hormone excretion. The nonhuman primate reproductive profiles described show similarities to human profiles. These are summarized in Fig. 8–5.

During the follicular phase of the ovarian cycle, the pattern of estrogen excretion is similar in all taxa. In the luteal phase, estrone conjugate excretion patterns are equivalent in the human, gorilla, chimpanzee, and orang-utan, whereas estrogen levels are lower in the gibbon and macaque. Intertaxon differences are, however, more marked in the progesterone metabolite excretion patterns seen during the luteal phase. Pregnanediol-3-glucuronide is the major metabolite excreted by the human, gorilla, chimpanzee, and orang-utan, with the human reaching levels 100 times greater than the orang-utan and slightly greater than the gorilla. The gibbon and macaque excrete no measurable pregnanediol, but they do excrete a compound similar to 20α-OHPo. While the gibbon exhibits a 20αOHPo cross-reactivity profile that is temporally patterned like the pregnanediol pattern of the great apes and human, its metabolite levels are lower than those of the orang-utan, but these levels are still higher than those of the macaque.

During pregnancy, the human and orang-utan have similar total estrogen profiles, which is reflective of the similarity of these hominoids in the fetal zones of their fetal adrenals. In contrast to humans and orang-utans, the gorilla and chimpanzee are themselves similar in having lower level total estrogen profiles and correspondingly smaller fetal zones in their fetal adrenals.

Among the large hominoids, two hormonal differences in reproduction are presently recognized. The first is a difference in progesterone metabolism in which the orang-utan is least like the human in the amount excreted. The second difference is in the estrogen profiles of pregnancy which reflect the relative size of the fetal adrenal component in each species. In this regard, the orang-utan is most similar to humans.

The decision as to which is the stronger evidence in arguing taxonomic relationships must be given to the structure-function argument. The morphological difference in the fetal adrenals between hominoids, with the subsequent variation in estrogen excretion during pregnancy, would suggest the orang-utan shares more characteristics with the human than do the gorilla or chimpanzee. Although hormone excretion by itself is good evidence, since we must assume either metabolic or morphological differences as the cause for hormone variation between individuals, at the present time without definitive metabolic or structural differences there is little evidence beyond the initial observation to argue this point.

REFERENCES

Czekala, N. M., Benirschke, K., McClure, H. and Lasley, B. L. 1983. Urinary estrogen excretion during pregnancy in the gorilla *(Gorilla gorilla),* orangutan *(Pongo pygmaeus),* and the human *(Homo sapiens). Biology of Reproduction, 28:*289–294.

Czekala, N. M., Meier, J. E., Gallusser, S. and Lasley, B. L. 1986. The development and application of an enzyme immunoassay for urinary estrone conjugates. *Zoo Biology, 5:*1–9.

Heger, H. W. and Neubert, D. 1983. Timing of ovulation and implantation in the common marmoset, *Callithrix jacchus,* by monitoring of estrogens and 6 beta-hydroxypregnanolone in urine. *Archives of Toxicology, 54:*41–52.

Lasley, B. L. 1985. Endocrinology and reproduction in exotic species. *In* Advances in Veterinary Science and Comparative Medicine, pp. 209–228, ed. C. E. Cornelius. Academic Press, Orlando.

Lasley, B. L., Stabenfeldt, G. H., Overstreet, J. W., Hanson, F. W., Czekala, N. M. and Munro, C. 1985. Urinary hormone levels at the time of ovulation and implantation. *Fertility and Sterility, 43:*861–866.

9

Physiological Development of Male Orang-utans and Gorillas

SUSAN R. KINGSLEY

Most studies on reproduction in great apes concentrate solely on the female, and as a result, little is known about the reproductive and physical development of the male. A study was therefore undertaken to examine the sexual and physical development of male orang-utans and gorillas.

Sexual maturity in males can be defined as the time of fertile sperm production and is reached during the later stages of pubertal development. Puberty itself is a graded series of hormonal and physical changes spanning about 5 years in humans (e.g., Kelch et al., 1972; Swerdloff and Rubin, 1978; Tanner, 1978) and chimpanzees (Martin et al., 1977; McCormack, 1971). Each stage of pubertal development occurs in an ordered manner so that the testes mature first, spermatogenesis occurs next (about 2 years later in humans and 1 year later in rhesus monkeys), and then other secondary sexual characteristics develop during the next year or two (Conaway and Sade, 1965; Tanner, 1978).

This relatively rapid, continuous sexual and physical development also seems to occur in the gorilla. Animals become sexually mature by about 8 years of age (personal observation), begin to develop the typical silver-backed appearance of the adult male at the age of 9 or 10 years, and complete development by about 12 years of age. These males also emit a pungent odor, have a prominent sagittal crest, and are larger than the adolescent black-backed males (Schaller, 1963).

However, this continuous development is not always so obvious in male orang-utans, who can become "fixed" in the subadult stage for many years. Sexually mature and fertile orang-utan males can be split into two groups on the basis of the degree of development of these secondary sexual characteristics. There are (1) the "subadult" males between the ages of 8 and 13–15 years in the wild (Rijksen, 1978) and 7 and 11–18 in captivity (Kingsley, 1981) and (2) the "mature" adult males of 13–15+ in the wild (Rijksen, 1978) and 11–18+ in captivity (Kingsley, 1981). The subadults, termed "nonflanged," retain many characteristics of the adolescent stage (between 5–6 and 8 years), such as the dark, flat face, short body hair, and general slimness of build. They have hard rims along the sides of their face, fully descended testes, and have begun to develop a beard. Their larynx is undeveloped and they emit no strong odor. Although capable of breeding, they have been known to remain in the nonflanged state for up to 9 years in captivity, with a captive-born male in the Philadelphia Zoo only starting to develop its flanges when it was 17½ years old (Ulmer, 1958). Examples of nonflanged males can be seen in Figs. 9–4A and 9–5A.

Eventually the subadult male begins to develop the cheek flanges (prominent pads of fat, connective tissue, and muscle; Deniker and Boulart, 1885) and throat pouch of the mature ("flanged") male. Once begun, this development has been reported to occur in a remarkably short time, often within a year (G. Brandes, 1932, 1939; Ulmer, 1958). Flange development is accompanied by the development of the larynx, a change in body proportions, a lengthening and thickening of the hair and, in the case of Bornean orang-utans, the appearance of a high fatty crown. There is also general growth: the nonflanged males weigh the same as adult females (30–59 kg) but grow to a weight of 50–90 kg once mature (Rijksen, 1978). The mature male has a wider range of vocalizations than the subadult (R. Brandes, 1931; MacKinnon, 1974; Rijksen, 1978), has a musty body scent not found in the less developed male, and has different behavior patterns. Figure 9–5C shows an example of a flanged male.

Physical and sexual maturity in humans and chimpanzees is correlated with increasing concentrations of plasma testosterone (August et al., 1972: Frasier et al., 1969; Kraemer et al., 1982; Martin et al., 1977; McCormack, 1971). Testosterone concentrations remain low in juveniles and increase rapidly throughout adolescence.

This rapid increase in circulating testosterone commences at 12 years of age or younger in boys, reaching adult levels at the age of 16 or more (August et al., 1972). In chimpanzees, increased testosterone secretion commences at the sixth or seventh year (Kraemer et al., 1982; Martin et al., 1977; McCormack, 1971). A second surge in secretion occurs at 11 years of age (Martin et al., 1977). In addition Collins et al. (1981) briefly reported an increase in plasma testosterone levels between juvenile gorillas (0–3 years old) and adults (10 years or more).

In order to ascertain whether developing orang-utan and gorilla males had hormonal changes throughout life similar to humans and chimpanzees, a study was undertaken to examine androgen levels (principally urinary testosterone) of juvenile, adolescent, and adult animals. In addition, in order to test whether flange development in orang-utans was associated with increasing steroid levels, urinary androgens and estrogens were measured in two individuals during the process of flange development. Estrogens as well as androgens were measured in these animals, as it is possible that these former hormones initiate the growth of some of the secondary sexual characteristics in humans. For example, estrogen levels double in boys between early and late puberty (Donovan and Werff ten Bosch, 1965).

MATERIALS AND METHODS

Animals

Twenty-four male orang-utans were used in the study. They belonged to four different age-status classes (juvenile, nonflanged adult, adults with flanges growing, flanged adult) and were maintained in six zoological collections in the United Kingdom and Germany. One of these male orang-utans, Saleh, was studied twice (before and after onset of flange growth), and thus the sample is regarded as comprising 25 individuals. The male gorillas belonged to three different age-status classes (juvenile, adolescent, adult) and were maintained in 10 zoological collections in the UK, Germany, and the Netherlands. Twenty-four individuals were studied, but one, Kum, was used twice (at ages 1 and 2 years); the data obtained are thus taken as representing 25 male gorillas. All animals were housed in visual, auditory, and olfactory contact with other members of the same species. All caging had indoor and outdoor areas and bars for climbing. The animals were fed several times a day with dried pellet food and fresh fruit and vegetables of the season, and most had water *ad libitum*.

Collection of Samples

Urine was collected either by syringe from the floor immediately after micturition, or in a special trap fitted into the cage drain. Normally six or more isolated samples were collected on different days from each male within a short time period. The routine collection of early morning samples minimized variation in hormone levels due to possible circadian changes in circulating testosterone (e.g., Goodman et al., 1974; Resko and Eik-Nes, 1966) or estrogens. Weekly samples were collected from two orang-utans undergoing flange development during the 2-year period of the study.

Samples were stored at −20°C for determination of salt concentration and analysis of androgens and estrogens by radioimmunoassay. The main androgen measured was testosterone, and therefore the term testosterone is used throughout when referring to androgens.

Radioimmunoassay of Urinary Steroids

Urinary estrogen and testosterone in both species were measured by the procedure described previously by Bonney et al. (1979) and Kingsley (1982).

Validations

The antiserum used for the total estrogen assay was raised in a goat to estradiol-17β-succinyl-bovine serum albumin (491/9) (supplied by Dr. Furr, ICI, Pharmaceutical Division, Macclesfield, UK) and diluted to give approximately 20.0% binding of radioligand [2,4,6,7(n)-^{3}H]estradiol (specific activity range 85–115 Ci/mmole: Amersham International, UK) in the absence of unlabeled steroid. Antiserum cross-reactivities with other estrogens were estradiol-17α, 50.3%; estrone, 27.9%; estriol, 2.8%. Cross-reactivity with androstenedione, testosterone, progesterone, and pregnanediol was <0.1%.

Intra- and interassay coefficients of variation were 4.7% (N = 10) and 9.4% (N = 7), respectively, and were calculated by repeated assay of a hydrolyzed urine pool in a single assay and in separate assays. Accuracy was determined by measuring added known amounts of estradiol (1.25, 0.625, 0.312, 0.156 ng/tube) to hydrolyzed urine (diluted to 1:10 and treated with charcoal to remove endogenous estradiol). A linear regression was found for estimated estradiol against added estradiol ($y = 0.93x - 0.023$), the slope of which was not significantly different from the expected value of 1.

The antiserum used for the testosterone assay was raised in a rabbit to testosterone-3-bovine serum albumin (supplies by Dr. Horth, G. D. Searle & Co. Ltd., High Wycombe, UK) and diluted to give approximately 30.0% binding of radioligand [1,2,6,7(n)-^{3}H]testosterone (specific activity range 81–93 Ci/mmole: Amersham International, UK) in the absence of unlabeled testosterone. Antiserum cross-reactivities were 5α-dehydrotestosterone (DHT), 35.3%; 19-nortestosterone, 5.1%; androstenedione, 3.4%; dehydroepiandrosterone, 0.3%. Cross-reactivity with 5β-DHT, estrone, estradiol-β, estradiol-α, estriol, and progesterone was <0.1%.

Intra- and interassay coefficients of variation were 6.5% ($N = 10$) and 10.2% ($N = 9$), respectively. These were calculated by the same procedure as the estrogens. Accuracy was also determined as above, adding known amounts of testosterone (1.0, 0.5, 0.25, 0.125 ng/tube) to the treated hydrolyzed urine (diluted 1:20). A linear regression was found for estimated testosterone against added testosterone ($y = 1.013\,x - 0.029$), the slope of which was not significantly different from the expected value of 1.

Assay of NaCl

Creatinine could not be detected in samples from one of the zoos (due to faulty refrigeration), and the normal procedure of relating the hormone levels of the isolated samples to urinary creatinine (to take into account the concentration of the sample) could not be followed. The hormone levels were instead related to the pure NaCl equivalent of urinary salts concentration, the salts level reflecting the concentration of the urine. The results obtained using NaCl were of the same pattern as those obtained using creatinine (calculated in samples suffering no creatinine breakdown) (Kingsley, 1981).

The freezing point depression of each sample (200 μl) was measured using an Osmometer (Advanced Instruments Inc.), allowing the osmolarity (the concentration of salts) of the sample to be determined. NaCl forms a major part of urinary salts, and therefore the osmolarity was converted totally to NaCl concentration (by means of pure salt standards) for ease of calculation.

Intra- and interassay coefficients of variation were 2.0% ($N = 8$) and 2.3% ($N = 7$), respectively, as established by repeated measurement of the untreated urine pool in a single assay, and in separate assays.

Statistics

Hormone levels were related to urinary salt concentration and the results averaged for each male and used in subsequent analysis. When fewer than three urine samples were collected from any one male, the results were excluded from the analysis, as the spread of values was such that their means may not have been typical.

The results from the two male orang-utans (Saleh and Dennis) undergoing flange development during the period of study were pooled into 3-monthly time periods and the means calculated. Results between the age-status groups of the individual species or between the 3-monthly time periods of the flange-developing orang-utans were compared by multiple comparison statistics (Fisher's LSD). The level of significance chosen was $p < 0.05$.

Measurement of Flange Growth Rate in Orang-utans

Full-face photographs were taken at monthly intervals, when possible, of the two study animals (Dennis and Saleh) in the process of flange development and the facial proportions examined. Since it was not possible to photograph the animals from a fixed distance (it was difficult to get full-face photographs at all), the photographs were of different sizes. The rate of flange growth was therefore calculated by dividing the total width of the face (TF) by the biorbital distance (BB) (the distance between the outer edges of the eyes) to obtain a ratio of flange size. The total width of the face increases with flange development, whereas the absolute postorbital breadth does not seem to do so (Biergert, 1979; Schultz, 1941). As a check on the method, facial photographs from 49 animals from zoological collections throughout the world were measured; the results showed that flanged males did indeed have the expected greater facial ratio than nonflanged males (Kingsley, 1982).

RESULTS

Hormonal Differences in Age-Status Classes

Orang-utan

Mean urinary concentrations of testosterone (±SD) in the 25 male orang-utans are presented in Fig. 9–1. Nonflanged adults had significantly higher testosterone levels (mean value 0.70 ng/mg NaCl) than juveniles (mean value 0.30 ng/mg NaCl) ($p < 0.05$, Fisher's LSD), while flanged adults had significantly higher testosterone concentrations (mean value 1.68 ng/mg NaCl) than nonflanged adults ($p < 0.01$, Fisher's LSD). Adults in the process of flange development had testosterone concentrations intermediate between those of flanged and nonflanged adults (mean value 1.33 ng/mg NaCl). One 5-year-old male, Dick, had appreciably higher testosterone concentrations than his peers. His testosterone levels were similar to those of nonflanged adults.

Gorilla

Figure 9–2 shows the mean urinary concentrations of testosterone in 24 gorillas. Mean testosterone levels in adolescents were 4 times higher than those in juveniles (mean values 0.86 and 0.21 ng/mg NaCl, respectively), but this difference did not reach statistical significance. However, mean urinary testosterone levels in adults (2.07 ng/mg NaCl) were significantly higher than those in adolescents ($p < 0.05$, Fisher's LSD).

Hormonal Correlation of Flange Development in Orang-utans

Figure 9–3 illustrates the pooled 3-monthly average levels of urinary estrogen and testosterone in the two males that were in the process of flange development.

Figure 9–3A shows the results from Saleh. Unfortunately, flange growth data are missing from periods 3 and 4 due to difficulties in obtaining full-face photographs, but half-face photographs show that flanges started to develop at the end of period 4. They rapidly continued to develop during periods 5 and 6 (Fig. 9–4 shows the preliminary flange growth at the beginning of period 5 and the relatively large flanges at the end of period 6). Testosterone concentrations increased during the 3-month period preceding the commencement of flange growth (period 3) and continued to increase in conjunction with flange development for the next 6 months (periods 4 and 5). Testosterone secretion then remained steady while the flange ratio continued to increase. Mean urinary testosterone levels were compared between time quarters using Fisher's LSD. The only statistically significant differences were be-

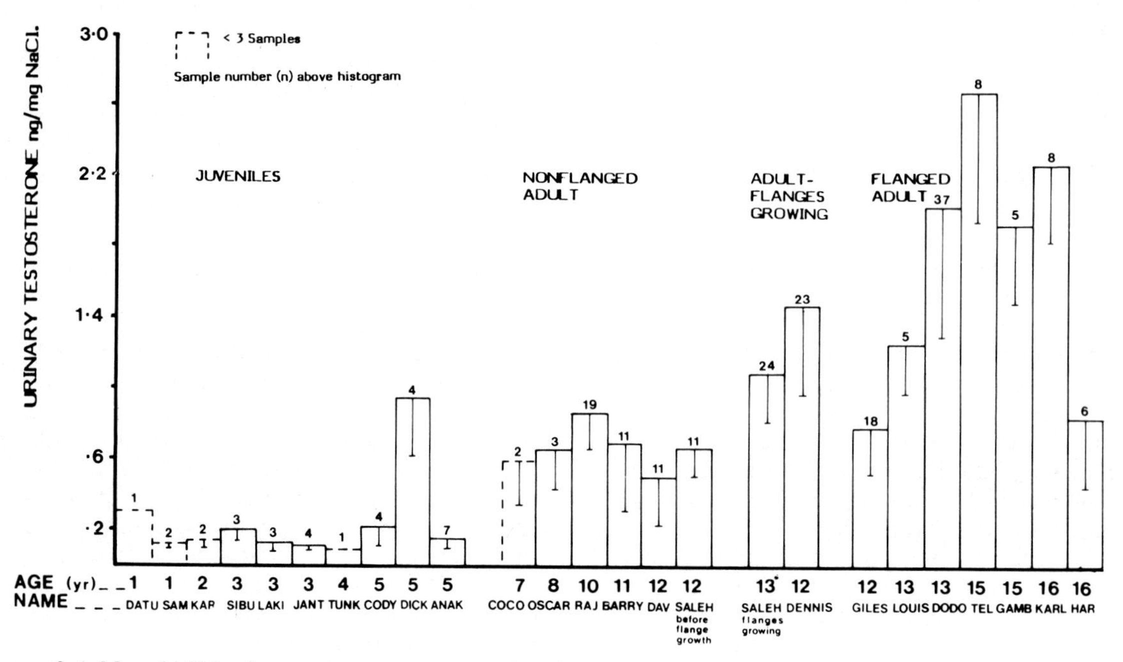

FIG. 9–1. Mean (±SD) urinary testosterone concentrations (ng/mg NaCl) in male orang-utans of all ages.

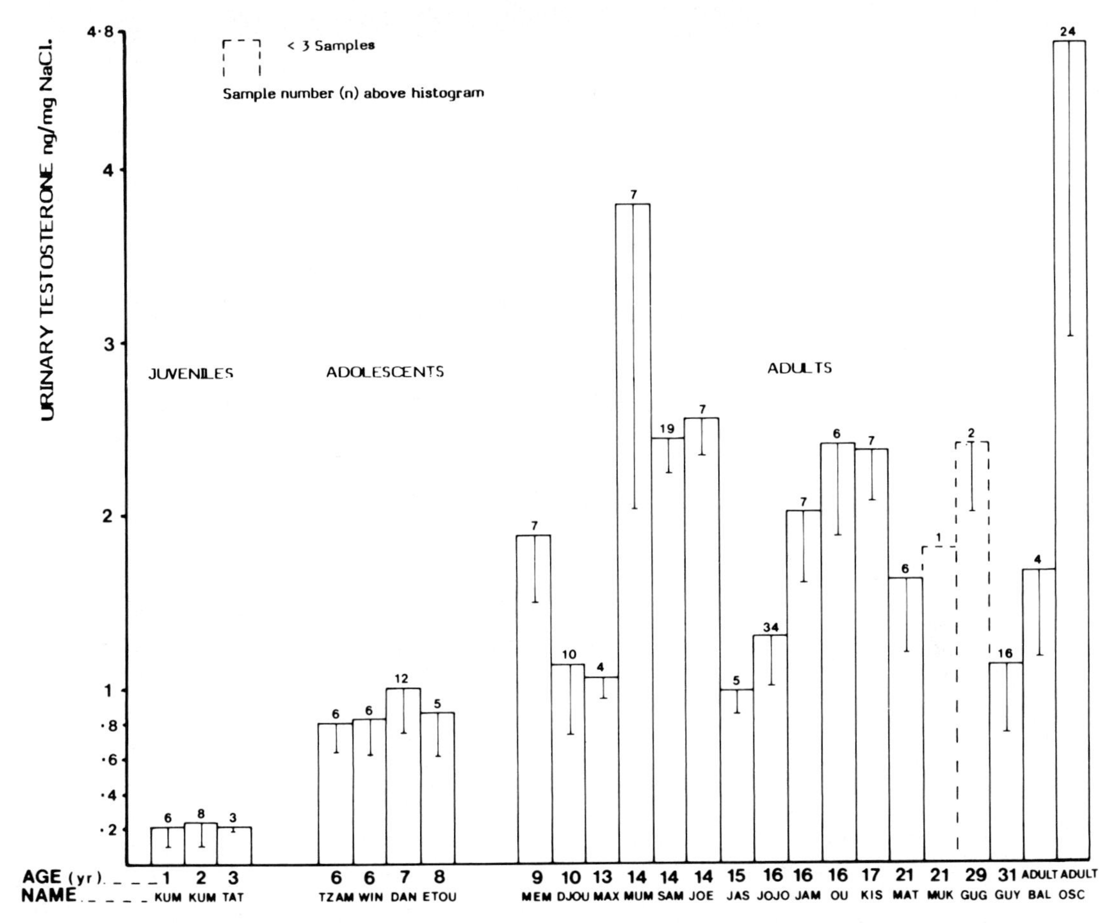

FIG. 9–2. Mean (±SD) urinary testosterone concentrations (ng/mg NaCl) in male gorillas of all ages.

tween period 1 (mean value 0.65 ng/mg NaCl) and period 4 (mean value 1.04 ng/ml NaCl) ($p < 0.02$) and period 1 and period 5 (mean value 1.11 ng/mg NaCl) ($p < 0.001$). Estrogen levels increased markedly during the 6-month period prior to flange development but then decreased rapidly to earlier levels during period 5. Mean levels between periods 1 and 3 (0.40 and 1.15 ng/mg NaCl, respectively), and 4 and 5 (1.42 and 0.48 ng/mg NaCl, respectively) were significantly different ($p < 0.01$). This later rapid decrease in estrogen levels coincided with Saleh's olfactory, auditory, and visual separation from a fully flanged male.

Figure 9–3B illustrates the results from Dennis. He had started to develop flanges during period 1 (see also Fig. 5A) and was almost fully flanged within a period of 1 year. Figures 9–5B and C show the flanges at the end of period 2 and the beginning of period 4, respectively. However, this rapid flange growth was not necessarily correlated with increasing levels of urinary testosterone, because although the mean levels between periods 1 and 8 (when flanges were fully developed) were significantly different from each other (1.39 and 2.08 ng/mg NaCl, respectively) ($p < 0.01$, Fisher's LSD), so were those between periods 4 and 8 (1.44 and 2.08 ng/mg NaCl) ($p < 0.05$, Fisher's LSD). Estrogen concentrations remained similar throughout the eight quarters.

DISCUSSION

In both the orang-utan and the gorilla, concentrations of urinary testosterone were higher in adolescents than in juveniles and higher in adults than in adolescents. These results are similar to those obtained for humans and chimpanzees (August et al., 1972; Frasier et al., 1969; Kraemer et al., 1982; Martin et al., 1977; McCormack, 1971). For example, Martin et al. (1977) found that adolescent chimpanzees (7–10 years of age) had serum testosterone levels 13 times higher than those of juveniles (1–6 years

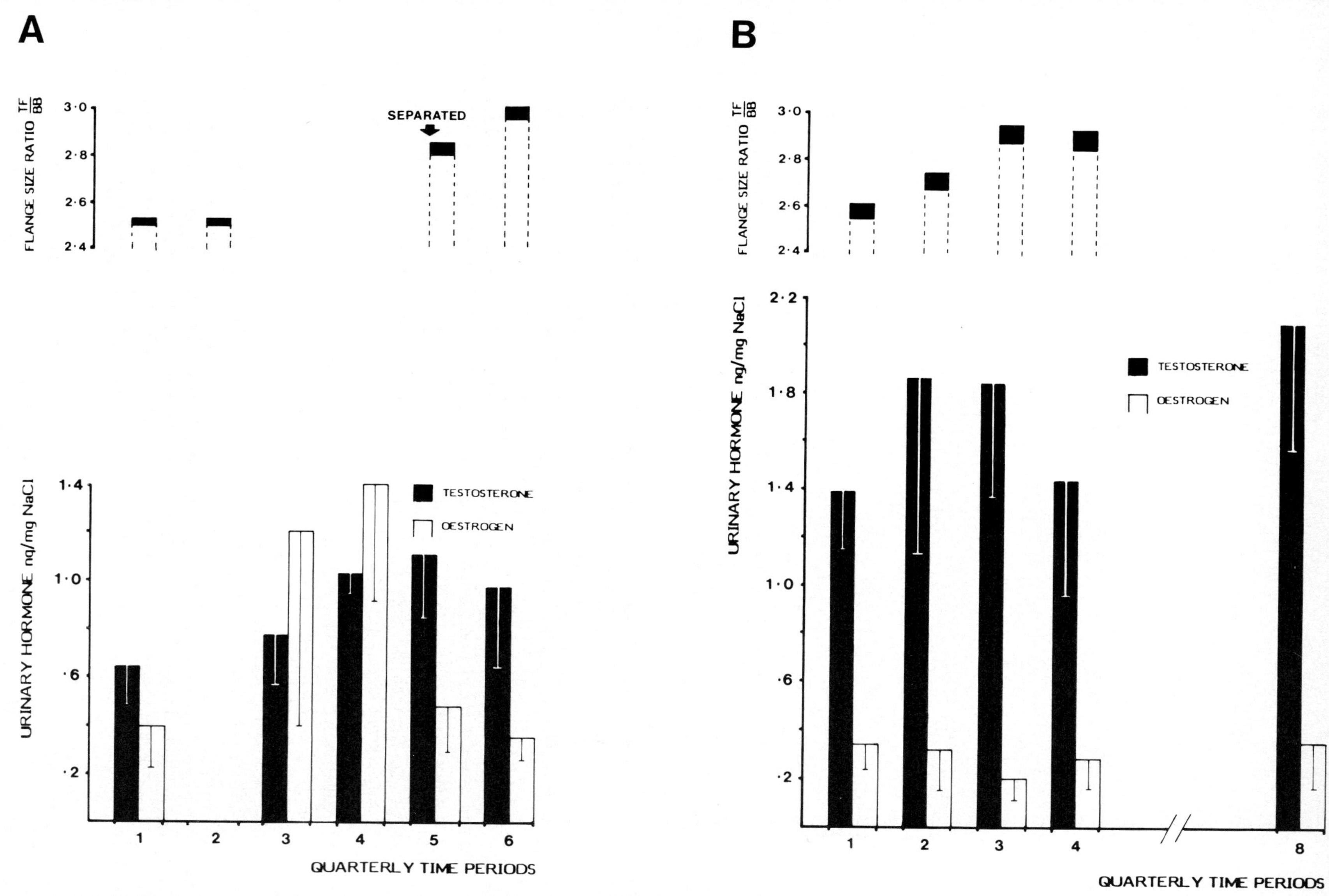

FIG. 9–3. Relationship between flange size and mean (±SD) quarterly urinary estrogen and testosterone concentrations in adult male orang-utans undergoing flange development: (A) Saleh, (B) Dennis.

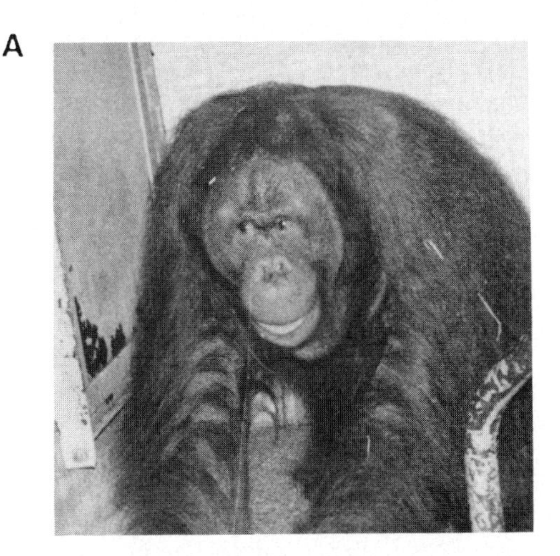

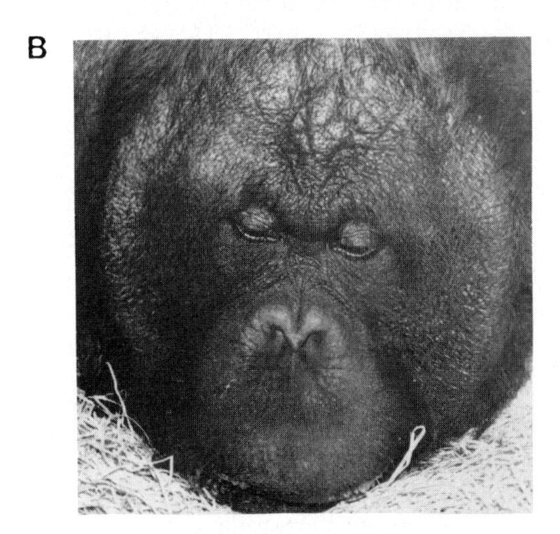

FIG. 9–4. Saleh: flanges at (A) beginning of period 5 (rudimentary), (B) end of period 6.

of age), while adults (11+ years) had over twofold increases in testosterone when compared with adolescents. In the present study, urinary testosterone levels in adolescent gorillas (6–8 years old) were 4 times those in juveniles (1–3 years old), while those in adults (9+ years) were 2.4 times those in adolescents. Orang-utan adults had urinary testosterone concentrations 2.4 times higher than subadult males (i.e., nonflanged), while subadult levels of urinary testosterone were 2.3 times higher than those of the juveniles. The relatively high levels of urinary testosterone in the juvenile orang-utans and gorillas, when compared with serum or plasma levels in juvenile chimpanzees or boys (August et al., 1972; Frasier et al., 1969; Martin et al., 1977), may be a function of the source of the urinary metabolites measured.

One 5-year-old orang-utan had urinary testosterone levels well within the range of those of nonflanged adult males, suggesting he had already reached the adolescent stage. Unfortunately, weight or testes size data were not available from him to confirm this suggestion. However, Rijksen (1978) has reported that the adolescent stage can be reached by 5 years of age.

Although none of the adolescent gorillas and only one of the nonflanged orang-utans (Saleh)

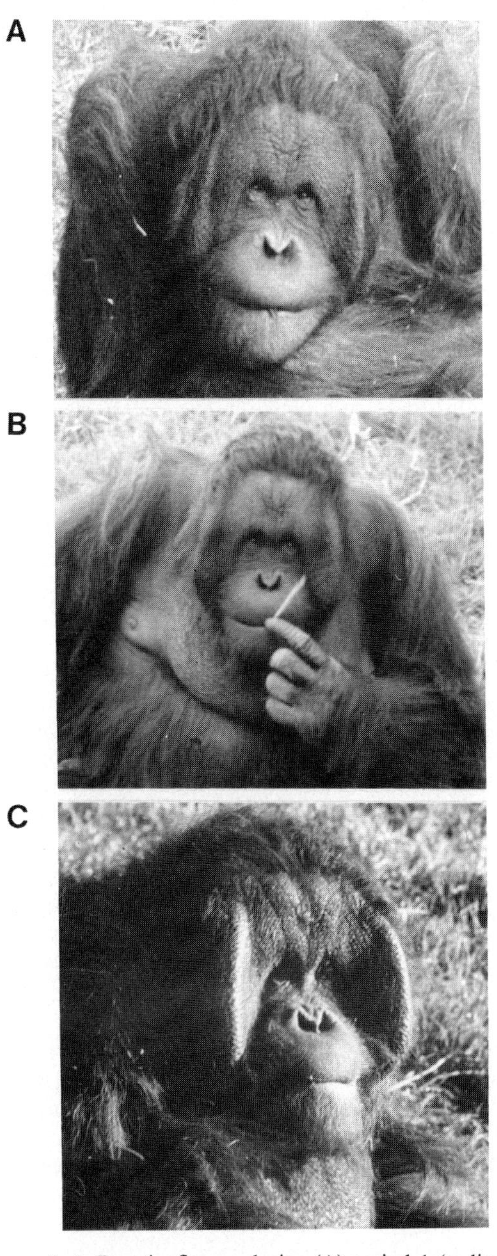

FIG. 9–5. Dennis: flanges during (A) period 1 (rudimentary), (B) period 2, (C) period 4 (almost fully developed).

had bred at the time of the study, it is possible that the increased testosterone levels in these animals signified that the animals were capable of producing sperm. Everts (personal communication) had reported that a 6½-year-old orang-utan had sired an infant at Hanover Zoo, while Dixson et al. (1982) reported spermatogenesis in three male orang-utans which were 6.4–7.7 years old. In addition, Fooden and Izor (1983) reported almost fully developed testes in a 6.8-year-old male orang-utan. Furthermore, a captive-bred gorilla at Bristol Zoo (Daniel, coded DAN in Fig. 2) sired an infant at 8 years of age when still black-backed and underdeveloped.

In this study, adulthood in gorillas was taken to be 9 years of age (after Schaller, 1963), although the 9- and 10-year-olds included in this category had not developed the typical silver back of fully adult male gorillas. Gorilla adults were found to have significantly higher levels of urinary testosterone than adolescents and, with the exception of the two animals cited above, all had silver backs and were sexually mature. However, only four of the animals had bred at the time of the study (MUM, SAM, JAM, and KIS), and these had an average testosterone concentration of 2.65 ng/mg NaCl (range 2.01–3.79). At least one other animal, the 10-year-old DJOU, subsequently bred at the age of 12 years.

The testes of two of the adult gorillas which died sometime after the study were examined by Dixson et al. (1980). These testes (of GUY and JOJO) were found to be grossly atrophied. Both animals had relatively low urinary concentrations in comparison to the breeding adults (1.13) and 1.28 ng/mg NaCl, respectively), and such low levels of androgens would be expected in animals with testicular dysfunction. It is possible that low urinary testosterone levels in some of the other gorillas in this cohort (e.g., MAX, JAS) are also indicative of testicular atrophy.

The flanged male orang-utans had significantly higher levels of urinary testosterone than nonflanged adults, while those in the process of flange development had intermediate levels. This suggests that increases in urinary testosterone concentrations are associated with flange development. Data from Saleh support this hypothesis, in that an increase in urinary testosterone levels preceded flange growth by a few months. Unfortunately, Dennis's flanges had started to develop at the beginning of the study and urine samples had not been collected prior to that time. It is possible that he also had lower levels of urinary androgens during the months preceding early flange development, but this will never be known. Once flanges started to grow rapidly, associated increases in testosterone no longer occurred in either animal. More males in the process of flange development need to be studied to confirm these results.

Urinary estrogen concentrations were also measured in these two males, as it had been found that flanged orang-utans had significantly higher levels of estrogens than nonflanged animals (Kingsley, 1981, 1982). The data on this hormone were equivocal, however. Dennis had relatively low levels of estrogens throughout the 2-year study, while Saleh experienced a peak of estrogen excretion during the 6 months prior to rapid flange development. It is possible that this surge of estrogen excretion was associated with the triggering of flange development and that concentrations decreased to preflange levels once flange growth started. This hypothesis would be consistent with the low levels of estrogens found in Dennis's urine. Alternatively, Kingsley (1982) has suggested that the high urinary estrogen levels found in Saleh were associated with stress produced by the presence of a fully flanged adult male in an adjacent cage. Only when Saleh was removed from the vicinity of this male did estrogen levels decrease and rapid flange development commence.

In both animals, flange development, once begun, proceeded rapidly. Both males reached full flange expression within a year or so. This rapid development is consistent with previous reports (Brandes, 1931; Brandes, 1939; Ulmer, 1958).

Unlike the case in the gorilla, lower levels of urinary testosterone in male orang-utans were not associated with infertility. The only non-breeding flanged adults in this study were TEL and LOU, and both were found to have relatively high urinary testosterone concentrations.

CONCLUSIONS

Male orang-utans and gorillas have changes in testosterone levels throughout life similar to those of humans and chimpanzees. Adolescents or subadults have higher urinary testosterone concentrations than juveniles, while adults have higher concentrations than adolescents/subadults. In addition, the actual process of flange development in the orang-utan may be accompanied by an increase in testosterone and estrogen excretion, but only during the early stages.

ACKNOWLEDGMENTS

I am very grateful to the directors, curators, and ape staff of Arp, Chester, Bristol, Flamingo Park, Frankfurt, Howletts, Jersey, London, Twycross, Wassenaar, and Weybridge zoological collections for their help and cooperation in the study. I would also like to thank Professor Bob Martin and Drs. Alan Dixson and Rosemary Bonney for their help on various aspects of the study, Dennis Tanner for preparing all the figures, and Lyn Arnold for typing the manuscript. This research was supported by the Medical Research Council, England, while the author was employed at the Wellcome Laboratories of Comparative Physiology, Institute of Zoology, Zoological Society of London, Regents Park, London NW1 4RY.

REFERENCES

August, G. P., Grumbach, M. M. and Kaplan, S. L. 1972. Hormonal changes in puberty: III. Correlation of plasma testosterone, LH, FSH, testicular size, and bone age with male pubertal development. *Journal of Clinical Endocrinology, 34:*319–326.

Biegert, J. 1979. Anatomy and morphology of the orang-utan. Paper read to the Workshop on the Conservation of the Orang-utan, Rotterdam, October 1979.

Bonney, R. C., Dixson, A. F. and Fleming, D. 1979. Cyclic changes in the circulating and urinary levels of ovarian steroids in the adult female owl monkey *(Aotus trivirgatus). Journal of Reproduction and Fertility, 56:*271–280.

Brandes, G. 1939. Buschi; Vom Orang-Saügling zum Backenwülster. Quelle and Meyer, Leipzig.

Brandes, R. 1931. Über den Kehlkopf des Orang-Utan in verschiedenen Altersstadien mit besonderer Berücksichtigung der Kehlsackfrage. *Morphologische Jahrbuch, 69:*1–61.

Collins, D. C., Nadler, R. D. and Preedy, J. R. K. 1981. Adrenarche in the great apes. *American Journal of Primatology, 1:*344.

Conaway, C. H. and Sade, D. S. 1965. The seasonal spermatogenic cycle in free ranging rhesus monkeys. *Folia Primatologica, 3:*1–12.

Deniker, J. and Boulart, R. 1885. Les sacs laryngiens, les excroissances adipeuses, les poumons, le cerveau etc des orang-utans. *Nouvelles Archives du Museum d'Histoire Naturelle (series 3), 7:*35–56.

Dixson, A. F., Knight, J., Moore, H. D. M. and Carman, M. 1982. Observations on sexual development in male orang-utans. *International Zoo Yearbook, 22:*222–227.

Dixson, A. F., Moore, H. D. M. and Holt, W. V. 1980. Testicular atrophy in captive gorillas *(Gorilla g. gorilla). Journal of Zoology, 191:*315–322.

Donovan, B. T. and Werff ten Bosch, J. J. Van de. 1965. *Physiology of Puberty.* Edward Arnold, London.

Fooden, J. and Izor, R. J. 1983. Growth curves, dental emergence norms, and supplementary morphological observations in known-age captive orang-utans. *American Journal of Primatology, 5:*285–301.

Frasier, S. D., Gafford, F. and Horton, R. 1969. Plasma androgens in childhood and adolescence. *Journal of Clinical Endocrinology and Metabolism, 29:*1404–1408.

Goodman, R. L., Hotchkiss, J., Karsch, F. J. and Knobil, E. 1974. Diurnal variations in serum testosterone concentrations in the adult male rhesus monkey. *Biology of Reproduction, 11:*624–630.

Kelch, R. P., Jenner, M. R., Weinstein, R., Kaplan, S. L. and Grumbach, M. M. 1972. Estradiol and testosterone secretion by human, simian and canine testes, in males with hypogonadism and in male pseudohermaphrodites with the feminizing testes syndrome. *Journal of Clinical Investigation, 51:*824–830.

Kingsley, S. R. 1981. The Reproductive Physiology and Behaviour of Captive Orang-utans *(Pongo pygmaeus).* Ph.D. Thesis, University of London.

Kingsley, S. R. 1982. Causes of non-breeding and the development of the secondary sexual characteristics in the male orang utan: a hormonal study. *In* The Orang-utan: Its Biology and Conservation, pp. 215–229, ed. L. E. M. de Boer. W. Junk, The Hague.

Kraemer, H. C., Horvat, J. R., Doering, C. and McGinnis, P. R. 1982. Male chimpanzee development focusing on adolescence: integration of behavioural with physiological changes. *Primates, 23:*393–405.

MacKinnon, J. 1974. The behaviour and ecology of wild orang-utans *(Pongo pygmaeus). Animal Behaviour, 22:*3–74.

Martin, D. E., Swenson, R. B. and Collins, D. C. 1977. Correlation of serum testosterone levels with age in male chimpanzees. *Steroids, 29:*471–481.

McCormack, S. A. 1971. Plasma testosterone concentration and binding in the chimpanzee: effect of age. *Endocrinology, 89:*1171–1177.

Resko, J. A. and Eik-Nes, K. B. 1966. Diurnal testosterone levels in peripheral plasma of human male subjects. *Journal of Clinical Endocrinology, 26:*573–576.

Rijksen, H. D. 1978. A Field Study on Sumatran Orang Utans (*Pongo pygmaeus abellii,* Lesson, 1827): Ecology, Behaviour and Conservation. H. Veenman and Zonen, Wageningen.

Schaller, G. B. 1963. The Mountain Gorilla: Ecology and Behavior. University of Chicago Press, Chicago.

Schultz, A. H. 1941. Growth and development of the orang-utan. *Contribution to Embryology, 29:*59–110.

Swerdloff, R. S. and Rubin, T. 1978. Psychological and endocrinological changes in puberty. *In* Perspectives in Endocrine Psychobiology, pp. 287–308, ed. F. Brambilla, P. K. Bridges, E. Endoczi and G. Heuser. John Wiley and Sons, London/Toronto.

Tanner, J. M. 1978. Foetus Into Man: Physical Growth From Conception to Maturity. Fletcher and Son, Norwich.

Ulmer, F. A. J. 1958. Rusty becomes a backenwulster! *America's First Zoo, 10:*7.

10

External Genitalia

JEREMY F. DAHL

The inadequacy in our knowledge of the external genitalia of orang-utans, especially of the adult female, derives from at least three major factors: the sparsity of both living and postmortem specimens, the problems posed in the examination of preserved genital tissues (see Matthews, 1946–47), and the restricted access to living specimens that might be available for study. This has had an impact not only on our ability to assess relatedness among the hominoids, but also on our ability to explore theoretical explanations for the diversity that is present. Particularly pertinent to the latter is the application of sexual selection theory to explain correlations between patterns of sexual behavior and genital anatomy (e.g., penile displays and conspicuous penes) as these patterns vary among the great apes (Harcourt, 1981; Harcourt and Stewart, 1977; Short, 1977, 1979, 1981). Verification of such explanations with respect to *Pongo* is dependent on additional evidence, since most available measurements of the penis are not for the entire pars libera and no direct measurements are available for the erect organ. In an attempt to clarify what is known of the external genitalia of orang-utans, and to extend our appreciation of their anatomical structure, I shall review the available literature and report observations made on 13 captive individuals maintained at the Yerkes Regional Primate Research Center.

PREVIOUS STUDIES

The Female

In a limited description of three adult orang-utans, Wislocki (1936:333) mentioned that in two, there were no well-defined elevations in the sparsely haired field flanking the vulva, but that the third had "pronounced, elongated cushions, containing fatty tissue" remarkably similar to the labia major illustrated for a juvenile chimpanzee (Wislocki, 1936, Fig. 26). He concluded that female orang-utans have variably conspicuous labia majora which, for the most part, lie anterior to the pudendal cleft; it is these structures that appear to swell during pregnancy, but not during intermenstrual intervals. This swelling was first noted by Fox (1929) and subsequently by Schultz (1938a), who failed to identify the edematous tissues. Graham-Jones and Hill (1962) report a swelling also, while Lippert (1974) provides both photographs and details of its presence along with an identification of the structures in question as labia majora; these structures become edematous 20–40 days after conception and remain variously enlarged for the remainder of pregnancy. The illustrations in Schultz (1938a) and Lippert (1974) appear to be the only representations of the *adult* female genitals available. The remaining literature pertaining to the adult female genitals concerns the identification by Pohl (1928, Plate III, Fig. 2) and Gerhardt (1909) of an os clitoridis.

The arrangement of the perivaginal structures is known from a number of accounts of immature individuals (Fisher, 1898; Friedenthal, 1910; Pocock, 1925; Sonntag, 1924; Wislocki, 1936; also see Rijksen, 1978, Fig. 127). In Fischer's illustrations (1898, Plate XVII, Fig. 13–15), large, fleshy structures lateral and ventral to the vaginal opening are identified as labia majora; they appear united ventrally to form a mons pubis. A relatively small conical structure between the labia majora is identified as the prepuce; it forms a sac-like entity about the clitoris. The tissues of the prepuce extend but a short distance dorsal to the clitoris and reunite at a labial commissure beyond the small vaginal aperture. A midsagittal raphe extends to the anus. Additional tissues extend from each side of the perivaginal tissues and are identified as "Querfalte zwischen grossen and kleinen Labien" (ibid.: 217). The clitoris shown is conspicuously grooved on its entire undersurface and is differentiated into a corpus and a glans of about equal length. Although the presence of a glans clitoridis was not reported, the descriptions by Sonntag (1924), Pocock (1925), and Friedenthal (1910) are in general agreement with Fischer's

(1898); Friedenthal's figure does not illustrate a conspicuous prepuce, but perivaginal folds (labia minora) are obvious, as are the plump lobes of the labia majora.

Wislocki (1936) described the structure of the external genitalia of a fetus and a newborn orang-utan. The fetus (sitting height 196 mm) had a small, almost circular rima pudendi that was flanked by protruding, puckered labia minora as well as a small, protruding clitoris. These structures are difficult to distinguish in the published illustration (Fig. 27), which more clearly figures two crescentic cushions of fatty tissue lying ventrolateral to the vulva, which Wislocki identified as the labia majora. The newborn female orang-utan (sitting height 264 mm) had a "puckered, rosette-like configuration of the labia minora" (Wislocki, 1936:333) identical to that of the fetus. In addition, the newborn also had an extension of skin on each side of the labia minora that traversed the pubic region, tapering imperceptibly into the tissue of the groin. Wislocki concluded that these structures (probably identical to those figured by Fischer) were homologous in position to a fold attendant to the scrotal area in an infant male.

Bischoff did not find labia majora in the immature specimen he examined and illustrated (Bischoff, 1879, Fig. XVII); consequently he argued that the presence of labia majora in human females but their absence in great ape females refuted Huxley's (1893) arguments applying Darwinian theory and establishing the close relatedness of apes and humans. Reviewing the subsequent debate, Bingham (1928) agreed with Schultz's (1927:26) conclusion that the labia majora "are laid down in fetal life, but, whereas in man they persist throughout growth, in the anthropoids they undergo a process of atrophy, which in many cases leads sooner or later to their complete disappearance." This suggestion was consistent with the notion that "each somatic characteristic of man is the result of a retardation or an arrest of a definite part of the developmental process" (Bolk, 1921:59) and, thus, was in keeping with the broader "fetalization" theory of human evolution (see Gould, 1977). It is surprising, however, that the atrophied labia majora in adult great apes became established as "fact," since definitive studies on adult female orang-utans did not predate the limited work of Wislocki (1936).

The Male

The anatomy of the external genitalia of the male orang-utan has received considerably more attention than that of the female (Chapman, 1880; Dixson et al., 1982; Fick, 1895; Friedenthal, 1910; Hill, 1939; Mayer, 1856; Pohl, 1928; Pousargues, 1894, 1895; Retterer, 1915; Sandifort, 1839; Wislocki, 1936). Clear reports on the arrangement of the internal tissues are provided by Pohl (1928) and Pousargues (1895).

Wislocki (1936) described the scrotal areas of four Bornean orang-utans: an infant (sitting height 300 mm), a juvenile (sitting height 550 mm), and two older specimens (sitting heights 687 and 770 mm). Although he did not take measurements, Wislocki did note the small size in all of the scrota and testes. The infant's scrotum was divided into two postpenial wrinkled areas that were separated by a stout median raphe extending from penis to anus; the small, flattened testes were found in low folds of skin just anterior to the borders of the scrotal area and lateral to the penis. The elevations of skin containing the partially descended testes were symmetrically placed on either side of the penis and tapered obliquely and superiorly toward the inguinal region. Wislocki thought these folds might be homologous to the lateral extensions of the labia minora of a young female (see above). In the older Bornean specimens the testes were completely descended into the postpenial scrotum, which was compact, bulged without being pendulous, and was similar in arrangement to those of the subadult individual figured by Friedenthal (1910).

That the testes are indeed relatively small in *Pongo* is supported by actual measurements of weight and volume (Table 10–1): for three *P. pygmaeus pygmaeus,* with a mean body weight of 101.3 kg, a mean testicular volume of 33.02 cm^3 is determined when volume is taken as one-sixth the width squared times length ($V = \frac{1}{6} W^2L$). For the subadult individuals ($N = 3$), with a mean body weight of 48.6 kg, the mean testicular volume was 18.39 cm^3. These data contrast with the volume of 7.48 cm^3 calculated from Short's (1977, 1979) data for an adult of a *P.p. abelii* weighing 114 kg and the volume of 6.3 cm^3 calculated from Hill's (1939) data for a large *P.p. abelii.* It may even be the case that the Sumatran form has smaller testes than the Bornean variety. When making a comparison between Bornean specimens and a Sumatran specimen (sitting height 625 mm), Wislocki (1936) remarked that the testes of the latter were different: they lay in a suprapubic position, were extremely small, and the scrotal area was unpigmented.

The external penis or pars libera is delineated into two components (Friedenthal, 1910; Pousargues, 1895): a proximal part enclosed within a highly folded prepuce when at rest, and a distal

TABLE 10–1. Measurements of the Penis and Testes of Orang-utans Derived from Previous Studies

Source	Subspecies/ Age Class	Body Weight (kg)	Testes Volume (cm³)	Testes Weight (g)	Testes Body Weight (%)	External Penis Length (cm)
Chapman (1880)	?/Juvenile?	—	—	—	—	4.4
Pousargues (1895)	?/Adult	—	17.1[1]	—	—	2.7[2]
Schultz (1938b)	?/Juvenile?	29.9	—	29.0	0.097	—
	?/Juvenile?	34.5	—	23.4	0.068	—
	?/Adult	72.2	—	32.4	0.045	—
	?/Adult	77.1	—	38.2	0.050	—
Hill (1939)	S/Adult	—	6.3	25.0	—	12.0
Short (1981)	B/Subadult	58.0	15.0	—	—	4.0[2]
	S/Adult	114.0	7.5	—	—	4.0[2]
Dixson et al. (1982)	B/Infant	1.6	—	1.6	0.097	1.4
	H/Subadult	41.8	21.0	17.3	0.041	2.5[2]
	B/Subadult	42.0	22.7	—	—	2.7[2]
	B/Subadult	45.7	17.5	—	—	3.0[2]
	B/Adult	85.0	33.8	—	—	—
	B/Adult	100.0	36.9	40.8	0.041	2.9[2]
	B/Adult	119.0	28.3	33.3	0.028	2.4[2]

B, Bornean; S, Sumatran; H, Hybrid.

[1]Measured with epididymus.

[2]Pars distalis only measured.

portion that extends from the reflected prepuce as a bilaterally compressed column. This latter component, referred to here as the pars distalis, bears a defined median raphe or longitudinal ridge caudally that could be homologous to the human frenulum (see Retterer, 1915). With the exception of its distal apex, the surface of the pars distalis is covered in small transverse ridges that Pousargues (1895) demonstrated were composed of epithelial tissue reinforced by small cornified plates; the apex is pierced by a simple, midsagittally oriented urinary meatus (contra Sandifort, 1839).

Internally, the central areolar tissue of the corpus cavernosum assumes a cylindrical configuration, is small in diameter, and is enclosed by a highly developed fibroelastic sheath; at its widest part, these tissues form a column 12 mm thick, but at the apex, the sheath is quite thin. The corpus cavernosum bears a marked indentation at its tip within which the base of the os penis (see below) is located. The corpus spongiosum encloses the urethra along its entire length from the muscular region to its terminus; it persists as a narrow band of tissue for most of its length, but it is expanded around the os penis and the terminal portion of the urethra, the latter being somewhat widened and thus reminiscent of, if not homologous with, the navicular fossa of humans (see Pousargues, 1895). Pousargues further believed that the pars distalis was equivalent to the glans penis and, while he noted the absence of a true corona, he did indicate in his diagram the presence of a "pseudo-cournonne du gland."

A clearly demarcated glans penis, as observed in *Gorilla* (Duvernoy, 1855–56) and humans (Dickinson, 1949; Hart, 1908), has not been reported for any specimens of *Pongo.* Rather, the portion of the pars libera that lies distal to its junction with the prepuce has been identified as the glans (e.g., Dixson et al., 1982). This opinion is substantiated by Retterer's (1915) discussion of the os penis, and is presumably based on homology with the pars libera of humans in which the entire corpus penis is covered with skin. In Pohl's photograph (1928, Plate II, Fig. 8) of a juvenile orang-utan, however, the distal apex appears to be partially demarcated from the transversely ridged portion by its smooth surface and the slightly larger diameter of its domed tip.

An os penis has been described for *Pongo* by Crisp (1865), Gerhardt (1909), Pousargues (1895), Pohl (1928), and Dixson et al. (1982), who demonstrated that the penile bone enlarges during maturity as it does in other primates (Fooden, 1975). The os penis described by Pousargues was covered thickly in periosteum and was 1.7 cm in length; the macerated length of the os was 1.5 cm as compared to lengths of 1.5 and 1.46 cm for two ossa measured from radiographs of adults by Dixson et al. (1982). The os penis of the orang-utan is irregular in shape, expanding from about 0.3 cm at its base to 0.5 cm below its apex. Retterer (1915) discussed the position of the bone and its probable development from the corpus cavernosum rather than the c. spongiosum. He was emphatic about the nature of this structure: "Cet osselet glandulaire est un segment squelettique, formé de tissu compact et pourvu d'un canal médulaire" (p. 384). Hill (1939) found a partially ossified structure in a large Sumatran male.

The size of the erect penis has not been measured, and measurements of postmortem materials or anesthetized individuals have been restricted primarily to the pars distalis (see Table 10–1). From published information, it would appear that the pars distalis ranges in length from 2.3 to 4.0 cm. Because the pars distalis has been mistaken for the entire pars libera, and the measurement by Hill ignored, the external penis of *Pongo* has been described as being remarkably small for such a large animal (as it is in *Gorilla*); this notion has apparently been corroborated by field observation (MacKinnon, 1974; Schürmann, 1982). This notion, however, is in conflict with an illustration in Friedenthal's (1910) paper of an erect penis in which the proximal shaft is enclosed in the reflected prepuce and it is longer than the pars distalis (= glans penis of Dixson et al., 1982). Moreover, the outline of the os penis has been superimposed on the drawing in its position near the distal apex. If the illustration is accurate, the erect penis would have been 9.4 cm in length, assuming a length for the os of 1.4 cm, and this is consistent with the measurement by Hill (1939).

OBSERVATIONS ON LIVING SPECIMENS

Method

Subjects

All animals studied are part of the collection maintained at the Yerkes Regional Primate Research Center (see Table 10–2 for details). W. C. O. Hill was responsible for the identification of original breeding group members as either Bornean or Sumatran orang-utans, and all but one of his designations have been confirmed karyologically.

Data Collection

Most of the information was gathered by close examination and photography of individuals

TABLE 10–2. Details of the Subjects

Accession Number	Subspecies	Origin	Age	Body Weight (kg)			Offspring
				N	$\overline{X}$	SD	
Females							
0-22	S[1]	W	24[2] years	8	45.6	2.6	8
0-26	S	W	26[2] years	10	59.0	5.8	6
0-38	S	C	18 years	8	53.2	2.0	3
0-76	S	C	63 months	1	27.4	—	0
0-80	S	C	50 months	2	21.8	—	0
0-90	H	C	12 months	1	5.8	—	0
Males							
0-7	S	W	27[2] years	10	109.0[3]	6.0	11
0-21	S[1]	W	27[2] years	10	82.1[3]	4.7	9
0-41	H[1]	C	16 years	6	100.0	5.8	3
0-93	B	C	55 months	1	26.2	—	0
0-95	S	C	42 months	2	19.5	—	0
0-97	H	C	34 months	1	15.1	—	0
0-101	S	C	14 months	1	6.0	—	0

S, Sumatran; B, Bornean; H, hybrid; W, wild-born; C, captive-born.

Weight of immature individuals are for the day or week of observation.

[1]Karyotyped.

[2]Approximation.

[3]Weight prior to ill health; weights at observation close to 67 kg.

while they were sedated with katamine hydrochloride during their annual medical checkup. Individuals were not anesthetized for the purpose of this study, and the length of time an animal was unconscious was not changed significantly in order to facilitate a prolonged examination of the external genitalia. Therefore, only 10–15 min were available for taking notes, measurements, and photographs; if time was too short, measurements were foregone rather than taken inadequately. Measurements of testicular width and length were estimated by applying spring calipers to each testis within the scrotum and taking a skinfold measurement. Two males were examined immediately postmortem, when their testes were removed, weighed, and their length and breadth measured with dividers. The lengths of the pars libera and the pars distalis were measured with a millimeter rule, and diameters were taken using dividers. These measurements were made when the pars libera was pulled outward until the prepuce was completely reflected. Additional information and photographs were obtained while working with fully conscious individuals who were either young nursery animals accustomed to being handled, or caged older individuals trained to present in their home cages.

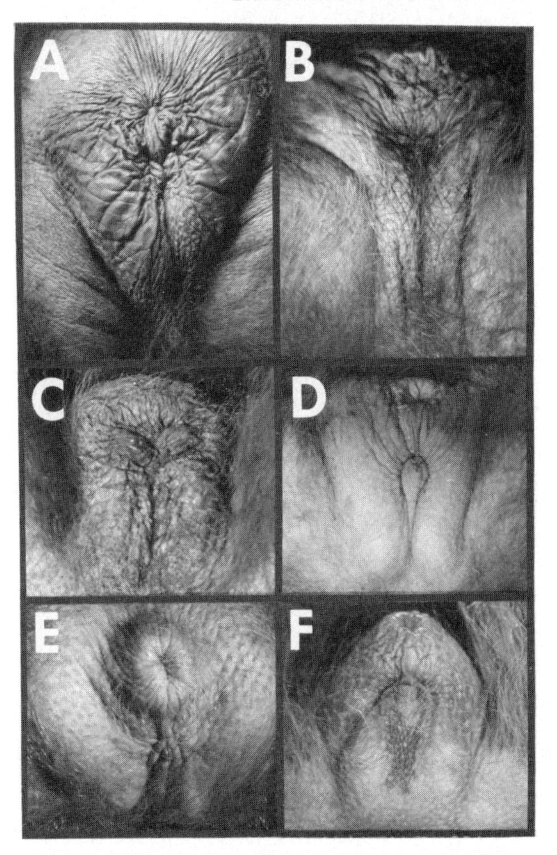

FIG. 10–1. General view of the perineum of six female orang-utans. (A) 0-26, adult, lying on side with thighs together. (B) 0-22, adult, lying on back with legs apart. (C) 0-38, adult. (D) 0-76, adolescent(?). (E) 0-80, juvenile(?). (F) 0-90, infant.

Results

The Female. The Sumatran orang-utan (accession number 0-26) was examined on two occasions when she was 26 years of age. A general view of her perineum is shown in Fig. 10–1A; the anus and vulval vestibule were 2.9 cm apart. Two large and sparsely wrinkled cushions of tissue, the labia majora, extended laterally and ventrally from the vulval vestibule, and blended into the circumanal tissues dorsally. The labia majora tapered between the thighs and were not accompanied by an anterior labial commissure resembling a mons pubis. There was a very sparse covering of hair but the variably wrinkled surface was essentially naked. With the labia majora of the vulval vestibule gently pulled back, a small vaginal aperture was exposed. The tissue immediately caudal to the aperture was a bright orange-pink but it was surrounded by a large unpigmented area. With the exception of the clitoris, which was a dull pink, the rest of the tissues were a uniform gray. The prepuce was a relatively small conical pouch that tapered ventrally and that appeared to be sharply demarcated from the perivaginal area posteriorly. The clitoris appeared adherent to the prepuce and was not manipulated during the observations; no evidence for its structure was obtained. No defined labia minora were recognizable.

A second adult wild-born Sumatran female (accession number 0-22) was examined once when she was 27 years old. A general view of her perinium is shown (Fig. 10–1B); the legs are separated in order to illustrate the labia majora and the absence of a mons pubis. Although the labia majora of 0-22 were hairier than those of 0-26, they were much less fleshy and tapered laterodorsally about the more obvious vulval vestibule. With the perivulval tissues gently pulled ventrolaterally, a glans clitoridis was exposed (Fig. 10–2A) and was clearly demarcated from a relatively long, darkly pigmented shaft (not shown). The prominent glans had a wide groove tapering to a point dorsal to the apex. The shaft of the clitoris was enclosed in what is best interpreted as a loose prepuce, because there were no demarcations between the tissue surrounding the clitoris and the perivaginal tissue; the pigmented area was not differentiated. In this and

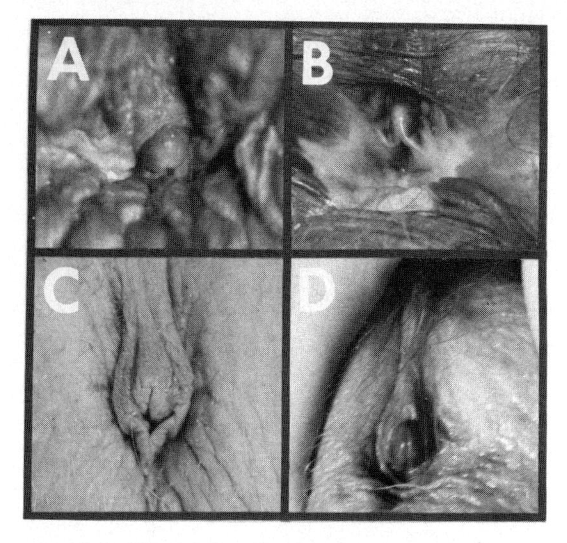

FIG. 10–2. View of four female orang-utans showing the structures immediately surrounding the vaginal aperture. (A) 0-22. (B) 0-38. (C) 0-76. (D) 0-80.

the previous specimen, the prepuce, clitoris, and vaginal aperature were located within a relatively deep vulval vestibule approximately 2–3 cm from the surface of the relaxed labia majora. Traces of frenulum or defined labia minora were not visible.

The first-born daughter of 0-22 (accession number 0-38) was trained to present through the mesh of her home cage so that her labia majora could be observed; she was studied when she was 19 years of age. This female had been diagnosed in February 1985 as pregnant for the third time. Her genitalia were observed for a total of 18 days distributed over the last 115 days of her full-term pregnancy. Her labia majora were variably tumescent but usually presented themselves as four large, pendulous lobes that sometimes lost the groove between dorsal and ventral lobes, but which always maintained a midsagittal demarcation. These lobes appeared to tumesce and partially detumesce as dorsal and ventral pairs independent of each other. The dorsal pair became maximally edematous 59 days prior to parturition; 2 days later each dorsal lobe had the appearance of a balloon completely filled and stretched taut by water, but this condition was maintained only for a few days. Three days after parturition, there was a marked reduction in tumescence but the ventral lobes remained large and their surfaces variably wrinkled. By the 10th day postpartum, tumescence was minimal. On the 17th day postpartum, 0-38 had her medical examination, at which time her external genitalia were examined in detail and photographed.

The perineal region of 0-38 (Fig. 10–1C) bears apparently normal labia majora which are notably hairy; a mons pubis is absent. The tissues immediately ventral and lateral to the vulval vestibule are more darkly pigmented than the surrounding slate-gray tissues and are clearly delineated as separate lobes. When the vaginal aperture was exposed, a cream-pink, cylindrical clitoris was revealed, from the apex and dorsolateral flanks of which two separate, thin extensions emanated. After a distance of a few millimeters these extensions fused with the ventrolateral walls of the vaginal aperture (Fig. 10–2B). The prepuce could not be differentiated clearly from the outer tissues but its stretched, posterior border is visible in the photograph. Most of the vaginal aperture of 0-38 was surrounded by an area devoid of dark pigment (similar to its mother, 0-26) but was suffused in parts by crimson red.

Another purebred daughter of 0-22 (accession number 0-76) was examined when she was 63 months old. Inspection of the perinium of 0-76 (Fig. 10–1D) revealed the presence of relatively large and sparsely haired labia majora that were clearly separated ventrally; an anterior labial commissure was not present. Dorsally, the labia majora appeared to extend beyond the circumanus and were imprecisely demarcated from it. Clearly demarcated from the labia majora, however, were perivaginal tissues, which presented themselves as a teardrop-shaped structure. The vaginal aperture itself was located toward the center of the apex of the "teardrop," but was occluded by the puckered, wrinkled mass of the surrounding tissue; when the labia majora were gently stretched ventrally, this aperture became distinct (Fig. 10–2C). The ventral border was formed by two small lobes, initially separated by a groove, which coalesced and then tapered ventrally to eventually fuse with the surrounding skin. On each side of this structure were folds of tissue (the labia minora) which continued past the vaginal aperture; the labia minora, which form the vaginal aperture's lateral and dorsal borders, fused together midsagittally to form a short raphe. The perineal region of 0-80 was examined when she was 50 months of age (Fig. 10–1E). The labia majora were much less conspicuous in this individual than in 0-76; they also were widely separated, and appeared as thickened folds ventral to the vaginal aperture. Between the labia majora there was a well-developed prepuce from which protruded a conspicuous, grooved clitoris (Fig. 10–2D) which occluded the vaginal aperture. The clitoris, however, was not differentiated into separate glans and corpus. Lateral to the vaginal aperture were two conspicuous folds, the labia minora, which fused midsagittally to form a raphe that extended toward the anus and fused with the raised circumanus.

A sixth female (accession number 0-90), a hybrid, was examined at age 12 months. She had a raised and wrinkled circumanus that extended ventrolaterally around the perivaginal structures and then gave rise to two raised and smooth cushions of tissue (Fig. 10–1H). These cushions were separated from each other by a darkly pigmented area, which was differentiated at its broad dorsal apex into a short, bell-shaped structure centrally, and two thin folds of tissue which met in the midline to form a short, midsagittal raphe that extended toward the anus. The vaginal aperture was located at the "open" end of the bell-shaped structure.

The Male. The adult male Sumatran orang-utan (accession number 0-7) was observed during his 27th year while he was enduring a protracted period of ill-health that was associated with a loss of weight (see Table 10–1). He had a parapenial scrotum that was not pendulous and which was delineated from the surrounding skin only by the darker and more finely rugose nature of its surface. Upon palpation it was apparent that the scrotal sheath was more vacuous than necessary for the small, loose testes within (Table 10–3). A relatively inconspicuous midsagittal raphe was present over the scrotum but faded into the surrounding skin caudally. The penis was completely withdrawn inside the scrotal sac, and the preputial folds projected slightly from the scrotum. With the pars libera gently pulled out until the junction of the prepuce at the corpus penis was exposed, the penis measured 9.0 cm in length. The base of the pars libera was cylindrical (diameter 1.05 cm) and indented into the general scrotal surface so that it was separated from it by a shallow, circular trough. The skin of the scrotum extended along the shaft for about 1 cm before disappearing under the reflected prepuce. The reflected surface of the prepuce was a dull but distinctive pink and was thrown into loose but numerous folds over its entire length (4.5 cm). Where the prepuce met the pars distalis the penis was bilaterally compressed (greatest diameter 1.2 cm, breadth 0.73 cm). The pars distalis was 3.5 cm in length, pale pink in color, and glossy. A clearly defined raphe originated 7–8 mm from the ventral apex of the urinary meatus and fused with the preputial tissue at its juncture with the pars distalis. Small, poorly differentiated transverse ridges, similar in general appearance to large dermatoglyphs, covered all but the apex of the pars distalis. The apex was rounded and smooth with a tightly appressed longitudinal urinary meatus (length ca. 1 cm). The base of the dome of the apex had a maximum diameter of 1.35 cm and a breadth of 0.7 cm. This individual was examined a second time immediately postmortem, and the same measurements for the penis were obtained. The testes were then dissected out and measured (Table 10–3). This individual had been a subject of behavioral study, and is illustrated in this volume (Nadler, chapter 7, Fig. 7–2) exhibiting a clearly visible penile display toward a female.

A second mature male Sumatran orang-utan (accession number 0-21) was examined immediately pre- and postmortem; this individual, as had 0-7, died as a consequence of glomerular nephritis. The scrotum of 0-21 was postpenial, and the relatively large testes within caused the scrotum to bulge outward. The penis had much the same structure as that of 0-7 but was not as long (Table 10–3). The testes were dissected out and measured (Table 10–3).

A third mature male (accession number 0-41) was the offspring of a female Bornean and a Sumatran male orang-utan and was 16 years old when examined (Fig. 10–3A). The scrotum presented itself as a raised, rounded parapenial structure covered relatively sparsely with single hairs and occasionally with long "clumps" of hair; its surface, like the surrounding tissue, was dark purple. The scrotum was filled completely by its two testes (Table 10–3). The prepuce was visible as a raised hillock of folded tissue. On manipulation, the pars libera (length ca. 10 cm) was liberated only with difficulty from the scrotal sac. The surface of the reflected prepuce was

TABLE 10–3. Size of the Testes and External Penis of Four Orang-utans

	Testes		Penis	
Accession number	Volume (cm^3)	Weight (g)	Pars distalis	Pars libera
0-7[1]	6.1	22.6	3.5	9.0
0-21	12.7	37.3	4.0	7.5
0-41	12.9	—	4.5	10.0
0-97[2]	0.62	—	3.5	6.5

Testes measurements for 0-41 and 0-97 were taken with spring calipers, which cause an underestimation of about 33% in the volume.

[1]This individual sired an infant approximately 1 year prior to these measurements.

[2]At 34 months of age.

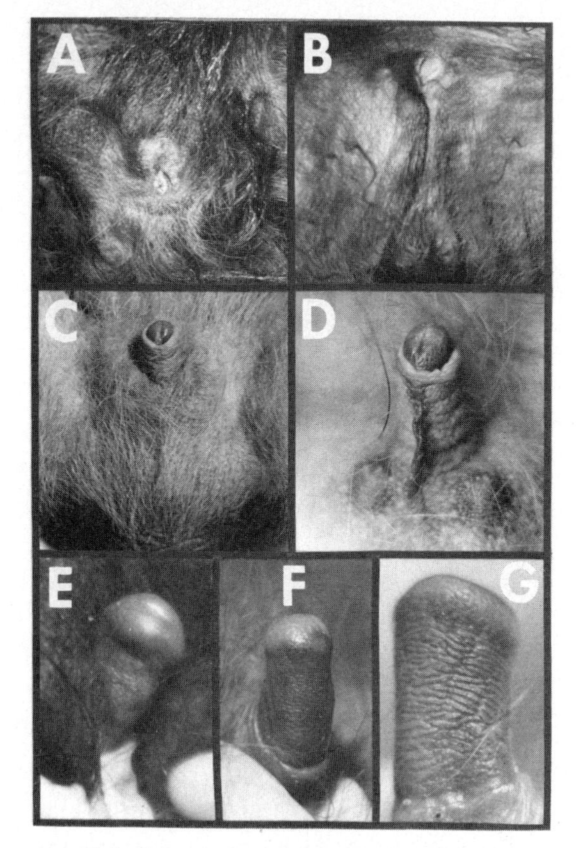

FIG. 10–3. General view of the external genitalia of four male orang-utans and view of the pars distalis of the penis: (A) and (E) 0-41, adult. (B) and (F) 0-93, juvenile/adolescent(?). (C) 0-95, juvenile. (D) and (G) 0-101, infant.

a lustrous white and extended for about two-thirds the length of the pars libera. The dark pink pars distalis was clearly differentiated into a slightly bilaterally compressed shaft that was surmounted by a large, smooth, domed glans (Fig. 10–3E); the shaft was covered in transversely organized dermatoglyph-like ridges and was separated from the glans by a corona.

A juvenile Bornean orang-utan (accession number 0-93) was examined when it was 55 months old. In Fig. 10–3B the pars libera is immediately evident because it was not fully retracted. No scrotal sac was obvious but two areas lying dorsolateral to the penis were pigmented a rich rufous color. These pigmented areas lie on either side of a prominent midsagittal raphe that extends up the thickly folded prepuce. At the apex of the prepuce the urinary meatus and the tip of the relaxed pars libera are visible. Dermatoglyph-like ridges are visible over the shaft of the pars distalis (Fig. 10–3F), but not the midsagittal raphe or the domed and slightly bulbous apex; the latter looked like a partially differentiated glans.

The scrotum of 0-95, examined when it was 42 months old, was identifiable as a parapenial bulge from which the penis protruded from the more ventral half to a distance of 2.0–2.5 cm. Dorsolateral to the pars libera were bilaterally symmetrical patches of rufous-colored skin, but the majority of the scrotal surface was unpigmented and sparsely covered with relatively long silvery-white hairs (Fig. 10–3C). The small testes (Table 10–3) were so loose within the scrotum that it was difficult to measure them. With the exception of its most distal apex, the pars libera was surrounded by a partially wrinkled prepuce on which a dorsal midsagittal raphe was not evident. The fully exposed pars libera measured 6.5 cm in length and the pars distalis was 3.5 cm. The shaft was covered in transverse, dermatoglyph-like ridges but its distal apex was smooth and no broader than the shaft. The distal centimeter of the penis was darkly pigmented.

At 34 months, the parapenial scrotum of a hybrid juvenile (accession number 0-97) protruded from the surrounding surface and was much more clearly defined than in 0-93. Ventrally, its surface was sparsely covered in hair but was nearly naked, pink, and crinkled dorsal to the protruding pars libera. A faint midsagittal raphe was observable on the prepuce but its presence over the scrotal sac was evident only by an interruption of the pigmentation. The prepuce surrounded the shaft of the pars libera in thick folds but the apex of the penis, with its urinary meatus, was visible at the tip. The exposed pars distalis was covered in transverse ridges; the tip did not appear to be differentiated into a glans.

An infant Sumatran orang-utan (accession number 0-101) was examined at 14½ months of age (coccygeal–crown height 385 mm). As seen in Fig 10–3D, protruding scrotal structures were not evident but the differentiated pigmentation dorsolateral to the protruding pars libera gave the appearance of two small elevated hillocks (although there was no protrusion). The tiny, loose testes were barely palpable under the thick skin of this region. A prominent midsagittal raphe was present on the dorsal surface of the prepuce; it continued a short distance between the paired patches of pigmented skin. The pars libera projected as much as 2 cm from the surrounding tissue. The shaft of the pars libera was covered in ridges (Fig. 10–3G); at its apex was a slightly expanded and smooth dome that was interrupted by the longitudinal slit of the urinary meatus.

DISCUSSION

Observations on six female and seven male captive orang-utans confirm much of the information previously reported on the external genitalia, particularly with regard to the structures of immature females (e.g., Fischer, 1898), and, in males, the inconspicuous nature of the scrotum, the small size of the testes (e.g., Short, 1977; Wislocki, 1936), and the basic appearance of the pars libera (e.g., Friedenthal, 1910; Hill, 1939; Pohl, 1928; Pousargues, 1895). Additional characterizations and some clarifications are also possible, but conclusions must be regarded as preliminary, since the total number of specimens examined remains small and there is evidence of considerable variability among individuals.

With respect to the female external genitalia, little evidence was obtained to support Schultz's (1927) conclusion that the labia majora of all great apes atrophy during development, and thus become inconspicuous or absent in the adult. From Wislocki's (1936) notes, as well as from the observations reported here, it appears that the labia majora of *Pongo* begin as barely differentiated cushions ventrolateral to the inner folds, but they continue to develop during infancy so that they extend caudally to fuse with circumanal extensions; in the adult, these labia surround and eventually envelop the inner folds. Defined labia majora are absent in adult *Gorilla* (Nadler, 1975, 1980; Nadler et al., 1979), but they do occur in *Pan paniscus* (Dahl, 1985), and their occurrence in at least some *P. troglodytes* is much more marked than previously thought (personal observations). The evidence for *Pongo* suggests that there may be differential development of these structures during ontogeny and that the variation in the adult forms reflects this phenomenon. For example, specimen 0-76 had large, clearly developed labia majora, but 0-80 did not. Only longitudinal, not transectional, study will clarify these points. The presence of a mons pubis was not confirmed here for *Pongo* (contra Fischer, 1898).

The labia minora and the associated prepuce can be clearly discerned in immature female *Pongo,* and together with the labia majora, bear a similarity with *Homo* in their arrangement. The labia minora of *Pongo* are small, however, surround a small vaginal aperture, and lose their identity in the adult; this distinguishes the orang-utan from all other known extant hominoids. In addition, tissues that might represent a frenulum were seen in only one adult individual. Moreover, in infant female *Pongo,* there are lateral extensions of the labia minora that have not been noted for other hominoids, but comparison with the transverse folds observed in *Macaca* by Wislocki (1936) may prove fruitful; these structures in *Macaca* appear to be responsive to hormonal influences (Hill, 1974), and, if this is also true for *Pongo,* it could account for the differential tumescence observed in the perivulval tissues noted in this study for a pregnant individual (0-38). These transverse structures in *Macaca* and *Pongo,* however, occupy different positions relative to the vaginal aperture.

In all but one individual the clitoris was relatively large and intimately associated with the relatively small vaginal aperture so that stimulation during intromission is highly likely. The structure of the clitoris, is, however, variable; a glans clitoridis may or may not be differentiated and is thus of little taxonomic relevance.

With respect to the external genitalia of the male, the structure of the pars libera of a hybrid male (0-41) provides an unexpected insight. Although previous studies have identified that portion of the pars libera distal to its junction with the prepuce as the glans penis, a bulb at the apex of the pars distalis is clearly differentiated in the hybrid (see Fig. 10–3E). In addition, at least some immature males (Pohl, 1928; Rijksen, 1978, Fig. 85; numbers 0-93 and 0-101 of this study) display some differentiation of the distal apex. Although these data support the observation that, in general, most *Pongo* fail to develop a glans, it is obvious that the species nevertheless retains some potential for developing the structure. Otherwise, the structure of the external penis of *Pongo* is remarkably similar to that of *Pan* (Hill, 1946–47; Izor et al., 1981) and unlike that of *Hylobates, Gorilla,* or *Homo.* An explanation for the curious variability in the distal apex of *Pongo* may be sought in Matthews's (1946–47) descriptions of hylobatids. Although adult *Hylobates* have a clearly differentiated glans, the glans of an immature individual detailed by Matthews is partially differentiated and in outline bears a striking resemblance to the configuration in immature *Pongo.* One could argue that the absence of a glans penis in *Pongo* and in *Pan* results from the retention into the adult of the "immature" configuration, the "mature" stage being inhibited (neoteny?); the occasional "release" of this inhibition would explain those orang-utans in which a glans does develop.

The process of "inhibition" of the glans can be associated with elongation of the pars libera; both species of chimpanzee have a long pars libera and the orang-utan's is longer than previously thought. The observations reported here tend to support the observation of Hill (1939) and the accuracy of Friedenthal's (1910) illustration which indicates that the erect organ can be at least 9 cm long, although in one of the males (0-21) the pars libera was only 7.5 cm long. Moreover, Dixson (personal communication) found that the length of the pars libera of an adult Bornean specimen was 8.5 cm. Thus, the penis of *Pongo* is at least as long as the 8-cm length reported for *Pan troglodytes* by Short (1981).

In order to check this *Pongo–Pan* comparison, the penes of six male common chimpanzees and one male pygmy chimpanzee were measured using the same technique as for the *Pongo*. The pars libera of the *Pan troglodytes* ($\overline{X}$ body weight 53.5 kg) had a mean length of 14.0 cm (SD 1.9 cm), and that of the *P. paniscus* (body weight 49.0 kg) was 17.0 cm long. No trace of a glans is reported for either species of *Pan*, and the tip is secondarily specialized; this is particularly the case for *P. paniscus* (Izor et al., 1981). The orang-utan's penis is, indeed, smaller than that of *Pan*, and this is consistent with an ability to retain a potential for developing a glans.

This amplification of an immature configuration to make elongation possible may be one of the few avenues by which the penis can (1) function as a visual stimulus to the female during sexual displays, and (2) in *Pan*, allow penetration and intromission past the highly edematous labia minora of the female. Originally it was believed that chimpanzees were the only great apes to exhibit penile displays, but penile displays by male Sumatran *Pongo* have been reported by Nadler (1982, and this volume) for captive individuals and by Schürmann (1982) for wild individuals. Although cyclical swelling during intermenstrual intervals does not occur in female *Pongo*, the vaginal aperture is set within a deep and occluded vulval vestibule; occlusion by the labia majora may preclude effective intromission with a pars libera that is less than 4 cm long. It may be the case that the pars distalis (possibly aided by the large os penis) is the only part of the penis to achieve intromission. The transverse, dermatoglyphic-like ridges on the shaft of the pars distalis might be sensory in function, enabling the male to find and penetrate the small, deep-set vaginal aperture. Such sensitivity would seemingly enhance the positive reinforcements of copulation for male *Pongo*, as copulations in *Pongo* are much longer than in other great apes (range 1–46 min; $\overline{X}$ 14 min; Nadler, 1977).

CONCLUSION

The arrangement of the external genitalia of adult female orang-utans is quite different from that of other hominoids in that the tissues of the labia minora lose their identity in the adult. Immature females of about 4 years of age, however, have a conspicuous prepuce and clearly demarcated labia minora; at this stage of development, the external genitalia resemble those of human females. The clitoris is highly variable in structure. Additional study is needed to clarify the ontogenetic differentiation of the glans clitoridis and the relationship it bears to the frenulum. Contrary to previous suggestions, the labia majora continue to develop after birth and form fairly large structures in the adult. A swelling of these tissues occurs during pregnancy, but other tissues (possibly derived from the labia minora) may be involved in this tumescence.

The scrotum and testes of orang-utans are relatively small and inconspicuous, but there may be some subspecific variation. There is some evidence that the Sumatran male form may have smaller testes than the Bornean variety, but many more specimens need to be examined. The external penis is similar to that of the chimpanzee in its configuration and, as Retterer (1915) emphasizes, is quite different from that of humans. The resemblance to *Pan* is no doubt a parallel development, since the surface of the pars distalis in *Pongo* is covered with highly specialized dermatoglyphic-like ridges as compared to the simple spicules of *Pan*. It is possible that similar ontogenetic changes have independently affected the penis in *Pongo* and *Pan* and that both taxa have emphasized an immature morphology in order to obtain a relatively long, filiform penis. This amplification of an immature configuration may not have extended so far in *Pongo* as in *Pan* because (1) a partially differentiated glans penis is found in some immature specimens; (2) a clearly defined glans penis is present in a hybrid individual; and (3) the distal apex of the penis in *Pan paniscus*, for example, appears to be secondarily specialized (Izor et al., 1981). Contrary to most previous assessments, the external penis of the orang-utan is relatively long, although it is not as long as that of *Pan;* this is consistent with recent documentation of penile displays by this ape, and appears to be functionally correlated with the structure of the female's genitalia. Numerous characters of the external genitalia of orang-utans are distinctive and are almost certainly unique to this genus.

ACKNOWLEDGMENTS

This paper would not have been possible without the expertise of the veterinary staff of the Yerkes Center, particularly Dr. R. B. Swenson, Dr. E. Strobert, Dr. J. L. Orkin, and Mr. E. Hunter, and the skillful assistance of Dr. D. C. Anderson, Mr. R. Collins, Ms. S. Mathis, Mr. S. Hughley, Mr. R. Mathis, Ms. G. Lillie, Mr. K. Hilliard, and Ms. L. Kinkaid. Input from Dr. A. Dixson was most valuable. The author thanks Mr. F. Kiernan for printing the photographs and providing greatly appreciated advice, and Ms. N. Johns, Ms. E. Jones, and Ms. P. Plant for their assistance with the literature search. Dr. M. Tigges gave generously of her time to explicate material in German. This work was supported in part by U.S. Public Health Service Grants RR-00165 and Division of Research Resources, National Institutes of Health HD-19060 (to Dr. R. D. Nadler. NICHD). The Yerkes Center is fully accredited by the American Association for Accreditation of Laboratory Animal Care.

REFERENCES

Bingham, H. C. 1928. Sex development in apes. *Comparative Psychology Monographs, 5:*1–165.

Bischoff, T. L. W. 1879. Vergleichend anatomische Untersuchungen über die äusseren weiblichen Geschechts- und Begattungs- organe des Menschen und der Affen, insbesondere der Anthropoiden. *Abhandlungen der Mathematisch-Physikalischen Klasse der Akademie der Wissenschaften zu München, 13:*207-274.

Bolk, L. 1921. The part played by the endocrine glands in the evolution of man. *Lancet, 2:*588–592.

Chapman, H. C. 1880. On the structure of the orang utang. *Proceedings of the Academy of Natural Sciences of Philadelphia, 1880:*160–175.

Crisp, E. 1865. On the os penis of the chimpanzee *(Troglodytes niger)* and of the orang *(Simia satyrus). Proceedings of the Zoological Society of London, 1865:*48–49.

Dahl, J. F. 1985. The external genitalia of female pygmy chimpanzees. *Anatomical Record, 211:*24–28.

Dickinson, R. L. 1949. Human Sex Anatomy, 2d ed. Williams and Wilkins, Baltimore.

Dixson, A. F., Knight, J., Moore, H. D. M. and Carman, M. 1982. Observations on sexual development in male orang-utans *Pongo pygmaeus. International Zoo Yearbook, 22:*222–227.

Duvernoy, M. 1855–56. Des caractères anatomiques des grands singes pseudo-anthropomorphes. *Paris Muséum National d'Historie Naturelle, Archives, VIII:*1–248.

Fick, R. 1895. Vergleichend anatomische Studien an einem erwachsenen Orang-Utang. *Archiv für Anatomie und Physiologie, Leipzig, 1895:*1–100.

Fischer, E. 1898. Beiträge zur anatomie der weiblichen Urogenitalorgane des Urang-Utan. *Morphologische Arbeiten, 8:*153–218.

Fooden, J. 1975. Taxonomy and evolution of liontail and pigtail macaques (Primates: Cercopithecidae). *Fieldiana, Zoology, 67:*1–169.

Fox, H. 1929. The birth of two anthropoid apes. *Journal of Mammalogy, 10:*37–51.

Friedenthal, H. 1910. Sonderformen der menschlichen Liebesbildung. *Beiträge zur Naturgeschichte des Menschen, 5:*1–100.

Gerhardt, U. 1909. Ueber des Vorkommen eines Penis- and Clitorisknochens bei Hylobatiden. *Anatomischer Anzeiger, 35:*353–358.

Gould, S. J. 1977. Ontogeny and Phylogeny. Harvard University Press, Cambridge.

Graham-Jones, O. and Hill, W. C. O. 1962. Pregnancy and parturition in a Bornean orang. *Proceedings of the Zoological Society of London, 139:*503–510.

Harcourt, A. H. 1981. Intermale competition and the reproductive behavior of the great apes. *In* Reproductive Biology of the Great Apes: Comparative and Biomedical Perspectives, pp. 301–318, ed. C. E. Graham. Academic Press, New York.

Harcourt, A. H. and Stewart, K. J. 1977. Apes, sex, and societies. *New Scientist, 76:*160–162.

Hart, D. B. 1908. On the role of the developing epidermis in forming sheaths and lumina to organs, illustrated specially in the development of the prepuce and urethra. *Journal of Anatomy, 42:*50–56.

Hill, W. C. O. 1939. Observations on a giant Sumatran orang. *American Journal of Physical Anthropology, 24:*449–510.

Hill, W. C. O. 1946–47. Note on the male external genitalia of the chimpanzee. *Proceedings of the Zoological Society of London, 116:*129–132.

Hill, W. C. O. 1974. Primates: Comparative Anatomy and Taxonomy, VII: Cynopithecinae. Halsted Press, New York.

Huxley, T. H. 1893. Man's Place in Nature and Other Anthropological Essays. H. L. Fowlie, New York.

Izor, R. J., Walchuk, S. L. and Wilkins, L. 1981. Anatomy and systematic significance of the penis of the pygmy chimpanzee, *Pan paniscus. Folia Primatologica, 35:*218–224.

Lippert, W. 1974. Beobachtungen zum Schwangerschafts- und Geburtsverhalten beim Orang-Utan *(Pongo pygmaeus)* im Tierpark Berlin. *Folia Primatologica, 21:*108–134.

MacKinnon, J. R. 1974. The behaviour and ecology of wild orang-utans *(Pongo pygmaeus). Animal Behaviour, 22:*3–74.

Matthews, L. H. 1946–47. Notes on the genital anatomy and physiology of the gibbon *(Hylobates). Proceedings of the Zoological Society of London, 116:*339–364.

Mayer, C. 1856. Zur anatomie des Orang-Utang und des Chimpanse. *Archive für Naturgeschichte, Berlin, 22:* 281–304.

Nadler, R. D. 1975. Cyclicity in tumescence of the perineal labia of female lowland gorillas. *Anatomical Record, 181:*791–797.

Nadler, R. D. 1977. Sexual behavior of captive orangutans. *Archives of Sexual Behavior, 6:*457–475.

Nadler, R. D. 1980. Reproductive physiology and behaviour of gorillas. *Journal of Reproduction and Fertility, Supplement, 28:*79–89.

Nadler, R. D. 1982. Reproductive behavior and endocrinology of orang utans. *In* The Orang Utan: Its Biology and Conservation, pp. 231–248, ed. L. E. M. de Boer. W. Junk, The Hague.

Nadler, R. D., Graham, C. E., Collins, D. C. and Gould, K. G. 1979. Plasma gonadotropins, prolactin, gonadal steroids, and genital swelling during the menstrual cycle of lowland gorillas. *Endocrinology, 105:*290–296.

Pocock, R. I. 1925. The external characters of the catarrhine monkeys and apes. *Proceedings of the Zoological Society of London, 1925:*1479–1579.

Pohl, L. 1928. Zur Morphologie der männlichen Kopulationsorgane der Säugetiere: insbesondere der Versuch einer vergleichend-anatomischen Studie über den Penis der Primaten, einschliesslich des Menschen. *Zeitschrift für Anatomie und Entwicklungsgeschichte, 86:*71–119.

Pousargues, E. de. 1894. Sur l'appereil genital mâle de l'Orang-Outan *Simia satyrus* (L.). *Comptes Rendus de l'Acadêmie des Sciences, 119:*238–240.

Pousargues, E. de. 1895. Note sur l'appareil genital male des orang-outans. *Muséum National d'Histoire Naturelle, Nouvelles Archives, Paris, Séries 3, 7:*57–82.

Retterer, E. 1915. Du gland de l'Orang-outan. *Comptes Rendus des Séances et Mémoires de la Société de Biologie, 78:*383–387.

Rijksen, H. D. 1978. A field study on Sumatran orang utans (*Pongo pygmaeus abelii* Lesson 1827): ecology, behaviour and conservation. H. Veenman & Zonen, Wageningen.

Sandifort, G. 1839. Ontleedkundige beschouwing van een' volwassen Orang-oetan (*Simia satyrus,* Linn.) van het mannelijk geslacht. *In* Verhandelingen over de natuurlijke geschiedenis der Nederlandische overzeesche bezittingen, door de leden den Natuurkundige Commissie in Indië en andere schrijvers, pp. 29–56, ed. S. Q. S. Luchtmans and C. C. van der Hoek. Leiden, The Netherlands.

Schultz, A. H. 1927. Studies on the growth of gorilla and of other higher primates with special reference to a fetus of gorilla, preserved in the Carnegie Museum. *Memoirs of the Carnegie Museum, 11:*1–86.

Schultz, A. H. 1938a. Genital swelling in the female orang-utan. *Journal of Mammalogy, 19:*363–366.

Schultz, A. H. 1938b. The relative weight of the testes in primates. *Anatomical Record, 72:*387–394.

Schürmann, C. 1982. Mating behaviour of wild orang utans. *In* The Orang Utan: Its Biology and Conservation, pp. 269–284, ed. L. E. M. de Boer. W. Junk, The Hague.

Short, R. V. 1977. Sexual selection and the descent of man. *In* Reproduction and Evolution, pp. 3–19, ed. J. H. Calaby and C. H. Tyndale-Biscoe. Australian Academy of Sciences, Canberra.

Short, R. V. 1979. Sexual selection and its component parts, somatic and genital selection, as illustrated by man and the great apes. *Advances in the Study of Behavior, 9:*131–158.

Short, R. V. 1981. Sexual selection in man and the great apes. *In* Reproductive Biology of the Great Apes: Comparative and Biomedical Perspectives, pp. 319–341, ed. C. E. Graham. Academic Press, New York.

Sonntag, C. F. 1924. On the anatomy, physiology, and pathology of the orang-outan. *Proceedings of the Zoological Society of London, 1924:*349–450.

Wislocki, G. B. 1936. The external genitalia of the simian primates. *Human Biology, 8:*309–347.

11

Morphology of the Placenta

HIROAKI SOMA

The orang-utan has for years been a focus of debate on the times of divergence among the hominoids. It is, in many ways, the most specialized of the large apes and has been at times thought of as a "living fossil," as well. As such, details of reproduction and parturition in the great apes are of particular interest in terms of questions involving not only the evolutionary history of the hominoids in general but also hominid origins. Although gestational lengths differ among mammals, the placenta has certain basic and vital functions for fetal development and adaptation *in utero* common to all. But the length of gestation in the orang-utan (which varies from 240 to 275 days) is quite similar to humans (Lippert, 1974; Soma, 1983)—and this makes even more intriguing the investigation of placental morphology and physiology in the large hominoids.

On a more practical and immediate level, the high rate of perinatal mortality among the large apes (22%) (Benirschke et al., 1980) and a stillbirth rate of almost 10 times that seen in humans dictates a need for further research and improved monitoring and management techniques for the periods of late pregnancy as well as early infancy (Graham and Bohn, 1985). In this regard, study of the placenta may yield information of diagnostic significance for infant great apes. However, little is known for these large hominoids of the relation of many placental lesions to fetal development because of difficulties in obtaining fresh, whole, delivered placentas, due in large part to the incidence of placentophagy, which occurs even in captivity.

I have been fortunate in being able to amass since 1974 nine orang-utan placentas, which were collected at the end of gestation under the supervision of the animals' keepers. I have studied these placentas using methods routinely applied to the analysis of human placentas and will report here the results of this investigation.

METHODS AND MATERIALS

Placentas were collected at term from four male and five female newborn orang-utans born in various zoos in Japan. The placentas were examined according to the methods outlined by Benirschke (1961). The umbilical cord, the rolled fetal membranes (the amnion and chorion), and the placental mass were fixed in a 10% formalin solution. Tissues were impregnated with hematoxylin and eosin (HE), periodic acid-Schiff (PAS), and Mallory stains and used subsequently in histological analyses.

Study of surface and transmission ultrastructure was undertaken on the placentas. The specimens were fixed with 25% glutaraldehyde and 1% osmic acid, dehydrated in a series of ethanol baths, embedded in Epon, and sectioned. The sections were double-stained with uranyl and lead citrate and examined using a HU-T1 DS transmission electron microscope.

The pathologic findings on the nine orang-utan placentas were then compared with those on 14 chimpanzee *(Pan troglodytes)* placentas (collected from Japanese zoo animals) as well as human placentas (obtained at the Tokyo Medical College Hospital) (see Tables 11–2 and 11–3).

RESULTS

Macroscopic Analysis

Of the nine orang-utan placentas, the gestational age of only one was determined with accuracy (see Tables 11–1 and 11–2 and the following discussion); the date of delivery of that particular placenta was 34 weeks. Seven of the infants were

TABLE 11–1. Placental Findings of the Orang-utan *(Pongo pygmaeus)*

Number and Name (Zoo)	Date of Delivery Gestational Age Newborn Weight and Sex	Placental Weight Dimensions Cord Length Insertion Site (no. of vessels)	Fetal Surface Membranes Subchorionic fibrin	Maternal and Cut Surfaces Infarct
1.— (Ritsurin Zoo)	12/16/74 ? 1,700 g male, alive	340 g 17 x 16 x 2 cm 28 cm eccentric (3)	Circummarginate Small fibrin	Infarct: 1 1 x 0.6 x 0.8 cm
2. Hatsuko (Ueno Zoo)	05/21/75 34 weeks 1,530 g female, alive	280 g 17 x 14 x 1.8 cm 3 + ? cm marginal (3)	Circummarginate Small fibrin	None
3. — (Tama Zoo)	12/02/77 ? 1,520 g female, stillborn	380 g 20 x 16.5 x 1.8 cm 50 cm eccentric (3)	Meconium-stained Patchy fibrin	Marginal hemorrhage
4. Eve (Japan Monkey Centre)	01/11/79 ? 1,900 g male, alive	340 g 17 x 15 x 1.8 cm 60 cm eccentric (3)	Circummarginate Small fibrin	Infarct: 1 1.4 x 1.2 x 1.0 cm Marginal

5. Sally (Tama Zoo)	06/21/79 ? over term 1,500 g female, alive	290 g 16 x 16 x 1.8 cm 4 + ? cm eccentric (3)	Circummarginate Patchy fibrin	Marked calcification Degeneration 2 x 0.4 cm
6. — (Tama Zoo)	09/22/81 ? ? male, alive	240 g 18 x 17 x 1.4 cm 30+ ? cm marginal (3)	Circummarginate Diffuse fibrin	Maternal floor 1.4 x 1.0 cm Marginal 1.6 x 1.7 × 0.8 cm
7. Hanako (Yokohama Zoo)	01/31/83 ? 1,340 g female, alive	185 g 16 x 14 x 1.2 cm 37 cm central (3)	Decidua necrosis Brown-dark Patchy fibrin	None
8. Norie (Japan Monkey Centre)	12/10/83 ? 1,950 g female, alive	360 g 16 x 15 x 1.9 cm 47 cm eccentric (3)	Circummarginate Yellow-white Patchy fibrin	Slight hemorrhage Marginal 2 2 x 1 x 2.1 cm 2.3 x 2.1 cm
9. Hanako (Yokohama Zoo)	4/22/84 ? 1,352 g male, died shortly after birth	240 g 16 x 15 x 1.4 cm 30 cm eccentric (3)	Circummarginate Brown-red	None

TABLE 11–2. Gross Findings of Orang-utan, Chimpanzee, and Human Placentas

	Orang-utan (N = 9)		Chimpanzee (N = 14)		Total (N = 23)		Human (N = 4,764)
Lesions	No.	%	No.	%	No.	%	
Infarct	4	44.4	5	35.7	9	39.1	32.8
Intervillous thrombosis			4	28.5	4	17.3	10.0
Extrachorialis							
Circumvallate			2				
Circummarginate	7	77.7	9	78.5	18	78.2	10.0
Marked subchorionic fibrin	1	11.1	4	28.5	5	21.7	1.6
Chorionic cyst			1	7.14	1	4.3	2.6
Marginal insertion of cord	1	11.1	1	7.14	2	8.6	1.8
Marked calcification	1	11.1			1	4.3	1.6
Decidua necrosis	2	22.2	5	35.7	7	30.4	9.8
Marginal or maternal surface hemorrhage	2	22.2	5	35.7	7	30.4	2.7
Meconium-stained	2	22.2			2	8.6	12.8
Abnormal length of cord (long cord)			1	7.14	1	4.3	4.1
True knot			1	7.14	1	4.3	

born alive and two were stillborn. The newborn infants ranged in weight from 1,340 to 1,900 g; the average weight for eight of the infants was 1,599 g. For a sample of seven newborn chimpanzees, the average weight was 1,840 g. The orang-utan placental weights ranged from 185 to 380 g, with an average placental weight of 295 g; the average placental weight for a sample of 14 chimpanzee newborns was 276 g.

The length of the umbilical cord in the newborn orang-utan varied from 28 to 60 cm, averaging 34.1 cm in length. The shortest distances of the umbilical cord's insertion ran from 0 to 7.0 cm, thereby demonstrating either a marginal or an eccentric insertion. The number of umbilical cord blood vessels was always three.

The length of the placentas of these orang-utans varied from 14 to 20 cm and the thicknesses from 1.2 to 2.0 cm. The average diameter of the orang-utan placentas was 17 × 15.4 cm compared to 15.6 × 12.5 cm for the chimpanzee. Interestingly, the diameter, size, and weight of the orang-utan placenta are usually greater than in the chimpanzee placenta, even though there is little difference between these two apes in average fetal weight.

Although the membranes were incomplete and lacerated, circummarginate placentas were present in seven (77.7%) of the orang-utans studied (Fig. 11–1). On the fetal surface of the orang-utan placentas, small, patchy or diffuse subchorionic fibrin could be recognized. Marginal hemorrhage as well as marked calcification occurred on the maternal side of the placenta (Fig. 11–2). On the cut surface, several marginal and maternal floor infarcts were easily identifiable in four of the orang-utan placentas (44.4%) (Fig. 11–3).

Microscopic Analysis

Pseudoinfarcts consisting of fibrin deposits as well as coagulation with degeneration are evident beneath the chorionic plate (see Table 1). In the umbilical cord and membranes of the pla-

FIG. 11–1. An orang-utan placenta with circummargination (No. 4).

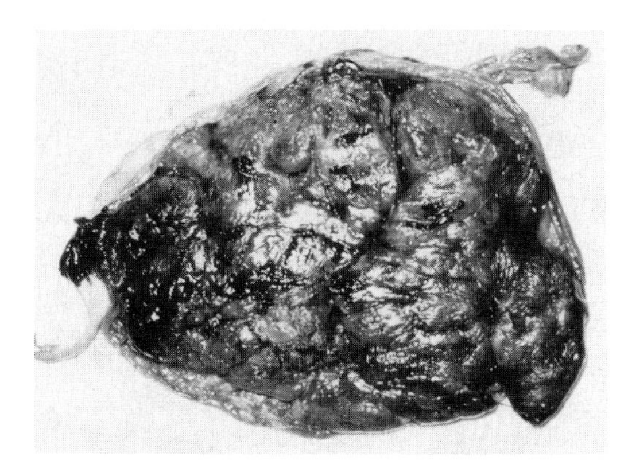

FIG. 11–2. An orang-utan placenta with maternal marginal hemorrhage (No. 3).

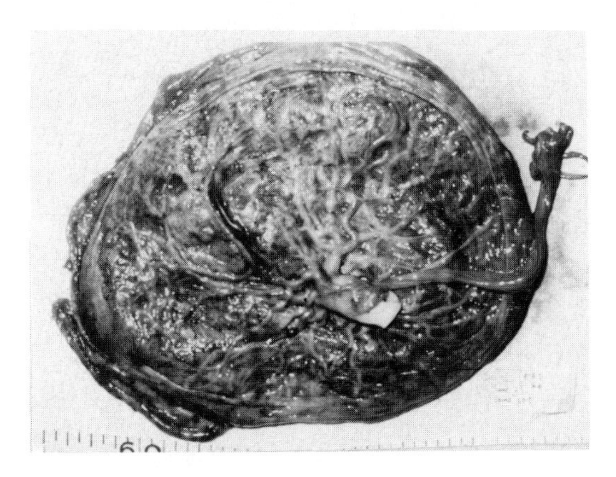

FIG. 11–4. An orang-utan placenta (No. 5) with circummargination and brown membrane. Placental vessels show a disperse type.

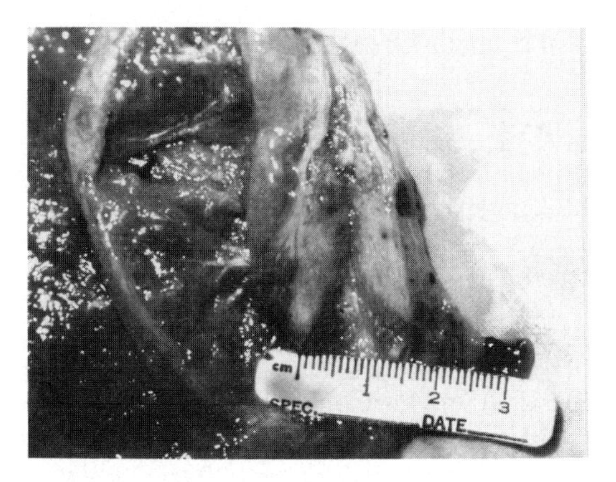

FIG. 11–3. Marginal infarcts in the orang-utan placenta (No. 4).

centa of the stillborn (case 3, Table 11–1), leucocyte infiltration was marked. The fetal surface of the placenta of case 5 (Table 11–1) was covered with a brownish, opaque membrane (Fig. 11–4) and, beneath the chorionic plate, round cell infiltration could be identified (Fig. 11–5). In addition, microinfarct with degenerative villi and round cell infiltration were discovered within the intervillous space; similar inflammatory lesions within the chorionic villi were also seen in the placenta of case 8 (Table 11–1, Fig. 11–6). With further regard to case 5, denuded trophoblasts with syncytial knots and syncytial bridges as well as dilated fetal vessels were found (Fig. 11–7) and fibrinoid necrosis of the stroma of the chorionic villi was often noted. In the basal plate, giant cells as well as X cells and decidual cells were scattered throughout, and fibrin layers were formed, even though degeneration was marked.

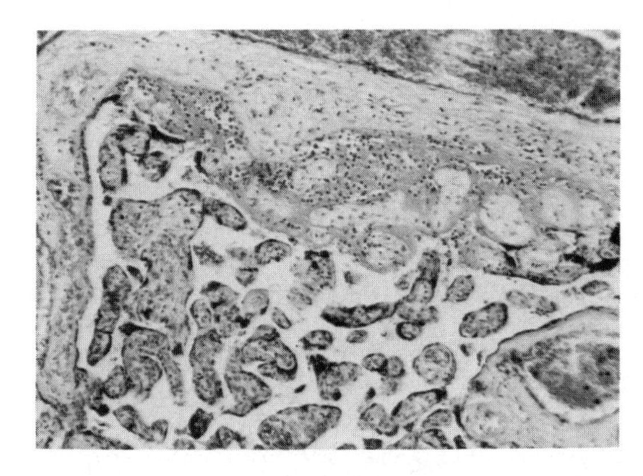

FIG. 11–5. Under the chorionic plate: round cell infiltration, fibrin, and condensed degenerated villi are noted (No. 5) (×50).

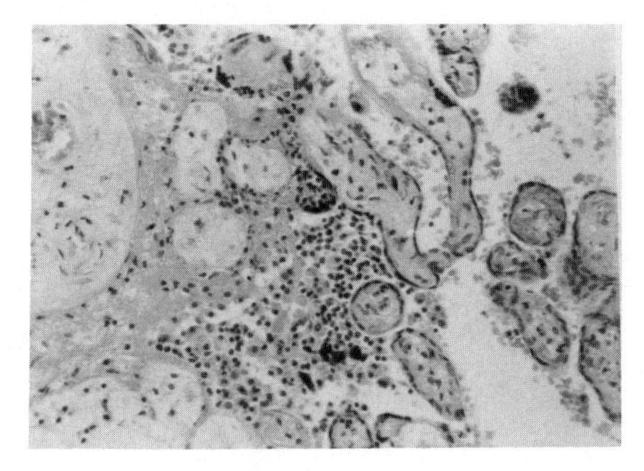

FIG. 11–6. Round cell infiltration in villi and intervillous space (No. 8) (×100).

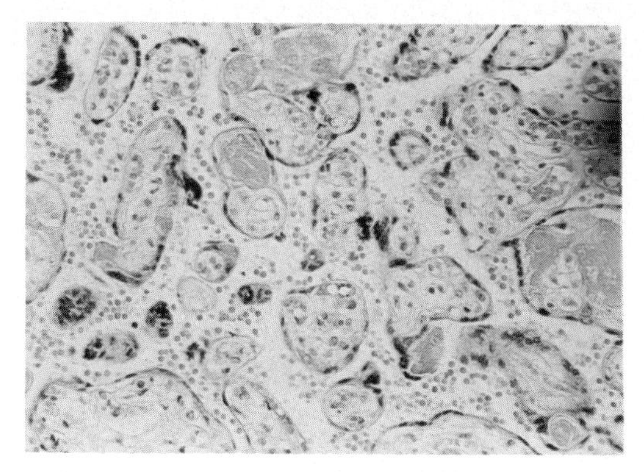

FIG. 11–7. Denuded trophoblast, syncytial knots and bridges, and fibrinoid necrosis (No. 5) (×50).

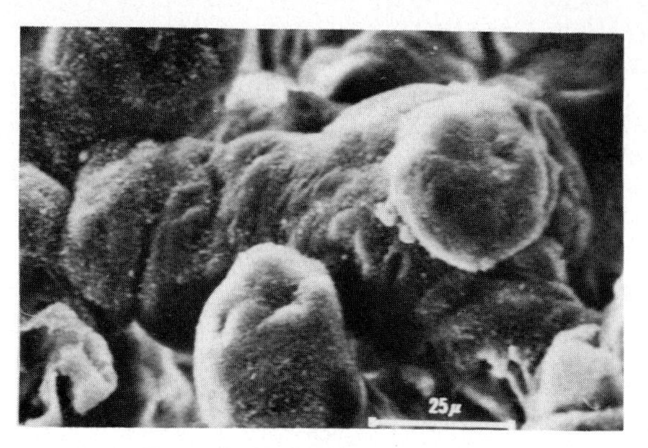

FIG. 11–8. Dome-shaped protrusions from the chorionic villous surface with microvilli.

Analysis of Ultrastructure

Characteristic of the surface ultrastructure of the orang-utan chorionic villi are dome-shaped protrusions from the villous surface, which consists of numerous microvilli (Fig. 11–8). With regard to the ultrastructural morphology of the trophoblast of the orang-utan chorionic villi, the syncytium is covered with numerous microvilli, which appear to be short and rodlike, and it contains irregular nuclei as well as dilated rough endoplasmic reticulum and numerous mitochondria (Fig. 11–9). Adjacent to the syncytium, the cytoplasm of the cytotrophoblast contains a large nucleus and a few mitochondria of relatively high density. Beneath the trophoblast lie a relatively thick basal membrane as well as endothelial cells associated with fetal capillaries.

Localization of Placental Proteins in the Orang-utan Placenta

Although there have been many studies of hormonal pattern during pregnancy in nonhuman primates, very few have dealt specifically with the great apes. Of these, the endocrinology of the orang-utan during pregnancy has been described by Bonney and Kingsley (1982) and Czekala et al. (1983) (see Czekala, Shideler, and Lasley, Chapter 8, as well as Graham, Chapter 6, for reviews). From these studies it appears that excretion of steroid hormones and chorionic gonadotropin during pregnancy in the orang-utan resembles that in the African apes and humans.

It is well known that several placental proteins—such as human chorionic gonadotropins (hCG) and human prolactin (hPL)—are pro-

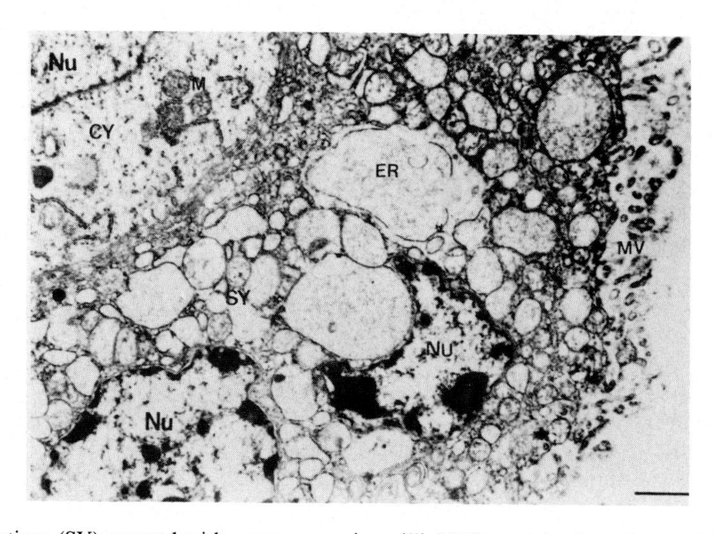

FIG. 11–9. The syncytium (SY) covered with numerous microvilli (MV) contains irregular nuclei (Nu), dilated rough endoplasmic reticulum (ER), and mitochondria (M); the cytotrophoblast (CY) has a large nucleus (Nu) and scanty mitochondria (M) (×2.8).

duced and excreted from the human placenta. Pregnancy-specific β_1-glycoprotein (SP_1), which is one of the placental proteins extracted from the human placenta by Bohn (1971), was identified here in the syncytium of the orang-utan and chimpanzee by its being strongly stained by an enzyme-immunohistochemical procedure (Fig. 11–10). Thus localization of placental proteins can be demonstrated in the synyctium of nonhuman hominoids.

DISCUSSION AND CONCLUSION

Placentophagy is widespread among nonhuman primates, but it does not always occur. Chimpanzee births have been observed frequently in captivity, but there is little information available on births in gorillas and orang-utans (Brandt and Mitchell, 1971). Since the orang-utan mother often eats the fetal membranes, it is difficult for the keeper to know whether or not the placenta has actually been expelled. Maple (1980) was able to obtain an orang-utan placenta, which he described as being discoid, 17 cm in diameter, and a few millimeters to 2.5 cm thick; the umbilical cord measured 60 cm in length, which is remarkable because the whole mass had dried by the time Maple got hold of the specimen. Prior to Maple, Graham-Jones and Hill (1962) had examined the placenta of an orang-utan after first observing its parturition.

Obviously, collecting whole placentas of great apes, even those in captivity, is not easily accomplished. This is what makes the data presented here—resulting from the analyses of nine orang-utan placentas collected over the past 12 years—so unique.

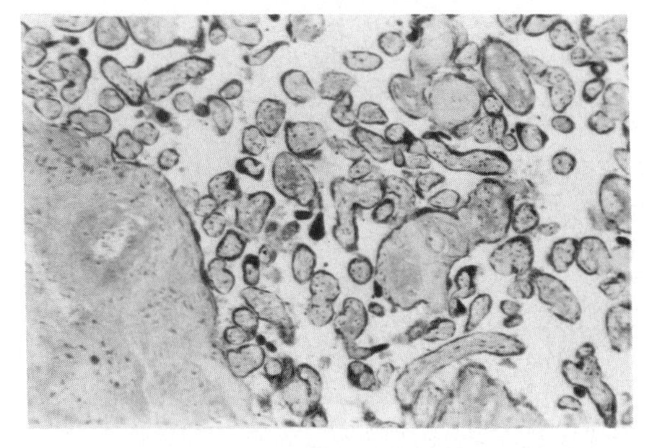

FIG. 11–10. Localization of SP_1 can be demonstrated in the syncytium of the chorionic villi as well as in the syncytial knots and cells in the basal plate of the orang-utan placenta by using enzyme-immunohistochemical staining (×50).

According to Wislocki (1933), placentation in humans and apes is so similar that, in their later stages of development, the placentas of these extant large hominoids are virtually indistinguishable from one another. Great similarity exists between humans and the orang-utan throughout a large part of placental development, but in later stages, the chorionic villi of the orang-utan become more slender and delicate than in humans (Strahl, 1903).

The average placental weight of the orang-utan (as judged from my sample) is 295 g and the average diameter is 17 × 15.4 cm. Thus the orang-utan placenta is usually larger than the placenta of a chimpanzee, even though the average body weight of a newborn orang-utan is actually less than that of a chimpanzee. In addition, the average length of the umbilical cord is shorter in the orang-utan than in the chimpanzee. But the site of insertion of the umbilical cord is most frequently eccentric or marginal in orang-utan as well as chimpanzee placentas (Ludwig and Baur, 1971; Soma, 1983; Young, 1972), although Benirschke and Miller (1982) did find five orang-utan placentas which exhibited a central insertion of the cord.

Following the pattern of placental vessels as described by Bacsich and Smout (1938), the human placenta can be divided into two groups, one representing a "magistral" type (in which the two arteries extend almost to the margin of the placenta) and the other a "disperse" type (in which the two arteries divide dichotomously and rapidly diminish in caliber). With regard to the arrangement of the placental arteries of the baboon, the magistral pattern is the most common (66%) (Houston, 1971). The magistral pattern was seen in 10 of the 14 chimpanzee placentas I studied. The magistral pattern was observed in six of nine orang-utan placentas; thus this pattern was more common than the dispersed type.

Prominent leucocyte infiltration of the umbilical cord and associated membranes was recognized in the placentas of two orang-utans and two chimpanzees. Foci of villous inflammation were present in each of the orang-utan and chimpanzee placentas, and the inflammatory lesions within the villi were focally overlain by a subtrophoblastic stroma.

Placenta extrachorialis (circumvallate and circummarginate) in the orang-utan and chimpanzee occurred at an extremely high frequency (78%) compared to the relatively low rate of 10% in humans (Table 11–2). Placenta extrachorialis has been thought to be associated with obstetric complications such as antepartum hemorrhage (Benirschke and Driscoll, 1967). In the rhesus macaque, extrachorial placenta occurs with a particularly high frequency (Myers et al. 1971). While it still needs to be determined if extrachorial growth of the placenta in the rhesus macaque results from the absence of interstitial implantation, it remains the case that placenta extrachorialis is one of the most interesting developmental aberrations involving the placenta (Benirschke, 1961).

The incidence of a marked development of subchorionic fibrin as well as of marginal and retroplacental hemorrhages is much higher in the placentas of orang-utans and chimpanzees (21.7% and 30.4%, respectively) than in humans (1.6% and 2.7%, respectively) (Tables 11–2 and 11–3).

Microscopically it was determined that, among the apes reported on here, villous changes such as fibrinoid necrosis, crowded villi, increased syncytial knots and bridges, microinfarcts, and villitis were prominent (Table 11–3). In the chorionic and basal plates of the apes, subchorionic fibrin, pseudoinfarct, and round cell infiltration proved to be higher than in the human placenta (Table 11–3).

The surface of the amniotic membrane of the orang-utan consists of an oval flat sheet that resembles the human amniotic membrane, in which the amniotic epithelial cells are regularly arranged and covered with dense microvilli. In the mature chorionic villi of the orang-utan and chimpanzee placentas, there are anuclear, thinned-out areas of trophoblast which directly overlie and appear to fuse with the fetal capillary wall. These areas are identified as vasculosyncytial membrane (Fox, 1979).

Scanning electron micrographs of mature chorionic villi in the orang-utan placenta illustrate that the surface of the syncytium is covered with microvilli that present themselves as dome-shaped swellings protruding into the intervillous space from the villous surface. Perhaps this area might be especially effective for gas transfer across the placenta. It is clear that, in contrast to the cytotrophoblast, the striking ultrastructural features of the syncytiotrophoblast of the orang-utan chorionic villi are richly defined with active microvilli as well as with organelles that include rough endoplasmic reticulum and mitochondria.

Recently, bio- and radioimmunoassays have demonstrated that there is a significant correlation between the placentas at term of humans and great apes in the concentration of chorionic gonadotropin (Bonney and Kingsley, 1982;

TABLE 11–3. Histological Findings of Orang-utan, Chimpanzee, and Human Placentas

Findings	Orang-utan (N = 9)		Chimpanzee (N = 14)		Total (N = 23)		Human Premature (N = 28)
	No.	%	No.	%	No.	%	%
Villous changes							
Fibrinoid necrosis	6	66.6	6	42.8	12	52.1	32
Fibrotic stroma	6	66.6	4	28.5	10	43.4	43
Crowded villi	1	11.1	2	14.2	3	13.0	7
Engorged villi	3	33.3	6	42.8	9	39.1	25
Increased syncytial knots and bridges	6	66.6	6	42.8	12	52.1	29
Microinfarct	5	55.5	3	21.4	8	34.7	11
Villitis	1	11.1	1	7.1	2	8.6	
Thrombosis	1	11.1	3	21.4	4	17.3	
Chorionic plate							
Subchorionic fibrin	7	77.7	5	35.7	12	52.1	
Pseudoinfarct	6	66.6	6	42.8	12	52.1	14
Infiltration	1	11.1	3	21.4	4	17.3	
Basal plate							
Fibrinoid degeneration	5	55.5	5	35.7	10	43.4	
Infiltration	4	44.4	4	28.5	8	34.7	18
Degeneration of vessel wall	3	33.3			3	13.0	18
Hemorrhage	1	11.1	2	14.2	3	13.0	36
Membranes and cord							
Chorioamnionitis	2	22.2	4	28.5	6	26.0	14
Vasculitis	2	22.2			2	8.6	4

Czekala et al., 1983; Hobson and Wide, 1981), and various features of the great ape syncytiotrophoblast are strongly suggestive of its being capable of protein synthesis in the placenta. Comparative biochemical studies on placental alkaline phosphatase indicate the presence in the orang-utan and chimpanzee (but apparently not the gorilla) of an isoenzyme resembling that in humans (Doellgast and Benirschke, 1979). Pregnancy-specific β_1-glycoprotein (SP_1), otherwise only isolated from the human placenta (Bohn, 1971), has now been localized in the syncytium of the orang-utan and chimpanzee placenta. The measurement during pregnancy of SP_1 levels in urine using enzymeimmunoassay might prove helpful for monitoring fetoplacental function in the great apes (cf. Yamabe et al., 1985).

In the basal plate of ape placentas, giant cells, X cells, and decidual cells are intermingled with the fibrin layer in a manner that is characteristic of the basal plate in humans (Soma, 1977).

Fetal death caused by abruptio placentae resulting from pregnancy toxemia associated with extensive retroplacental hematomata has been reported for the gorilla (Benirschke et al., 1980). Placental infarcts have been noted in the patas monkey's placenta (Gille et al., 1977), as has placenta previa in the orang-utan (Kingsley and Martin, 1979). Marked infarcts as well as intervillous thrombi were detected in the orang-utan and chimpanzee placentas in this study (Table 11–2). Obviously, continued study of the ape placenta is needed in order to understand the causes and high incidences of perinatal death among these large hominoids. And, equally obviously, given the possible distinctions among the hominoids in the nature of fetoplacental function (cf. Czekala et al., 1983), further study of the ape placenta may contribute to a better understanding of the evolutionary relationships of the African apes, humans, and orang-utans.

REFERENCES

Bacsich, P. and Smout, C. F. V. 1938. Some observations on the foetal vessels of the human placenta with an account of the corrosion technique. *Journal of Anatomy, 72:*358–364.

Benirschke, K. 1961. Examination of the placenta. *Obstetrics and Gynecology, 18:*309–333.

Benirschke, K., Adams, F. D., Black, K. L. and Gluck, L. 1980. Perinatal mortality in zoo animals. *In* Comparative Pathology of Zoo Animals, pp. 471–481, ed. R. Moxtali. Smithsonian Institution, Washington, DC.

Benirschke, K. and Driscoll, S. G. 1967. The Pathology of the Human Placenta. Springer-Verlag, New York.

Benirschke, K. and Miller, C. J. 1982. Anatomical and functional differences in the placenta of primates. *Biology of Reproduction, 26:*29–53.

Bohn, H. 1971. Nachweis and Charakterisierung von Schwangerschaftproteinen in der menschlichen Plazenta sowie ihre quantitative immunologische Bestimmung im Serum von schwangeren Frauen. *Archives für Gynäkologie, 210:*440–457.

Bonney, R. C. and Kingsley, S. 1982. Endocrinology of pregnancy in the orang-utan *(Pongo pygmaeus)* and its evolutionary significance. *International Journal of Primatology, 3:*431–444.

Brandt, E. M. and Mitchell, G. 1971. Parturition in primates: behavior related to birth. *In* Primate Behavior: Developments in Field and Laboratory Research, pp. 178–223, ed. L. A. Rosenblum. Academic Press, New York.

Czekala, N. M., Benirschke, K., McClure, H. and Lasley, B. L. 1983. Urinary estrogen excretion during pregnancy in the gorilla *(Gorilla gorilla),* orangutan *(Pongo pygmaeus),* and the human *(Homo sapiens). Biology of Reproduction, 28:*289–294.

Doellgast, G. J. and Benirschke, K. 1979. Placental alkaline phosphatase in Hominidae. *Nature, 280:*601–602.

Fox, H. 1979. The correlation between placental structure and transfer function. *In* Placental Transfer, pp. 15–29, ed. G. V. P. Chamberlain and A. W. Wilkinson. Pitman Press, London.

Gille, J. H., Moore, D. G. and Sedwick, C. J. 1977. Placental infarction: a sign of preeclampsia in a patas monkey *(Erythrocebus patas). Laboratory Animal Science, 27:*119–121.

Graham, C. E. and Bowen, J. A. 1985. Preface. *In* Clinical Management of Infant Great Apes, pp. ix–xi, ed. C. E. Graham and J. A. Bowen. Alan R. Liss, New York.

Graham-Jones, O. and Hill, W. C. O. 1962. Pregnancy and parturition in a Bornean orang. *Proceedings of the Zoological Society of London, 139:*503–510.

Hobson, B. M. and Wide, L. 1981. The similarity of chorionic gonadotrophin and its subunits in term placentae from man, apes, Old and New World monkeys and a prosimian. *Folia Primatologica, 35:*51–64.

Houston, M. L. 1971. Placenta. *In* Embryology of the Baboon, pp. 153–172, ed. A. G. Hendrickx. University of Chicago Press, Chicago.

Kingsley, S. B. and Martin, R. D. 1979. A case of placenta previa in an orangutan. *Veterinary Record, 106:*56–57.

Lippert, W. 1974. Beobachtungen zum Schwangerschafts- und Geburtsverhalten beim Orang-utan *(Pongo pygmaeus)* im Tierpark Berlin. *Folia Primatologica, 21:*108–134.

Ludwig, K. S. and Baur, R. 1971. The chimpanzee placenta. *Chimpanzee, 4:*349–372.

Maple, T. L. 1980. Orang-utan Behavior. Van Nostrand, New York.

Soma, H. 1977. The dynamic relationship of the morphologic changes of the placenta. *In* Biological and Clinical Aspects of the Fetus, pp. 81–100, ed. Y. Notake and S. Suzuki. Igaku Shoin, Tokyo.

Soma, H. 1983. Notes on the morphology of the chimpanzee and orang-utan placenta. *Placenta, 4:*270–290.

Strahl, H. 1903. Uteri gravidi des Orangutan. *Anatomischer Anzeiger, 22:*170–175.

Wislocki, G. B. 1933. Gravid reproductive tract and placenta of the chimpanzee. *American Journal of Physical Anthropology, 18:*81–92.

Yamabe, S., Saito, T., Kikuchi, K., Takayama, M., Sayama, S. and Soma, H. 1985. The concentration of SP_1 using enzyme immunoassay during pregnancy. *Journal of Obstetrics, Gynecology, Neonatology and Haematology, 9:*207–213 (in Japanese).

Young, A. 1972. The primate umbilical cord with special reference to the transverse communicating artery. *Journal of Human Evolution, 1:*345–359.

III
NEUROANATOMY

The two chapters included here—by Karl Zilles and Gerd Rehkämper on the brain, with special reference to the telencephalon, and by Este Armstrong and G. Thomas Frost on the diencephalon—represent major advances in the compilation of basic information on the brain of the orang-utan. I do not mean to imply that there has not been any research of note on the neuroanatomy of the orang-utan. Much of the focus of past endeavors, however, has been either largely volumetric or descriptive on the level of overall shapes and size, and not equally on cytoarchitectural characteristics. Indeed, the amount of research that has been done on the brain of the orang-utan pales in the face of similar efforts with regard to the chimpanzee and even the gorilla. Nevertheless, Marjorie LeMay and Norman Geschwind's discovery (cited in chapter 5) that the orang-utan develops cerebral and Sylvian sulcus asymmetries that are as marked as those in humans—and thus that these two hominoids are distinct from the African apes—provides just enough of the unexpected to warrant further research on the neuroanatomy of the Asian ape.

Zilles and Rehkämper begin by reviewing the basic appearance of the orang-utan brain, which they find to be most similar in shape to the gorilla brain when viewed ventrally, but a better match for the chimpanzee in lateral aspect. Although it had been well known that the positions of the primary motor and sensory areas of the cortex are similar in humans and the African apes, Zilles and Rehkämper point out that the favorable comparison also applies to the orang-utan. In fact, their cytoarchitectural analyses do not reveal differences of any significance among the large hominoids in the primary cortical areas. In the calculation of encephalization indices, Zilles and Rehkämper do find differences, with the gibbon, chimpanzee, and *Homo* being separated from the gorilla and orang-utan; humans, of course, are markedly separated from all other hominoids. In various other features (e.g., slender frontal poles, low degree of development of the telencephalon), the orang-utan also emerges with the gorilla as distinct not just from humans and chimpanzees, but from the gibbon as well, which is similar to the latter two large hominoids. Rather than attribute these possible groupings to the phylogenetic relatedness of the relevant taxa, Zilles and Rehkämper suggest that the similarities noted represent convergent solutions to individual evolutionary demands.

Armstrong and Frost undertake the study of the diencephalon, which, although small, is actually a key element in that it comprises the (dorsal) thalamus, the hypothalamus, the epithalamus, and the subthalamus, each of which is associated with a discrete function. Since there are absolutely no data on this region for the orang-utan, Armstrong and Frost must begin with the basics, first demonstrating qualitative similarities with other hominoids in the diencephalic nuclei. With regard to quantitative measurements, Armstrong and Frost predominantly find either that the orang-utan's nuclei are equivalent in absolute size to homologous nuclear regions in the African apes or that these nuclei are of the size expected for an ape of their brain weight. The exception they discover involves the ventrobasal nuclear complex, which is rather enlarged. Since hypertrophy of the ventrobasal nuclear complex predominantly affected the ventrobasal pars medialis, Armstrong and Frost hypothesize that the presumed increase in fiber number in this region would be correlated with the development in male orang-utans of cheek flanges, as their specimens are male. Armstrong and Frost predict that this nuclear complex may be sexually dimorphic in the orang-utan and that the answers to this and other questions must await future investigations.

12

The Brain, with Special Reference to the Telencephalon

KARL ZILLES AND GERD REHKÄMPER

Although there are numerous reports on brain structure in *Pan* (in the following discussion *Pan* refers to *Pan troglodytes,* the common chimpanzee) and *Gorilla,* there are relatively few that deal with this subject in *Pongo.* Tiedemann (1827) appears to have been one of the first to study the brain of *Pongo,* followed by Bischoff (1870, 1876), who provided detailed information from the study of two juvenile specimens. Tilney and Riley (1928) dealt at considerable length with especially the gross anatomy of the orang-utan brain. Cyto- and myeloarchitectonic details have rarely been reported (e.g., Campbell, 1905; Filimonoff, 1933; Mauss, 1908, 1911), and rigorous quantitative analyses are essentially nonexistent. There are, however, data on brain and body weights and occasionally on percentage brain composition (e.g., Bischoff, 1876; Tilney and Riley, 1928). Bauchot and Stephan (1969) used these data for their formulation of an encephalization index.

Here, we shall present new qualitative and quantitative data on the brain of *Pongo* and make comparisons with other apes as well as with *Homo* in order to delineate similarities between *Pongo* and other genera in brain structure.

MATERIALS AND METHODS

Two orang-utan brains were available for this study. The first was that of an adult female originally from Sumatra. She died in the zoological garden of Cologne and was subsequently transferred to our institution's collection. The whole animal had been fixed in 70% alcohol for several years. The brain was removed from the skull, photographed, embedded in paraffin, and cut into serial sections (20 μm). One out of each group of 20 sections was mounted and stained with cresyl fast violet for Nissl stain. The overall histological preservation is very poor, but sufficient for the delineation of basic components of the brain (cf. pages 170–2). The volumes of these brain components were determined, using the planimetric technique of Zilles et al. (1982a) (Table 12–2).

The second brain came from the collection of the Netherlands Institute of Brain Research. It had been in different fixation fluids (the last years in formalin) over a period of 75 years. Only the telencephalon and diencephalon were present, the brainstem and cerebellum having earlier been removed. This brain was embedded by us in paraffin and cut into several sections. Instead of using a Nissl stain with cresyl fast violet, which gives poor results with old museum material, we employed a new silver staining technique, which marks the cytological details of the perikarya and the most proximal part of their process (modified after Merker, 1983). This partial brain was used for the analysis of cytoarchitecture, the delineation of different parts of the telencephalon (cf. pages 163–7), and for the determination of volumes (Table 12–2).

The structure of the different cortical areas was analyzed with a computer-controlled image analyzer (Micro-Videomat 2, C. Zeiss, F.R.G.) by measuring the gray level index (GLI). This parameter is the proportion of the profiles of stained particles (predominantly nerve cell bodies) on a measuring field of 18 × 18 μm (for further information cf. Schleicher et al., 1978; Zilles et al., 1978a,b, 1979a,b, 1982b, 1984). The GLI was measured in neocortical areas 3 (part of the primary somatosensory cortex), 4 (the primary motor cortex), 17 (the primary visual cortex), and 41/42 (auditory cortex) by a complete scanning of a representative area (at least 6,500 measuring fields) in each cortical field. Mean values of GLIs in each cortical layer of each of the four areas were calculated. This permits interspecific comparison of the laminar structure after standardizing the actual mean GLIs of the layers to the mean GLI of the complete cortical area under consideration. The mean GLI of the complete area was defined as 1. Layer-specific GLIs

>1 were found in layers which had cell-packing densities higher than the mean of all layers of a respective area; a GLI < 1 indicates a lower cell-packing density. The volume proportion of each layer in a cortical area was determined by the ratio between the number of measuring fields in a layer and the total number of measuring fields in all cortical layers of the respective area.

MACROSCOPIC APPEARANCE OF THE BRAIN

Remarks on the macroscopic anatomy of the orang-utan brain can be found in Bolk (1902) and Connolly (1950). Very useful, high-quality photographs of the orang-utan brain (as well as of other primate brains) have been published by Retzius (1906). The brain of the orang-utan is very compact; this is primarily due to the development of the telencephalon (Fig. 12–1). In dorsal view, the brain of an orang-utan is nearly egg-shaped, although the occipital poles are somewhat prominent and separated from each other by a caudally diverging interhemispherical fissure. In lateral view, the rostral part of the telencephalon appears to be especially well developed. The height of an orang-utan's brain is remarkable, reaching nearly three-quarters of its total length. The temporal lobe is well developed and its blunt tip is directed rostrally, thereby creating a protrusion of the outer contour of the brain. The cerebellum is quite prominent later-

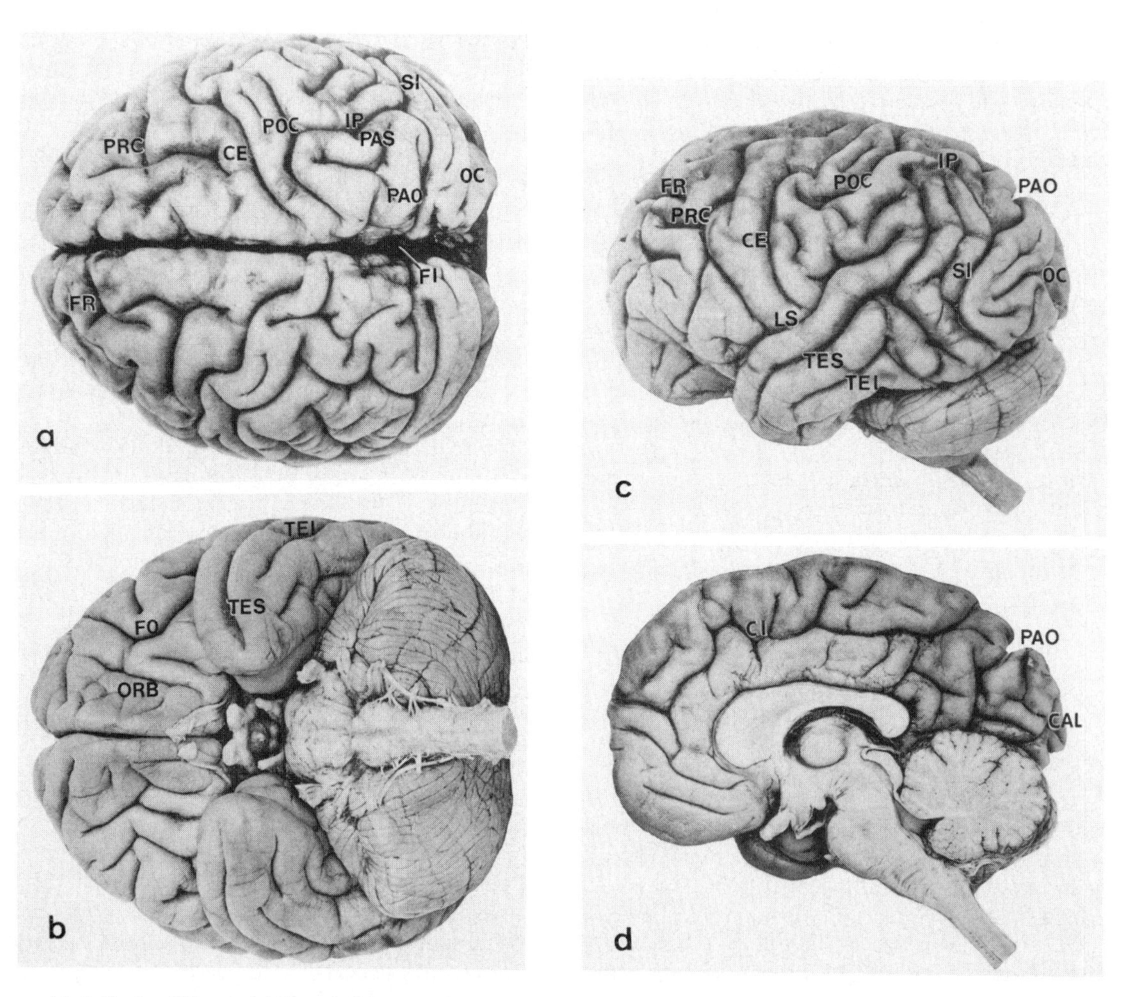

FIG. 12–1. Brain of *Pongo.* (a) Dorsal, (b) ventral, (c) lateral, and (d) medial views. CAL = calcarine sulcus; CE = central sulcus; CI = cingulate sulcus; FI = interhemispheric fissure; FO = frontoorbital sulcus; FR = frontal sulcus; IP = interparietal sulcus; LS = lateral (Sylvian) fissure; OC = occipital sulcus; ORB = orbital sulcus; PAO = parietooccipital sulcus; PAS = superior parietal sulcus; POC = postcentral sulcus; PRC = precentral sulcus; SI = simian sulcus; TEI = inferior temporal sulcus; TES = superior temporal sulcus (From Retzius, 1906).

ally and especially ventrally, although it remains inconspicuous when viewed from the dorsal aspect. A typical feature of the orang-utan brain is the keel-like form of the orbital part of the frontal lobe, which is best seen when viewed rostrocaudally (cf. Retzius, 1906). Additional special brain characteristics need to be described in comparisons with other anthropoid brains, i.e., especially between cercopithecids and the other apes.

Comparison with Cercopithecidae

Compared with cercopithecids (e.g., *Cercopithecus, Papio, Macaca,* and *Cercocebus*), the telencephalon of *Pongo* has a very elaborate sulcal pattern (Fig. 12–1). This detail is most conspicuous in the occipital pole, which is nearly smooth in cercopithecids. A significant part of the difference in the development of the orangutan's telencephalon can be attributed to the enlargement of the frontal pole, which appears oversized when viewed laterally, in contrast to the frontal pole of a cercopithecid, which tapers markedly and remains clearly smaller than the occipital pole. Apparently, the cortical region involved in the increased size in the rostral pole of the *Pongo* hemisphere is located much more caudally in cercopithecids. The cerebellum of *Pongo* also differs from that of cercopithecids in that it is so large, especially ventrally, so that it is nearly as broad as the entire brain. In medial aspect, the interhemispheric fissure appears to be relatively deeper in *Pongo* than in cercopithecids, the former primate also having a more elaborate sulcal pattern in its medial hemispherical surface.

Comparison with Other Apes

In features of its brain, *Pongo* appears more similar to *Gorilla* than to *Pan.* In *Pongo* and *Gorilla,* the brain in dorsal outline is egg-shaped, the thicker end lying at the occipital pole. In *Pan,* as well as in *Homo,* the anterior part of the brain is much broader in dorsal view, its rostral end being quite blunt compared to *Pongo* and *Gorilla.* In lateral view, however, some similarities between *Pongo* and *Pan* become apparent. In these two genera, the dorsal contour of the telencephalon is not semicircular, as in *Gorilla,* but exhibits a remarkable dorsal protrusion of the frontal lobe. We could not determine whether the olfactory bulbs in *Pongo* extend beyond the rostral tip of the telencephalon, as in *Gorilla,* or if they are more truncated, as in *Pan,* because these structures were not preserved in our material and are not discernible in the illustrations published by other investigators (e.g., Retzius, 1906). Viewed ventrally, the shape of the orangutan brain is again similar to the gorilla's. In both *Pongo* and *Gorilla,* the temporal lobe appears to be less pronounced than in *Pan.*

SULCAL PATTERN

Earlier descriptions of sulcal patterns in *Pongo* demonstrate an enormous amount of variation from individual to individual (e.g., Bischoff, 1876; Connolly, 1950; Fick, 1895a, b; Kükenthal and Ziehen, 1895; Retzius, 1906). However, some major sulci do occur regularly among different specimens and can be described using the nomenclature applicable to human brain anatomy.

The frontal lobe is caudally delineated by a central sulcus, which originates near the sulcus lateralis Sylvii and then courses toward the medial border of the hemisphere, but which does not extend onto the medial hemispherical surface (Fig. 12–1). Rostral to the central sulcus, a precentral sulcus and a frontal sulcus can be identified. These two sulci give rise to secondary sulci. In lateral and ventral views, there is an orbital sulcus and a frontoorbital sulcus (Retzius, 1906). The frontal pole is divided by several smaller sulci (e.g., the sulcus rectus of Connolly, 1950) which vary from individual to individual in shape and size.

Caudal to the central sulcus, a postcentral sulcus can be found, which can be subdivided into an inferior and a superior segment. The inferior segment may become confluent with the intraparietal sulcus, which, in turn, may be in close contact with the "sulcus simiarum" (Retzius, 1906). The homology of the "sulcus simiarum" with the sulcus lunarum in *Homo* has been the subject of much discussion (see Kuhlenbeck, 1978).

The parietal lobe exhibits an interparietal sulcus. It is further subdivided by a sulcus parietalis superior (Retzius, 1906), which runs nearly parallel to the interhemispherical fissure.

The occipital lobe displays an occipital sulcus and its derivatives, together with a very variable system of minor sulci.

Most of the temporal lobe (with the exception of the temporal pole) is separated into two parts by the long sulcus temporalis superior. The caudal and larger part of the temporal lobe bears a sulcus temporalis inferior as well as some minor, variable sulci.

Three major sulci can be seen on the medial aspect of the brain. The first of these is the cingulate sulcus, which anteriorly may be accompanied by a parallel sulcus. The second is the parietooccipital sulcus, which crosses the medial border of the hemisphere and may contact the sulcus simiarum on the lateral aspect of the brain. The third is the calcarine sulcus, which may not even extend onto the lateral surface of the occipital lobe (Connolly, 1950; Retzius, 1906; our own observations).

CORTICAL AREAL PATTERN

Only two investigations of the cortical areal pattern of *Pongo* have been published. Campbell (1905) produced a map based on the study of cytoarchitecture, and Mauss (1908, 1911) analyzed myelin-stained material. In spite of some differences in identifying details of the borders on nonprimary cortical areas, these two studies are in nearly perfect agreement on the position and extent of the primary areas (Figs. 12–2 and 12–3).

Mauss's (1908, 1911) area 4 and the respective area in Campbell's (1905) study can be identified as the "motor" area of Grünbaum and Sherrington (1901, 1903) and Leyton and Sherrington (1917). On the basis of physiological studies, the latter authors identified a motor area immediately rostral to the central sulcus, on the precentral gyrus, which extends over the whole length of this sulcus; motor reactions can also be elicited deep within the sulcus. Beevor and Horseley (1890–1891) had earlier reported motor reactions to electrical stimuli of *Pongo*'s cortex, but although they observed a somatotopic order, their description does not allow precise topographical identification of the stimulation sites. Leyton and Sherrington (1917) identified two additional motor regions, which lie apart from this "primary" motor field. One of these regions, on the lateral surface of the hemisphere and rostral to the precentral gyrus, leads to conjugated eye movements and thus resembles the frontal eye field in other primates, including *Homo* (Smith, 1944; Wagman, 1964). The second motor region occurs on the medial surface of the

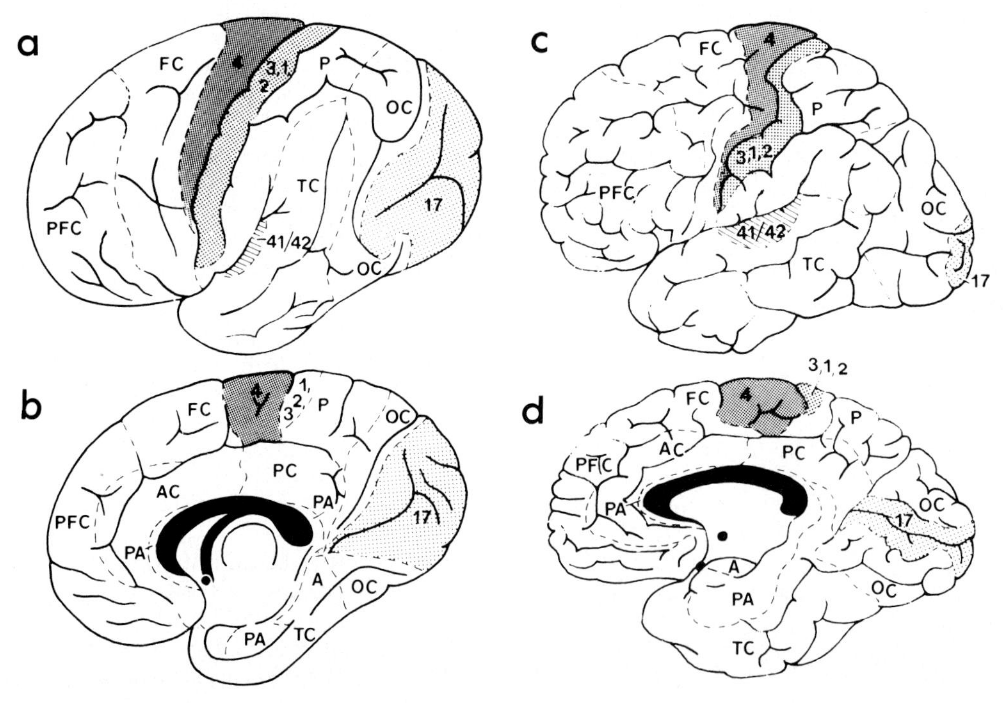

FIG. 12–2. (a) Lateral and (b) medial view of areal mapping of the cortex in *Pongo*, based on studies by Mauss (1908, 1911). (c and d) Corresponding views of cortex of *Homo*, based on studies by Brodmann (1909). 3, 1, 2 = Primary somatosensory cortex; 4 = primary motor cortex; 17 = primary visual cortex; 41/42 = auditory cortex; A = archicortex; AC = anterior cingulate cortex; FC = frontal cortex; OC = occipital cortex; P = parietal cortex; PA = periarchicortex; PC = posterior cingulate cortex; PFC = prefrontal cortex; TC = temporal cortex. All hemispheres are standardized to equal lengths of the hemispheres.

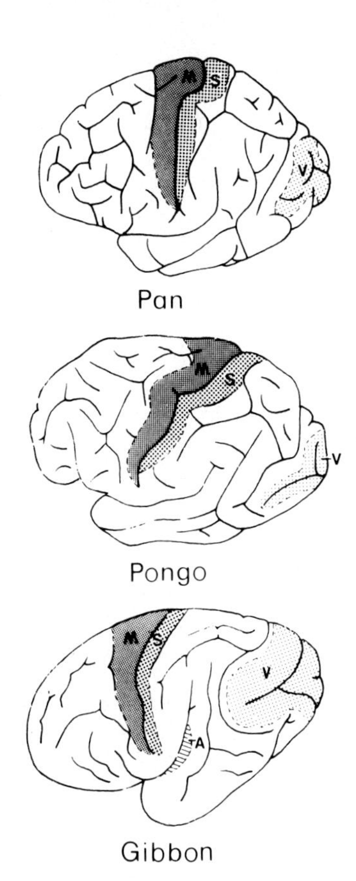

FIG. 12–3. Areal mapping of the cortex in *Pan, Pongo,* and gibbon *(Hylobates)* based on studies by Campbell (1905). A = auditory cortex; M = primary motor cortex; S = primary somatosensory cortex; V = primary visual cortex. All hemispheres are shown in lateral view and are standardized to equal lengths of the hemispheres.

hemisphere near the cingulate sulcus and is responsible for movements of the shoulder, trunk, wrist, and fingers. This region may be analogous to the supplementary motor area in *Homo* (Penfield and Welch, 1951).

Upon stimulation of the precentral gyrus, *Pan* and *Gorilla* show reactions similar to those of *Pongo* (Grünbaum and Sherrington, 1901, 1903; Leyton and Sherrington, 1917). Area 4 in *Pongo* is therefore comparable in position, extent, and function with area 4 in *Pan* and *Gorilla.* As such, area 4 in *Pongo* is also similar to *Homo* with respect to topography (Brodmann, 1909; Förster, 1936; von Economo and Koskinas, 1925) as well as function (Penfield and Rasmussen, 1950).

With the exception of the enormously enlarged finger representation area in *Homo,* the somatotopy in the primary motor area (Fig. 12–4) is very similar among hominoids. The topology of the different parts of the pyramidal tract is also identical, as is evident in the crura cerebri of the hominoids (Förster 1936; Leyton and Sherrington, 1917). For example, fibers for toe and lower extremity movements are positioned most laterally in the crura cerebri. These fibers are followed medially by the fibers for the upper extremity, which, in turn, are followed by those for the fingers. The fibers for the face and tongue lie most medially.

Physiological observations of the primary sensory areas (Figs. 12–2 and 12–3) of *Pongo* have not been published. A comparison of the positions and extents of these areas in Campbell's (1905), Mauss's (1908, 1911), and Filimonoff's (1933) cortical maps of *Pongo* with those of the primary somatosensory regions (areas 3, 1, 2), visual region (area 17), and auditory regions (areas 41/42) in *Homo* (Brodmann, 1909) reveals such a striking resemblance that equivalent functions can be inferred for these areas in the orang-utan. This topological equivalence is corroborated by the cytoarchitectural comparison of their laminar pattern (see page 163). The cor-

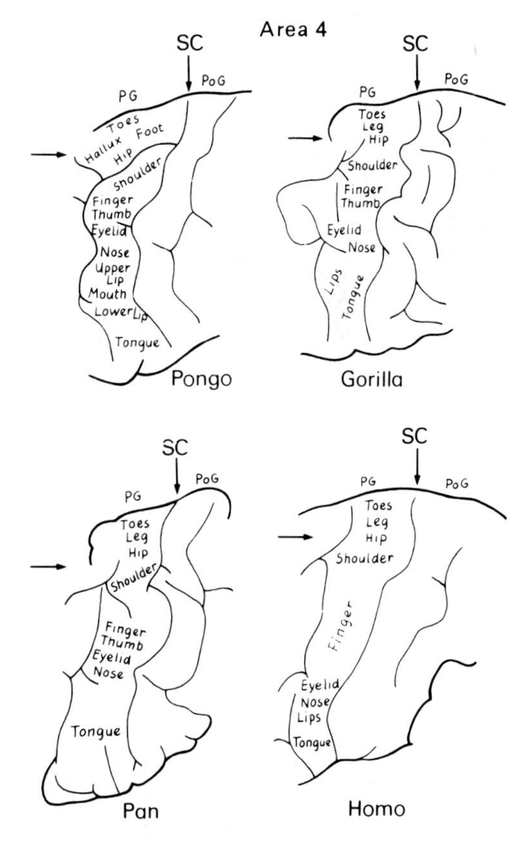

FIG. 12–4. Somatotopic representation in area 4 of *Pongo, Gorilla, Pan,* and *Homo* according to the results of Grünbaum and Sherrington (1901, 1903), Leyton and Sherrington (1917), and Penfield and Rasmussen (1950). SC = central sulcus; PG = precentral gyrus; POG = postcentral gyrus.

tical map of *Pan* (Bailey et al., 1950; Campbell, 1905) also displays comparable topological features for the primary motor, somatosensory, and visual areas.

As a rule, the positions of the primary cortical areas in the apes and *Homo* are very similar; nevertheless, slight differences are obvious. When viewed laterally, area 17 covers only a very small part of the hemisphere in *Homo,* a greater part in *Pan* and *Gorilla,* and the greatest part in the gibbon. The situation in *Pongo* is very similar in this respect to *Pan* and *Gorilla.*

The secondary areas (Fig. 12–2) of the *Pongo* neocortex can be subdivided into many different cortical fields (Campbell, 1905; Mauss, 1908, 1911), but the methodological approaches used in these cyto- and myeloarchitectural studies, as well as the lack of any physiological analysis, do not allow for a detailed comparison of both maps in these parts of the neocortex. Therefore, we have only delineated the main groups of the secondary areas, i.e., the prefrontal cortex (PFC), the nonprimary motor cortex (FC), the parietal cortex (P), the nonprimary visual cortex (OC), and the temporal cortex (TC) (Figure 12–2).

From a structural perspective, the anterior (AC) and posterior (PC) cingulate cortices, which are found on the medial surface, although representing transitional zones between the iso- and allocortex, are more similar to the isocortex (Vogt, 1976). A very small cortical region, identified as the periarchicortex (PA), is located between the cingulate cortex and the corpus callosum. The PA surrounds the allocortex on the medial surface of the hemisphere. The allocortex is represented in this aspect by the archicortex (A), whose primary constituent is the hippocampus.

The paleocortex, a further part of the allocortex, is one of the smallest parts of the cerebral cortex in apes and is extremely reduced in comparison to prosimians and insectivores (Stephan, 1975). The peripaleocortex, which represents another transitional zone, is found on the lateral aspect, between the paleocortex and the isocortex.

The cyto- and myeloarchitectonical maps (Figs. 12–2 and 12–3, respectively) of the orangutan and the gibbon are also similar with regard to the position and extent of the primary cortical areas (Campbell, 1905; Mauss, 1908, 1911). Thus, this topological similarity appears as a fundamentally common structure in all primates (cf. Brodmann, 1909; Mauss, 1908, 1911; Zilles et al., 1979a,b, 1982b). Apart from this general similarity, however, some peculiarities are typical for special groups. In prosimians, the elongated primary motor and sensory areas are oriented along a roughly frontooccipital axis. The shapes of these areas are similar in platyrrhines and in lower catarrhines, but their positions are increasingly divergent from the frontooccipital axis. These areas are oriented nearly perpendicularly to the frontooccipital axis in apes and in *Homo.* Such differences may be influenced by different factors, such as skull structure, degree of encephalization (see pages 167–170), neocortical development, and formation of the central sulcus. *Pongo* is more similar to *Pan, Gorilla,* and *Homo* than to any other catarrhine with respect to the orientation of the primary motor and sensory areas.

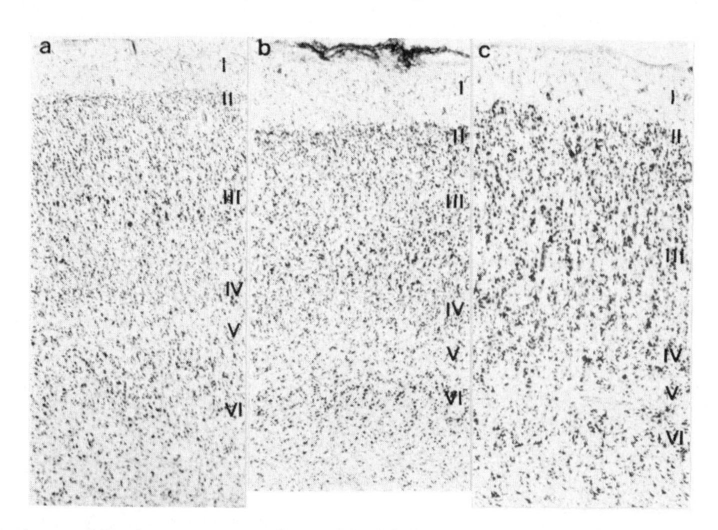

FIG. 12–5. Cytoarchitecture and laminar structure of area 3 in (a) *Pongo,* (b) *Pan,* and (c) *Homo.* I = molecular layer; II = outer granular layer; III = outer pyramidal layer; IV = inner granular layer; V = inner pyramidal layer; VI = multiform layer. The magnification is identical for all examples.

LAMINAR STRUCTURE OF PRIMARY CORTICAL AREAS

The primary sensory areas (Figs. 12–5, 12–7, and 12–8) have the typical basic six-layered isocortical structure in all hominoids. This structure reaches the highest degree of differentiation in the primary visual cortex (area 17), where layer IV is divided into at least three sublayers. On the other hand, a clearly visible layer IV cannot be detected in the primary motor cortex (area 4), because this region has such a reduced

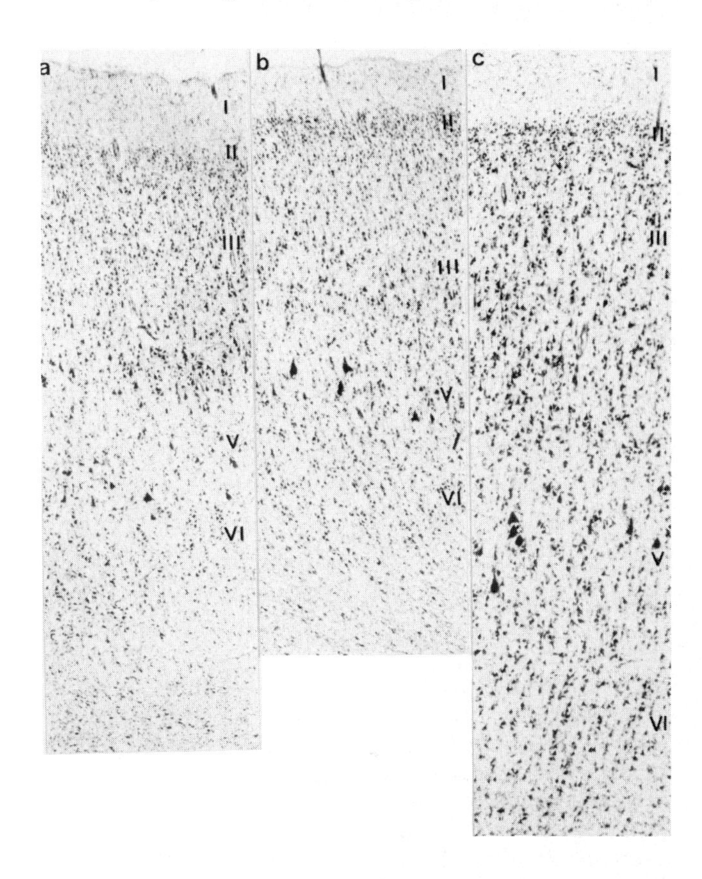

FIG. 12–6. Cytoarchitecture and laminar structure of area 4 in (a) *Pongo,* (b) *Pan,* and (c) *Homo.* For further explanation see Fig. 5.

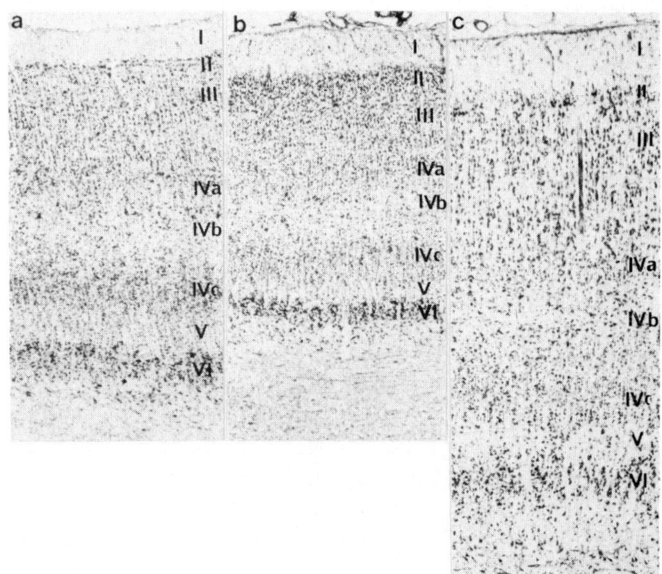

FIG. 12–7. Cytoarchitecture and laminar structure of area 17 in (a) *Pongo,* (b) *Pan,* and (c) *Homo.* For further explanation see Fig. 5.

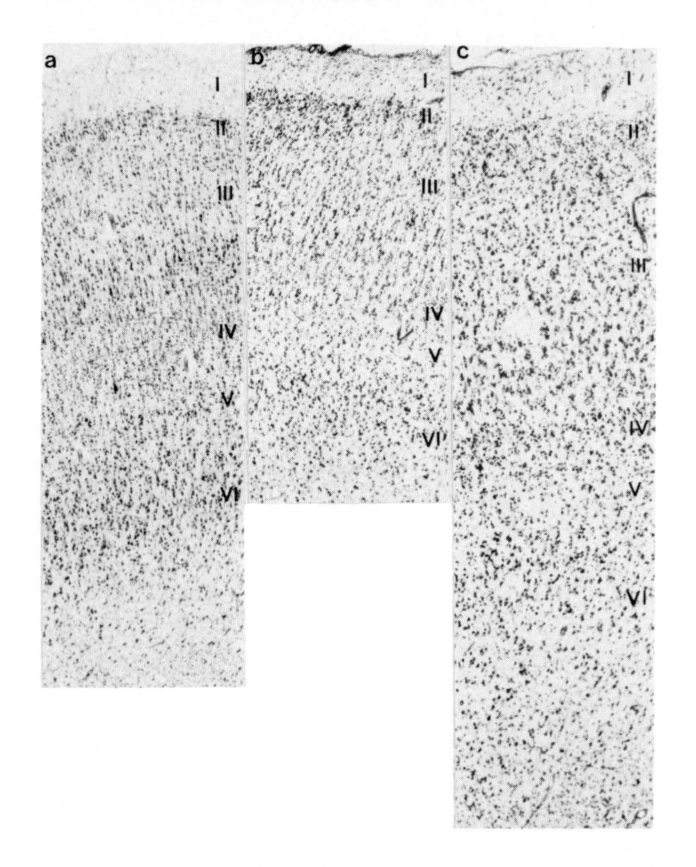

FIG. 12–8. Cytoarchitecture and laminar structure of areas 41/42 in (a) *Pongo,* (b) *Pan,* and (c) *Homo.* For further explanation see Fig. 5.

density of small granular cells (Fig. 12–6). However, since one can find the basic six-layered structure ("homotypische Formation," Brodmann, 1909), as well as a varying number of layers ("heterotypische Formation," ibid.), in all primates and in other mammals (Braak, 1980; Brodmann, 1909; Hubel and Wiesel, 1977; Jones, 1975; Sanides, 1968; Zilles et al., 1982b), the equivalent areas in hominoids must be distinguished with respect to several parameters, such as structural differences in cell shape and in such quantitative aspects as layer thickness and cell-packing density. Because the quantitative aspects can be objectified by measurement, we decided to use this approach for a comparison between the cortical structure in the primary areas of *Pongo, Gorilla, Pan,* and *Homo.* Since difficulties in obtaining well-preserved *Pongo* brains prohibit the analysis of a statistically sufficient sample, we attempted to minimize the effect of variation on absolute size: the variable under consideration (layer thickness or GLI as a measure of the cell-packing density) was set at 1 for the whole cortical thickness as well as for the mean GLI of each of the different cortical areas. Using this approach, the parameter can readily be interpreted as a ratio, i.e., layer thickness can be expressed as the proportion of a single layer to the cortex in percent. The motor cortex of *Gorilla* was not available for this analysis.

The relative thicknesses of the different cortical layers *(layer proportions)* in area 3 (Fig. 12–9) show a similar pattern in *Pongo* and *Pan. Gorilla* and *Homo* differ in this respect, because their layer III is markedly thicker. Measurement of the same parameter in areas 4 and 17 reveals a high degree of similarity in the laminar pattern of *Pongo* and the other hominoids (Figs. 12–10 and 12–11). *Pongo* does have a higher layer IV proportion in area 41/42 than *Gorilla, Pan,* and *Homo.* A comparable feature is, however, evident in the overall pattern of the other layer proportions (Fig. 12–12).

The standardized GLIs of the different layers show comparable patterns in the same cortical field of apes and humans (Figs. 12–13 to 12–16). The sensory cortical areas are roughly characterized by high GLI values in layers II–IV (Figs. 12–13, 12–15, and 12–16). Area 3 in *Pongo* differs to a minor degree from the other apes and *Homo* in that, compared with layers III and IV, it has a rather low GLI in layer II and a high GLI in layers V and VI. In all large hominoids, the maximum values for the primary motor cortex (Fig. 12–14) are found in layer II followed by (in descending order) layer III, layer V, and layer

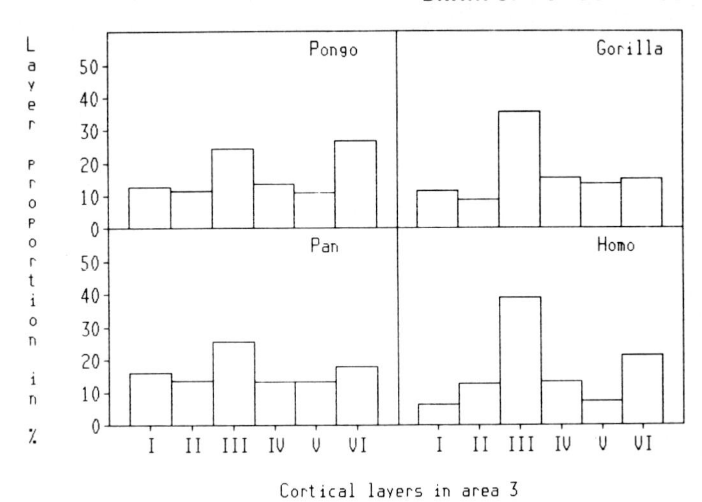

FIG. 12–9. Layer proportions in area 3 of the large hominoids (*Pongo, Gorilla, Pan,* and *Homo*) for the different cortical laminae.

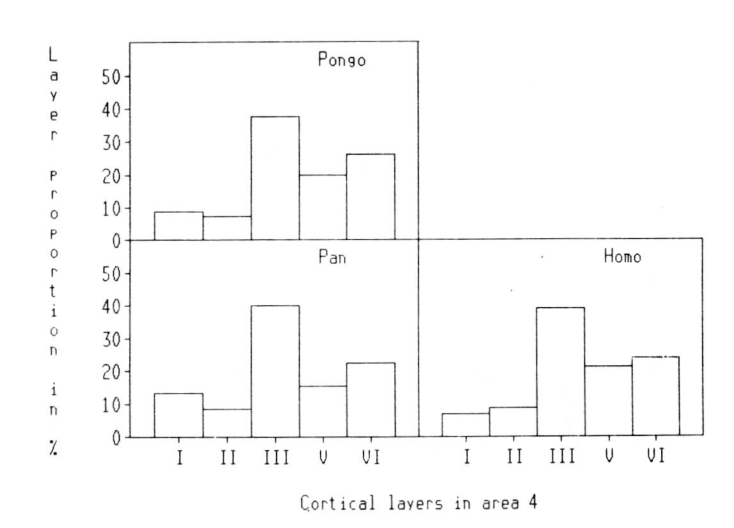

FIG. 12–10. Layer proportions in area 4 of the large hominoids (data unavailable for *Gorilla*) for the different cortical laminae.

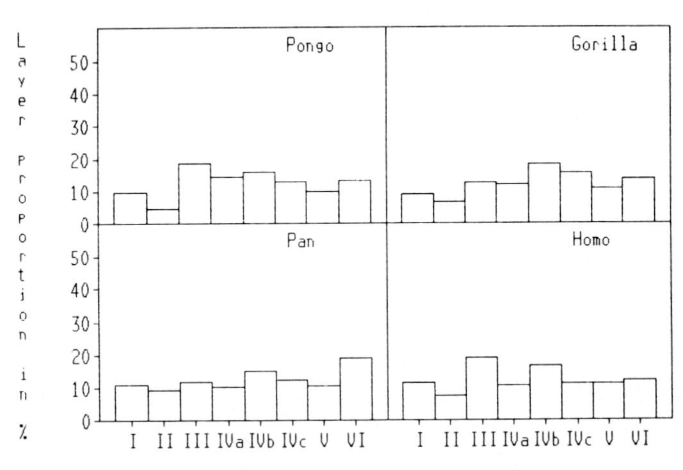

FIG. 12–11. Layer proportions in area 17 of the large hominoids for the different cortical laminae.

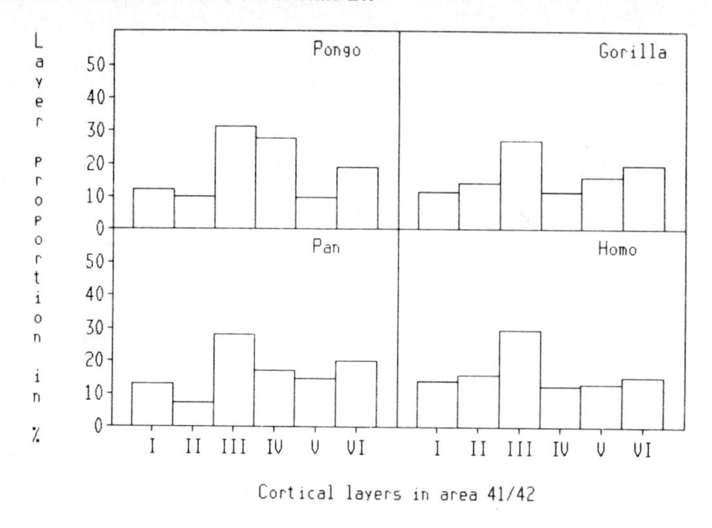

FIG. 12–12. Layer proportions in areas 41/42 of the large hominoids for the different cortical laminae.

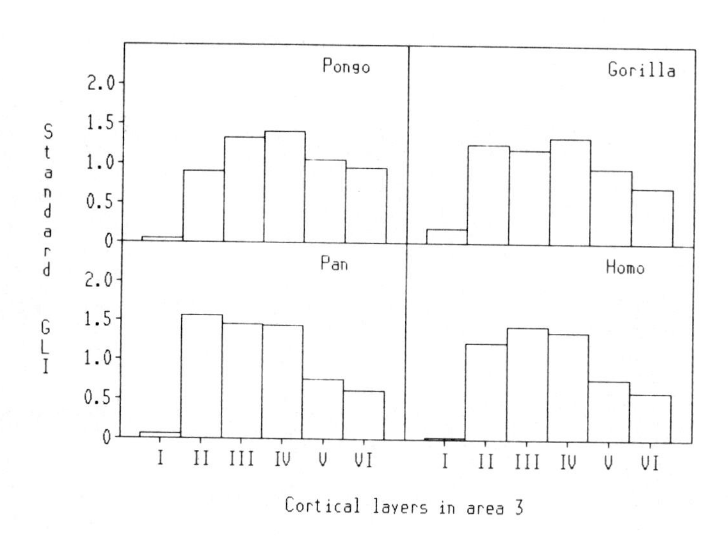

FIG. 12–13. Standardized GLIs of the different layers in area 3 of the large hominoids (*Pongo, Gorilla, Pan,* and *Homo*).

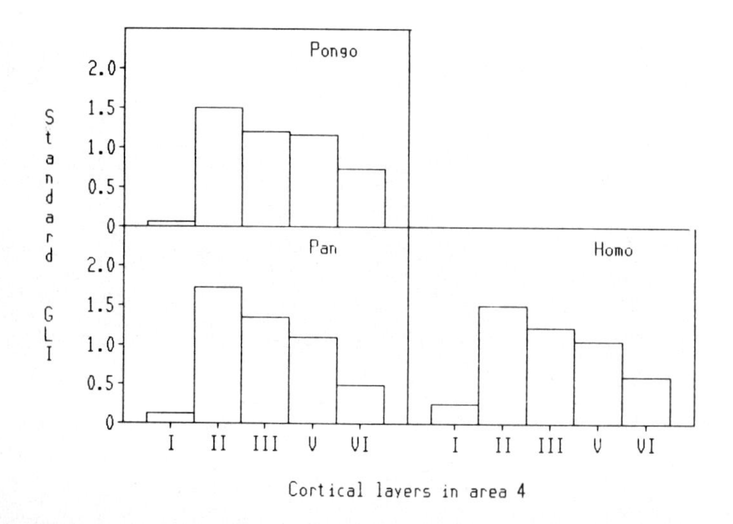

FIG. 12–14. Standardized GLIs of the different layers in area 4 of the large hominoids (data unavailable for *Gorilla*).

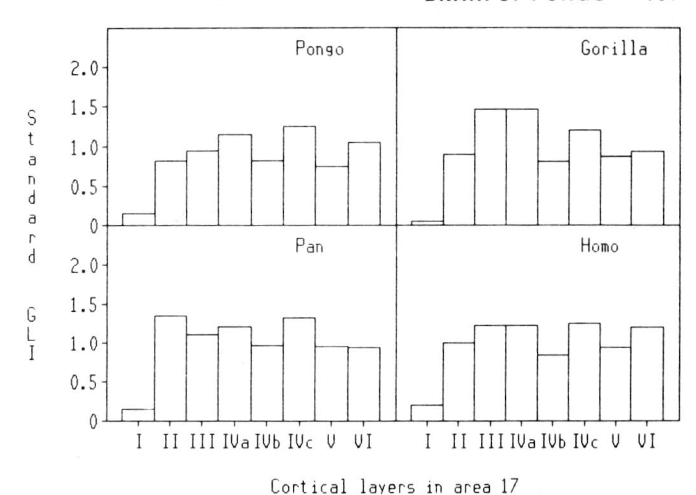

FIG. 12–15. Standardized GLIs of the different layers in area 17 of the large hominoids.

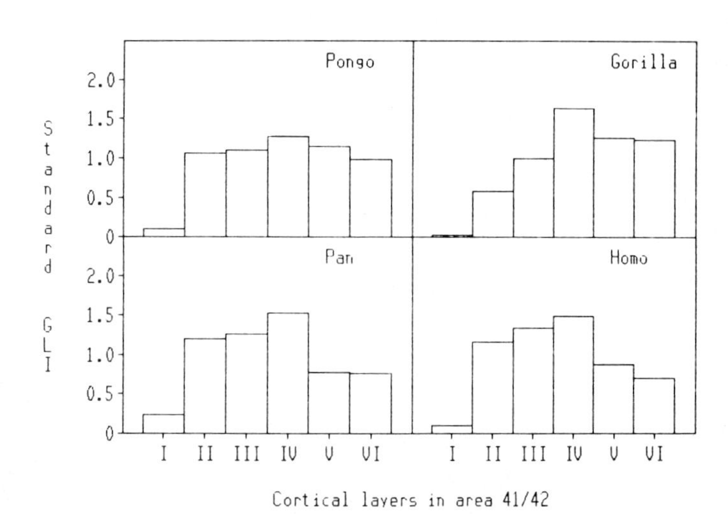

FIG. 12–16. Standardized GLIs of the different layers in areas 41/42 of the large hominoids.

VI. In summary, the cytoarchitectural structure of the primary neocortical areas is quantitatively similar in *Pongo, Gorilla, Pan,* and *Homo,* if layer thickness and cell packing density are considered as relative (standardized) values.

We can, therefore, state that the evolution of the neocortex in at least the large hominoids has led to a comparable basic structure in primary sensory cortical areas on one side and to a different laminar pattern in the primary motor areas. Our analysis of the cytoarchitectural pattern of the primary cortical areas with quantitative methods corroborates Campbell's (1905) and Brodmann's (1909) findings, which failed to uncover any fundamental differences between the apes and *Homo* in these parts of the neocortex.

ENCEPHALIZATION

The "level" of brain development, as reflected by a brain–body relationship, can be quantitatively expressed as an encephalization or progression index (Stephan, 1967a,b). This index is designed to reflect the relationship between the brain weight of a given species and that of a hypothetical insectivore of the same body size. Insectivores are taken as the basal comparison because insectivore-like animals are thought to most closely approximate the primitive mammalian condition and because their level of brain development is the lowest among extant mammals. The latter is particularly true of the Malagasy tenrecines, e.g., *Tenrec* and *Echinops* (Stephan, 1967c). The brain-body relationship in this tenrecid subfamily, which subsumes species of a variety of body sizes, can be described allometrically in a double logarithmic system with brain weight on the ordinate and body weight on the abscissa. In this system, the brain–body weight relation can be approximated by a regression line with a slope of 0.63 and a *y*-intercept of 1.6128 (Stephan et al., 1981; Stephan et al., 1986, in press). This regression line appears suitable

TABLE 12–1. Brain and Body Weights from Literature

Author	Sex	Body Weight (g)	Brain Weight (g)	Origin of the Data	
Crisp 1865	j	9,157	297.7		
Deniker and Boulart 1894	?	73,000	400	Milne-Edwards?	
	Mj	7,500	365		
Keith 1895	M	7,600	340.2	Manouvrier	
	M	18,600	325.1	Rolleston 1861	
	M	73,500	400	Owen	
	F	3,170	248	Milne-Edwards	
	M		283.5		
	M		425.8	Chapman 1880–81	
	M		470	Owen	Cranial capacities (cm³)
	F		395	Milne-Edwards	
	?		288.5	Owen	
Weber 1897	Mj	5,925	334.5	Owen	
	M	7,600	340		
	Mj	8,830	339	Rolleston, 1861	
	M	11,275	375		
	M	73,500	400	Deniker and Boulart 1895	
	M	76,500	395	Fick 1895a	
	F	18,593.5	315.5		
	F	20,200	306		
Warncke 1908	?	18,000	333	Kohlbrugge	
	Mj	7,665	365	Manouvrier	
	Mj		334	Kohlbrugge	
	M	53,600	400	Milne-Edwards	
	M	73,200	400	Deniker and Boulart 1895	
	M		283.5	Chapman 1880–81	
	Mj		325.1	Owen	
	M		339	Weber 1896	
	Mj		340.2	Rolleston 1861	
	Mj		381		
	Fj		225	Bischoff	
	Fj		248	Keith	
	F		306	Weber 1896	
Lapicque 1907	?	73,500	400	Deniker and Boulart, also Dubois 1914, Anthony 1928, Brummelkamp 1940. See Bauchot and Stephan 1969	
Filimonoff 1933	?		350		
	?		350		
	?		390		
	?		360		
Henneberg 1910	?		79.7		
Hrdlička 1925	M	49,895	347.5		
	M	54,431	371		
	M	63,503	368		
	M	72,575	326		
	M	79,379	422		
	M	83,915	384		
	M	86,183	359		
	M	90,720	395		
	F	32,659	322		
	F	32,659	326		
	F	34,019	291		

TABLE 12–1. (*Continued*)

Author	Sex	Body Weight (g)	Brain Weight (g)	Origin of the Data
	F	34,474	267	
	F	35,834	279	
	F	36,287	274.5	
	F	36,288	283	
	F	36,741	304	
	F	37,195	344	
	F	39,917	340.5	
	F	44,452	287.5	
Kappers 1926	?		293	See Bauchot and Stephan 1969
Kappers 1927	?		295	Weber, Deniker, and Boulart, Milne-Edwards, Fick
Putnam 1927	?		212.86	See Bauchot and Stephan 1969
	?		222.2	
Leboúcqú 1928	M		370	See Bauchot and Stephan 1969
Tilney and Riley 1928	?		246	
Bonin 1937	?	63,400	400	
Schultz 1940	M	79,300	444	
	F	1,500	129.15	
	F	9,980	359	See Bauchot and Stephan 1969
	F	36,900	343	
	F?	32,500	341	
	?	4,060	308.6	
Kennard and Willner 1941	M	4,500	325	
	M	6,700	353	See Bauchot and Stephan 1969
	F	25,200	340	
	F	30,000	315	
Count 1947	M	72,575	371.6	See Bauchot and Stephan 1969
	F	36,439	299.8	
Connolly 1950	Mj	5,415	337.5	
	Mj	7,554	284.0	Presumably the same material as Hrdlička 1925
	M	56,782	361.2	
	Fj	27,745	262.6	
	F	33,972	302.9	
Hopf 1965	M	73,000	400	
Napier and Napier 1967	M	69,000		
	F	37,000		
Walker 1975	M	75,000		
	F	40,000		

M = male; F = female; j = juvenile.

for making comparisons among orders of mammals (for methodological problems, see Stephan et al., 1986). A brain has an index of 1 if its brain–body weight relation lies on the regression line. A position above the regression line reflects an index greater than 1, while an index of less than 1 would be indicated by a position below the regression line. The progression index of the whole brain or of a region of the brain is the ratio between the mean brain weight of a species and the brain weight of a hypothetical insectivore of the same body weight as the species under consideration. The brain weight of the hypothetical insectivore can be estimated by extrapolation of the above-mentioned regression line.

Since we were unable to collect a sufficient sample of actual body weights and brain weights for *Pongo,* we had to rely on published data (see Table 12–1). These sources yield a mean body weight of 54,000 g and a fresh brain weight of 333 g. Sexual dimorphism is obvious: mean body weight in males is 72,000 ± 14,000 g (N= 11) and in females is 36,000 ± 4,000 g (N=14); mean brain weight in males is 359 ± 13.9 g and in females is 306 ± 14 g. Despite these sexual differences, we worked with one mean, including data from both sexes; the means can, therefore, be regarded as representative for the species without reflecting sexual dimorphism. The progression index, which was calculated with species- but not sex-specific brain and body weights in an interspecific comparison, is thought to be reflective of the evolutionary level of brain development. Treating the sexes separately is problematic in an interspecific comparison, since this differentiation would question the evolutionary unity of sexually reproducing species. An intraspecific comparison would first have to solve the problem of defining the regression line which would be used as a reference. Many more brain–body weight pairs than have been published are required before such regression lines, which would be specific for each sex, could be calculated on a sound basis.

Potential differences in body weight (Lyon, 1911, but see Napier and Napier, 1967) between the two subspecies from Borneo and Sumatra (Jacobshagen, 1979; Röhrer-Ertl, 1982, 1983) were not considered because of the lack of precise data.

It is unclear whether the brain weight data in the literature (Table 12–1) refer to fresh weights (brains weighed immediately after death without meninges) or not. Although we could not be certain about this in any of our references, we were forced to use them for lack of better data.

Given our data base, we calculated an encephalization index (cf. page 167) of 8.48 for *Pongo* (Table 12–3), i.e., the *Pongo* brain is 8.48 times larger than the brain of a hypothetical insectivore with the same body size. Bauchot and Stephan (1969) arrived at the slightly different index of 8.90, but they relied on different brain and body weight data. We also used available data on brain and body weights (Stephan et al., 1981, in press) to calculate the encephalization index of the other extant hominoids (Table 12–3). Interestingly, the orang-utan and the gorilla emerge as having similar indices of encephalization and are, in this regard, separated from *Hylobates, Pan,* and *Homo,* with *Homo* being demonstrably the most distinguished. These results are in general agreement with those of Hofman (1983), who employed a different methodological approach.

DEVELOPMENT OF DIFFERENT BRAIN COMPONENTS

The different parts of the brain (see below) were delineated on photographs of serial histological sections and the areas measured with a digitizer. The volume of each brain component was calculated as outlined in Zilles et al. (1982a). We analyzed the medulla oblongata, the cerebellum (with and without the pons), the mesencephalon including tectum and ventral parts of the tegmentum (nucleus ruber, substantia nigra, nucleus subthalmicus, griseum centrale), the diencephalon, and the telencephalon. Although the delineation of components within the tegmentum is essentially artificial, such delineations are reproducible across analyses and have been followed here in order to make use of the already published data (Stephan et al., 1981). The basis for the analysis of these major brain parts has been the histological series of the brain in our institution's collection (see Materials and Methods section). The volumetric values obtained from this series must be corrected for the effect of shrinkage during histological processing. The correction factor is the quotient of the species-specific fresh volume of the brain and the brain volume in our histological series. The species-specific fresh volume of the brain is given by the species-specific brain weight of 333 g (see page 167) divided by the specific gravity of brain tissue (1.036 g/cm^3, Stephan, 1960). This procedure not only corrects for shrinkage, but it gives brain part volumes which are more representative for the species because it uses species-specific brain weights, rather than individual brain weights, as its basis.

Within the telencephalon, we measured the neocortex (gray plus white matter without gray of area striata), the area striata, the hippocampus, the regio entorhinalis, the paleocortex (regio prepiriformis, regio retrobulbaris, tuberculum olfactorium, and corpus amygdaloideum), the septum, and the corpus striatum (nucleus caudatus and putamen). These areas were delineated in the Netherlands brain series (see Materials and Methods section) and had to be corrected for the effect of shrinkage as well. In this case the correction factor was the ratio of the fresh volume of the telencephalon of our specimen (77.2% of total brain, see Table 12–2) and the volume of the telencephalon in the Netherlands brain series.

Table 12–2 lists the fresh volumes of the different structures in *Pongo* along with their proportion in the whole brain. The calculated indices are given in Table 12–3. The corresponding indices for *Gorilla, Pan, Hylobates,* and *Homo* were calculated from the data of Stephan et al. (1981, in press).

The index for the medulla oblongata is similar in *Pongo* and *Gorilla.* The higher values in *Pan, Hylobates,* and *Homo* are probably due to an enlarged pyramidal system, which passes through the medulla before entering the spinal cord. Therefore, this result may not necessarily point out relevant differences in the function of the medulla oblongata proper, which controls various basic vegetative functions such as breathing, but it could be an indication of a more highly developed pyramidal system in *Pan, Hylobates,* and *Homo* as compared with *Pongo* and *Gorilla.*

The indices of the cerebellum are most similar among *Pan, Pongo,* and *Gorilla,* with the lowest value occurring in *Gorilla.* The index is somewhat higher in *Hylobates,* but it does not reach that of *Homo.* Since one of the main functions of the cerebellum is the coordination of motor activity, the low cerebellar volume index of *Gorilla* may well be reflective of the primarily terrestrial nature of this hominoid. Since both are actively arboreal, one would expect the index of *Pongo* to approach that of *Hylobates* most closely, but the index of *Pan* is slightly higher than the orang-utan's. The higher cerebellar index in *Pan* may, however, be related to the higher degree of neocorticalization of this hominoid (see below), since such development leads to elaborate connections of the neocortex with the cerebellum (especially the neocerebellum). Neocorticalization surely contributes in part to the high index in *Homo.*

The mesencephalic index is lowest in *Gorilla,* somewhat higher in *Pongo* and *Pan,* and highest in *Hylobates* and *Homo.* In addition to the nu-

TABLE 12–2. Volumes of Brain Components and Their Percentages of Total Brain Volume

Structure	Fresh Volume (cc^3)	Percentage of Total Brain Volume[1]
Medulla oblongata	5.5	1.8
Cerebellum (without pons)	42.9	13.9
Pons	4.3	1.4
Mesencephalon	4.0	1.3
Diencephalon	13.5	4.4
Telencephalon	238.3	77.2
Neocortex	219.8	71.2
Gray (without area striata)	129.9	42.1
Gray area striata	8.4	2.7
White matter	81.5	26.4
Hippocampus	2.7	0.9
Regio entorhinalis	1.3	0.4
Paleocortex	2.4	0.8
Regio praepiriformis	1.0	0.3
Corpus amygdaloideum	1.4	0.5
Septum	0.6	0.2
Corpus striatum	11.5	3.7
Globus pallidus	1.8	0.6

[1]Excluding ventricles and nerves.

TABLE 12–3. Indices of Brain Structure Size in *Pongo* and Other Hominoidea

Structure	Slope	*Y*-Intercept	*Pongo*	*Gorilla*	*Pan*	*Hylobates*	*Homo*
Whole brain	0.63	1.6128	8.48	8.37	11.41	10.71	30.12
Medulla oblongata	0.60	0.7930	1.28	1.17	1.49	2.02	2.01
Cerebellum (including pons)	0.67	0.5690	8.60	8.08	8.85	9.92	22.10
Mesencephalon	0.54	0.5690	2.99	2.28	3.06	3.69	5.49
Diencephalon	0.60	0.4765	6.52	6.28	8.19	10.64	14.40
Telencephalon	0.63	1.3570	10.93	11.16	15.91	14.38	43.41
Neocortex	0.67	0.4790	49.25	49.00	72.73	66.50	199.15
Hippocampus	0.57	0.5870	1.39	1.70	2.15	5.00	4.81
Regio entorhinalis	0.60	0.0805	1.51	2.20	2.67	5.26	6.61
Paleocortex (+ corpus amygdaloideum)	0.57	1.0391	0.43	0.61	0.55	0.84	1.49
Septum	0.57	−0.0346	1.34	1.75	2.03	2.36	5.10
Corpus striatum (nucleus caudatus/putamen)	0.6	0.2570	9.18	7.83	10.80	14.77	20.56

Indices are based on a regression line through the corresponding data in Tenrecinae. Slopes and *y*-intercepts are from Stephan et al., 1987a; data from genera other than *Pongo* are from Stephan et al., 1981.

cleus ruber and the substantia nigra, which are parts of the extrapyramidal system, the mesencephalon contains the colliculi superior and inferior. The superior colliculus is associated with the visual system, especially the coordination of eye movement, while the inferior colliculus is associated with the auditory system; the extrapyramidal system plays a role in movements and postural adjustments. It is tempting to try to interpret differences among the hominoids in their mesencephalic indices with regard to their respective degrees of arboreality vs. terrestriality, but at present the necessary physiological data are lacking.

The orang-utan's diencephalic index is somewhat higher than in *Gorilla,* but it is clearly lower than in *Pan; Hylobates* and *Homo* have much higher indices. Most of the diencephalon connects the telencephalon with sensory afferents and lower centers of the brain; there are elaborate reciprocal connections with the neocortex. As indicated in other studies, the development of the diencephalon appears to be correlated with that of the telencephalon, e.g., as seen in insectivores (Stephan, 1967b,c), primates (Stephan, 1972; Stephan et al., in press), and bats (Stephan and Pirlot, 1970). This also seems to be the case in the orang-utan. A detailed qualitative description of the thalamus in *Pongo* can be found in Simma (1957) and some quantitative data in Hopf (1965) and Kraus and Gihr (1965) (see also Chapter 13, this volume).

With regard to the telencephalon, *Pongo* yields the lowest value of the hominoids, even though the difference between it and the gorilla is small; *Pan* and *Hylobates* have higher indices, and *Homo* greatly surpasses all. Brain evolution in vertebrates has generally been characterized by an increased development of the telencephalon, a region which claims at least representative areas of all sensory systems in addition to wielding some control over all other brain functions; this enlargement is very obvious in mammals, especially primates (Stephan et al., 1981). The differences among the hominoids in telencephalic enlargement might be correlated with differences not only in diet, but in food procurement activity. For example, *Pongo* and *Gorilla* are essentially herbivorous and "lethargic" feeders, at least in comparison with the more omnivorous and "frenetic" chimpanzee; *Hylobates* is an active arborealist. The Old World monkeys may provide an analogous situation: the more "lethargic," herbivorous colobine monkeys have a lower telencephalic index than their more omnivorous cercopithecine counterparts (Stephan et al., 1981).

The allometric data on the neocortex are of interest because of the important role of this telencephalic component (Stephan, 1972). Neocortical volume has been regarded to be the most suitable parameter for defining the evolutionary level of brain development (Stephan, 1967b). In this regard, *Pongo* is just barely higher than *Gorilla* in neocortical volume, whereas *Pan* and, somewhat less so, *Hylobates* are clearly distinguished by higher indices; the index in *Homo* again surpasses all other hominoids. The primary representation areas of motor and sensory systems are located in a relatively small part of the neocortex. This part seems to decrease as the level of encephalization increases (Zilles et al., 1982b). Apparently, most of the remainder of the neocortex has the function of integrating information from different cortical areas. It can be justifiably argued that a stronger development of the neocortex is due, perhaps primarily, to an enlargement of the integrative apparatus (Passingham, 1973; Passingham and Ettlinger, 1974; Starck, 1982), which, in turn, could be expected to be somewhat less differentiated in *Pongo* and *Gorilla* than in the other genera because of their lower indices.

The index of the hippocampus is lowest in *Pongo,* followed by *Gorilla* and then the other genera, with *Hylobates* surpassing *Homo.* Stephan (1975) and Stephan and Manolescu (1980) pointed out that the hippocampus behaves allometrically very differently among various mammals; for example, extremely high indices are found not only in *Hylobates* and *Homo,* but in *Daubentonia* and Macroscelididae, as well. Although the hippocampus is usually associated with learning and memory, this function may be secondary to its role in activity, attentiveness, and awareness. Since *Pongo* and *Gorilla* appear to be less active than *Pan* and *Hylobates,* it would not be unjustified to interpret *Pongo*'s and *Gorilla*'s low hippocampal indices in this context. The function of the hippocampus is complex and far from fully understood, so that interpretation of the data needs further support.

The indices of the regio entorhinalis and of the septum proceed from *Pongo* (the lowest) to *Gorilla, Pan, Hylobates,* and then *Homo* (the highest). With the exception of *Hylobates,* which has a slightly higher hippocampal index than *Homo,* the sequence of indices is the same in the regio entorhinalis and septum as in the hippocampus. The parallelism of these three areas of the limbic system results from their close interconnection (Siegel et al., 1975; Steward, 1976; for review see Stephan and Andy, 1982).

The index of the paleocortex is lowest in *Pongo,* which is followed by *Pan, Gorilla, Hylobates,* and then *Homo.* The paleocortex, as delineated here, contains part of the olfactory system as well as parts of the limbic system. Since the single components of these parts seem to have developed in very different and in partly divergent directions (Stephan, 1975; Stephan and Andy, 1977), functional interpretations are not possible until data on the single components are available.

The striatum (the nucleus caudatus and the putamen) yields a lower index in *Gorilla* than in *Pongo; Pongo* is followed by higher indices in *Pan* and *Homo.* The striatum appears to have undergone an enlargement from insectivores to primates (Stephan, 1979), but there are conflicting reports on the details. Harman and Carpenter (1950) suggest that the nucleus caudatus decreased in size relative to the putamen. Hopf (1965), however, came to the opposite conclusion at least in terms of a comparison between *Pongo* and *Homo* (see also Bonin and Shariff, 1951). The striatum exhibits two main features: it is reciprocally interconnected with almost the whole neocortex and it is also an important link in the extrapyramidal system. The latter function of the striatum is of particular interest in analyzing the difference between *Pongo* and *Gorilla* in their striatum indices. The index for *Pongo* is clearly higher and lies closer to *Pan* than to *Gorilla,* and this could reflect a fundamental difference in locomotory behavior between the arboreal *Pongo* and terrestrial *Gorilla.* Differences in general among taxa in striatal indices may well be reflective of the need to control body movement in space via a well-developed extrapyramidal system.

CONCLUSIONS

In many details, the brain of *Pongo* is very similar to those of other hominoids. This is especially true with regard to cytoarchitecture and the areal arrangement of the primary areas. In particular, however, there are some features in which *Pongo* shows some special similarities primarily with *Gorilla.* For example, the brains of both genera are slender at the frontal pole in contrast to *Pan, Hylobates,* and *Homo,* in which the opposite condition prevails. Similarly, the indices of encephalization and the development of several brain parts, e.g., the telencephalon and the neocortex, are often as low as those of *Gorilla,* which tends to have the lowest indices of the hominoids. These latter similarities between *Pongo* and *Gorilla* probably reflect convergent solutions to similar biological demands.

ACKNOWLEDGMENTS

Because it was exceedingly difficult to obtain well-preserved *Pongo* material, this study would have been considerably restricted without the generosity of the Netherlands Institute of Brain Research and especially that of Dr. H. Uylings, who gave us a fixed *Pongo* brain. We also thank Dr. E. Armstrong, Louisiana State University, New Orleans, for her histological slides of a *Gorilla* brain. We would, furthermore, like to express our thanks to Dr. H. Stephan and Dr. H. Frahm (Max Planck Institute of Brain Research, Frankfurt) as well as Prof. Dr. G. Baron (Université de Montréal) for their support and helpful discussions on the delineation of problems and for their permission to use data from a forthcoming paper (Stephan et al., in press). The English was smoothed by Johannes Bernbeck. This work has been supported by the Deutsche Forschungsgemeinschaft (grant Zi 192/4-5).

REFERENCES

Anthony, R. 1928. Anatomie comparée du cerveau. Doin, Paris.

Bailey, P., Bonin, G. von and McCulloch, W. S. 1950. Isocortex of the Chimpanzee. University of Illinois Press, Urbana.

Bauchot, R. and Stephan, H. 1969. Encephalisation et niveau evolutif chez les simiens. *Mammalia, 33:*225–275.

Beevor, C. E. and Horseley, V. 1890–1891. A record of the results obtained by electrical excitation of the so-called motor cortex and internal capsule in an Orang Outang *(Simia satyrus). Proceedings of the Royal Society of London 47:*159–160.

Bischoff, T. L. W. 1870. Die Großhirnwindungen des Menschen mit Berücksichtigung ihrer Entwicklung bei dem Foetus und ihrer Anordnung bei den Affen. *Abhandlungen der Mathematisch-physikalischen Klasse der Koeniglisch bayerische Akademie der Wissenschaften, München, 10:*391–497.

Bischoff, T. L. W. 1876. Über das Gehirn eines Orang-Outan. *Sitzungsberichte der Mathematisch-physikalischen Klasse der bayrischen Akademie der Wissenschaften, München, VI:*193–205.

Bolk, L. 1902. Beiträge zur Affenanatomie. 2. Über das Gehirn von Orang-Utan. *Peterus Camper Nederlandsche Bijdragen tot se Anatomie, 1:*25–84.

Bonin, G. von. 1937. Brain weight and body weight of mammals. *Journal of General Psychology, 16:*379–389.

Bonin, G. von and Shariff, G. A. 1951. Extrapyramidal nuclei among mammals. A quantitative study. *Journal of Comparative Neurology, 94:*427–439.

Braak, H. 1980. Architectonics of the Human Telencephalic Cortex. Springer-Verlag, New York.

Brodmann, K. 1909. Vergleichende Lokalisationslehre der Großhirnrinde in ihren Prinzipien dargestellt auf Grund des Zellenbaues. J. A. Barth, Leipzig.

Brummelkamp, R. 1940. Brain weight and body size. A study of the cephalization problem. *Nederlandsche Akademie van Wetenschappen Verhandelingen, 39:*1–57.

Campbell, A. W. 1905. Histological Studies on the Localisation of Cerebral Function. Cambridge University Press, Cambridge.

Chapman, H. C. 1880–81. On the structure of the Orang Outang. *Proceedings of the Academy of Natural Sciences, Philadelphia, 1880–1881:*160–175.

Connolly, G. J. 1950. External Morphology of the Primate Brain. Charles C Thomas, Springfield, IL.

Count, E. W. 1947. Brain and body weight in man: their antecedents in growth and evolution. *Annals of the New York Academy of Sciences, 46:*993–1122.

Crisp, E. 1865. On the relative weight of the brain, and on the external form of this organ, in relation to the intelligence of the animal. 35th Meeting of the British Association for the Advancement of Science, *1865:*84–85.

Deniker, K. J. and Boulart, R. 1894. Sur divers points de l'anatomie de l'Orang-Outan. *Comptes Rendus de l'Academie de Sciences, Paris, 119:*235.

Deniker, J. and Boulart, R. 1895. Notes anatomiques sur les sacs laryngiens, les excroissances adipeuses, les poumons, les cerveaux etc. de Orang-Outans. *Nouveaux Archives de Muséum d'Histoire Naturelle, 7:*35–56.

Dubois, E. 1914. Die gesetzmäßige Beziehung von Hirnmasse zur Körpergröße bei den Wirbeltieren. *Zeitschrift für Morphologie und Anthropologie, 18:*323–350.

Economo, C. von and Koskinas, G. N. 1925. Die Cytoarchitektonik der Hirnrinde des erwachsenen Menschen. Springer-Verlag, Berlin.

Fick, R. 1895a. Vergleichend-anatomische Studien an einem erwachsenen Orang-Utang. *Archiv für Anatomie und Physiologie, Leipzig, 1895:*1–100.

Fick, R. 1895b. Beobachtungen an einem zweiten erwachsenen Orang-Utang und einem Schimpansen. *Archiv für Anatomie und Physiologie, Leipzig, 1895:*289–318.

Filimonoff, I. N. 1933. Über die Variabilität der Großhirnrindenstruktur. Mitteilung III. Regio occipitalis bei den höheren und niederen Affen. *Journal fuer Psychologie und Neurologie, 45:*69–137.

Förster, O. 1936. Motorische Felder und Bahnen. *In* Handbuch der Neurologie, vol. 6, pp. 1–357, ed. O. Bumke and O. Förster. Springer-Verlag, Berlin.

Grünbaum, A. S. F. and Sherrington, C. S. 1901. Observations on the physiology of the cerebral cortex of some of the higher apes. *Proceedings of the Royal Society of London, 69:*206–208.

Grünbaum, A. S. F. and Sherrington, C. S. 1903. Observations on the physiology of the cerebral cortex of the anthropoid apes. *Proceedings of the Royal Society of London, 72:*152–155.

Harman, P. J. and Carpenter, M. V. 1950: Volumetric comparison of the basal ganglia of various primates including man. *Journal of Comparative Neurology, 94:*427–439.

Henneberg, R. 1910. Messungen der Oberflächenausdehnung der Großhirnrinde. *Journal für Psychologie und Neurologie, 17:*144–158.

Hofman, M. A. 1983. Encephalization of hominids: evidence for the model of punctuationalism. *Brain, Behavior and Evolution, 22:*102–117.

Hopf, A. 1965. Volumetrische Untersuchungen zur vergleichenden Anatomie des Thalamus. *Journal für Hirnforschung, 8:*25–38.

Hrdlička, A. 1925. Weight of the brain and internal organs in American monkeys (with data on brain weight of other apes). *American Journal of Physical Anthropology, 8:*201–211.

Hubel, D. and Wiesel, T. 1977. Functional architecture of macaque visual cortex. *Proceedings of the Royal Society of London, series* **B**, *198:*1–59.

Jacobshagen, B. 1979. Morphometric studies in the taxonomy of the Orang Utan (*Pongo pygmaeus* L. 1760). *Folia Primatologica, 32:*29–34.

Jones, E. G. 1975. Lamination and differential distribution of thalamic afferents within the sensory motor cortex of the squirrel monkey. *Journal of Comparative Neurology, 160:*167–204.

Kappers, C. U. A. 1926. The relation of the cerebellum weight to the total brain weight in human races and in some animals. *Proceedings of the Koninklijke Nederlandsche Akademie van Wetenschappen, Amsterdam, 29:*129–141.

Kappers, C. U. A. 1927. The influence of the cephalization coefficient and body size upon the form of the forebrain in mammals. *Proceedings of the Koninklijke Nederlandsche Akademie van Wetenschappen, Amsterdam, 31:*65–80.

Keith, A. 1895. The growth of brain in men and monkeys, with a short criticism of the usual method of stating brain-ratios. *Journal of Anatomy, 29:*282–303.

Kennard, M. A. and Willner, M. D. 1941. Findings at autopsies of 70 anthropoid apes. *Endocrinology, 28:*967–976.

Kraus, C. and Gihr, M. 1965. Statistischer Beitrag zur Gliederung des Nucleus anterior thalami. *Journal für Hirnforschung, 8:*39–45.

Kuhlenbeck, H. 1978. The Central Nervous System of Vertebrates, vol. 5/II. S. Karger, Basel.

Kükenthal, W. and Ziehen, T. 1895. Untersuchungen über die Großhirnfurchen der Primaten. *Jena Zeitschrift für Naturwissenschaften, 29:*1–122.

Lapicque, L. 1907. Tableau géneral des poids somatique et encephalique dans les espèces animales. *Bulletin et Mémoire du Société d'Anthropologie, 8:*248–262.

Leboúcqú, G. 1928. Le rapport poid-surface dans le cerveau des singes. *Comptes Rendus de l'Association des Anatomistes, Prague,* 1–6.

Leyton, A. S. F. and Sherrington, C. S. 1917. Observations on the excitable cortex of the Chimpanzee, Orang-Utan, and Gorilla. *Quarterly Journal of Experimental Physiology, 11:*135–222.

Lyon, M. W. 1911. Mammals collected by Dr. W. L. Abbott on Borneo and some of the small adjacent islands. *Proceedings of the U.S. National Museum of Natural History, 40:*53–146.

Mauss, T. 1908. Die faserarchitektonische Gliederung der Großhirnrinde bei den niederen Affen. *Journal für Psychologie und Neurologie, 13:*263–325.

Mauss, T. 1911. Die faserarchitektonische Gliederung des Cortex cerebri von den anthropomorphen Affen. *Journal für Psychologie und Neurologie 18:*410–467.

Merker, B. 1983. Silver staining of cell bodies by means of physical development. *Journal of Neuroscience Methods 9:*235–241.

Napier, J. R. and Napier, P. H. 1967. A Handbook of Living Primates. Academic Press, New York.

Passingham, R. E. 1973. Anatomical differences between the neocortex of man and other primates. *Brain, Behavior and Evolution, 7:*337–359.

Passingham, R. E. and Ettlinger, G. 1974. A comparison of cortical functions in man and other primates. *International Review of Neurobiology, 16:*233–299.

Penfield, W. and Rasmussen, T. 1950. The Cerebral Cortex of Man. Macmillan, New York.

Penfield, W. and Welch, K. 1951. The supplementary motor area of the cerebral cortex. A clinical and experimental study. *Archives of Neurology and Psychiatry, 66:*289–317.

Putnam, I. K. 1927. The proportion of cerebellar to total brain weight in mammals. *Proceedings of the Academy of Science, Amsterdam, 31:*155–168.

Retzius, G. 1906. Das Affenhirn. Gustav Fischer, Jena.

Röhrer-Ertl, O. 1982. Über Subspecies bei *Pongo pygmaeus. Spixiana, 5:*317–321.

Röhrer-Ertl, O. 1983. Zur Erforschungsgeschichte und Namensgebung bei Orang Utan mit Kurzbibliographie. *Spixiana, 6:*301–332.

Rolleston, G. 1861. On the affinities of the brain of the Orang-Utang. *Natural History Record, London, 1861:*201.

Sanides, F. 1968. The architecture of the cortical taste nerve areas in squirrel monkey *(Saimiri sciureus)* and their relationships to insular and prefrontal regions. *Brain Research, 8:*97–124.

Schleicher, A., Zilles, K. and Kretschmann, H.-J. 1978. Automatische Registrierung und Auswertung eines Grauwertindex in histologischen Schnitten. *Anatomischer Anzeiger, 114:*413–415.

Schultz, A. H. 1940. The relative size of the cranial capacity in primates. *American Journal of Physical Anthropology, 28:*273–287.

Siegel, A., Ohgami, S and Edinger, H. 1975. Projections of the hippocampus to the septal area in the squirrel monkey. *Brain Research, 99:*274–260.

Simma, K. 1957. Der Thalamus der Menschenaffen. Eine vergleichend-anatomische Untersuchung. *Psychiatria et Neurologia, Basel, 134:*145–175.

Smith, W. K. 1944. The frontal eye fields. *In* The Precentral Motor Cortex, pp. 307–342, edited by P. C. Bucy. University of Illinois Press, Urbana.

Starck, D. 1982. Vergleichende Anatomie der Wirbeltiere, Bd. 3. Springer-Verlag, Berlin.

Stephan, H. 1960. Methodische Studien über den quantitativen Vergleich architektonischer Struktureinheiten des Gehirns. *Zeitschrift für Wissenschaftliche Zoologie, 164:*143–172.

Stephan, H. 1967a. Zur Entwicklungshöhe der Primaten nach Merkmalen des Gehirns. *In* Neue Ergebnisse der Primatologie, pp. 108–119, ed. D. Starck, R. Schneider, and H.-J. Kuhn. Fischer, Stuttgart.

Stephan, H. 1967b. Quantitative Vergleiche zur phylogenetischen Entwicklung des Gehirns der Primaten mit Hilfe von Progressionsindices. *Mitteilungen des Max-Planck-Gesselschaft, 2:*63–86.

Stephan, H. 1967c. Zur Entwicklungshöhe der Insektivoren nach Merkmalen des Gehirns und die Definition der "Basalen Insektivoren." *Zoologischer Anzeiger, 179:*177–199.

Stephan, H. 1972. Evolution of primate brains: a comparative anatomical investigation. *In* The Functional and Evolutionary Biology of Primates, pp. 155–174, ed. R. Tuttle. Aldine/Atherton, Chicago.

Stephan, H. 1975. Allocortex. *In* Handbuch der mikroskopischen Anatomie des Menschen, Bd. IV/9, ed. W. Bargmann. Springer-Verlag, Berlin.

Stephan, H. 1979. Comparative volumetric studies on striatum in insectivores and primates: evolutionary aspects. *Applied Neurophysiology, 42:*78–80.

Stephan, H. and Andy, O. J. 1977. Quantitative comparison of the amygdala in insectivores and primates. *Acta Anatomica, 98:*130–153.

Stephan, H. and Andy, O. J. 1982. Anatomy of the limbic system. *In* Stereotaxy of the Human Brain, pp. 269–292, ed. G. Schaltenbrand and A. E. Walker. Thieme, Stuttgart.

Stephan, H., Baron, G. and Frahm, H. in press. Comparative size of brain and brain components. *In* Comparative Primate Biology: Neuroscience, ed. H. D. Steklis. Alan R. Liss, New York.

Stephan, H., Baron, G., Frahm, H. and Stephan, M. 1986. Größenvergleiche an Gehirnen und Hirnstrukturen von Säugern. *Zeitschrift fuer Mikroskopisch-Anatomische Forschung, Leipzig, 100:*189–212.

Stephan, H., Frahm, H. and Baron, G. 1981. New and revised data on volumes of brain structures in insectivores and primates. *Folia Primatologica, 35:*1–29.

Stephan, H. and Manolescu, J. 1980. Comparative investigations on hippocampus in insectivores and primates. *Zeitschrift für Mikroskopisch-Anatomische Forschung, Leipzig, 94:*1025–1050.

Stephan, H. and Pirlot, P. 1970. Volumetric comparison of brain structures in bats. *Zeitschrift für Zoologische Systematik und Evolutionsforschung, 8:*200–236.

Steward, O. 1976. Topographic organization of the projections from the entorhinal area to the hippocampal formation in the rat. *Journal of Comparative Neurology, 167:*285–315.

Tiedemann, F. 1827. Das Hirn des Orang-Outangs mit dem des Menschen verglichen. *Zeitschrift für Physiologie, 2:*17–28.

Tilney, F. and Riley, M. A. 1928. The Brain from Ape to Man. Hoeber, New York.

Vogt, B. A. 1976. Retrosplenial cortex in the rhesus monkey: a cytoarchitectonic and Golgi study. *Journal of Comparative Neurology, 169:*63–98.

Wagman, I. H. 1964. Eye movements induced by electrical stimulation of cerebrum in monkeys and their relationship to bodily movements. *In* The Oculomotor System, pp. 18–39, ed. M. B. Bender. Hoeber, New York.

Walker, E. P. 1975. Mammals of the World, 3d edition. Johns Hopkins University Press, Baltimore.

Warncke, P. 1908. Mitteilung neuer Gehirn- und Körpergewichtsbestimmungen bei Säugern, nebst Zusammenstellung der gesamten bisher beobachteten absoluten und relativen Gehirngewichte bei den verschiedenen Species. *Journal für Psychologie und Neurologie, 12:*355–403.

Weber, M. 1897. Vorstudien über das Gehirngewicht der Säugetiere. *Festschrift Gegenbaur, 3:*105–123.

Zilles, K., Rehkämper, G. and Schleicher, A. 1979a. A quantitative approach to cytoarchitectonics. V. The areal pattern of the cortex of *Microcebus murinus* (E. Geoffroy, 1828), (Lemuridae, Primates). *Anatomy and Embryology, 157:*269–289.

Zilles, K., Rehkämper, G., Stephan, H. and Schleicher, A. 1979b. A quantative approach to cytoarchitectonics. IV. The areal pattern of the cortex of *Galago demidovii* (E. Geoffroy, 1796), (Lorisidae, Primates). *Anatomy and Embryology, 157:*81–103.

Zilles, K., Schleicher, A. and Kretschmann, H.-J. 1978a. A quantitative approach to cytoarchitectonics. I. The areal pattern of the cortex of *Tupaia belangeri. Anatomy and Embryology, 153:*195–212.

Zilles, K., Schleicher, A. and Kretschmann, H.-J. 1978b. A quantitative approach to cytoarchitectonics. II. The allocortex of *Tupaia belangeri. Anatomy and Embryology, 154:*335–352.

Zilles, K., Schleicher, A. and Pehlemann, F. W. 1982a. How many sections must be measured in order to reconstruct the volume of a structure using serial sections? *Microscopica Acta, 86:*339–346.

Zilles, K., Stephan, H. and Schleicher, A. 1982b. Quantitative cytoarchitectonics of the cerebral cortices of several prosimian species. *In* Primate Brain Evolution, pp. 177–201, ed. E. Armstrong and D. Falk. Plenum Press, New York.

Zilles, K., Wree, A., Schleicher, A. and Divac, I. 1984. The monocular and binocular subfields of the rat's primary visual cortex. A quantitative morphological approach. *Journal of Comparative Neurology, 226:*391–402.

13

The Diencephalon: A Comparative Review

ESTE ARMSTRONG AND G. THOMAS FROST

An orang-utan brain has not been the subject of as many studies as have those of the chimpanzee or gorilla. While descriptions of its major diencephalic nuclei exist (Feremutsch, 1963; Hopf, 1965; Simma, 1957), these data have not been quantitatively compared to those of the better-known pongids. Descriptions and comparisons are important for learning about the orang-utan, an interesting animal in its own right, and also for shedding some light on the evolution of the hominoid brain. A better appreciation of human brain evolution can also be acquired if the ape pattern, and not just the African ape pattern, is understood.

A major problem for comparative work on an endangered species like *Pongo* is that a limited number of specimens are available for study. Such is the case for this review: we had access to the brains of only two orang-utans. One of these, from an immature male, contained the diencephalon and was used only for qualitative observations. The serial sections obtained from this brain were stained with cresyl violet as well as a silver stain which colored individual neuronal perikarya (chapter 12, this volume). Use of a Nissl stain in this specimen permitted us to compare the neuronal appearance with those reported for other orang-utans and apes. The second orang-utan brain is from the collection of the Netherlands Institute of Brain Research and is also described by Zilles and Rehkämper (chapter 12, this volume).

Previous comparative studies have established that quantitative rather than qualitative differences are expected among hominoid brains (Armstrong and Falk, 1982). However, quantitative comparisons are difficult when both brain and body weights are unknown. Thus, for our comparisons, we report the absolute volumes of particular areas as well as their size in relation to that of the lateral geniculate body.

Absolute size is a critical variable for comparisons, but shrinkage produced during the fixation and processing of tissue can cause marked quantitative variations. Since most of the shrinkage occurs in white matter and since most of the nuclei of this study are predominantly gray matter, the absolute figures reported here are of value. Nuclei in specimens of the same species which were prepared in different laboratories show ranges of individual variation similar to those of specimens prepared in the same laboratory (personal observation). In both situations the coefficient of variation is close to that reported for individual variation determined in one laboratory (Stephan et al., 1970).

To correct for the effects of possible shrinkage, we compared our volumes to that of the lateral geniculate body. This nucleus can be used as a morphometric standard because its boundaries are extremely well defined and its volume remains relatively constant among hominoids, especially among the great apes (Armstrong, 1979, 1982). Wherever possible, both volumes and these ratios are used to compare our orang-utan data with values reported in the literature. In Hopf's (1965) quantitative study of an orang-utan thalamus, only ratios are compared. Because he corrected his volumes for shrinkage, but did not publish the correction factor, his absolute values are not directly comparable to ours.

The diencephalon, a small but central component of the central nervous system, is divisible into four regions: the (dorsal) thalamus, the hypothalamus, the epithalamus, and the subthalamus. The central location of these structures, many of which are associated with a discrete function, makes the diencephalon a key region for comparative studies. In this initial comparative review of the orang-utan diencephalon, we shall examine major nuclear divisions within all four regions.

THALAMUS

The thalamus is by far the most completely studied component of the diencephalon. It is the largest part of the diencephalon and, as the major gateway to the cortex, its structure–function relationships are thought to be critical for understanding the cortex (Walker, 1966). An integrative center, the thalamus can be subdivided many ways: our subdivisions are the metathalamus, the lateral tier, the intralaminar nuclei, the medial tier, and the anterior portion. In each instance the size and major cytoarchitectural features as well as the primary functions or connections as they are presently understood from studies in anthropoids will be described.

Metathalamus

The lateral and medial geniculate bodies are two sensory relay nuclei of the dorsal thalamus which constitute the metathalamus. They are the same approximate size, irrespective of brain weight, among great apes and humans.

The dorsal nucleus of the lateral geniculate body is a laminated structure which has four outer parvicellular layers and two inner magnocellular layers (Figs. 13–1 and 13–2). The neurons take a medium stain in pars parvicellularis and a dark stain in pars magnocellularis. The latter portion contains not only large neurons, as its name implies, but also small-to-medium-sized ones. Connections from the retina and to and from the striate cortex are well established in primates, including hominoids (Hendrickson et al., 1970; Jackson et al., 1969; Kaas et al., 1977; Walker and Fulton, 1938). The lateral geniculate body of this orang-utan has a volume of 60.7 mm^3, a value which is similar to those of the other great apes (gorilla, 51.0 mm^3; chimpanzee, 56.3 mm^3) and human (73.6 mm^3). Hopf (1965) also observed that orang-utans and humans have similarly sized lateral geniculate bodies.

A small pars ventralis or grisea pregeniculatus (Walker and Fulton, 1938) is dorsolateral to the

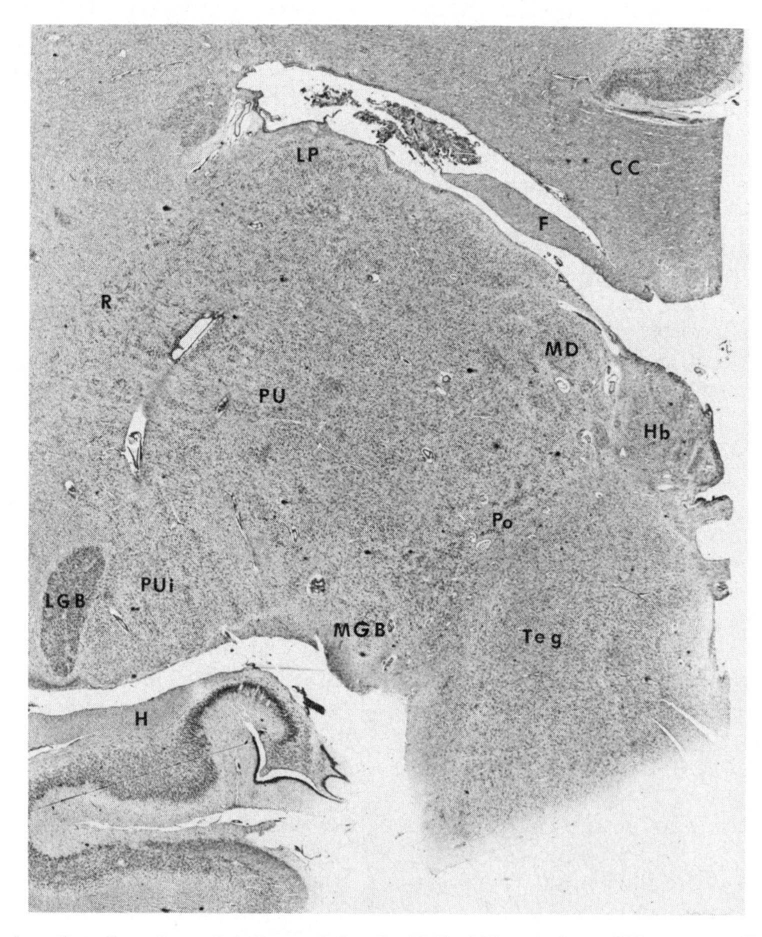

FIG. 13–1. Coronal section through caudal diencephalon (×10.1). Abbreviations: CC, corpus callosum; F, fornix; H, hippocampus; Hb, habenular nuclei; LGB, lateral geniculate body; LP, lateral posterior nucleus; MD, mediodorsal nucleus; MGB, medial geniculate body; PO, posterior complex; PU, pulvinar; PU_i, pulvinar, pars inferior; R, reticular nucleus; Teg, brainstem tegmentum.

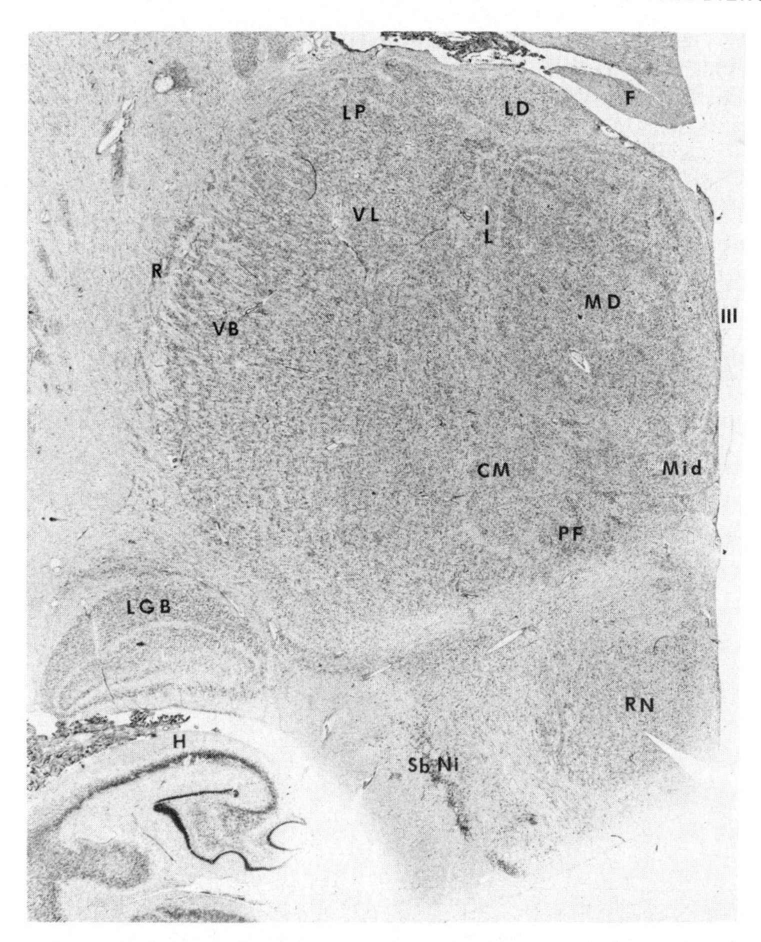

FIG. 13–2. Cornonal section at level of midthalamus (×10.1) Abbreviations: CM, centromedian nucleus; F, fornix; H, hippocampus; IL, intralaminar nuclei; LGB, lateral geniculate body; LD, lateral dorsal nucleus; LP, lateral posterior nucleus; MD, mediodorsal nucleus; MID, midline nuclei; PF, parafasciculus; R, reticular nucleus; RN, red nucleus; SbNi, substantia nigra; VB, ventrobasal complex; VL, ventrolateral complex; III, third ventricle.

lateral geniculate body (Fig. 13–2). This nucleus has been noted in all primates, including these orang-utans.

The medial geniculate body, which is intimately associated with the auditory system, lies medial to the lateral geniculate body (Fig. 13–1). In hominoids, a division into a pars parvicellularis (= pars ventralis [Morest, 1964]) and a pars magnocellularis can be seen. The former component is the larger and its major connections to the auditory cortex are tonotopic. This organization is thought to be similar among all anthropoids (Imig et al., 1977). The smaller pars magnocellularis lies dorsomedial to pars parvicellularis and, in addition to auditory information, it receives afferents from the spinal cord and brainstem tegmentum (Casseday et al., 1976). The boundary between the pars parvicellularis and pars magnocellularis is not clear and thus the entire medial geniculate body was measured. The volume of 28.4 mm^3 is similar to that of another orang-utan (33.9 mm^3) (Blinkov and Zvorykin, 1950) and to other great apes (gorilla, 31.9 mm^3, $N = 1$; chimpanzee, 34.4 mm^3, $N = 2$) (Armstrong, 1979; Blinkov and Zvorykin, 1950). The medial-to-lateral geniculate body ratio among great and lesser apes is approximately 0.5; it is only slightly higher (0.8) in humans. The volumes of both the lateral and medical geniculate bodies have been evolutionarily conserved among all hominoids.

Lateral Tier

The ventral subdivision of the lateral thalamus consists of the ventroanterior, ventrolateral, and ventrobasal complexes, each of which can be further subdivided. The ventroanterior complex, the most rostral component of the ventral group, has two nuclei. The major portion, pars parvicellularis, contains clusters of large to medium-sized cells, while the smaller component, pars magnocellularis, is formed by small clusters of large neurons caudal to the mamillothalamic tract. These cytoarchitectural characteristics of the orang-utan thalamus are shared by other anthropoids. In nonhominoid primates, the ventroanterior nuclei receive information from the motor nuclei, the globus pallidus, and substantia nigra, as well as from the intralaminar thalamic nuclei and the frontal cortex.

Caudal to the ventroanterior nucleus is the ventrolateral complex (Figs. 13–2 and 13–3), a group of structures also involved with motor systems, particularly the cerebellum (Asanuma et al., 1983) and the motor cortex (Strick, 1975). It, too, is made up of several components. Similar to the arrangement in other hominoids, the orang-utan ventrolateral complex is divided into a pars caudalis, medialis, and oralis. In pars caudalis and oralis, the neurons are heterogeneous in appearance and arrangement; the ventrally located pars medialis is cytoarchitecturally distinct, having small and evenly distributed neurons. The orang-utan pars oralis has a more clustered neuronal arrangement than does its pars caudalis.

Boundaries between the ventroanterior and ventrolateral nuclei were not determined in this study. The caudal boundary between the ventrolateral and the ventroposterior lateral nucleus could be delimited by the appearance of larger cells in the latter. Typically, the transition in cell types between these two regions is gradual, so that in places the identification of a boundary is somewhat arbitrary. We drew the boundary around the large-celled ventroposterior lateral nucleus and placed transitional regions within the ventrolateral nucleus. The ventroanterior-

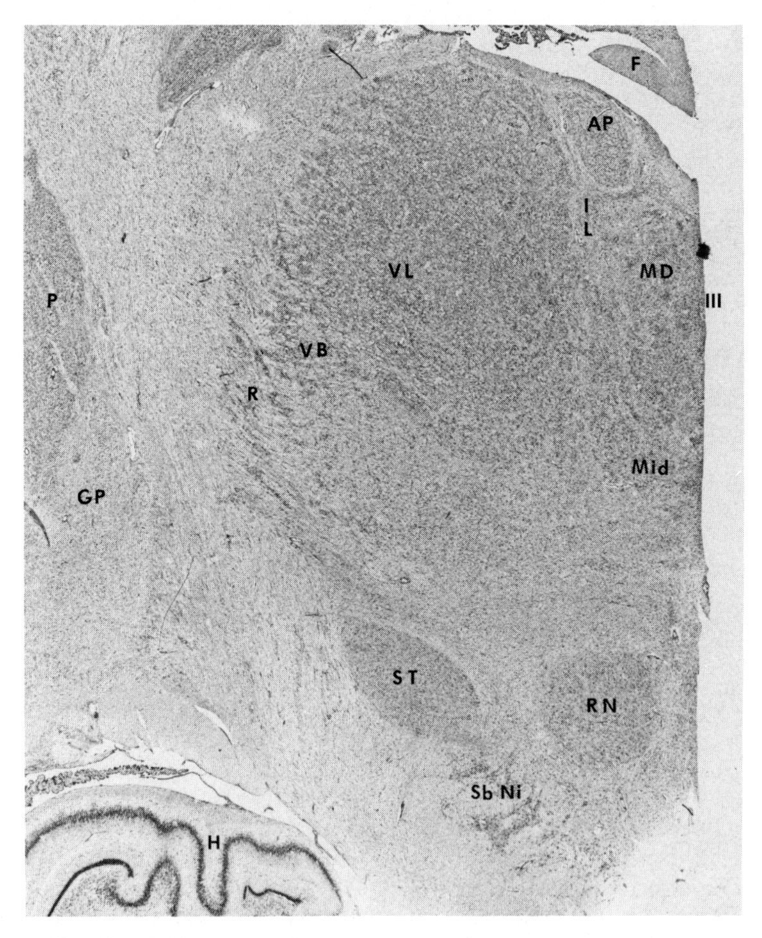

FIG. 13–3. Coronal section at level of subthalamus (×10.1). Abbreviations: AP, anterior principal nucleus; F, fornix; BP, globus pallidus; H, hippocampus; IL, intralaminar nuclei; MD, mediodorsal nucleus; mid, midline nuclei; P, putamen; R, reticular nucleus; RN, red nucleus; SbNi, substantia nigra; ST, subthalamus; VB, ventrobasal complex; VL, ventrolateral complex; III, third ventricle.

lateral complex was measured and compared as a unit.

In all hominoids the ventroanterior-lateral complex increases in size as a correlate of brain enlargement (Armstrong, 1980b, 1982). Consequently, the motor to lateral geniculate body ratio becomes larger as one proceeds from gibbons (4:1) to the great apes (chimpanzee, 6:1; gorilla, 8:1) and then humans (11:1) (Armstrong, 1982). The orang-utan ventroanterior-lateral complex of this study has a mass of 353 mm^3, which is six times the size of its lateral geniculate body. This ratio matches that of the chimpanzee, probably a result of their equivalent brain sizes. The slightly higher ratio of the gorilla reflects its somewhat larger brain (Tobias, 1971).

The ventrobasal complex (Figs. 13–2 and 13–3) is the major somatosensory and kinesthetic relay center for the thalamus. This complex is composed of ventroposterolateral and medial portions, which project information from the body and face, respectively (Jones, 1983). Our measurements incorporated only the large-celled portion of pars medialis; the small-celled subdivision of pars medialis, which relays taste information, was omitted, as was pars inferior. In the orang-utan, the small cells of the lightly staining pars inferior and the homogeneously staining medium-sized cells of pars medialis qualitatively resemble the homologous nuclei of other hominoids. A large-celled anterior portion of the ventroposterior lateral complex, the pars oralis, relays vestibular and cerebellar information (Asanuma et al., 1983), and is within the ventrobasal boundaries of this study.

The large-celled ventrobasal complex in this orang-utan measured 261 mm^3 and was four times as large as its lateral geniculate body, a size which is slightly larger than those observed in the African apes (chimpanzee, 144.8 mm^3, 2.5:1; gorilla, 178.9 mm^3, 3.5:1). It appeared to us that the large-celled portion of pars medialis was especially large in the orang-utan specimen. An enlargement in this region may be the result of more neurons and/or neuropil for the integration of sensory afferents from the surface of the enlarged cheek pads of the male orang-utan. Such an enlargement would be expected, since animals with enlarged body parts have appropriately enlarged ventrobasal components (Welker and Johnson, 1965).

The pulvinar and lateral posterior complex (Figs. 13–1 and 13–2), which is the largest thalamic complex in hominoids (Armstrong, 1981), occupies the posterior pole of the thalamus in orang-utans as it does in all other primates (Harting et al., 1972; Ogren, 1982). The neurons in this region are small-to-medium in size and moderate in staining characteristics. While neuronal morphology and arrangement appear more homogeneous in the pulvinar than in the lateral posteriour nucleus, the differentiation of the two nuclei is not clear and they were measured as a unit. The considerably larger cells of the ventral and intralaminar nuclei clarify most of the medial borders, but the boundary between the mediodorsal nucleus and pulvinar is less clear and sometimes arbitrary.

The pulvinar lies posterior to the lateral posterior nucleus and can be subdivided further. Subnuclei pars oralis, medialis, lateralis, and inferior have been defined more by topography than by architecture (Crouch, 1934; Van Buren and Borke, 1972; Walker, 1938), but studies of their connections support the recognition of these internal divisions. The pars inferior receives afferents from the superior colliculus and is intimately associated with the visual system (Ogren and Hendrickson, 1976). The rest of the pulvinar is richly, but differentially, interconnected with association cortices in the occipital, parietal, temporal, and frontal lobes (Bos and Benevento, 1975; Markowitsch et al., 1985). The lateral posterior nucleus is also interconnected with the parietal lobe (Pearson et al., 1978).

The pulvinar appears to be an outgrowth of the lateral posterior nucleus (Walker, 1938) and is more prominent in primates than in other mammalian orders (Diamond and Hall, 1969). The hominoid pulvinar-lateral posterior complex is large and is absolutely largest in the human brain (Armstrong, 1981, 1982; Harting et al., 1972; Kuhlenbeck, 1954). However, since the size of this complex scales with brain size, neither the human nor the orang-utan pulvinar-lateral posterior region deviates from its expected size. The pulvinar-lateral-posterior complex in the orang-utan specimen measured 371 mm^3, which falls between the values for the gorilla (414 mm^3) and chimpanzee (323 mm^3). While the pulvinar-lateral posterior region of the gibbon is about 4.5 times bigger than the lateral geniculate body, in the great apes this region is approximately six to eight times larger than the lateral geniculate body (Armstrong, 1982; Hopf, 1965).

Recent evidence links the primate laterodorsal nucleus (Fig. 13–2) with limbic and association circuitries, particularly to the posterior cingulate and parietal cortices (Mikol et al., 1977). The moderately staining neurons of this nucleus are medium-sized. While fibers of the internal medullary lamina clearly define its medial borders, its lateral border with the lateral posterior-pulvinar complex is in places unclear. The orang-utan's laterodorsal volume of 11.7 mm^3 resembles that of the African apes absolutely and relatively (gorilla, 11.6 mm^3, 22.7%; chimpanzee, 12.9 mm^3, 22.9%). the human laterodorsal nucleus, similar to the anterior principal nucleus, is relatively enlarged, being about half (52.7%) as large as its lateral geniculate body.

The posterior thalamic nuclei (Fig. 13–1) include the large-celled suprageniculate and limitans nuclei. This complex is found along the ventral border of the thalamus, medial to the medial geniculate body and ventral to the ventroposterior and pulvinar nuclei. Recent evidence suggests that this region may be associated with pain mechanisms (Albe-Fessard et al., 1985). In hominoids, the suprageniculate and limitans nuclei both have large and heterogeneous neurons, but the more medial part, the nucleus limitans, has an increased number of fusiform cells. The cytoarchitectural appearance of the neurons in this region resembles that of other hominoids.

Intralaminar Nuclei

Intralaminar nuclei (Fig. 13–2 and 13–3) are neuronal populations within the internal medullary lamina, fibers which separate the medial and lateral portions of the thalamus. As in other hominoids and primates in general, the neurons in these regions are generally large, dark, and multipolar or spindle-shaped. Several nuclear groups are recognized within this aggregation; they include the nuclei paracentralis, centralis lateralis, centromedianus, and parafascicularis. Sometimes the nucleus centrolateralis is subdivided with an additional nucleus, the centralis superior lateralis, being identified (= n. cuccularis [Hassler, 1959]). As a group, the intralaminar nuclei have major afferents from the brainstem reticular formation, spinothalamic tracts, and cerebellum, and project to the basal ganglia as well as to the cortex (Bovie, 1979; Walker, 1938). Because of the patchy nature of the rostral intralaminar nuclei, only the centromedianum and parafascicular nuclei were morphometrically measured, and they were measured as a single unit.

The centromedianum-parafascicular complex (Fig. 13–2) is prominent in all primates (Mehler, 1966; Niimi et al., 1969), including the orang-utan. The centromedianum is a large nucleus which is almost entirely encapsulated by fibers; at its medial boundary, it merges with the parafascicular nucleus. The centromedianum nucleus has been divided into a lateral pars parvicellularis and a medial pars magnocellularis (Vogt and Vogt, 1941), but the distinction is not very clear. In the orang-utan the neurons in pars parvicellularis are predominantly fusiform-to-multipolar and take a lighter Nissl stain than the neurons in pars magnocellularis. These latter neurons are typically multipolar and their medium-sized perikarya are larger than the cells in pars parvicellularis, but they are smaller than the cells in n. parafascicularis. Parafascicular neurons surround the fasciculus retroflexus or the habenulointerpeduncular tract. Similar to the more rostral intralaminar nuclei, the centromedianum-parafascicular complex is interconnected with the striatum, in particular the putamen (Jones and Leavitt, 1974) and with the cortex, particularly the motor areas (Petras, 1969). Its functions are also quite diverse, and it has been implicated in the modulation of pain, cortical recruiting responses, and the integration of motor activities.

This orang-utan centromedianum-parafascicular complex has a volume of 61.3 mm^3, which is the same approximate size as its lateral geniculate body. Similar morphometric relationships were observed in the African apes (Armstrong, 1982) and by Hopf (1965) in another orang-utan specimen. The size of the centromedianum-parafascicular complex is, however, tied closely to brain weight. Consequently, gibbons have smaller centromedianum-parafascicular to lateral geniculate body ratios than any of the great apes, while humans have larger ratios.

Medial Tier

The large, spherical mediodorsal nucleus (Figs. 13–1 to 13–3) is almost completely surrounded by the "nonspecific" thalamic nuclei; posteriorly it abuts the medial portion of the pulvinar, and this transition is indistinct. Three subdivisions are identified within the mediodorsal nuclear complex. The pars parvicellularis is the largest component and has medium to small-sized neurons which take a lighter Nissl stain than those of the pars magnocellularis. The pars magnocellularis contains large cells intermingled with smaller ones and is typically seen as clusters of cells in the more rostromedial portions of the mediodorsal complex. A very small pars paralamellaris or pars multiformis has large, multipolar cells, which are scattered along the posterior ventrolateral edge of the mediodorsal complex. None of these divisions, however, is distinct in the orang-utan specimens. Lack of definitive boundaries within this region has also been noted in other hominoids (Heiner, 1960; Kanagasuntheram and Wong, 1968; Simma, 1957; Van Buren and Borke, 1972; Walker, 1938). A more detailed division, similar to those on the human mediodorsal nucleus (Hassler, 1959; Namba, 1958) was not attempted here; rather, the mediodorsal nucleus was analyzed as a unit.

The primate mediodorsal nucleus has major reciprocal connections with the prefrontal cortex and has been associated with memory, learning complex tasks, and emotions, particularly aggression (Girgis, 1971; Schulmann, 1964; Squire, 1982). This nuclear complex is large in all hominoids. Its volume is 159.4 mm^3 in the orang-utan specimen; in the gorilla it is 123.2 mm^3; and in the chimpanzee it is 145.9 mm^3 (Armstrong, 1982). In all great apes, including Hopf's (1965) specimen, the mediodorsal nucleus is approximately 2.5 times larger than the lateral geniculate body. This ratio is just slightly larger than that found in gibbons (2.0:1), but much less than that found in humans (7.2:1). Despite vast differences in social organization and behavior between orang-utans and African apes, the volumetrics of the mediodorsal nucleus in all are similar.

Nuclei of the midline are identified as groups of cells lying near the third ventricle. Since the orang-utan hemisphere had been cut sagittally, the presence or absence of a massa intermedia could not be determined. The massa intermedia is variably present in both ape and human brains (Shantha and Manocha, 1969; Walker, 1938) and has been observed in orang-utan brains (Simma, 1957).

A volumetric comparison of the midline nuclei was not attempted, since some of that region had been removed. The nuclear groups, however, could be recognized. As a whole, the midline nuclei form a relatively small component of the orang-utan thalamus. The paratenial nucleus lies ventrolateral to the stria terminalis and contains medium-sized neurons. Caudal to this nucleus is the nucleus paraventricularis, whose cells in the orang-utan are somewhat smaller than those of the paratenial nucleus. Further caudally, the small-celled nuclei rhomboidalis and reuniens could be recognized. Qualitatively, the midline nuclei appear similar among extant hominoids.

Anterior Nuclei

The anterior nuclei (Fig. 13–3) are part of the limbic system and, in particular, Papez's circuit (Papez, 1937). In the orang-utan brain, as in those of other hominoids, the anterior complex contains two distinct nuclei: the anterior principal and the anterior dorsal. The anteroprincipal nucleus is composed of a large anteroventral and a smaller anteromedial nucleus, which are not as morphologically distinct in hominoid brains as they are in other primates and mammals (Feremutsch, 1963; Toncray and Krieg, 1946; Van Buren and Borke, 1972; Walker, 1938). Its medium-sized, moderately stained neurons are homogeneously distributed. With the exception of its inferior edge, the anterior principal complex is surrounded by fibers, and is thus an anatomically distinctive nucleus in the thalamus.

Although limbic structures are frequently thought of as being conservative neural structures, morphometric studies show that the human anterior principal nucleus is actually relatively enlarged compared to other anthropoids (Armstrong, 1980a, 1986). The size of the orang-utan anterior principal nuclear complex (31.2 mm^3) is comparable to those in the gorilla (31.3 mm^3; N = 3) and chimpanzee (32.1 mm^3; N = 5), and each is about half the size of its respective lateral geniculate body. Hopf's (1965) orang-utan specimen has the smallest anterior to lateral geniculate body ratio (0.40:1), a ratio which is close to that of gibbons (0.43:1). The gorilla (0.61:1) and chimpanzee (0.57:1) ratios are slightly higher than the one we observed in the orang-utan specimen (0.51:1). All of these ratios are much lower than those of humans, where the anterior principal has greatly expanded (1.4:1). Although the anterior principal nucleus has undergone a marked evolutionary change in the hominids, it has remained morphometrically conservative among the apes.

The small anterior dorsal nucleus is situated dorsomedial to the principal component. It is surrounded by the fibers of the stria medullaris and its neurons are moderately sized, spindle-shaped, and darker than those in the anterior principal. The small size of the anterodorsal nucleus precludes morphometric comparisons, but qualitatively it appears similar in all observed hominoids.

HYPOTHALAMUS

The hypothalamus is an important limbic and autonomic integrative area. Although it can be divided into nuclei, the nuclear boundaries are much less clear than in the thalamus. Consequently, with the exception of the medial mamillary bodies, only the orang-utan's cytoarchitecture is compared.

Following Kuhlenbeck (1954) and Keyser (1979), we treat the preoptic region as part of the hypothalamus. The preoptic region lies rostral to the optic chiasm and contains a periventricular nucleus as well as medial and lateral nuclear divisions. In the orang-utan specimens, the periventricular nucleus is composed of very small neurons. The area preopticus medialis has small and medium-sized cells; the lateral group has similarly sized neurons, but they are distributed less densely.

In the orang-utan, the lateral nucleus extends from the preoptic to the rostral mamillary region. Its cells are heterogeneous in size, ranging from small to large, and they are sparsely distributed. Fibers from the medial forebrain bundle course through the ventral part of this region. Functionally, the lateral hypothalamus is thought to modulate the limbic and brainstem information which traverses this tract. Fibers to the habenular nuclei have also been noted (Saper et al., 1978).

The supraoptic region sits just caudal to the preoptic area and, as its name implies, it lies superior to the optic chiasm. In addition to a small-celled periventricular column and the large-celled lateral region, four nuclei can be identified in the orang-utan specimens: the anterior, suprachiasmatic, supraoptic, and paraventricular nuclei.

The anterior hypothalamic nucleus contains small and medium-sized cells which are predominantly round or multipolar neurons, although some spindle-shaped neurons may also be present. While this nucleus is relatively large, its boundaries, especially with the more rostral preoptic areas, are indistinct. Similar mergings of borders have been noted in other hominoids (Heiner, 1960; Shantha and Manocha, 1969). The primary connections of the anterior hypothalamic nucleus are to the septum, central gray, and other hypothalamic regions (Saper et al., 1978).

The suprachiasmatic nucleus contains small cells. Contrary to some early anatomic studies (e.g., Le Gros Clark, 1938), this nucleus can be identified as a separate entity in hominoids, but its boundaries are less distinct than those reported for monkeys, prosimians, or nonprimates (Kanagasuntheram and Wong, 1968; Shantha and Manocha, 1969; Simmons, 1976). Unclear suprachiasmatic borders apparently characterize hominoids. The suprachiasmatic nucleus receives retinal fibers and may be important in regulating circadian rhythms (Tigges et al., 1977).

The large-celled supraoptic and paraventricular nuclei distinguish this hypothalamic region. The supraoptic nucleus of the orang-utan is a well-defined, compact nucleus situated dorsolateral to the optic tract. The second large-celled nucleus, the nucleus paraventricularis (n. filiformis of Le Gros Glark, 1938), is also discernible in the orang-utan hypothalamus. Paraventricular neurons stain like those of the supraoptic nucleus, but they are vertically arranged just lateral to the third ventricle. The cellular architecture of these structures in the orang-utan resembles that of other apes and primates in general. The supraoptic and paraventricular neurons are neurosecretory and release vasopressin (antidiuretic hormone) and oxytocin in the neurohypophysis.

The tuberal region occupies the middle third of the orang-utan hypothalamus. Both Heiner (1960) and Kesarev (1966) suggest that this area is enlarged in the great apes (as seen in chimpanzees) compared to monkeys, but that it is smaller than in the human brain. We did not take any measurements in this area. In addition to the small-celled periventricular region and the large-celled lateral area, several nuclear groups were also identified in the tuberal region of the orang-utan brains. The ventromedial nucleus, part of the tuber cinereum, is prominent here, as it is in other hominoids (Shantha and Manocha, 1969); its small to medium-sized neurons have connections with the amygdala and other limbic structures (Mehler, 1980).

The nucleus dorsomedialis lies dorsal to the ventromedial nucleus and spans the distance from the anterior to the posterior hypothalamic nuclei. The nucleus dorsomedialis is bounded medially by the periventricular zone, laterally by the lateral hypothalamus, and dorsally by the area dorsalis. It is composed of small and medium-sized neurons, but it is not as well delimited as the ventromedial group. A lack of clear demarcation is also noted in other anthropoids (Shantha and Manocha, 1969; Simmons, 1976). Heiner (1960) did not separate this area from the posterior nucleus, which lies caudal to the dorsomedial nucleus.

The arcuate or infundibular nucleus appears as a slight expansion of the small-celled periventricular region ventral to the ventromedial nucleus. The axons that form these neurons make up the tuberoinfundibular tract.

Toward the caudal end of the tuberal region, the orang-utan hypothalamus contains a loosely structured, heterogeneous area, the posterior hypothalamic nucleus. Most of its cells are small to medium in size with a few large cells scattered throughout. The cytoarchitecture of this nucleus resembles that described for other primates (Shantha and Manocha, 1969; Simmons, 1976).

The posterior region of the hypothalamus is the mamillary region; the largest nucleus here is the medial mamillary body. In the orang-utan, as in other anthropoids, several subdivisions can be observed. A large pars medialis contains medium-sized cells, the pars basalis has very small, lightly staining neurons, and the pars intercalatus has a few larger cells. The medial mamillary complex is intimately connected with the hippocampus, the tegmental nuclei of Gudden, and the anterior thalamic nucleus (Veazey et al., 1982). It is a large nucleus in hominoid brains and its size scales to that of the whole brain (Armstrong, 1986). The orang-utan specimen has a volume of 8.4 mm^3, a size which resembles that found in the other great apes (gorilla, 8.1 mm^3, $N = 3$; chimpanzee, 9.5 mm^3, $N = 5$) and larger than that found in the lesser apes (gibbons, 3.9 mm^3, $N = 3$).

Lateral to the ventral portion of the medial mamillary complex, the orang-utan hypothalamus has a small cluster of large cells, the lateral mamillary region. Neurons of similar size and shape are found dorsal to the medial mamillary complex, and these, as in other hominoids, are the neurons of the supramamillary complex. A very small premamillaris nucleus with small neurons lies just rostral to the medial complex. These perimamillary nuclei appear similar in size and configuration to those reported for other hominoids (Diepen, 1962; Heiner, 1960; Shantha and Manocha, 1969).

EPITHALAMUS

The pineal gland, the habenular complex, and the related stria medullaris constitute the epithalamus. Baron (1979) compared the epithalamus in a series of primates and determined that it is a progressive structure when body weight is used as the standard. On the other hand, Simmons (1976) found the anthropoid habenular complex to be regressive when the total epithalamic volume was used as the standard.

The habenular complex consists of two nuclei, the medial and the lateral; the lateral habenular nucleus is the larger of the two. This complex is located medial to the posterior portion of the dorsomedial thalamic nucleus, from which it is separated by a thick fiber bundle. Its neurons are small to medium in size and take a dark Nissl stain. Fibers of the fasciculus retroflexus (habenulointerpeduncular tract) course through the lateral portion. The medial habenular nucleus, which lies between the lateral nucleus and the third ventricle, contains smaller, more faintly staining neurons. No qualitative features distinguish the orang-utan habenular complex from that of other hominoids.

SUBTHALAMUS

The subthalamic region of the diencephalon consists of the subthalamic nucleus of Luys, the prerubral or Forel's field, and the zona incerta. Only the subthalamic nucleus was morphometrically measured in the orang-utan diencephalon.

The orang-utan subthalamic nucleus has reciprocal connections with the pallidum and contains medium-sized neurons which are evenly arrayed and surrounded by fibers. The cellular architecture is thought to be conservative among primates, including hominoids (Yelnik and Percheron, 1979). While this nucleus varies in size, both absolutely and relative to body size, it has enlarged among primates and is especially large in New World monkeys and humans (Baron, 1979; Bauchot, 1982). The orang-utan subthalamic nucleus has a volume of 26.4 mm^3, which is about 40% of the lateral geniculate body's volume.

CONCLUSIONS

The cytoarchitecture of the orang-utan brains in this study is qualitatively similar to the configuration in other hominoids. The topographic relationships of components and the appearance of their constituent cells are stable among all extant hominoids. These characteristics are thus considered to be conservative or primitive retentions for hominoids and may very well be typical of other anthropoids, as well.

The primary observation here is that the nuclei in the orang-utan diencephalon closely resemble these structures in other hominoids. Morphometric differences among taxa were most apparent in nuclei which scale according to brain size. One exception to this is the enlarged ventrobasal complex of the orang-utan, where an increase in size appears to stem from changes within the pars medialis. An increased size in this nuclear subregion could reflect an increased number of fibers carrying information from the enlarged cheek pads. If this is the case, presumably female orang-utans would have a smaller ventroposterior medial nucleus, and thus a smaller ventrobasal complex. Future studies must resolve the question of whether such sexual dimorphism exists. The other sensory relay nuclei of the orang-utan's thalamus closely resemble those of the African apes.

Several complexes attained sizes which appear to covary with brain size. Among these were the motor nuclei of the ventral thalamic tier and the intralaminar centromedianum-parafascicular complexes. Behavioral motor differences between orang-utans and other apes suggest that structural differences are to be expected within these regions. The seeming lack of difference among the great apes in this study is most probably the result of our focus on large complexes rather than on subcomponents thereof.

Among the regions that were quantified, four were components of the limbic system: the anterior principal, lateral dorsal, mediodorsal thalamic nuclei, and the mamillary bodies. Their sizes approximate those reported for the chimpanzee and gorilla. Neurological differences supporting different social behaviors are thus to be found in other limbic structures and/or smaller neuroanatomical units.

Association regions in the orang-utan thalamus are the size expected for an ape. In particular, the scaling of the large lateral-posterior-pulvinar complex is a conservative feature in all hominoids, including the orang-utan. Differences in complex behaviors mediated by these regions are likely to be found in shifts within this, and other, complexes.

Obviously, the species-specific behavior exhibited by any taxon will be reflected in its central nervous system, and thus differences separating the orang-utan from other apes are to be expected in major integrative centers, such as the diencephalon. These differences are most likely quantitative shifts representing enlargements or diminutions in the size of various neuronal populations within the major regions compared here. To understand what role the size of various neuronal nuclei or complexes may have for a species, it is necessary to study the absolute as well as relative sizes of the regions and to determine the functional correlates of each region. It is therefore highly unlikely that the structural-functional relationships of an endangered species like *Pongo pygmaeus* will ever be known with exactitude. However, assuming that the general aspects of these relationships are similar to those of other anthropoids—and this is a very probable assumption given the similarity in cytoarchitecture (Zilles et al., 1984)—morphometric comparisons among various nuclear populations should expand our understanding of the brain's functioning in rare species (Bauchot, 1982). Future studies, by asking questions about particular nuclei or even intranuclear subdivisions, will help clarify which neural differences covary with behavioral shifts.

ACKNOWLEDGMENTS

We thank the Audubon Zoo, New Orleans, the Netherlands Institute of Brain Research and Karl Zilles, Anatomisches Institut zu Köln, F.R.G. for access to the orang-utan brains used in this study. We thank H. Stephan, Max-Planck-Institut, Frankfort, F.R.G.; C. Noback, Columbia University; W. Welker, University of Wisconsin; and the Yakovlev Collection, Armed Forces Institute of Pathology, Washington, D.C., for access to the other hominoid brains used in this study. The research was supported by NSF grant BNS-8317819.

REFERENCES

Albe-Fessard, D., Berkley, K. J., Kruger, L., Ralston, H. J. III and Willis, W. D., Jr., 1985. Diencephalic mechanisms of pain sensation. *Brain Research Reviews, 9:*217–296.

Armstrong, E. 1979. A quantitative comparison of the hominoid thalamus: I. Specific sensory relay nuclei. *American Journal of Physical Anthropology, 51:*365–381.

Armstrong, E. 1980a. A quantitative comparison of the hominoid thalamus: II. Limbic nuclei anterior principalis and lateral dorsalis. *American Journal of Physical Anthropology, 52:*43–54.

Armstrong, E. 1980b. A quantitative comparison of the hominoid thalamus: III. A motor substrate—the ventrolateral complex. *American Journal of Physical Anthropology, 52:*405–419.

Armstrong, E. 1981. A quantitative comparison of the hominoid thalamus: IV. Posterior association nuclei—the pulvinar and lateral posterior nucleus. *American Journal of Physical Anthropology, 55:*369–383.

Armstrong, E. 1982. Mosaic evolution in the primate brain: differences and similarities in the hominoid thalamus. *In* Primate Brain Evolution: Methods and Concepts, pp. 131–161, ed. E. Armstrong and D. Falk. Plenum Press, New York.

Armstrong, E. 1986. Enlarged limbic structures in the human brain: the anterior thalamus and medial mamillary body. *Brain Research, 362:*394–397.

Armstrong, E. and Falk, D. 1982. Primate Brain Evolution: Methods and Concepts. Plenum Press, New York.

Asanuma, C., Thach, W. T. and Jones, E. G. 1983. Distribution of cerebellar terminations and their relation to other afferent terminations in the ventral lateral

thalamic region of the monkey. *Brain Research Reviews, 5*:237–265.
Baron, G. 1979. Quantitative changes in the fundamental structural pattern of the diencephalon among primates and insectivores. *Folia Primatologica, 31*:74–105.
Bauchot, R. 1982. Brain organization and taxonomic relationships in insectivora and primates. *In* Primate Brain Evolution: Methods and Concepts, pp. 163–175, ed. E. Armstrong and D. Falk. Plenum Press, New York.
Blinkov, S. and Zvorykin, V. P. 1950. Dimensions of the auditory cortex and the medial geniculate body in man and monkeys. Referred to *In* The Human Brain (Figures and Tables, pp. 225–226 and 410–412), ed. S. M. Blinkov and I. I. Glezer. Plenum Press, New York, 1968.
Bovie, J. 1979. An anatomical reinvestigation of the termination of the spinothalamic tract in the monkey. *Journal of Comparative Neurology, 186*:17–48.
Bos, J. and Benevento, L. A. 1975. Projections of the medial pulvinar to orbital cortex and frontal eye fields in the rhesus monkey. *Experimental Neurology, 49*:487–496.
Casseday, J. H., Diamond, I. T. and Harting, J. K. 1976. Auditory pathways to the cortex in *Tupaia glis. Journal of Comparative Neurology, 166*:303–340.
Crouch, R. L. 1934. The nuclear configuration of the thalamus of *Macacus rhesus. Journal of Comparative Neurology, 59*:451–487.
Diamond, I. T. and Hall, W. C. 1969. The evolution of neocortex. *Science, 184*:251–262.
Diepen, R. 1962. Der Hypothalamus. *In* Handbuch der Mikroskopischen Anatomie des Menschen, pp. 1–430, ed. W. von Mollendorf and W. Bargmann. Springer-Verlag, Berlin.
Feremutsch, K. 1963. Thalamus. *Primatologica, 2*:1–226.
Girgis, M. 1971. The role of the thalamus in the regulation of aggressive behavior. *International Journal of Neurology, 8*:327–351.
Harting, J. K., Hall, W. C. and Diamond, I. T. 1972. Evolution of the pulvinar. *Brain, Behavior, and Evolution, 6*:424–452.
Hassler, R. 1959. Anatomy of the thalamus. *In* Introduction to Stereotaxis with an Atlas of the Human Brain, pp. 230–290, ed. G. Schaltenbrand and D. Bailey. Grune and Stratton, New York.
Heiner, J. R. 1960. A reconstruction of the diencephalic nuclei of the chimpanzee. *Journal of Comparative Neurology, 114*:217–238.
Hendrickson, A., Wilson, M. E. and Toyne, M. J. 1970. The distribution of optic nerve fibers in *Macaca mulatta. Brain Research, 23*:425–427.
Hopf, A. 1965. Volumetrische Untersuchungen zur vergleichenden Anatomie des Thalamus. *Journal für Hirnforschung, 8*:25–38.
Imig, T. J., Ruggero, M. A., Kitzes, L. M. Javel, E. and Brugge, J. F. 1977. Organization of auditory cortex in the owl monkey *(Aotus trivirgatus). Journal of Comparative Neurology, 171*:111–128.
Jackson, W. J., Reite, M. L. and Buxton, D. F. 1969. The chimpanzee central nervous system: a comparative review. *Primates in Medicine, 4*:1–51.
Jones, E. G. 1983. Distributional patterns of individual medial lemniscus axons in the ventrobasal complex of the monkey thalamus. *Journal of Comparative Neurology, 215*:1–16.
Jones, E. G. and Leavitt, R. Y. 1974. Retrograde axonal transport and the demonstration of non-specific projections to the cerebral cortex and striatum from thalamic intralaminar nuclei in the rat, cat, and monkey. *Journal of Comparative Neurology, 154*:349–378.
Kaas, J. H., Huerta, M. F. and Weber, J. T. 1977. Patterns of retinal terminations and laminar organization of the lateral nucleus of primates. *Journal of Comparative Neurology, 182*:517–554.
Kangasuntheram, J. and Wong, W. C. 1968. Nuclei of the diencephalon of Hylobatidae. *Journal of Comparative Neurology, 134*:265–286.
Kesarev, V. S. 1966. Structural features of hypothalamus in man and other primates (chimpanzee, macaque). *Federal Proceedings, 25*:T243–T247.
Keyser, A. 1979. Development of the hypothalamus in mammals: an investigation into its morphological position during ontogenesis. *In* Handbook of the Hypothalamus, Vol. 1, Anatomy of the Hypothalamus, pp. 65–135, ed. P. J. Morgane and J. Panksepp. Marcel Dekker, New York.
Kuhlenbeck, H. 1954. The Human Diencephalon. Karger, New York.
Le Gros Clark, W. E., Beattie, J., Riddoch, G. and Dott, N. 1938. The Hypothalamus. Oliver and Boyd, Edinburgh.
Markowitsch, H. J., Emmans, D., Irle, E., Streicher, M. and Preilowski, B. 1985. Cortical and subcortical afferent connections of the primate's temporal pole: a study of rhesus monkeys, squirrel monkeys and marmosets. *Journal of Comparative Neurology, 242*:425–458.
Mehler, W. R. 1966. Further notes on the center median nucleus of Luys. *In* The Thalamus, pp. 109–127, ed. D. P. Purpura and M. D. Yahr. Columbia University Press, New York.
Mehler, W. R. 1980. Subcortical afferent connections of the amygdala in the monkey. *Journal of Comparative Neurology, 190*:733–762.
Mikol, J., Brion, F., Derome, P., DePommery, J., and Gallissot, M. C. 1977. Connections of laterodorsal nucleus of the thalamus. II. Experimental study of *Papio papio. Brain Research, 138*:1–16.
Morest, D. K. 1964. The neuronal architecture of the medial geniculate body of the cat. *Journal of Anatomy, 98*:611–638.
Namba, M. 1958. Über die feineren Structuren des Medio-dorsalen Supranucleus und der Lamella Medialis des Thalamus beim Menschen. *Journal für Hirnforschung, 4*:1–42.
Niimi, K., Katayama, E., Karaseki, T. and Morimoto, K. 1969. Studies on the derivation of the centremedian nucleus of Luys. *Tokushima Journal of Experimental Medicine, 6*:261–268.
Ogren, M. P. 1982. The development of the primate pulvinar. *In* Primate Brain Evolution: Methods and Concepts, pp. 113–129, ed. E. Armstrong and D. Falk. Plenum Press, New York.
Ogren, M. P. and Hendrickson, A. E. 1976. Pathways be-

tween the striate cortex and subcortical regions in *Macaca mulatta* and *Saimri sciureus:* evidence for a reciprocal pulvinar connection. *Experimental Neurology, 53:*780–800.

Papez, J. W. 1937. A proposed mechanism of emotion. *Archives of Neurology and Psychology, 38:*725–744.

Pearson, R. C. A., Brodal, P. and Powell, T. P. S. 1978. The projection of the thalamus upon the parietal lobe in the monkey. *Brain Research, 144:*143–148.

Petras, J. M. 1969. Some efferent connections of the motor and somatosensory cortex of simian primates, and felid, canid, and procyonid carnivores. *Annals of the New York Academy of Sciences, 167:*469–505.

Saper, C. B., Swanson, L. W. and Cowan, W. M. 1978. The efferent connections of the anterior hypothalamic area of the rat, cat, and monkey. *Journal of Comparative Neurology, 182:*575–600.

Schulmann, S. 1964. Impaired delayed response of thalamic lesions. *Archives of Neurology, 11:*477–499.

Shantha, T. and Manocha, S. 1969. The brain of chimpanzee *(Pan troglodytes). In* The Chimpanzee, Vol. 1, pp. 238–305, ed. G. Bourne. Karger, New York.

Simma, K. 1957. Der Thalamus der Menschenaffen: Eine vergleichendanatomische Untersuchung. *Psychiatrie und Neurologie, 134:*145–175.

Simmons, R. M. T. 1976. The diencephalon of the vervet monkey *(Cercopithecus aethiops). South African Journal of Medical Sciences, 41:*109–136.

Squire, L. R. 1982. The neuropsychology of human memory. *Annual Review of Neuroscience, 5:*241–273.

Stephan, H., Bauchot, R. and Andy, O. J. 1970. Data on size of brain and of various parts of insectivores and primates. *In* The Primate Brain, pp. 289–297, edited by C. R. Noback and W. Montagna. Appleton-Century-Crofts, New York.

Strick, P. 1975. Multiple sources of thalamic input to the primate motor cortex. *Brain Research, 88:*372–377.

Tigges, J., Bos, J. and Tigges, M. 1977. An autoradiographic investigation of the subcortical visual system in the chimpanzee. *Journal of Comparative Neurology, 172:*367–380.

Tobias, P. V. 1971. The Brain in Hominid Evolution. Columbia University Press, New York.

Toncray, J. E. and Krieg, W. 1946. The nuclei of the human thalamus: a comparative approach. *Journal of Comparative Neurology, 85:*421–461.

Van Buren, J. and Borke, R. 1972. Variations and Connections of the Human Thalamus, vols. I and II. Springer-Verlag, New York.

Veazey, R. B., Amaral, D. G. and Cowan, W. M. 1982. The morphology and connections of the posterior hypothalamus in the cynomolgous monkey *(Macaca fascicularis).* I. Cytoarchitectonic organization. *Journal of Comparative Neurology, 207:*114–132.

Vogt, C. and Vogt, O. 1941. Thalamusstudien, I–III. *Journal für Psychologie und Neurologie, 50:*32–152.

Walker, A. E. 1938. The Primate Thalamus. University of Chicago Press, Chicago.

Walker, A. E. 1966. Internal structure and afferent-efferent relations of the thalamus. *In* The Thalamus, pp. 1–11, ed. D. P. Purpura and M. D. Yahr. Columbia University Press, New York.

Walker, A. E. and Fulton, J. F. 1938. The thalamus of the chimpanzee. III. Methathalamus normal structure and cortical connections. *Brain, 61:*250–268.

Welker, W. I., and Johnson, J. I. 1965. Correlation between nuclear morphology and somatotopic organization in ventrobasal complex of the raccoon thalamus. *Journal of Anatomy, 99:*761–790.

Yelnik, J. and Percheron, G. 1979. Subthalamic neurons in primates: a quantitative and comparative analysis. *Neuroscience, 4:*1717–1743.

Zilles, K., Wree, A., Schleicher, A. and Divac, I. 1984. The monocular and binocular subfields of the rat's primary visual cortex: a quantitative morphological approach. *Journal of Comparative Neurology, 226:*391–402.

IV

CRANIOFACIAL ANATOMY, GROWTH, AND FUNCTIONAL MORPHOLOGY

Although the anatomies of interest in the first two chapters of this section differ, their common thrust is the development of variation in the orang-utan and the degrees to which sexual dimorphism is manifested. In this regard, Linda Winkler begins the section with a contribution on variation in the suboccipital anatomy of the orang-utan. Although the region itself is relatively small, the anatomy is quite complicated; its muscles serve to rotate the head as well as the head and the atlas laterally to the same side of the body and to pull the head backward. Winkler dissected a sample that included individuals of various ages and of both sexes and found, for instance, that the degree of variation that exists in suboccipital anatomy is far greater than had been recognized previously, even though patterns of innervation and vascularization appear to be fairly constant. Muscular variation is seen in size, morphology, and relationship, and seems to be expressed largely on an individual basis, being correlated neither with sex nor age class. However, variation in some aspects of the *m. obliquus capitis superior* and *m. rectus capitis posterior major* and *minor* might be related to age, sex, and the degree to which craniofacial prognathism is developed.

Olav Röhrer-Ertl continues the theme of variation in his exhaustive study of cranial growth in Bornean and Sumatran orang-utans of both sexes. In general, he finds that the extremes of morphological variation that develop in the orang-utan parallel those seen in modern humans. The midfacial region of the orang-utan is, however, relatively stable. In comparing samples from Borneo and Sumatra, Röhrer-Ertl discriminates between the two on the basis of the distribution of single characteristics as well as the craniofacial indices used in the study, especially that of facial width. Although cranial growth (as expected) is correlated with overall body growth, Röhrer-Ertl emphasizes the fact that the pre- and posteruptive states of the dentition, as well as the development and use of muscles of mastication, greatly influence final cranial form. In this regard, one cannot overlook the observation that brain volume continues to increase even after cranial sutures have closed. And finally, Röhrer-Ertl provides evidence of marked sexual dimorphism cranially in the orang-utan, with the skulls of females typically being higher and shorter than males of the same age.

The delineation of sexual dimorphism in the orang-utan is pursued further by Linda Winkler, Glenn Conroy, and Michael Vannier in their analysis of exo- and endocranial dimensions in the orang-utan. Their study is quite complementary to Röhrer-Ertl's in that they assess a series of 37 craniometric exocranial dimensions. For a sample of 130 orang-utan crania, representing individuals of both sexes and all ages, 34 of these measures demonstrate statistically significant differences between the sexes. When only adult individuals are analyzed, the level of significance is even greater and applicable to all exocranial dimensions. The second focus of their chapter is the analysis of endocranial structures, which they pursue using three-dimensional CT-scans on a male and female adult orang-utan and an adult male gorilla. The results, although preliminary, indicate that sexual dimorphism in the orang-utan is also reflected in endocranial volume and in the degree to which airorhynchy (dorsiflexion of the face) is developed; the development of cranial sexual dimorphism may also be influenced by the nuchal-occipital and masticatory musculature, tooth size, the presence of a throat sac and cheek flanges, as well as by orbital and cerebral volume. Comparisons of the orang-utans with the gorilla specimen indicate not only that dorsiflexion is restricted to the former, but also that the inner and outer tables of the frontal bone are approximated in orang-utans, whereas in the gorilla, they are well separated. Winkler et al. suggest that the greater disparity of frontal bone components in the gorilla results from its having a relatively low endocranial volume in conjunction with a large orbital volume, and is reflected in the development of massive browridges and frontal sinuses, features lacking in the orang-utan.

The contributions by Brian Shea and by Barbara Brown and Steven Ward attempt to incorporate data on relevant fossils into a framework on which to interpret the evolution of the extant orang-utan. In both chapters, two elements that emerge as pivotal in the assessment of the relationships of the orang-utan to and among fossil and extant hominoids are (as Winkler et al. also touched upon) supraorbital morphology and craniofacial hafting. But beyond recognizing these features, and being consistent in morphological description, the last two, and even the antepenultimate, contributions to this section illustrate divergent approaches to inference and interpretation, and, thus, to suggestions of potential relatedness of the taxa involved.

Shea's review of fossil and extant hominoids leads him to conclude that dorsiflexion of the facial skeleton and minimal browridge development are primitive features, and largely retained as such in the orang-utan. In fact, overall skull form in the orang-utan may be less specialized in certain respects than has been previously recognized. Thus, the similar features one finds in *Sivapithecus* are not reflective of the relatedness of the fossil and extant taxa. Indeed, Shea suggests that all "ramapiths" *(Ouranopithecus, Gigantopithecus, Ramapithecus/Sivapithecus)* may be primitive relative to the extant large-bodied hominoids and may constitute a group among whose members lies the ancestor of the living forms. Shea continues the argument by contrasting facial position, supraorbital morphology, and other cranial features in the African apes and hominids, which he concludes represent synapomorphies of the three taxa.

Brown and Ward review the skeletal and soft tissue correlates of airohynchy in the orang-utan and make comparisons with many of the same hominoid crania Shea discusses. They conclude that *Sivapithecus* possesses an airorhynchous condition similar to that of the orang-utan, as evidenced by the topography of the frontal bone, anterior cranial base, interorbital septum, palate, and premaxilla. Brown and Ward discover, however, that the oral region of *Sivapithecus* retains a more "generalized" configuration: the orang-utan's unique anatomy of the suprahyoid, for example, is definitely not present in the fossil form. They suggest that the difference in suprahyoid morphology may be due to hypertrophy in the orang-utan's laryngeal air sac system. Thus, while apparently closely related to the extant orang-utan, *Sivapithecus* should not be viewed as a Miocene orang-utan, but only as a member of the same radiation. For *Sivapithecus* to be related to the orang-utan at all implies, as Brown and Ward argue, that the features of airorhynchy shared by these two hominoids are indeed due to synapomorphy and not primitive retentions, as Shea concluded. The logical extension of Brown and Ward's interpretation is that craniofacial similarity between hominids and the African apes is due to primitive retention and is not reflective of the relatedness of these hominoids.

14

Variation in the Suboccipital Anatomy

LINDA A. WINKLER

The assessment of evolutionary relationships among taxa based on anatomical data has become increasingly common. The complexity and diversity of anatomical structures would appear to contribute to an understanding of hominoid phylogenetic relationships. However, individual, sexually dimorphic, and age-related differences may produce significant variation in anatomy, particularly in the soft tissues. Therefore, an appreciation of anatomical variation in each of the hominoids is essential before one can use anatomical features to reconstruct phylogenetic relationships. Dissections of only one or two animals cannot attempt to reveal or delineate the range of potential variation. Since descriptions of the suboccipital region of the orang-utan (Hill, 1939; Sakka, 1973; Sonntag, 1924) differ in their reports of the size and morphology of the relevant musculature, I shall here attempt to assess variation in the anatomy of this area.

MATERIALS AND METHODS

The present study is based on the dissection of the suboccipital region of eight orang-utans, representing a variety of age categories. Given the rarity of orang-utan specimens, and the even greater paucity of those suitable for dissection, the sample reported on here is actually quite large. Included are a female fetus of approximately 6 gestational months; a male newborn; a 9-month-old female; an 8-year-old female; a 9-year-old male; an 11-year-old female; a 17-year-old female; and a 21-year-old male. All specimens were at least raised, if not born, in captivity.

The general structure of each muscle, its attachments, relationships, and nerve supply within the suboccipital region were recorded. The positions and relationships of nerves, arteries, and veins within the area were also documented. The muscles were defined on the basis of attachment points—bony origin and insertion.

GENERAL SUBOCCIPITAL ANATOMY

The suboccipital region is an area at the dorsal base of the cranium; its apices are the spinous process of the axis, the transverse process of the atlas, and the occipital bone below the superior nuchal line at midline. Anatomical structures within these boundaries (Fig. 14–1) include: *m. rectus capitis posterior major, m. obliquus capitis superior, m. obliquus capitis inferior, m. rectus capitis posterior minor,* a descending branch of the occipital artery, the deep cervical arterial and venous plexuses, and the suboccipital and greater occipital nerves. *M. rectus capitis posterior major, m. obliquus capitis superior,* and *m. obliquus capitis inferior* are superficial, and their adjacent borders form the suboccipital triangle. *M. rectus capitis posterior minor* is deep to, and generally concealed by, *m. rectus capitis posterior major. M. rectus capitis posterior major* and *m. rectus capitis posterior minor* rotate the head and draw it backwards. *M. obliquus capitis superior* and *m. obliquus capitis inferior* rotate the head and atlas laterally toward the same side of the body. These muscles are innervated by the suboccipital nerve.

M. rectus capitis posterior major originates from the superior part of the spinous process of the axis. It travels superiorly and broadens before inserting on the occipital bone inferior to the superior nuchal line. The insertion extends laterally from the midline. *M. obliquus capitis inferior* originates from the lateral side of the spinous process of the axis. It courses laterally, maintaining a fairly constant width throughout, and inserts on the medial part of the transverse process of the atlas. *M. obliquus capitis superior* originates from the superior portion of the transverse process of the atlas, and proceeds superiorly, fanning somewhat as it does, to insert on the occipital bone along the occipitomastoid suture. *M. rectus capitis posterior minor* originates from the posterior tubercle of the atlas; it runs superiorly, broadening from its attachment near midline to its lateral extent, to insert on the oc-

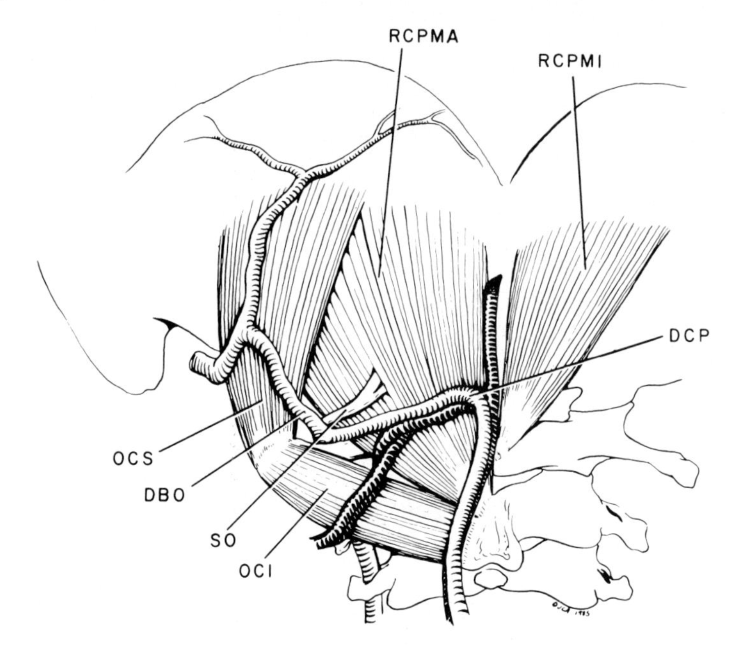

FIG. 14–1. General suboccipital anatomy. RCPMA, *m. rectus capitis posterior major;* RCPMI, *m. rectus capitis posterior minor;* OCS, *m. obliquus capitis superior;* OCI, *m. obliquus capitis inferior;* DBO, descending branch of the occipital artery; DCP, deep cervical arterial and venous plexus; SO, suboccipital nerve.

cipital bone inferior, as well as deep, to the insertion of *m. rectus capitis posterior major.*

The suboccipital nerve emerges through the atlantooccipital membrane; and in the suboccipital triangle, it divides into branches that supply the muscles on their deep surfaces. The nerve then continues superficially and emerges onto the superficial face of *m. rectus capitis posterior major* which it innervates and over which it courses medially.

The occipital artery gives off its descending branch as it crosses *m. obliquus capitis superior.* The descending branch courses inferiorly and medially across *m. obliquus capitis superior* to enter the center of the suboccipital area where it anastomoses with the deep cervical artery and with the vertebral artery deep to the atlantooccipital membrane (Fig. 14–1).

The deep cervical artery and veins, which comprise the deep cervical plexus, travel through the deep cervical fat and fascia of the dorsal midline. Slightly above the level of the vertebral axis, the deep cervical artery anastomoses with the descending branch of the occipital artery as well as with the vertebral artery. The veins of the deep cervical plexus travel medially and inferiorly across the superficial face of both *m. rectus capitis posterior major* and *m. obliquus capitis inferior.*

The anatomy of the suboccipital region is directly related to the *planum nuchale* of the occipital bone upon which all the suboccipital muscles, with the exception of *m. obliquus capitis inferior,* insert. The *planum nuchale* is defined as that portion of the occipital bone that is bounded superiorly by the superior nuchal line, inferiorly by the posterior border of the foramen magnum, and laterally by the mastoid region of the temporal bone.

VARIATION IN ANATOMY

There is notable variation between these eight specimens in their occipitovertebral musculature. In all specimens, the only muscle that remains relatively constant in both size and position is *m. obliquus capitis inferior.*

There is variation among the eight specimens in the number of heads of *m. rectus capitis posterior major* and the morphology of this muscle. In the fetus, the newborn, the 9-month-old, and the 11-year-old, this muscle possesses two heads (Fig. 14–2A,B). In the 11-year-old, the two heads of the *m. rectus capitis posterior major* are joined for a short distance above their origin, while in the remaining animals the two heads are totally separate. In all cases, both heads, the *caput mediale* and *caput laterale,* originate from the axis. *Caput laterale* runs laterally and superiorly, widening extensively from its origin to its insertion

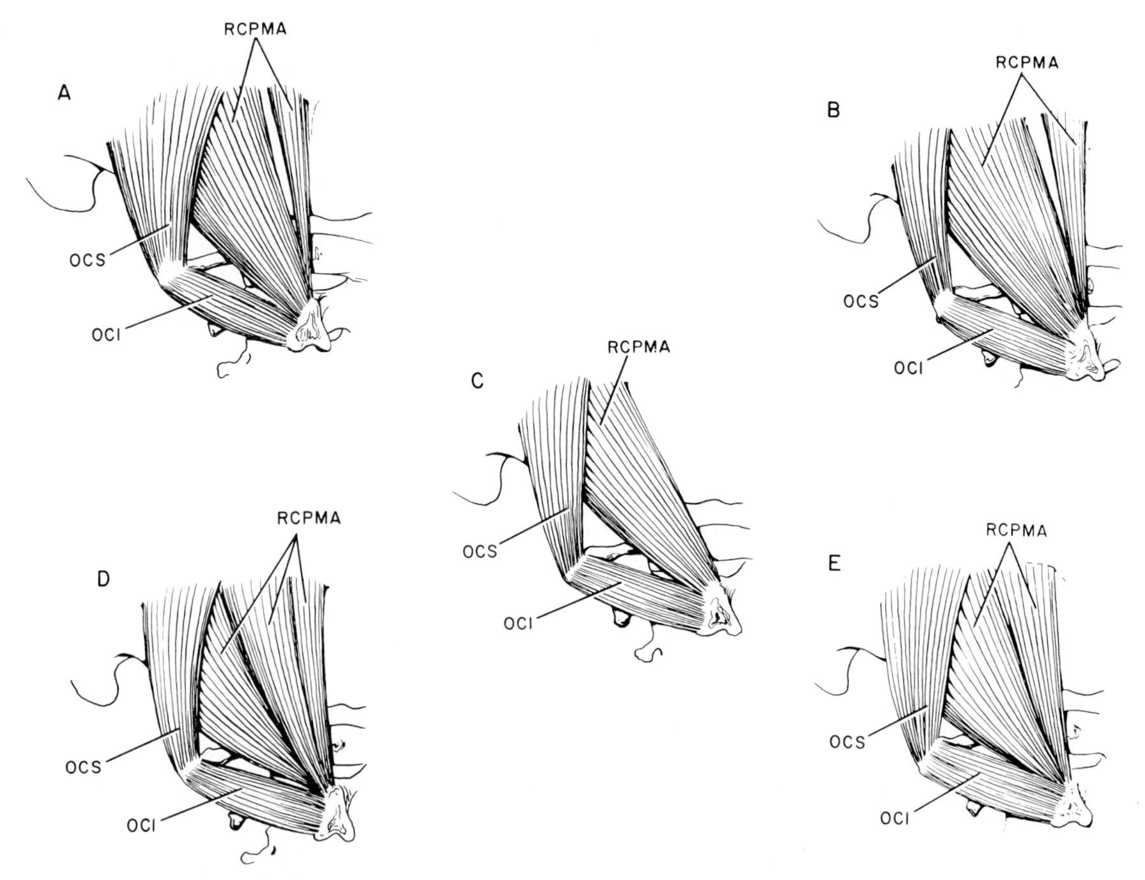

FIG. 14–2. Variation in the superficial musculature. (A) Fetus, newborn, and 9-month-old. (B) Eleven-year-old. (C) Eight-year-old. (D) Nine-year-old and 17-year-old. (E) Twenty-one-year old. RCPMA, *m. rectus capitis posterior major;* OCS, *m. obliquus capitis superior;* OCI, *m. obliquus capitis inferior.*

on the occipital bone deep to and along the medial border of *m. obliquus capitis superior. Caput mediale,* which is noticeably narrower than the lateral head, lies near midline of the occiput and is clearly separated from the more extensive lateral head. *Caput mediale* runs superiorly, fanning slightly as it proceeds from its origin to its insertion on the occipital bone.

In the remaining specimens, *m. rectus capitis posterior major* presents itself as a single functional unit or head (Fig. 14–2C–E). In the 9-year-old, the 17-year-old, and the 21-year-old, this single head may actually be composed of smaller muscular components that have fused; in fact, the *m. rectus capitis posterior major* of the 9- and the 17-year-olds can be dissected into three smaller, closely apposed components. At the midline of the occiput, there is a medial component that runs superiorly and that appears morphologically similar to the *caput mediale* seen in the fetus, newborn, 9-month-old, and 11-year-old. Lateral to this medial component lies an intermediate as well as a more lateral component, both of which are very well developed. The intermediate component runs superiorly to insert upon the occipital bone medial to *m. obliquus capitis superior.* The lateral component travels laterally and superiorly, inserting along the medial border and deep to *m. obliquus capitis superior.* The intermediate and lateral components are approximately equal in size, are closely apposed, and are bound by a fascia.

In the 21-year-old, *m. rectus capitis posterior major* possesses a fascial sheet that runs through the muscle and appears superficially to separate the medial one-third from the lateral two-thirds of the muscle. The two muscular components in this adult male are tightly apposed and cannot be dissected into smaller muscle units. However, the presence of a fascial sheet within the muscle suggests that the muscle may, at some point, have had two components. *M. rectus capitis posterior major* in the 21-year-old is particularly well developed and has a total muscle mass substantially greater than that seen in any other specimen.

In the remaining specimen, an 8-year-old, *m. rectus capitis posterior major* has a distinctive single head which cannot be dissected into smaller units and which shows no evidence whatsoever of possible fusion from multiple muscle components. The mass of this single head is relatively small and it covers a much more limited area of the occiput than the muscle in the other specimens. From its origin, the muscle runs laterally and superiorly, widening only slightly in comparison to the fanning seen in other specimens; it inserts deep and medial to *m. obliquus capitis superior.*

The relationship between the insertion of *m. rectus capitis posterior major* and *m. obliquus capitis superior* appears to vary with the size and morphology of both muscles. *M. obliquus capitis superior* is relatively narrow in the 8-year-old, 9-year-old, 11-year-old, and 17-year-old, widening but minimally as it runs superiorly onto the occipital bone (Fig. 14–2B–D). In the 9-, 11-, and 17-year-olds, less than half of the superior border width of *m. rectus capitis posterior major* (including both muscular heads in the 11-year-old) inserts deep to *m. obliquus capitis superior.* The remainder of *m. rectus capitis posterior major* in these individuals inserts medial to *m. obliquus capitis superior,* between this latter muscle and the dorsal midline of the occiput. The 8-year-old, however, differs in that its *m. rectus capitis posterior major* is rather poorly developed and, therefore, although *m. obliquus capitis superior* is narrow in this individual, more than half of the width of the superior border of the relatively narrow *m. rectus capitis posterior major* inserts deep to *m. obliquus capitis superior.*

M. obliquus capitis superior in the remaining four specimens—the fetus, the newborn, the 9-month-old, and the 21-year-old—broadens more extensively from its origin to its insertion and, thus, it covers a relatively greater amount of the occipital bone (Fig. 14–2A,E). *M. rectus capitis posterior major* is well developed in these four individuals, but the larger relative size of *m. obliquus capitis superior* results in the overlapping by this latter muscle of more than half of the insertion of *m. rectus capitis posterior major,* which lies medial and deep to it.

Variation in the size and degree of overlap of the three superficial muscles—*m. obliquus capitis superior, m. obliquus capitis inferior,* and *m. rectus capitis posterior major*—determines the extent to which the suboccipital triangle is obliterated. The suboccipital triangle is more delineated in the perinatal specimens, in which *m. rectus capitis posterior major* and *m. obliquus capitis inferior* are relatively small and are in less extensive contact along their adjacent borders. Conversely, the suboccipital triangle is less well demarcated and may even be obliterated in the adolescent and adult individuals as a result of the extensive contact along their adjacent borders of the relatively larger and well-developed *m. rectus capitis posterior major* and *m. obliquus capitis inferior.* There are, however, two exceptions to these general patterns: in the newborn, the suboccipital triangle is nearly obliterated due to extensive contact between the superficial muscles, whereas the superficial muscles in the 8-year-old appear relatively less developed and, thus, the adjacent borders of *m. rectus capitis posterior major* and *m. obliquus capitis inferior* are in limited contact and the suboccipital triangle is clearly defined.

The deep muscle of the suboccipital triangle, *m. rectus capitis posterior minor,* also varies significantly in size and shape, and variation exists, as well, in the relationship between the insertion of this muscle and the insertions of *m. rectus capitis posterior major* and *m. obliquus capitis superior.* The insertion of *m. obliquus capitis superior* extends along the lateral edge of the occipital bone, abutting the occipitomastoid suture and the digastric notch lateral to this suture; the insertion of this muscle lies lateral to the insertion of *m. rectus capitis posterior major* and *m. rectus capitis posterior minor.* The insertion of *m. rectus capitis posterior major* abuts *m. rectus capitis posterior minor* laterally and superiorly.

The insertion of *m. obliquus capitis superior* is oval-shaped and concave medially in the fetus, the newborn, the 9-month-old, the 9-year-old, the 11-year-old, and the 17-year-old (Fig. 14–3A,B). It extends inferosuperiorly and curves slightly medially. In the 8-year-old, the insertion of *m. obliquus capitis superior* is also oval-shaped, but it lacks a medial concavity (Fig. 14–3C) and merely extends inferosuperiorly. The attachment site of *m. obliquus capitis superior* covers a relatively broader area in the 21-year-old (Fig. 14–3D); this muscle runs inferosuperiorly and becomes broadened along its superior face.

Variation in the size and shape of *m. obliquus capitis superior* corresponds to differences in the bony morphology related to its insertion. The occipitomastoid suture separates the lateral border of *m. obliquus capitis superior* from the origin of *m. digastricus* on the temporal bone. Sakka (1972) described a bony crest, the *linea m. obliquus capitis superior,* which courses along the lateral border of the insertion of *m. obliquus capitis superior,* as separating it from the origin of *m. digastricus* and as delineating the occipital from the temporal bone. The presence of a *linea*

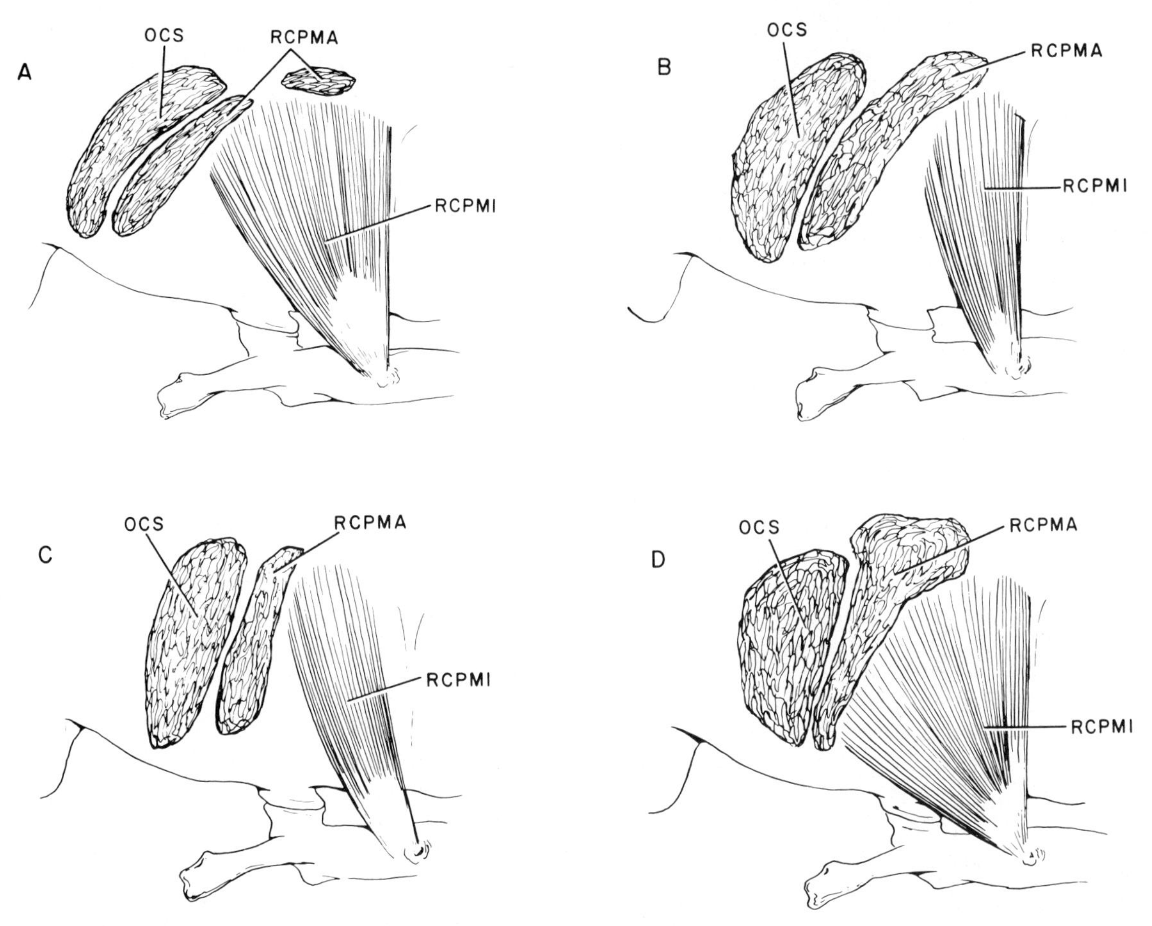

FIG. 14–3. Variation in *m. rectus capitis posterior minor* and the insertions of *mm. rectus capitis posterior major* and *obliquus capitis superior.* (A) Fetus, newborn, 9-month-old, and 11-year-old. (B) Nine-year-old and 17-year-old. (C) Eight-year-old. (D) Twenty-one-year old. OCS, *m. obliquus capitis superior* insertion; RCPMA, *m. rectus capitis posterior major* insertion; RCPMI, *m. rectus capitis posterior minor.*

m. obliquus capitis superior appears to be related to the development of both *m. obliquus capitis superior* and the origin of the *m. digastricus* and associated digastric notch.

In the fetus, the newborn, and the 9-month-old, the digastric notch is nonexistent or slight and, although *m. obliquus capitis superior* is well shaped, it is of relatively small size with a limited insertion area compared to older specimens. Consequently, there is little or no evidence of a *linea m. obliquus capitis superior.* The occipitomastoid suture remains patent in these specimens along the lateral border of the insertion of *m. obliquus capitis superior.*

With the exception of the 11-year-old and the 21-year-old, the remaining specimens possess a distinct digastric notch with a *linea m. obliquus capitis superior* delineating the origin of *m. digastricus* laterally from the lower portion of the insertion of *m. obliquus capitis superior* medially. The *linea m. obliquus capitis superior* merges into the *planum nuchale* of the occipital bone above the origin of *m. digastricus.* The origin of *m. digastricus* and the corresponding digastric notch are related only to the lower third of *m. obliquus capitis superior,* since the insertion of the latter muscle continues for a significant distance above the origin of *m. digastricus* on the dorsal cranium. *Linea m. obliquus capitis superior* is clearly evident only along the medial border of the digastric notch and the origin of *m. digastricus.* In these specimens, *m. obliquus capitis superior* is again well developed but does not possess a significantly large insertion area, which accounts for the absence of a well-developed *linea m. obliquus capitis superior* along the superior lateral portions of this muscle.

The 11-year-old and the 21-year-old possess a distinct *linea m. obliquus capitis superior* along the entire lateral border of the insertion of *m. ob-*

liquus capitis superior. As in previously described specimens, *m. obliquus capitis superior* is well developed in the 11-year-old but, in contrast to the 21-year-old, does not insert over a relatively broad area. However, in the former individual, the origin of *m. digastricus* has migrated superiorly, its upper border coinciding with the upper border of the insertion of *m. obliquus capitis superior.* Thus, the extensive *linea m. obliquus capitis superior* in the 11-year-old appears related to the relatively larger and more superior origin of *m. digastricus* along the lateral border of *m. obliquus capitis superior* rather than to the size of the insertion of the latter muscle.

In contrast, in the 21-year-old, the origin of *m. digastricus* is related only to the inferior portion of *m. obliquus capitis superior* and cannot account for the extensive *linea m. obliquus capitis superior.* However, in this individual, *m. obliquus capitis superior* is extremely well developed, possesses a relatively larger mass than the other specimens, and inserts over a broad area. Therefore, the development of a distinct bony crest along the entire lateral border of *m. obliquus capitis superior* in this individual must correspond to size-related changes in this muscle.

Because it develops two muscular heads, *m. rectus capitis posterior major* has two insertions on the occiput in the fetus, the newborn, the 9-month-old, and the 11-year-old (Fig. 14–3A). The insertion of *caput laterale* is oval-shaped and bears a slight medial concavity; it courses inferosuperiorly, curving medially as it proceeds, and its lateral border maintains broad contact with the insertion of *m. obliquus capitis superior.* The upper half of the medial border of *caput laterale* abuts the superior portion of the lateral third of *m. rectus capitis posterior minor.* The insertion of *caput mediale* lies medial to the insertion of *caput laterale;* it is oval-shaped, extending mediolaterally above the medial third of *m. rectus capitis posterior minor.*

The insertion of *m. rectus capitis posterior major* is relatively broader in the 9- and 17-year-olds (Fig. 14–3B). This insertion is oval-shaped, concave medially, and runs inferosuperiorly with substantial medial curvature. It is in broad contact laterally with the insertion of *m. obliquus capitis superior* and extends medially across the occiput, bounding the superior lateral two-thirds of *m. rectus capitis posterior minor.*

In the 8-year-old, the insertion of *m. rectus capitis posterior major* is oval-shaped and lacks the medial concavity or curvature seen in the previously discussed specimens (Fig. 14–3C). It runs inferosuperiorly, abutting the insertion of *m. obliquus capitis superior* laterally and the superior portion of the lateral half of *m. rectus capitis posterior minor* medially.

The 21-year-old possesses the most extensive attachment site of *m. rectus capitis posterior major* (Fig. 14–3D). The insertion of this muscle is somewhat T-shaped, with its base along the medial edge of the insertion of *m. obliquus capitis superior.* The muscle broadens mediolaterally along the upper portion of its insertion, extending laterally above the superior edge of *m. obliquus capitis superior* and medially above the superior border of the lateral three-fourths of *m. rectus capitis posterior minor.*

M. rectus capitis posterior minor is well-developed in the 21-year-old (Fig. 14–3D). It is relatively larger than in the other specimens, broadening extensively as it passes from origin to insertion, and covering a substantial portion of the occiput inferior and medial to the insertion of *m. rectus capitis posterior major. M. rectus capitis posterior minor* also flares extensively from origin to insertion in the fetus, the newborn, the 9-month-old, and the 11-year-old (Fig. 14–3A). However, in these specimens, it is significantly smaller and covers less of the occipital bone than in the 21-year-old. In the 8-, 9-, and 17-year-olds, *m. rectus capitis posterior minor* is less well developed than in the 21-year-old and broadens but slightly from its origin to insertion in the 9-year-old and 17-year-old (Fig. 14–3B). In the 8-year-old, there is limited flare from origin to insertion (Fig. 14–3C).

In the 21-year-old, the extensive size and insertion areas of *m. rectus capitis posterior major* and *m. rectus capitis posterior minor* are directly related to a distinct, well-developed inferior nuchal line. This line has two noncontinuous portions which correspond to the two muscles inserting on it. The portion associated with the insertion of *m. rectus capitis posterior minor* commences at the midline of the occiput and runs laterally, terminating before reaching the second portion of this line which corresponds to the insertion of *m. rectus capitis posterior major.* The second portion begins along the inferior border of the insertion of *m. rectus capitis posterior major* medial to the insertion of *m. obliquus capitis superior* and courses superiorly, terminating in a raised tubercle superior and lateral to the insertion of *m. rectus capitis posterior minor* and the first portion of the inferior nuchal line related to the latter muscle.

In the remaining specimens, the inferior nuchal line is poorly developed or nonexistent, thus reflecting the relatively smaller size and insertion areas of *m. rectus capitis posterior major* and *m. rectus capitis posterior minor* in these individuals compared to the 21-year-old.

DISCUSSION

The degree of variation in the suboccipital anatomy of these eight specimens clarifies the apparent discrepancy among previous descriptions of this region (Hill, 1939; Sakka, 1973; Sonntag, 1924). Sonntag, from the dissection of a young female orang-utan specimen, noted that the suboccipital muscles are not only similar overall to those of humans, but they also have the same attachments; the muscles in the orang-utan are well developed and in broad contact, obliterating the suboccipital triangle. From the dissection of an adult male, Hill described the suboccipital muscle of the orang-utan as being small and thin and bounding a clearly demarcated suboccipital triangle. In contrast, Sakka concluded from his dissection of an adult female that the orang-utan differs markedly from humans in its possession of a *m. rectus capitis posterior major* which is divided into two heads. He reported that these two heads are joined by their investing fascia but can be separated easily from each other.

Since each of these descriptions derives from the dissection of only a single individual, the differences between accounts could be the consequence of normal variation in musculature patterns. And indeed, the present study indicates that, with the exception of *m. obliquus capitis inferior,* a nontrivial amount of variation does exist in the suboccipital anatomy of the orang-utan. Furthermore, as is indicated by the differences in the shape of (degree of fanning, curvature, etc.) and relationships between the muscles and their respective insertions, much of this anatomical variation appears to be between individuals and not clearly related to age or sex.

In contrast, differences in the morphology and mass or size of *m. rectus capitis posterior major,* as well as differences in the relative sizes of *m. obliquus capitis superior* and *m. rectus capitis posterior minor,* may be tentatively correlated with the individual orang-utan's age, sex, and the degree to which its craniofacial complex has developed and become prognathic. The younger specimens possess two clearly separate heads of *m. rectus capitis posterior major,* whereas subadult and adult specimens have either a single head or some degree of fusion or apposition of multiple heads.

M. rectus capitis posterior minor and the combined heads of *m. rectus capitis posterior major* of the fetus, newborn, and 9-month-old possess the smallest relative total mass among the eight specimens. In the 8-year-old and the 11-year-old, these muscles are relatively larger than those in the younger specimens but the latter muscle is smaller than in the 9-year-old and the 17-year-old. Compared to all other specimens, the 21-year-old possesses both *m. rectus capitis posterior major* and *m. rectus capitis posterior minor,* which are significantly larger in relative size and insertion area. The relatively larger size of these muscles in this adult male correlates with the development of a distinct inferior nuchal line along the muscles' attachment sites on the occiput. As previously discussed, the presence of a substantial inferior nuchal line occurs only in this specimen.

Comparisons of the relative mass of *m. obliquus capitis superior* and the related development of a *linea m. obliquus capitus superior* also reveal age related patterns. The neonatal specimens possess a relatively small *m. obliquus capitis superior* with a limited insertion area compared to older specimens; concurrently, *linea m. obliquus capitis superior* is virtually nonexistent in these individuals. Older individuals with a relatively larger *m. obliquus capitis superior* show the development of at least a partial *linea m. obliquus capitis superior* along the muscle insertion's lateral border; an extensive *linea m. obliquus capitis superior* is seen in the 21-year-old, which possesses the largest *m. obliquus capitis superior* among the specimens.

Although age-related patterns in the relative sizes of and the bony morphology related to *m. rectus capitis posterior major, m. rectus capitis posterior minor,* and *m. obliquus capitis superior* are clearly indicated, variability between the adult specimens, the 21-year-old male and the 17-year-old female, also suggests the possible influence of sexual dimorphism. The relatively smaller mass of these muscles in the adult female compared to the male, as well as the lack in the former individual of a substantial inferior nuchal line and extensive *linea m. obliquus capitis superior,* may be a function of sexually dimorphic differences in cranial development and prognathism in the adults.

Schultz (1941, 1962, 1963) described the rapid and profound changes in relative head size of the orang-utan after birth, noting that some of the most significant changes in facial development and prognathism occurred at the time of the eruption of the secondary dentition. He commented that, concurrent with the development of the face and prognathism typical of the orang-utan, the relative position of the occipital condyles and foramen magnum shifts posteriorly. This backward relocation of the occipital condyles results in the heads of older individuals becoming less well balanced on the spine. Schultz also noted that such facial prognathism and the corresponding backwards shift of the skull base are not only age-related but are also correlated with sexual dimorphism; i.e., both facial prognathism and dorsal migration of the foramen magnum are more marked in adult males than in adult females. Schultz concluded that there is a much greater shift in males than in females of the center of gravity of the head, which consequently requires in males a stronger nuchal musculature for supporting the head on the spine.

Zuckerman (1955) also described age-related changes in the position of the foramen magnum. Like Schultz, he noted that the backward migration of the foramen magnum in older individuals occurs concurrently with the eruption of the permanent teeth and the development of facial prognathism. Zuckerman related this relative shift of the foramen magnum to a remodeling of the *planum nuchale* of the occipital bone due to growth-related changes in the neck muscles.

The osteological growth-related changes described by Schultz and Zuckerman offer an explanation for the age-related changes in the muscles of the suboccipital region as well as for the sexually dimorphic differences seen in the adult specimens. The development of larger muscle masses in *m. rectus capitis posterior major, m. rectus capitis posterior minor,* and *m. obliquus capitis superior,* as well as the fusion of multiple heads of *m. rectus capitis posterior major,* enhances muscle function and leverage, counterbalancing age-related and sexually dimorphic changes in facial prognathism and cranial development.

CONCLUSION

In contrast to the innervation, arterial, and venous patterns, which are fairly constant, the musculature of the suboccipital anatomy of the orang-utan may vary significantly between individuals as well as because of age and sex. With the exception of *m. obliquus capitis inferior,* there is considerable variability in the size, morphology, and relationships among the suboccipital muscles. The bony morphology related to the insertions of *m. rectus capitis posterior major, m. rectus capitis posterior minor,* and *m. obliquus capitis superior* is also variable in size, shape, and development of attachment area. Morphological differences among individuals in the shape of the muscles and their attachment areas, as well as in relationships between the muscles, are not clearly related to the age or sex of the individual. However, variations in relative size and the bony morphology related to areas of insertion of *m. rectus capitis posterior major, m. rectus capitis posterior minor,* and *m. obliquus capitis superior* appear to be age- and sex-related in the adult individuals. In addition, the morphology of *m. rectus capitis posterior major* demonstrates definite age-related trends with regard to the fusion or apposition of multiple heads. Age- and sex-related changes in the development of the face and of prognathism in the orang-utan, which result in a shift of the cranium's center of gravity and the remodeling of the occiput, appear to be the underlying reasons for similar categories of variability in the suboccipital muscles. The amount of variation seen in the suboccipital anatomy of the orang-utan should highlight the caution one should bring toward reconstructing the phylogenetic relationships of taxa on the basis of muscular anatomy derived from the study of small samples. Knowledge of variation in both soft tissues and related bony morphology must precede the pursuit of phylogenetic reconstruction.

ACKNOWLEDGMENTS

I wish to thank Dr. Jeffrey Schwartz for inviting me to participate in this volume, to express my sincere appreciation to Drs. Nikolajs Cauna, Jeffrey Schwartz, and Michael Siegel for their comments on this manuscript, and to express my gratitude to Mrs. Cynthia Strickland for typing it. The figures were drawn by John Anderton. This research was supported by funds or logistic support from the National Science Foundation (grant no. BNS-831252); Yerkes Regional Primate Center, Emory University (NIH RR-00165); the L.S.B. Leakey Foundation; Sigma Xi; the Anthropological Institute, University of Zürich; Departments of Anthropology and of Anatomy, Washington University, St. Louis; and the St. Louis Zoological Park.

REFERENCES

Hill, W. C. O. 1939. Observations on a giant Sumatran orang. *American Journal of Physical Anthropology, 24*:449–505.

Sakka, M. 1972. Comparative anatomy of the *squama oc-*

cipitalis and of the nape muscles in man and the pongids. pt. 1. Osteology. *Mammalia, 36:*696–750.

Sakka, M. 1973. Comparative anatomy of the *squama occipitalis* and of the nape muscles in man and the pongids. pt. 2. Myology. *Mammalia, 37:*126–191.

Schultz, A. H. 1941. Growth and development of the orangutan. *Contributions to Embryology, Carnegie Institute, 23:*57–110.

Schultz, A. H. 1962. Metric age changes and sex differences in primate skulls. *Zeitschrift für Morphologie und Anthropologie, 52:*239–255.

Schultz, A. H. 1963. Age changes, sex differences, and variability as factors in the classification of Primates. *In* Classification and Human Evolution, pp. 85–115, ed. S. L. Washburn. Aldine Publishing Co., Chicago.

Sonntag, C. F. 1924. On the anatomy, physiology, and pathology of the orang-outan. *Proceedings of the Zoological Society of London, 24:*349–451.

Zuckerman, S. 1955. Age changes in the basicranial axis of the human skull. *American Journal of Physical Anthropology, 13:*521–539.

15

Cranial Growth

OLAV RÖHRER-ERTL

A persistent theme throughout the scientific literature on the orang-utan is the problem of extreme variability in the morphological data (cf. Röhrer-Ertl, 1982a,b, 1983, 1984a). In fact, the degree of such morphological variability is similar to, if not in some nonmetrical features greater than, that recorded for modern *Homo* (cf. Martin, 1928). And in this regard, *Pongo* appears to differ from all other nonhuman primates.

Again similar in degree to modern *Homo,* the orang-utan responds to changes, even minor ones, in the environment; the corresponding changes in morphology can be measured accurately (Röhrer-Ertl and Frey, 1987). Since previous methods used to record and quantify such physical alterations have produced relatively unclear results, I have developed an approach to such analyses which better elucidates intrapopulational variation, both cross-sectionally and longitudinally. Since the correct species name of the orang-utan appears to be *satyrus* and not *pygmaeus* (see Röhrer-Ertl, this volume) I shall refer to this hominoid as *Pongo satyrus;* the Bornean subspecies is, I have argued, *P. s. borneensis* von Wurm, 1784 and the Sumatran variety *P. s. satyrus* (Linnaeus, 1758).

MATERIALS AND METHODS

Cranial growth was studied in two random samples of orang-utans, one from the subspecies *Pongo satyrus borneensis* from Skalau, West Borneo ($N = 219$), and the other from *P. s. satyrus* from Sumatra ($N = 59$). Each sample was subdivided according to age and sex. The method of sampling, the nature of the specimens used, and the representativeness of the samples for Skalau and Sumatra have been discussed elsewhere (Röhrer-Ertl, 1984a).

Following the procedures developed for *Homo sapiens* by Martin (1928) and for apes by Oppenheim (1927), standardized cranial indices were established in which the particular index is calculated as $(X/Y) \times 100$, where X is the mean of the smaller value or dimension and Y is the mean of the larger value or dimension (see Tables 15–5 and 15–6). In random samples, the calculation of indices can smooth out the variations introduced by single measurements. Further smoothing occurs when index averages are calculated for subsamples, e.g., age cohorts.

Cranial indices have been used in the past in attempts to quantify differences among populations, especially of *Homo sapiens,* but the results of such pursuits have been largely unsatisfactory. In this study, cranial indices were evaluated in order to determine if the dimensions measured are more or less correlated in their growth. In *Homo sapiens,* for instance, we could look at cranial length and breadth (width) or cranial length and height in the respective indices LBI and LHI (Rörher-Ertl, 1982b). The more independent in growth each of the individual bones is, the more the values of the indices should be skewed. As a general precaution, the use of cranial indices demands a stringent selection of data according to the specific problem involved.

Measurements were taken and indices calculated according to Martin (1928). The length-breadth (width) (LBI) and transverse-frontotemporal indices (TFTI) of Oppenheim (1927) were also calculated; the TFTI was used as a check against the data obtained by the LBI. Three other indices were also used. One was the palatine-basal-prosthion index (PBIpr), in which the landmark prosthion (pr) was substituted for orale (ol) in the measurement of distance from staphylion (sta); the measure staphylion-prosthion yields a better approximation of the development of the snout, especially in older age groups. The second index, LBI/Mand, measures mandibular length and breadth and can monitor the developing masticatory system. The third index, the facial-breadth index (GBI), focuses on changes in width of the middle face. Differences in N among the calculated indices are a reflection of the state of preservation of the cranial samples (cf. Röhrer-Ertl, 1984a).

If the number of samples is large enough to avoid error in the mean, it is sufficient to use the standard deviation(s) to evaluate the average values ($\overline{X}$). When the number of samples is small, the measured range of variation must be taken into account. In such cases it has proved helpful to use the difference calculated for the two extreme values (VB 1) as well as the absolute values of the two extremes themselves (VB 2).

DATA BASE FOR CRANIAL GROWTH

In the following subsections, growth parameters for 17 cranial indices are given for both samples, which, in turn, are subdivided according to sex and age (groups from infancy to complete maturity).

Length–Breadth Index (LBI) (Fig. 15–1 and Tables 15–1 to 15–4)

Skalau Females (N = 113)

Growth in cranial breadth exceeds length through suture closure. Index values indicate a rapid increase in cranial width until the juvenile phase, followed by a period up to the adult phase of equilibrium in growth of cranial length and breadth, and then an increase in breadth until full maturity, paralleling the development of the masticatory apparatus. The tympanic region of the temporal bone (around the landmark porion), becomes very stout in response to increasing masticatory pressures.

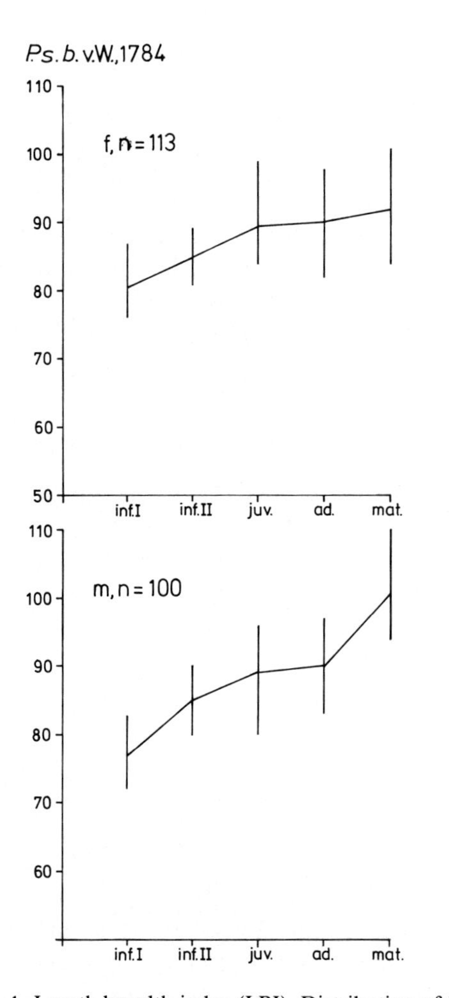

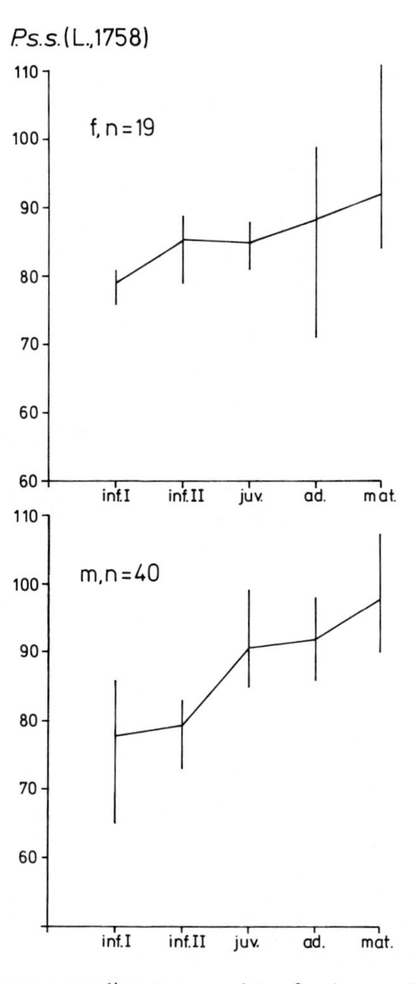

FIG. 15–1. Length-breadth index (LBI). Distribution of index averages according to sex and age for the samples from Skalau (western Borneo; *P.s.b., Pongo satyrus borneensis*) and Sumatra *(P.s.s., Pongo satyrus satyrus).* Range of variation per subsample (age group; cf. Röhrer-Ertl, 1984a) is also indicated. Ordinate, index averages with ranges of variation in index values; abscissa, age group.

TABLE 15–1. Cranial Indices and Cranial Capacity for *Pongo satyrus borneensis* von Wurmb, 1784. Skalau (West Borneo), Females, N = 116

	LBI (8:1)	LHI (17:1)	LBI/Mand (66:68)	LHI/Mand (70:68)	ZBI (45:40)	PBI (62:40)	PBIpr (62′:40)	GI/K (47:45)	TFTI (9:11(1))
Infant I, N = 11									
N	10	8	9	10	8	8	8	9	9
$\overline{X}$	80.7000	78.0000	82.4444	73.3000	99.7500	47.7500	52.6250	92.7778	70.1111
s	3.8887	3.1623	4.2753	5.4579	4.0620	2.0529	1.7678	5.2387	5.1828
VB 1	11	9	13	20	12	7	5	17	17
VB 2	87–76	84–75	88–75	83–63	107–95	51–44	55–50	102–85	80–64
Infants II, N = 6									
N	6	6	6	6	6	6	6	6	6
$\overline{X}$	85.1667	76.0000	82.6667	76.3333	97.8333	50.0000	54.5000	92.1667	61.5000
s	2.7896	3.6878	3.5024	2.5820	2.9269	1.5492	1.0488	3.8166	4.0866
VB 1	8	9	8	6	7	3	3	9	11
VB 2	89–81	82–73	87–79	80–74	102–95	52–49	56–49	98–89	65–54
Juvenile, N = 28									
N	28	28	27	27	28	28	27	27	28
$\overline{X}$	89.5000	75.6071	83.2222	75.8889	97.5000	48.1429	54.1111	93.5555	56.7143
s	3.3500	3.0833	5.3445	3.7245	4.6627	2.0132	1.5021	2.4089	3.8476
VB 1	15	12	25	16	17	8	6	22	17
VB 2	99–84	82–70	96–71	84–68	107–90	51–43	57–51	105–83	65–48
Adult, N = 62									
N	60	60	59	60	58	59	59	57	60
$\overline{X}$	9.2667	76.4667	84.4915	75.4333	95.2241	49.1864	54.4407	93.8421	52.7333
s	3.5361	3.3621	5.6090	5.1925	4.0351	2.0127	2.0109	5.8211	4.4029
VB 1	16	16	23	23	16	10	11	27	22
VB 2	98–82	86–70	96–73	87–64	102–86	54–44	60–49	107–80	64–42
Mature, N = 9									
N	9	8	6	7	9	9	9	7	8
$\overline{X}$	92.3333	77.3750	89.3333	79.7143	96.5556	49.2222	54.2222	93.7143	51.8750
s	5.5453	2.1998	3.0111	2.8115	4.5308	1.7873	2.1082	6.0474	5.0267
VB 1	17	6	8	7	12	5	6	14	18
VB 2	101–84	81–75	95–87	83–76	102–90	52–47	57–51	102–88	61–43

	JMI (66:45)	TCFI (11(1):45)	PI (63:62)	PHI (64:63)	OI (52:51)	OTI (53:52)	NI (54:55)	GBI (44:46)	CC (38)
Infant I, N = 11									
N	7	9	9	10	10	9	10	10	10
$\overline{X}$	68.000	98.3333	60.6556	29.6000	108.4000	113.4444	116.0000	85.4000	348.0000
s	3.0000	6.4031	4.1866	4.8808	12.7348	6.4829	10.3816	5.2957	33.2666
VB 1	8	21	13	14	45	18	28	18	110
VB 2	72–64	107–86	66–53	35–21	142–97	124–106	138–110	94–76	400–290
Infant II, N = 6									
N	6	6	6	6	6	6	6	6	6
$\overline{X}$	72.8333	92.0000	53.0000	31.8333	107.3333	118.1667	111.1667	76.3333	355.0000
s	2.1370	2.2804	4.1473	3.1885	7.2847	4.0702	13.0295	3.4448	30.1662
VB 1	4	6	9	7	17	11	34	10	80
VB 2	75–71	96–90	58–49	34–27	114–97	123–112	130–96	83–73	400–320
Juvenile, N = 28									
N	27	28	28	28	28	28	28	28	28
$\overline{X}$	73.4444	89.3214	53.9286	33.9286	106.6429	117.8571	104.3067	75.0714	370.5357
s	3.5337	3.9068	5.1992	4.2245	5.5990	7.7159	11.3561	5.1491	42.7815
VB 1	13	13	22	18	22	35	48	23	150
VB 2	81–68	95–82	69–47	44–26	119–97	135–100	129–81	86–63	460–310

TABLE 15–1 (*Continued*)

	JMI (66:45)	TCFI (11(1):45)	PI (63:62)	PHI (64:63)	OI (52:51)	OTI (53:52)	NI (54:55)	GBI (44:46)	CC (38)
Adult, N = 62									
N	55	58	59	59	59	59	59	59	60
$\overline{X}$	76.0000	85.2241	52.4576	35.2208	107.2034	117.9831	115.0678	74.2373	365.1667
s	4.0323	2.8100	4.0229	5.4774	6.4401	8.7975	13.7388	5.0184	39.9042
VB 1	19	12	19	19	21	38	71	19	170
VB 2	84–65	90–78	62–43	47–28	118–97	138–100	156–85	84–64	465–295
Mature, N = 9									
N	6	9	9	9	9	9	9	9	8
$\overline{X}$	77.6667	84.6667	56.2222	34.7778	111.2222	117.2222	105.2222	75.2222	384.3750
s	5.1251	3.7081	5.6519	3.8006	9.8841	11.5734	10.9747	3.9299	39.5002
VB 1	13	12	19	13	26	38	36	14	115
VB 2	86–73	93–81	66–47	43–30	126–100	134–98	129–93	84–70	465–350

Indices in index values, cranial capacity (CC) in cm^3. Further explanation in text and Figs. 15–1 to 15–15.

N, Number; $\bar{X}$, mean; s, standard deviation; VB 1, interval (range); VB 2, extremes. Age groups are given according to Röhrer-Ertl (1984a); measurement scale according to Martin (1928). Deviations from this scale in measurement No. 62′ (pr - sta), which is here introduced instead of 62 (ol - sta).

TABLE 15–2. Cranial Indices and Cranial Capacity for *Pongo satyrus borneensis* von Wurmb, 1784. Skalau (West Borneo), Males, N = 103

	LBI (8:1)	LHI (17:1)	LBI/Mand (66:68)	LHI/Mand (70:68)	ZBI (45:40)	PBI (62:40)	PBIpr (62′:40)	GI/K (47:45)	TFTI (9:11(1))
Infant I, N = 15									
N	14	12	15	15	11	11	11	14	14
$\overline{X}$	77.0000	76.6364	84.4000	70.9333	99.3636	46.8182	52.6364	93.000	72.2857
s	5.1441	2.1106	8.9267	8.0220	4.6926	3.3710	2.0136	7.6561	8.2409
VB 1	11	7	37	31	16	6	7	29	36
VB 2	83–72	79–72	100–63	91–60	108–92	50–39	55–48	105–76	94–58
Infant II, N = 6									
N	6	6	6	6	6	6	6	6	6
$\overline{X}$	85.3333	75.6667	86.0000	77.3333	94.6667	47.5000	53.6667	99.0000	58.8333
s	3.4448	4.7188	5.5498	7.5011	3.5024	1.0488	1.3663	4.0000	5.2313
VB 1	10	14	15	20	10	3	4	11	13
VB 2	90–80	81–67	96–81	89–69	99–89	49–46	56–52	103–92	66–53
Juvenile, N = 17									
N	17	16	17	17	16	16	16	16	17
$\overline{X}$	89.1765	77.1875	81.9333	75.2500	94.6875	47.1875	53.4375	95.0625	57.3529
s	4.3046	3.3110	6.2274	3.7859	5.4494	1.7595	2.1282	4.6543	4.1373
VB 1	16	9	22	13	20	6	8	13	14
VB 2	96–80	82–73	90–68	82–69	108–88	50–44	57–49	101–88	62–48
Adult, N = 38									
N	37	37	36	36	36	37	37	35	37
$\overline{X}$	90.9730	74.2703	82.3333	75.0278	93.3056	48.5676	54.1892	96.2000	52.7297
s	3.9755	2.9026	5.9952	4.1781	4.7136	1.5730	1.8233	5.9941	4.9702
VB 1	14	13	22	16	21	9	11	24	24
VB 2	97–83	82–69	94–72	83–67	107–86	55–46	61–50	109–85	63–39
Mature, N = 27									
N	26	26	26	27	26	26	26	27	26
$\overline{X}$	100.7692	77.1538	84.9231	75.6296	96.8077	48.1923	52.8077	90.6667	39.1538
s	4.3663	3.7703	6.0261	5.2341	6.4561	1.8766	1.7668	5.7513	7.8266
VB 1	16	15	22	19	30	7	6	28	23
VB 2	110–94	84–69	99–77	85–66	115–85	52–45	56–50	108–80	50–27

TABLE 15–2 (*Continued*)

	JMI (66:45)	TCFI (11(1):45)	PI (63:62)	PHI (64:63)	OI (52:51)	OTI (53:52)	NI (54:55)	GBI (44:46)	CC (38)
Infant I, N = 15									
N	14	14	14	14	14	14	14	14	15
$\overline{X}$	69.7143	101.5000	64.4286	28.0714	107.2143	112.6429	122.8571	83.2143	357.0000
s	6.3419	3.8779	6.5365	4.9995	10.3491	13.0479	19.3306	6.8296	38.9505
VB 1	21	17	24	19	39	51	67	23	125
VB 2	80–59	109–92	81–57	38–19	131–92	138–87	157–90	96–73	400–275
Infant II, N = 6									
N	6	6	6	6	6	6	6	6	6
$\overline{X}$	77.1667	94.8333	54.5000	33.6667	108.5000	112.3333	114.5000	77.6667	365.2381
s	3.4303	3.4303	4.4159	3.5590	6.0249	11.3078	11.2916	7.1181	47.2355
VB 1	9	10	11	9	16	26	30	19	185
VB 2	82–73	99–89	60–49	38–29	116–100	129–103	126–96	89–70	460–275
Juvenile, N = 17									
N	15	16	17	17	17	17	17	16	17
$\overline{X}$	73.7333	88.0625	53.8824	34.3529	105.3529	118.2353	111.6471	74.3125	398.8235
s	4.4955	1.9822	5.1220	4.7558	8.1236	8.5259	16.5225	5.6299	27.1299
VB 1	18	7	17	14	31	33	67	17	110
VB 2	82–64	91–84	64–47	41–27	118–87	135–102	146–79	82–65	440–330
Adult, N = 38									
N	35	36	36	36	37	37	37	36	37
$\overline{X}$	75.7143	84.7778	51.4722	35.5278	108.7027	114.6486	111.8919	73.3056	381.4865
s	4.7932	3.7577	4.3719	5.4010	17.5450	9.1355	12.3845	4.8156	42.7341
VB 1	22	14	22	25	23	35	58	19	180
VB 2	86–64	91–77	60–38	49–24	123–100	132–97	131–73	81–62	490–310
Mature, N = 27									
N	26	26	27	27	27	26	27	26	26
$\overline{X}$	76.2692	81.2308	49.0000	38.8889	107.5556	118.0000	100.1852	74.8462	431.1538
s	4.2101	5.3539	4.1231	5.6250	5.8266	11.1176	12.9289	5.1201	43.1349
VB 1	8	23	17	25	24	49	53	18	145
VB 2	88–72	96–73	61–44	50–25	122–98	151–102	129–76	86–68	510–365

Abbreviations as in Table 15–1.

TABLE 15–3. Cranial Indices and Cranial Capacity for *Pongo satyrus satyrus* (Linnaeus, 1958). Sumatra, Females, N = 19

	LIB (8:1)	LHI (17:1)	LBI/ Mand (66:68)	LHI/Mand (70:68)	ZBI (45:40)	PBI (62:40)	PBIpr (62:40)	GI/K (47:45)	TFTI (9:11(1))
Infant I, N = 3									
N	3	3	3	3	3	3	3	3	3
$\overline{X}$	79.0000	79.0000	92.0000	66.6667	105.6667	50.0000	57.0000	86.6667	78.6667
s	2.6458	6.2450	6.0000	6.4291	2.8868	7.0000	3.4641	6.6583	0.5774
VB 1	5	12	12	12	5	13	6	12	1
VB 2	81–76	86–74	98–86	74–62	109–104	58–45	61–55	91–79	79–78
Infant II, N = 3									
N	3	3	3	3	3	3	3	3	3
$\overline{X}$	85.3333	84.3333	90.0000	73.0000	94.0000	47.6667	53.6667	96.0000	60.0000
s	5.5076	3.5119	12.1244	4.3589	5.2915	2.5166	1.5275	4.3589	6.0828
VB 1	10	7	21	8	10	5	3	8	11
VB 2	89–79	88–81	104–83	78–70	100–90	50–45	55–52	99–91	67–56

TABLE 15–3 (*Continued*)

	LIB (8:1)	LHI (17:1)	LBI/ Mand (66:68)	LHI/Mand (70:68)	ZBI (45:40)	PBI (62:40)	PBIpr (62:40)	GI/K (47:45)	TFTI (9:11(1))
Juvenile, $N = 4$									
N	4	4	4	4	4	4	4	4	4
$\overline{X}$	85.0000	78.2500	78.5000	70.7500	88.7500	52.0000	56.7500	99.2500	48.5000
s	2.9439	2.8723	2.8868	6.7515	3.3040	2.4495	2.3629	6.0208	12.3693
VB 1	7	6	7	15	7	5	5	13	28
VB 2	88–81	82–76	82–75	77–62	92–85	54–49	60–55	107–94	59–31
Adult, $N = 4$									
N	4	4	3	3	4	4	4	3	4
$\overline{X}$	88.5000	80.5000	78.6667	77.3333	96.7500	52.2500	57.5000	93.0000	47.2500
s	12.5565	2.3805	3.2146	18.0093	2.8723	0.9574	1.7321	10.1489	3.8622
VB 1	28	5	6	33	6	2	3	20	8
VB 2	99–71	82–77	81–75	98–65	99–93	53–51	59–56	104–84	51–43
Mature, $N = 5$									
N	5	5	5	5	5	5	5	5	5
$\overline{X}$	92.0000	79.0000	81.2000	77.4000	95.0000	50.6000	56.4000	93.2000	48.6000
s	11.1580	3.0822	7.4067	3.2863	4.3012	1.9494	1.5166	3.3466	6.6182
VB 1	27	7	18	8	11	5	3	8	16
VB 2	111–84	83–76	91–73	83–75	102–91	54–49	58–55	97–89	54–38

	JMI (66:45)	TCFI (11(1):45)	PI (63:62)	PHI (64:63)	OI (52:51)	OTI (53:52)	NI (54:55)	GBI (44:46)	CC (38)
Infant I, $N = 3$									
N	3	3	3	3	3	3	3	3	3
$\overline{X}$	70.0000	101.0000	65.0000	32.6667	102.3333	114.0000	110.6667	80.3333	291.6667
s	1.7321	7.5498	10.5357	2.5166	2.0817	7.9373	3.7859	4.0415	97.7667
VB 1	3	15	21	5	4	15	7	7	195
VB 2	71–68	109–94	75–54	35–30	104–100	123–108	115–108	85–78	385–190
Infant II, $N = 3$									
N	3	3	3	3	3	3	3	3	3
$\overline{X}$	75.6667	99.3333	53.0000	29.6667	105.3333	124.0000	102.3333	73.0000	366.6667
s	5.0332	3.2146	9.8489	10.4083	4.0415	4.3589	10.3333	0.0000	33.2916
VB 1	10	6	19	20	7	8	21	0	65
VB 2	81–71	103–97	64–45	38–18	110–103	127–119	111–90	73–73	395–330
Juvenile, $N = 4$									
N	4	4	4	4	4	4	4	4	4
$\overline{X}$	78.2500	87.5000	49.000	32.7500	110.5000	118.7500	105.5000	66.2500	363.7500
s	6.1305	7.8528	9.3452	5.3151	7.9373	8.0156	10.2794	1.5000	31.1916
VB 1	13	17	19	13	19	18	25	3	65
VB 2	87–74	99–82	63–44	39–26	119–100	129–111	118–93	68–65	395–330
Adult, $N = 4$									
N	3	4	4	4	4	4	4	4	3
$\overline{X}$	75.0000	85.5000	54.5000	31.2500	110.2500	110.5000	92.2500	69.5000	333.3333
s	2.6458	2.0817	2.6458	5.8523	16.6608	8.9629	9.1788	7.8951	34.0343
VB 1	5	5	6	14	37	21	21	16	65
VB 2	78–73	88–83	58–52	38–24	131–94	119–98	100–84	79–63	360–295
Mature $N = 5$									
N	5	5	5	5	5	5	5	5	5
$\overline{X}$	73.0000	84.8000	53.6000	30.8000	110.2000	109.8000	107.2000	68.4000	380.0000
s	6.1237	9.8843	1.1402	7.5961	5.3572	2.6833	13.3679	2.7019	34.0955
VB 1	17	24	3	19	13	6	34	7	95
VB 2	82–65	102–78	55–52	38–19	118–105	114–108	130–96	73–66	430–335

Abbreviations as in Table 15–1.

TABLE 15–4. Cranial Indices and Cranial Capacity for *Pongo satyrus satyrus* (Linnaeus, 1758). Sumatra, Males, $N = 40$

	LBI (8:1)	LHI (17:1)	LBI/Mand (66:68)	LHI/Mand (70:68)	ZBI (45:40)	PBI (62:40)	PBIpr (62′:40)	GI/K (47:45)	TFTI (9:11(1))
Infant I, $N = 6$									
N	6	6	5	4	5	6	6	4	6
$\overline{X}$	78.0000	81.1667	86.8000	73.2500	97.8000	49.5000	55.3333	100.7500	73.8333
s	8.0250	3.4303	10.5214	4.6458	19.1102	3.3912	3.8297	15.7560	11.2146
VB 1	21	8	28	11	52	9	10	35	32
VB 2	86–65	85–77	100–72	78–67	128–76	54–45	59–49	124–89	93–61
Infant II, $N = 3$									
N	3	3	3	3	3	3	3	3	3
$\overline{X}$	79.6667	76.0000	75.0000	70.3333	95.0000	51.3333	54.6667	109.3333	72.0000
s	5.7735	4.3589	3.6056	5.0332	8.7178	1.5275	1.5275	13.5031	13.0000
VB 1	10	8	7	10	16	3	3	27	23
VB 2	83–73	79–71	79–72	75–65	105–89	53–50	56–53	123–96	87–64
Juvenile, $N = 8$									
N	8	7	8	8	7	7	7	8	8
$\overline{X}$	91.1250	78.5714	82.3750	74.2500	95.4286	50.0000	55.8571	97.2500	56.1250
s	2.2763	3.1547	6.4573	3.0119	11.0151	1.5275	2.5166	12.7410	10.5040
VB 1	14	9	18	9	20	3	5	23	21
VB 2	99–85	85–76	89–71	79–70	108–88	52–49	59–54	103–80	56–35
Adult, $N = 3$									
N	3	3	3	3	3	3	3	3	3
$\overline{X}$	92.6667	75.6667	90.6667	72.0000	95.3333	50.3333	56.6667	88.3333	45.3333
s	6.1101	1.1547	10.5987	7.5498	11.0151	1.5275	2.5166	12.7410	10.5040
VB 1	12	2	21	15	20	3	5	23	21
VB 2	98–86	77–75	102–81	79–64	108–88	52–49	59–54	103–80	56–35
Mature, $N = 20$									
N	20	19	19	19	19	19	19	18	20
$\overline{X}$	98.0000	78.1579	86.5263	77.0000	97.3158	50.4211	54.6842	91.1667	37.1500
s	5.0158	3.4199	8.2420	7.0317	4.7147	2.1684	2.1097	5.9532	5.8694
VB 1	17	14	32	24	18	9	9	23	22
VB 2	107–90	89–75	104–72	85–61	107–89	55–46	58–49	99–76	48–26

	JMI (66:45)	TCFI (11(1):45)	PI (63:62)	PHI (64:63)	OI (52:51)	OTI (53:52)	NI (54:55)	GBI (44:46)	CC (38)
Infant I, $N = 6$									
N	4	5	6	6	6	6	6	6	6
$\overline{X}$	78.0000	101.8000	63.6667	25.5000	104.0000	117.8333	116.8333	80.8333	341.6667
s	17.3781	17.1231	14.0097	2.6646	7.1833	7.6004	7.5211	8.0850	65.4726
VB 1	41	42	38	7	18	22	23	24	170
VB 2	101–60	132–90	80–42	30–23	110–92	131–109	127–104	94–70	400–230
Infant II, $N = 3$									
N	3	3	3	3	3	3	3	3	3
$\overline{X}$	67.0000	94.6667	53.3333	24.3333	102.3333	120.6667	115.0000	78.0000	345.0000
s	1.7321	5.0332	12.7410	12.3423	4.0415	7.6376	14.1067	5.2915	37.7492
VB 1	3	10	23	24	7	15	28	10	75
VB 2	69–66	100–90	68–45	38–14	107–100	129–114	128–100	84–74	385–310
Juvenile $N = 8$									
N	8	8	8	8	8	8	8	8	8
$\overline{X}$	73.2500	88.2500	51.000	32.8750	107.2500	117.8750	92.6250	68.1250	392.5000
s	4.4641	3.0589	5.3984	6.9372	5.9940	8.3570	9.1329	2.9970	55.9336
VB 1	13	10	18	21	19	24	29	9	155
VB 2	79–66	94–84	62–44	41–20	116–97	132–108	107–78	73–64	460–305

TABLE 15–4 (*Continued*)

	JMI (66:45)	TCFI (11(1):45)	PI (63:62)	PHI (64:63)	OI (52:51)	OTI (53:52)	NI (54:55)	GBI (44:46)	CC (38)
Adult, N = 3									
N	3	3	3	3	3	3	3	3	3
$\overline{X}$	80.3333	83.3333	50.6667	27.0000	115.0000	109.3333	109.3333	64.3333	423.3333
s	4.6188	4.0415	4.7258	13.0000	11.5326	14.3643	14.3643	5.5076	14.4338
VB 1	8	8	9	23	22	27	27	10	25
VB 2	83–75	87–79	56–47	42–19	128–106	120–93	120–93	68–58	440–415
Mature, N = 20									
N	19	20	19	19	19	19	20	20	20
$\overline{X}$	75.7895	78.8500	49.0000	36.4211	112.0000	111.1579	94.5500	64.2500	408.5000
s	5.6232	6.1923	4.8419	5.1566	9.8319	10.1065	9.8166	4.6326	34.2629
VB 1	20	29	15	18	45	35	34	18	115
VB 2	84–64	98–69	55–40	42–24	129–84	129–94	112–78	69–51	475–360

Abbreviations as in Table 15–1.

Skalau Males (N = 100)

While there is some change from the infant to juvenile phases (77–85–89), the most marked growth occurs from the adult to the fully mature stage (91–101), especially in increased cranial breadth, a dimension that changes much more drastically during that period than it does in females; until the adult stage, males have a longer and narrower cranium than do females. The marked growth spurt in males between the adult and fully mature phases could be related to their being generally more "active" during that period (see below).

Sumatran Females (N = 19)

Index values exhibit parallels with the Skalau female sample. The noted range of variation could reflect the possibility that individuals from Atjeh (northern Sumatra) and Deli (northwestern Sumatra) are included in the sample (see below for further discussion).

Sumatran Males (N = 40)

The sample of Sumatran males parallels in general that from Skalau. Sumatran males, however, do also show an increase in cranial breadth up to the fully mature stage, but it is unlikely that this expansion is influenced by an increase in brain size at that time.

Length-Height Index (LHI) (Fig. 15–2 and Tables 15–1 to 15–4)

Skalau Females (N = 110)

Differences in index values indicate that cranial length increases relative to breadth from infant stages I–II, that the two dimensions grow in parallel during the juvenile phase, and that an emphasis on cranial length resumes throughout adulthood. The latter period of increasing cranial length is no doubt coupled with the development of the nuchal and masticatory musculature as well as the associated nuchal crest (crista occipitotemporalis). It would also appear that, as in *Homo sapiens,* there is a correlation between the length and height of the skull and the volume of the cranial cavity (i.e., cranial capacity) (see Rörher-Ertl, 1982).

Skalau Males (N = 97)

There does not seem to be any reason, other than sampling bias, for the growth patterns of males to differ so significantly from those of females.

Sumatran Females (N = 57)

Aside from sampling bias, there is seemingly no reason for there to be a growth pattern different from that observed in the Skalau female sample.

Sumatran Males (N = 38)

Again, there is no reason to expect that changes in these parameters should differ so much from those in the Skalau males.

Mandibular Length-Breadth Index (LBI/Mand) (Fig. 15–3 and Tables 15–1 to 15–4)

Skalau Females (N = 107)

Index averages indicate that the lower jaw widens (as one would expect) in tandem with the upper jaw. Such growth is particularly apparent

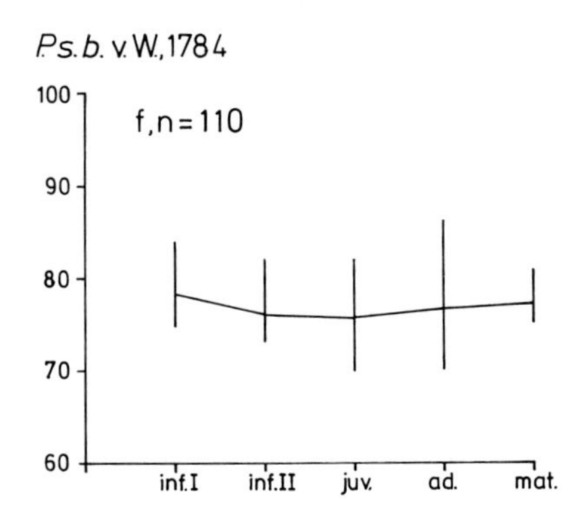

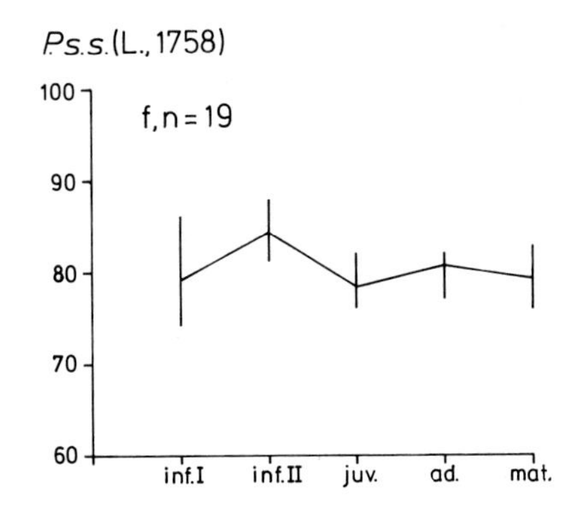

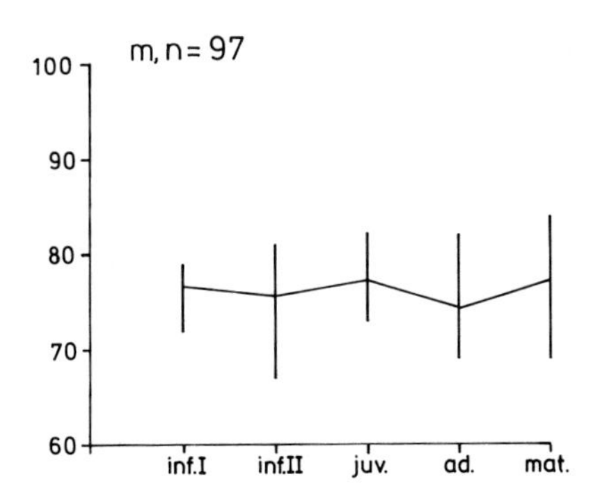

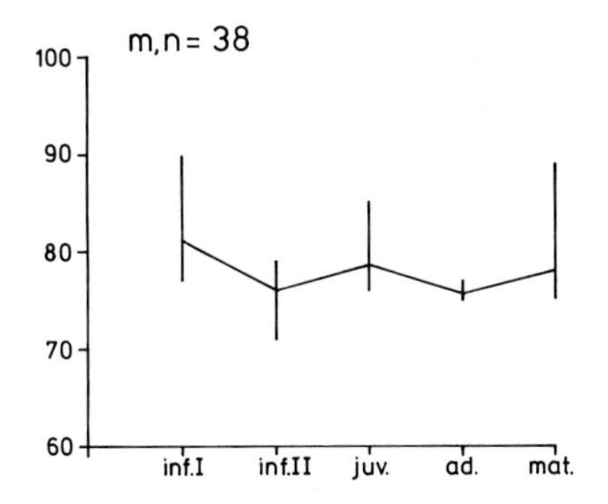

FIG. 15–2. Length-height index (LHI). See Fig. 15–1 for explanation.

from the juvenile phase onward, presumably as a functional correlate of the eruption and subsequent use of the teeth.

Skalau Males (N = 100)

The irregularity of the curve may be related to biases in sampling, since a curve similar to that of females is expected. In general, increase in breadth is not as apparent in males as it is in females, which may be due to the fact that females have smaller mandibles.

Sumatran Females (N = 18)

There is no significant difference from the curve obtained for Skalau females.

Sumatran Males (N = 38)

The index averages are similar to those for Skalau males.

Mandibular Length-Height Index (LHI/Mand) (Fig. 15–4 and Tables 15–1 to 15–4)

Skalau Females (N = 110)

It appears that, until the juvenile stage, growth proceeds slowly, with mandibular length being a stronger component than ramal height. During the juvenile phase, mandibular length and height increase at the same rate, but in the final phase, length is once more the predominant vector of mandibular growth. These changes are correlated with changes in the dentition, such as tooth replacement, addition of posterior teeth, and alterations in occlusion, as well as with the development of the muscles of mastication (Röhrer-Ertl, 1984a).

Skalau Males (N = 101)

The curve for males is generally similar to and parallels that of females. It would appear that the

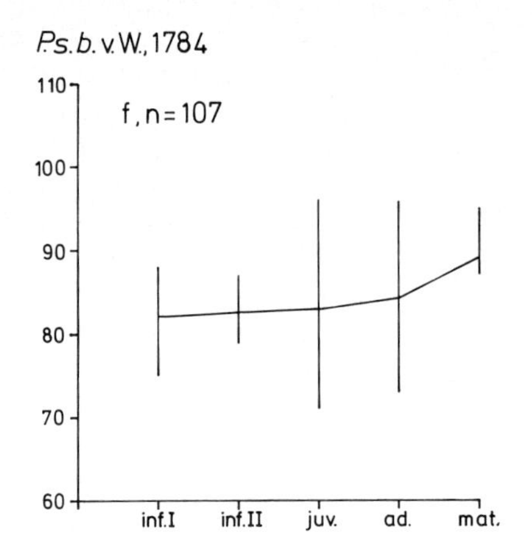

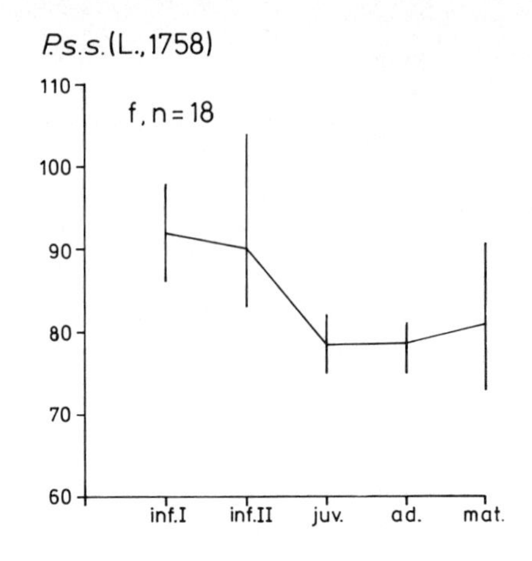

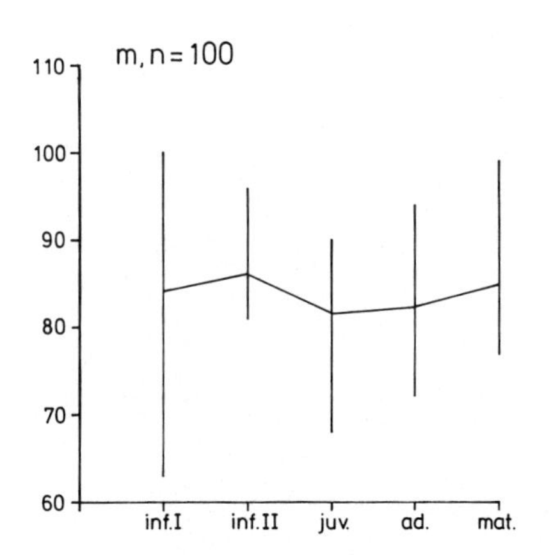

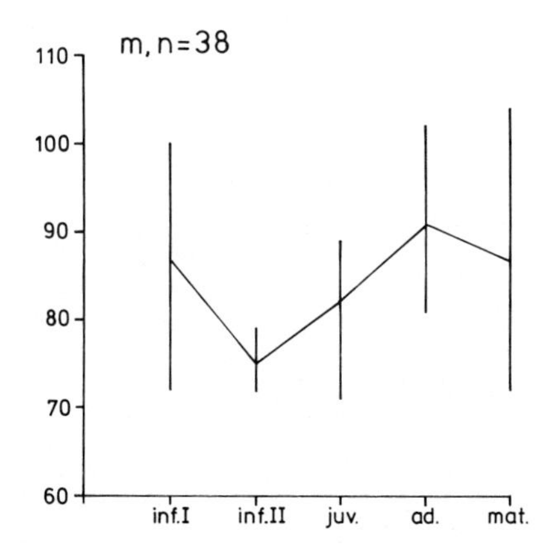

FIG. 15–3. Mandibular length-breadth index (LBI/Mand). See Fig. 15–1 for explanation.

visual impression one gets that the female mandible is relatively higher than in males is actually due to the smaller size overall of the female mandible and to its smoother and less muscle-scarred surfaces.

Sumatran Females (N = 18)

The curve is not significantly different from that for Skalau females.

Sumatran Males (N = 37)

Index values, although differing in some detail from those for Skalau males, are confined within the same general parameters and reflect the same pattern of mandibular growth.

Zygomaticobasal Index (ZBI) (Fig. 15–5 and Tables 15–1 to 15–4)

Skalau Females (N = 109)

Throughout most of growth facial length increases at a more rapid pace than does facial breadth, but in the final phase, the reverse relation occurs.

Skalau Males (N = 95)

The curve actually resembles that for LHI/Mand for Skalau females and thus reflects facial lengthening. The increase is relatively much greater in males.

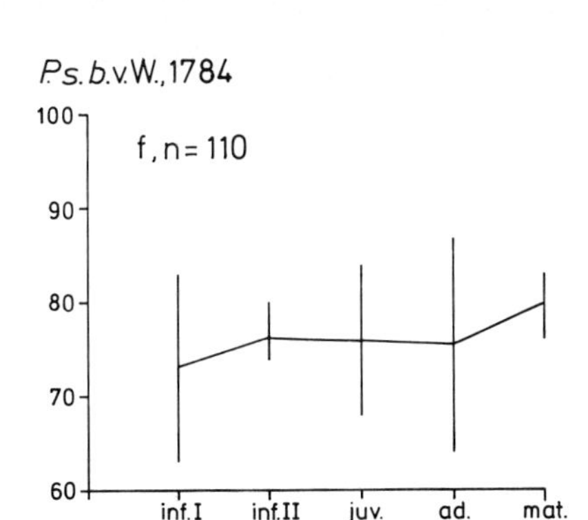

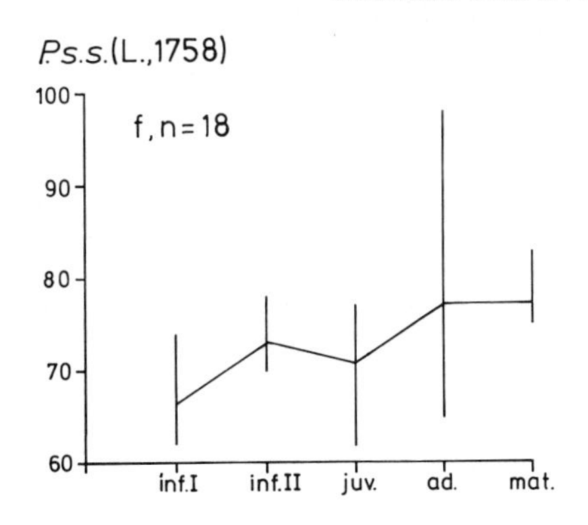

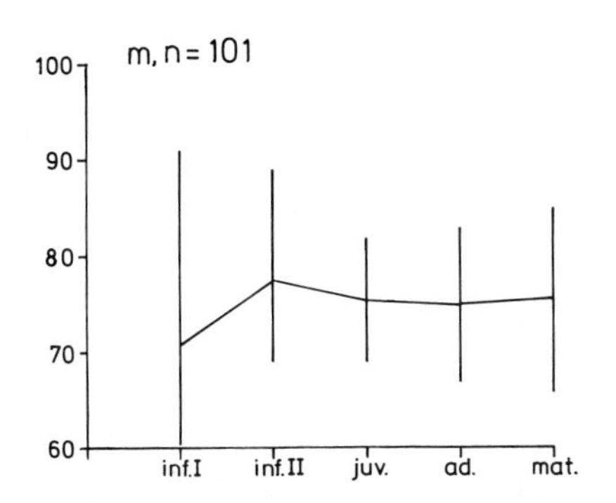

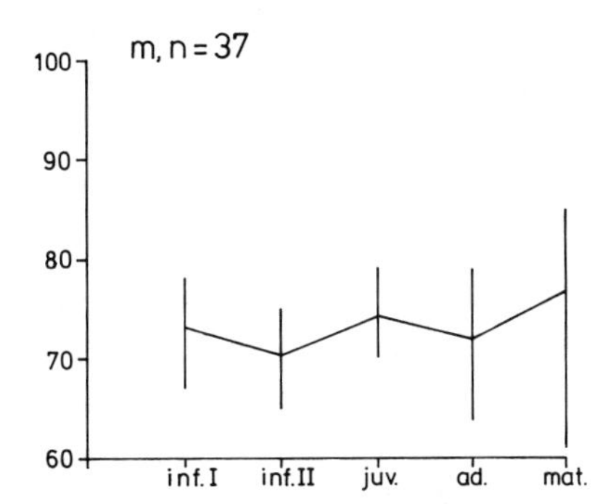

FIG. 15–4. Mandibular length-height index (LHI/Mand). See Fig. 15–1 for explanation.

Sumatran Females (N = 19)

Although the pace at which events occur is somewhat different, the emphasis on length versus width is similar to that seen in the sample of Skalau females.

Sumatran Males (N = 37)

There is greater similarity than difference overall between Skalau and Sumatran males in their growth curves, which emphasize increase in length.

Palatine-Basal Index (PBI) and Palatine-Basal Index/Prosthion (PBIpr) (Fig. 15–6 and Tables 15–1 to 15–4)

Skalau Females (N = 109)

The PBI remains relatively unchanged with age and thus indicates that, although there is a very minimal emphasis on elongation of the snout, relative growth of the palate and cranial base is nonetheless rather stable and correlated. The index averages of PBIpr parallel those of PBI.

Skalau Males (N = 96)

The index averages of PBI reflect a slight increase with age in growth of the muzzle. Although there is minor fluctuation in the PBIpr curve, a similar growth trend emerges. The absolute measurements of muzzle development indicate a continuous progress dependent on changes in the cranial base.

Sumatran Females (N = 19)

Although there is a bit more fluctuation in the PBI curve than that for Skalau females, a similar correlation between growth of the palate and the cranial base is demonstrated. The same holds true for PBIpr.

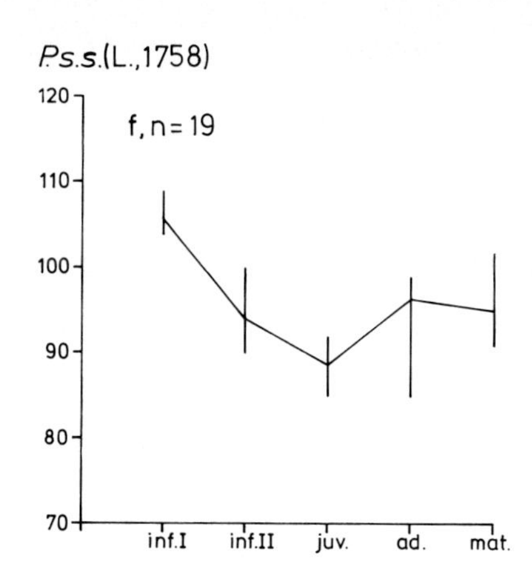

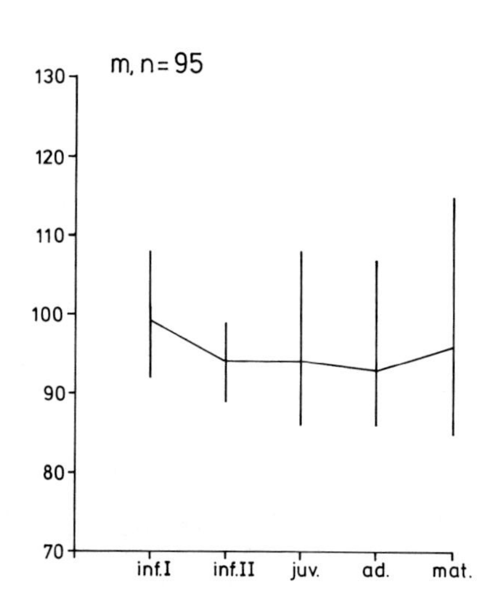

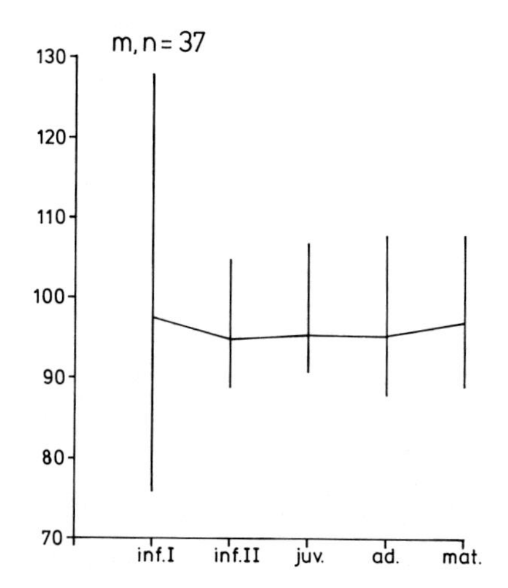

FIG. 15–5. Zygomaticobasal index (ZBI). See Fig. 15–1 for explanation.

Sumatran Males (N = 38)

For all practical purposes, the curves obtained for PBI and PBIpr are similar to those for Skalau males, reflecting the concerted growth of the palate and cranial base.

Facial Index (according to Kollmann) (GI/K) (Fig. 15–7 and Tables 15–1 to 15–4)

Skalau Females (N = 106)

The index average drops from 93 (infant I) to 92 (infant II) but then rises to 94, where it remains constant through full maturity as a consequence of the development of the jaws, which, in turn, affects the snout and the muscles of mastication.

Skalau Males (N = 98)

The curve is irregular and reflects a widening of the face until the adult stage, followed by an increase in facial length and height (and thus an enlargement of the snout), which is correlated with the development of the jaws. In addition to differences in absolute measurements of the splanchnocranium, the pronounced snout in the male is one of the most striking manifestations of craniofacial sexual dimorphism in the orang-utan.

Sumatran Females (N = 18)

The index averages indicate a phase of facial widening followed by an increase in facial length and height. The end result is not significantly different from Skalau females.

Sumatran Males (N = 36)

First there is a period of increasing facial breadth, then a long phase of increasing length and height, terminating in additional facial widening. The resulting sexual dimorphism is similar to that seen in the Skalau orang-utans.

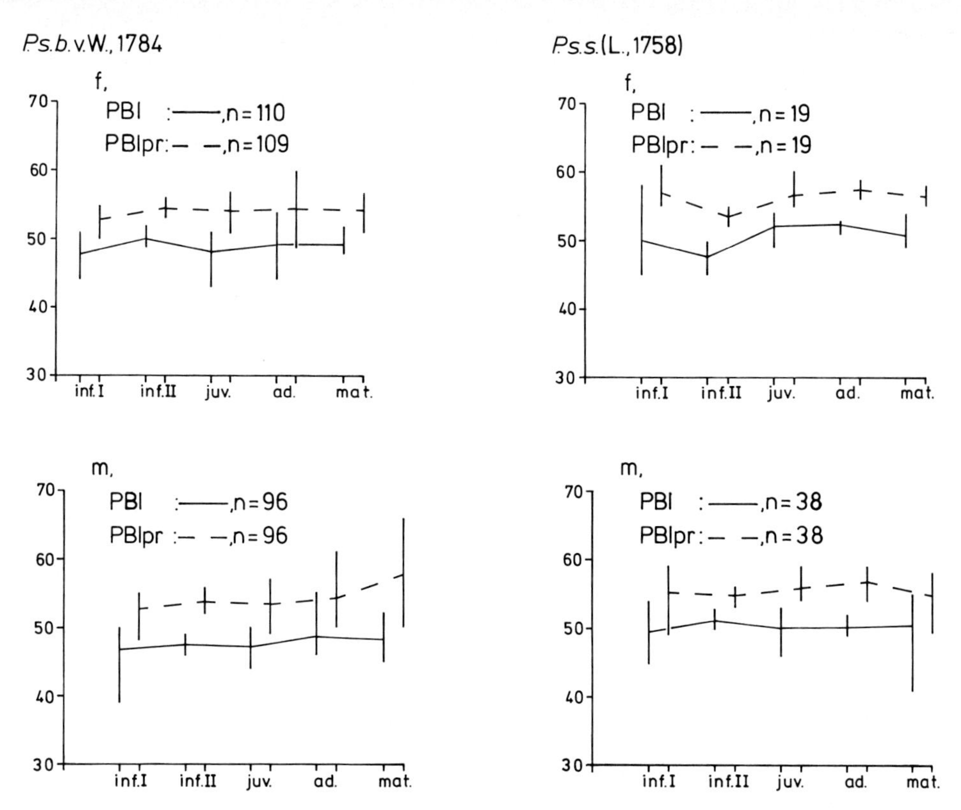

FIG. 15–6. Palatine-basal index (PBI) and palatine-basal index/prosthion (PBIpr). See Fig. 15–1 for explanation.

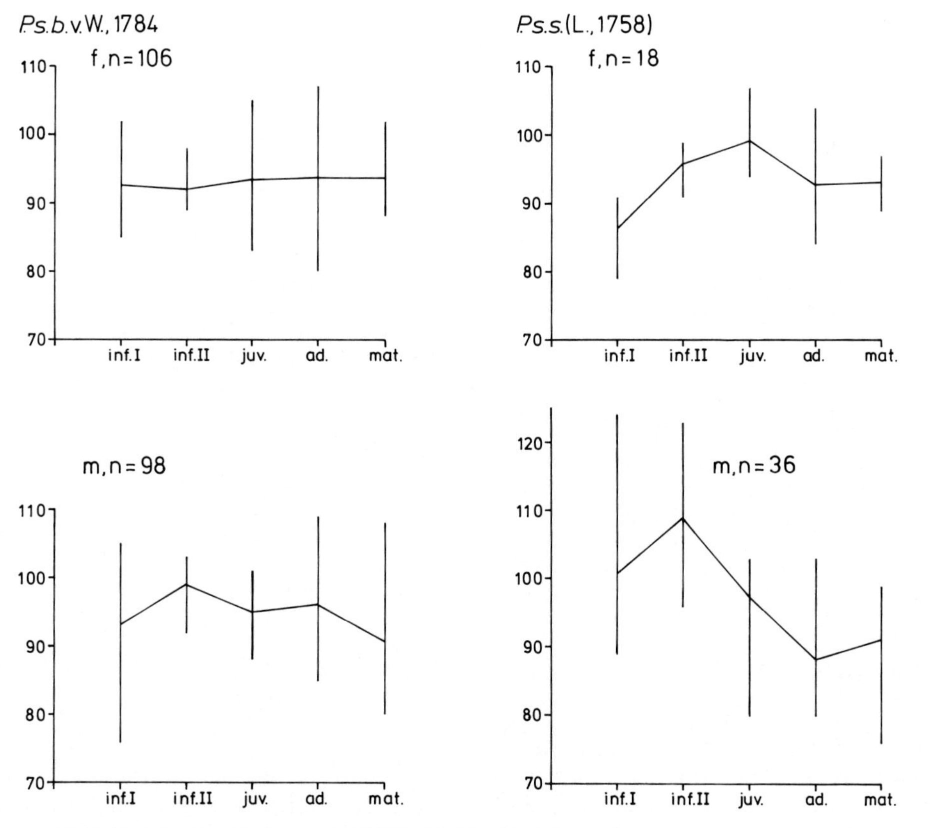

FIG. 15–7. Facial index (Kollmann) (GI/K). See Fig. 15–1 for explanation.

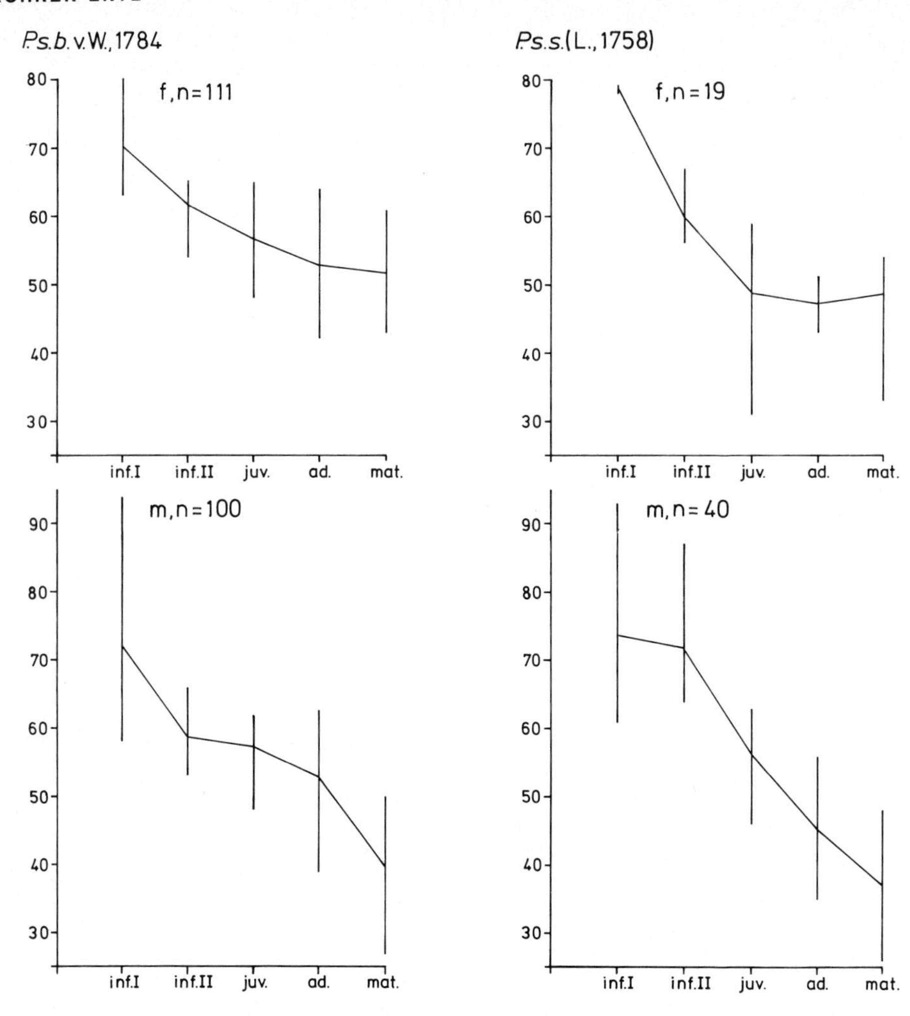

FIG. 15–8. Transverse frontal temporal index (TFTI). See Fig. 15–1 for explanation.

Transverse Frontotemporal Index (TFTI) (Fig. 15–8 and Tables 15–1 to 15–4)

Skalau Females (N = 211)

The index averages drop consistently from infant I to fully mature, reflecting continual growth in biauricular width as measured against the narrowest width of the suraorbital region. This is associated with the development of the muscles of mastication.

Skalau Males (N = 100)

Although the curve is steeper, the same tendencies of increasing biauricular width seen in Skalau females are seen in the males. For males and females, as well, one would expect a correlation with the growth factors reflected in the LBI.

Sumatran Females (N = 19)

The curve for Sumatran females drops more sharply than that for Skalau females, but the general tendency toward increasing biauricular width is apparent.

Sumatran Males (N = 40)

The curve for Sumatran males is steeper than that for Skalau males, but the resulting increase in biauricular width is not significantly different.

Jugomandibular Index (JMI) (Fig. 15–9 and Tables 15–1 to 15–4)

Skalau Females (N = 116)

The index average rises from infant I to fully mature as a reflection of an increase in bimalar or

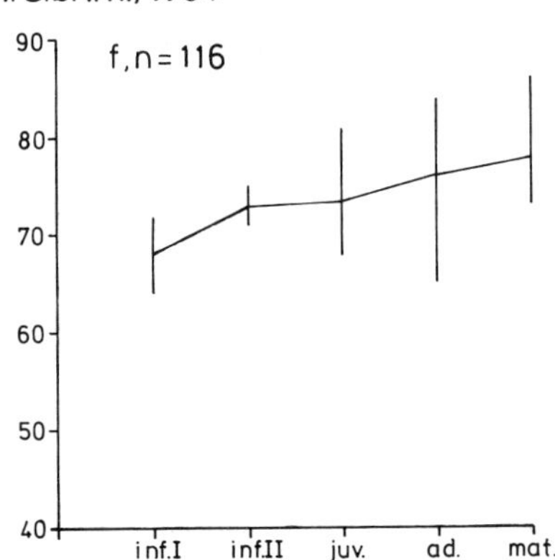

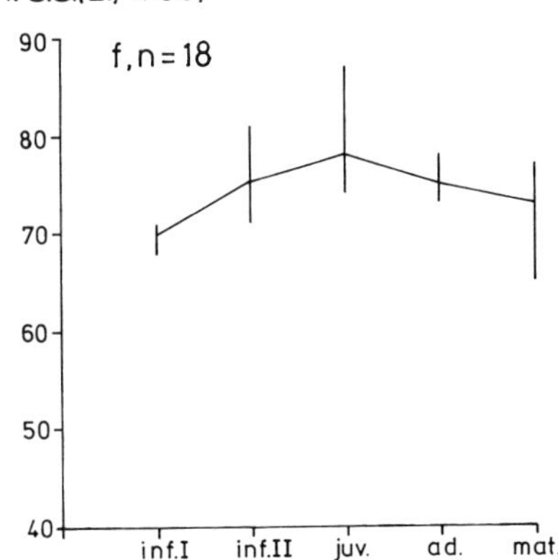

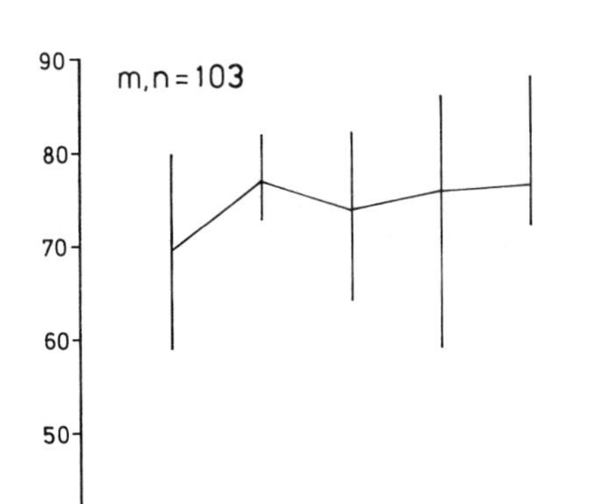

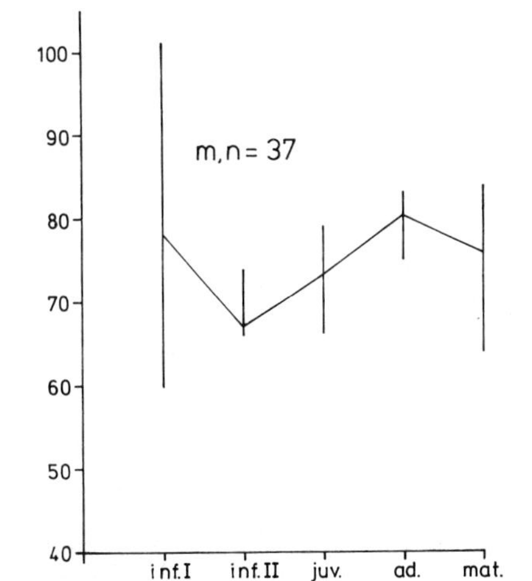

FIG. 15–9. Jugomandibular index (JMI). See Fig. 15–1 for explanation.

bizygomatic (bijugulare) width in conjunction with bigonial width and thus highlights the influence of the masticatory system on cranial development (see Röhrer-Ertl, 1984a).

Skalau Males (N = 103)

The pattern is in general similar to that of Skalau females, but at the beginning, bimalar (bizygomatic) growth exceeds bigonial growth.

Sumatran Females (N = 18)

The index averages reflect an early period of greater bimalar (bizygomatic) expansion followed by an increase in bigonial width.

Sumatran Males (N = 37)

Here it appears that bigonial width increases prior to and then subsequent to a phase of increasing bimalar width.

Transverse Craniofacial Index (TCFI) (Fig. 15–10 and Tables 15–1 to 15–4)

Skalau Females (N = 110)

The index averages indicate that, with age, the pace of bimalar or bizygomatic (bijugulare) widening decelerates as the rate of biauricular expansion increases. This pattern runs parallel to the development of the masticatory apparatus.

Skalau Males (N = 98)

The curve for Skalau males is steeper than that for females, but the overall effect of decreasing bimalar (bizygomatic) growth in inverse proportion to increasing biauricular growth is the same.

Sumatran Females (N = 18)

Although the initial drop in the curve is not as steep as it is for Skalau females, the general effect

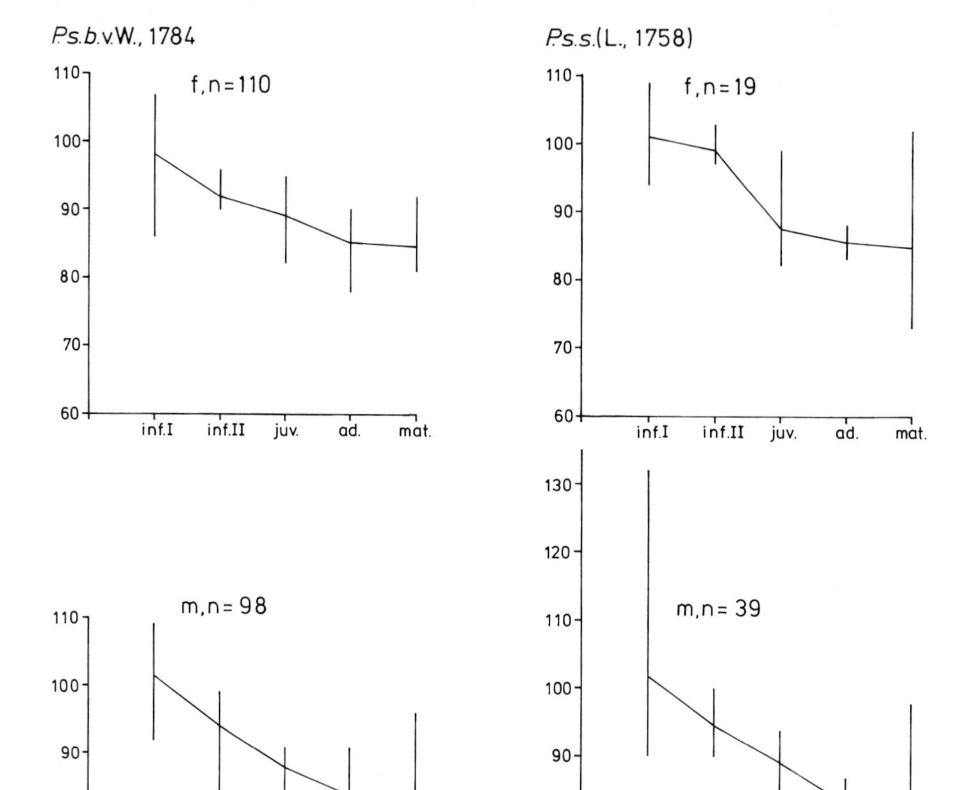

FIG. 15–10. Transverse craniofacial index (TCFI). See Fig. 15–1 for explanation.

of bimalar (bizygomatic) versus biauricular rates is essentially the same.

Sumatran Males (*N* = 39)

The average index values are slightly higher in Sumatran males than in Skalau males, but the basic slope of the growth curve is similar.

Palatine Index (PI) and Palatine-Height Index (PHI) (Fig. 15–11 and Tables 15–1 to 15–4)

Skalau Females (PI, *N* = 111; PHI, *N* = 112)

The index averages of PI illustrate that, during growth, the palate essentially becomes longer and narrower as a result of the development of the muzzle, whereas those of PHI indicate that width increases more than height during palatal growth.

Skalau Males (PI, *N* = 100; PHI, *N* = 100)

Similar to Skalau females, the male palate becomes long with age but it also becomes relatively narrower, as would be expected because males have relatively longer snouts. The index averages of PHI rise from infant I to fully mature, reflecting a greater increase in width than in height of the palate, with the sharp rises in the curve being especially due to the development of the masticatory apparatus and snout.

Sumatran Females (PI, *N* = 19; PHI, *N* = 19)

Although the PI curve is much steeper in the early phases of development than it is for Skalau females, Sumatran females also achieve long and

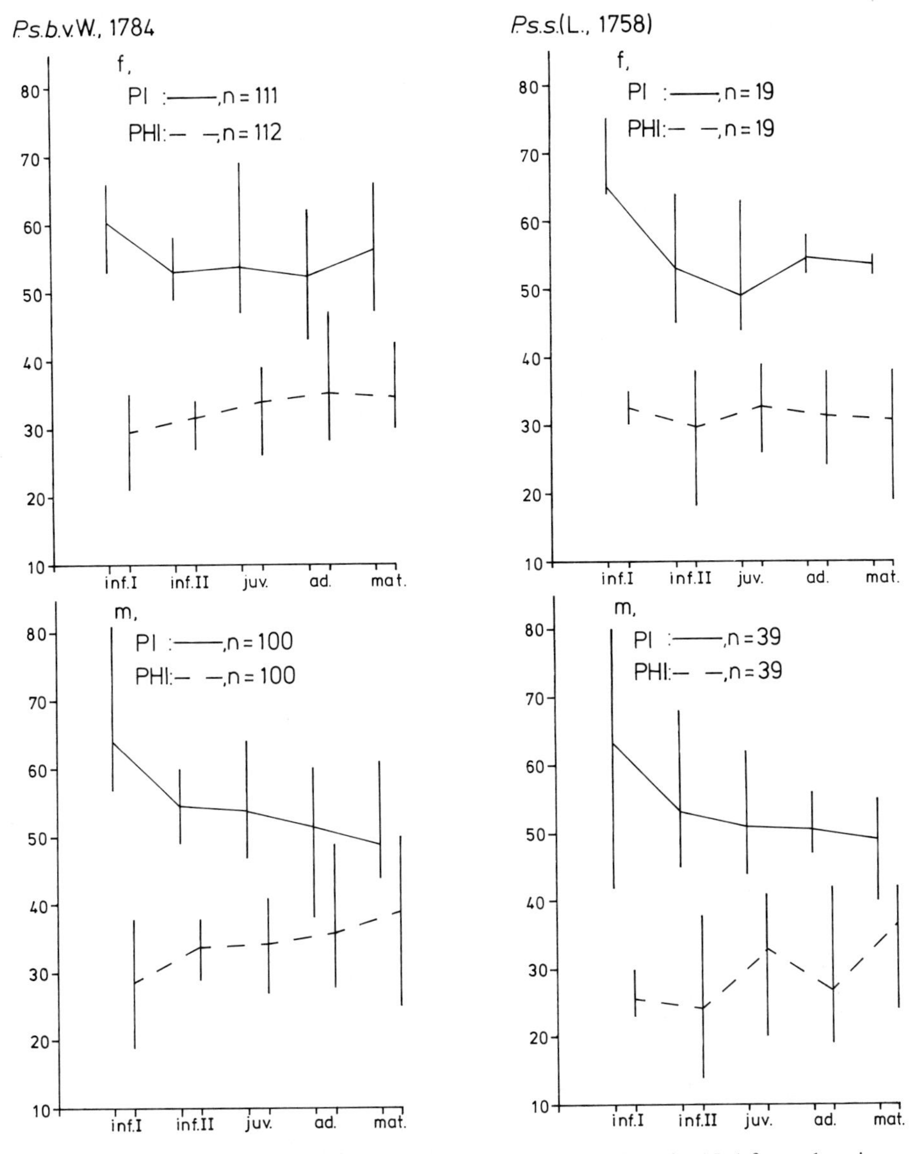

FIG. 15–11. Palatine index (PI) and palatine-height index (PHI). See Fig. 15–1 for explanation.

narrow palates. The net assessment of the PHI curve is that, during palatal growth, width is emphasized over height.

Sumatran Males (PI, N = 39; PHI, N = 39)

The PI curve for Sumatran males is essentially the same as for Skalau males, reflecting the long and relatively narrow palate developed by this sex. The PHI curve of Sumatran males has more definite spurts of increasing palatal width and height, but the net effect is the same as in Skalau males, with emphasis being on the former dimension.

Orbital Index (OI) and Orbital-Depth Index (OTI) (Fig. 15–12 and Tables 15–1 to 15–4)

Skalau Females (OI, N = 112; OTI, N = 112)

The relatively minor fluctuations in the OI averages do not alter the interpretation of a rather closely integrated and stable growth in orbital height and width. Concomitantly, and although there is a slight rise in the OTI (to 118 [infant II]), this index reflects the parallel growth rates of orbital height, width, and depth, which are presumably determined by the development of the eye. Although these growth rates may be re-

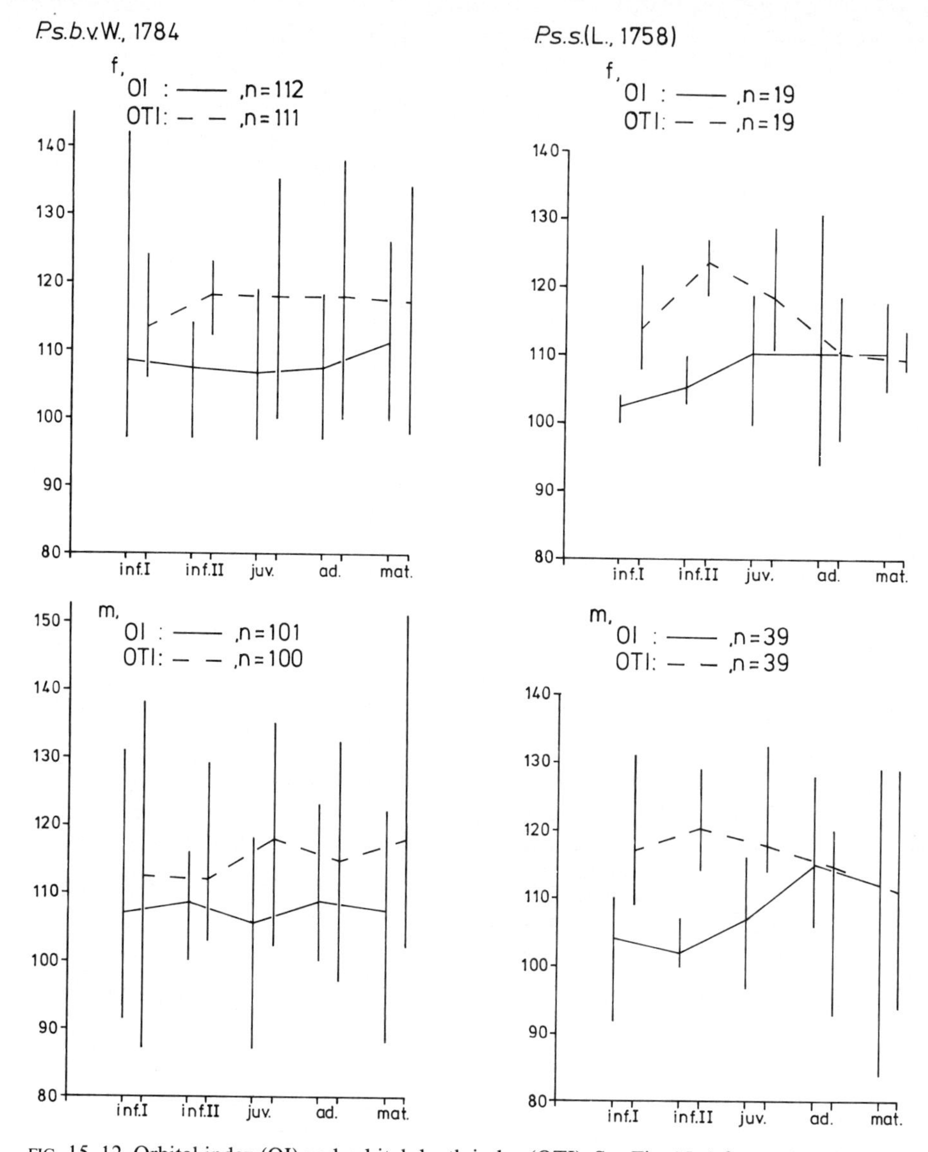

FIG. 15–12. Orbital index (OI) and orbital-depth index (OTI). See Fig. 15–1 for explanation.

latively synchronized, the observed ranges of variation point out the differences in form that the orbits of different individuals may attain.

Skalau Males (OI, N = 101; OTI, N = 101)

The OI fluctuates narrowly between 107 and 109 and mirrors the relatively coordinated growth in orbital height and width. The OTI is a bit more irregular but, nonetheless, reflects a reasonably correlated growth in orbital height, width, and depth. Similar to the Skalau females, individual males may differ in the details of orbital shape, but on the average, the orbits of males are typically higher than those of females. The latter seems to be another expression of craniofacial sexual dimorphism in the orang-utan.

Sumatran Females (OI, N = 19; OTI, N = 19)

The OI essentially levels off at the juvenile phase at similar growth rates of orbital height and width. The OTI fluctuates much more markedly, indicating, on first impression, a different rate of increasing orbital depth. However, the ranges of variation fall within those obtained for the Skalau females. The differences between curves for Sumatran and Skalau females are most likely due to the small sample for the former.

Sumatran Males (OI, N = 39; OTI, N = 39)

The OI and OTI averages plot more irregularly than those of the Skalau males, but the ranges of variation for the Sumatran males are overlapped by those for the Skalau males. One would expect there to be no significant difference between these two populations in rates of orbital growth.

Nasal Index (NI) (Fig. 15–13 and Tables 15–1 to 15–4)

Skalau Females (N = 112)

The index averages indicate that (1) up to the juvenile stage, nasal aperture width increases faster than height; (2) this relation is reversed in the adult stage; and (3) with the fully mature phase, growth in the width of the nasal aperture predominates.

Skalau Males (N = 101)

In the males, however, there is, with advancing age, a decreasing accent on nasal aperture width and an increasing emphasis on height. In males, therefore, elongation of the snout and development of the masticatory system may affect the width but not the height of the nasal aperture.

Sumatran Females (N = 19)

The plot of the index averages fluctuates more widely than that for Skalau females, but the ranges of variation for Sumatran females fall within those for Skalau females, thus indicating overall similarity between the two populations. Differences between the index averages are probably due to differences in sample size and are not biologically significant.

Sumatran Males (N = 40)

In general, the plot of index averages reflects a trend similar to that observed for Skalau males: i.e., with age, growth in nasal aperture width decreases while height increases. The unevenness of the curve for Sumatran males (especially the severe drop to 93 at the juvenile phase) is probably due to sampling biases.

Facial-Breadth Index (GBI) (Fig. 15–14 and Tables 15–1 to 15–4)

Skalau Females (N = 112)

The curve from 85 (infant I) to 74 (adult) reflects a greater increase in biorbital width than in bimalar (bizygomatic) width; the slight rise to 75

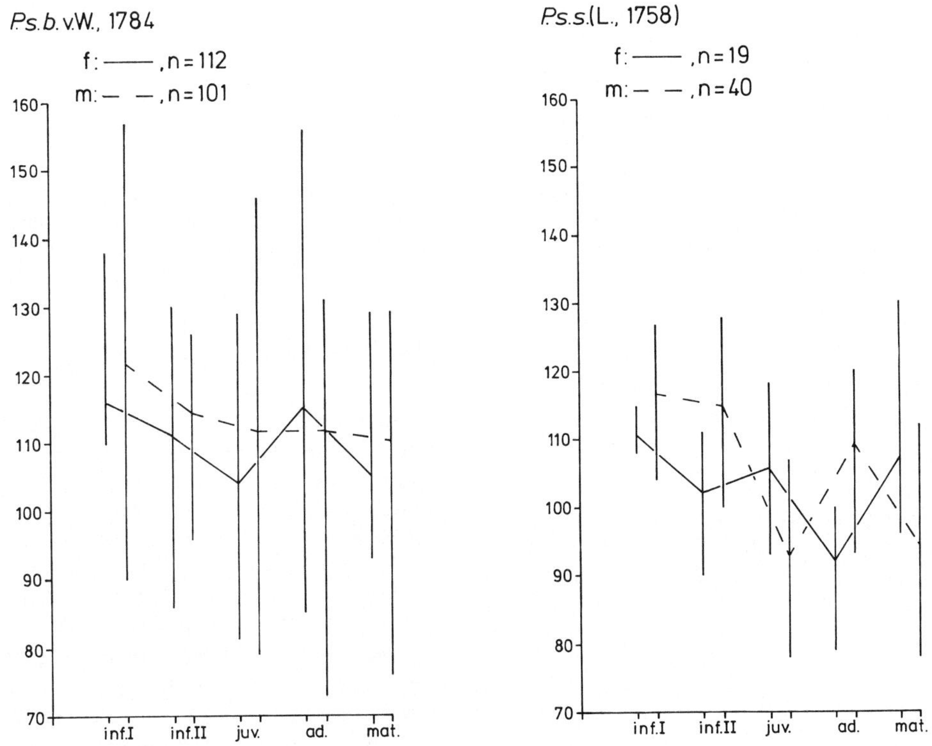

FIG. 15–13. Nasal index (NI). See Fig. 15–1 for explanation.

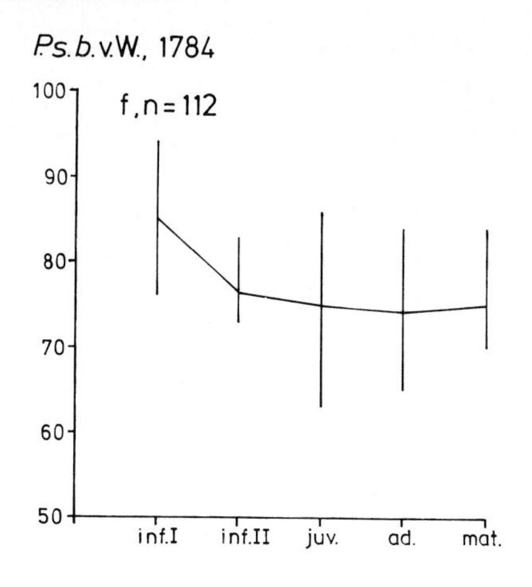

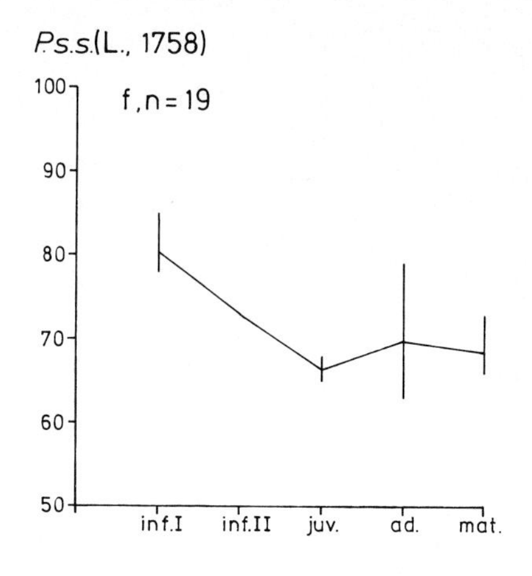

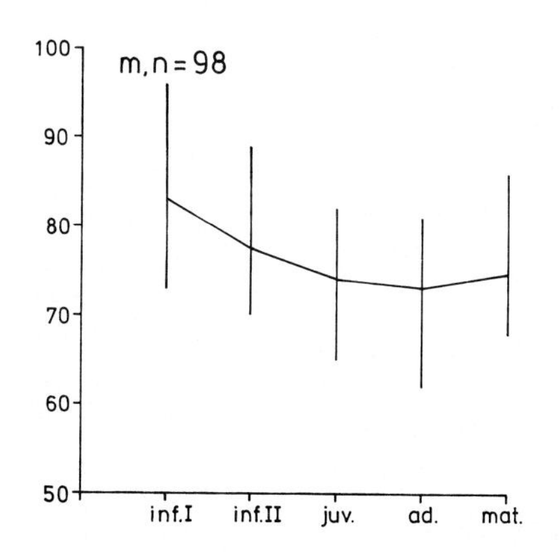

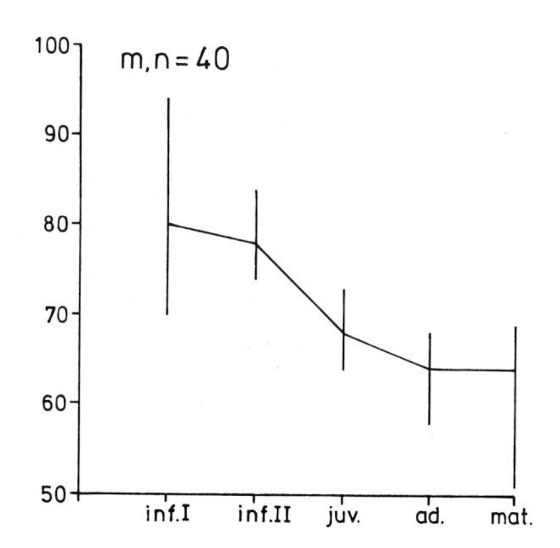

FIG. 15–14. Facial-breadth index (GBI). See Fig. 15–1 for explanation.

in the fully mature phase indicates that the relation is reversed somewhat. This shift is presumably correlated with the development of the masticatory apparatus.

Skalau Males (N = 98)

The curve for Skalau males essentially parallels that for females.

Sumatran Females (N = 19)

Even though the sample size is quite small, one can still detect a pattern somewhat similar to that determined for the Skalau females: there is an early phase of greater biorbital than bimalar (bizygomatic) widening followed by a reversal of emphases.

Sumatran Males (N = 40)

Although differing in slight detail from the curve for Skalau males, that for Sumatran males reflects a generally similar pattern. Perhaps the greatest deviation is in less of a final emphasis on bimalar (bizygomatic) expansion.

DISCUSSION

There are differences at different ages in the emphasis of cranial growth. First, there are simple changes, such as rates of increase in overall size, that diminish with age. Second, there are changes related to the development of other systems, e.g., the gradient of mandibular growth is correlated with the developing dentition. Third, there are functional changes related to the development of the masticatory apparatus which cannot be reduced simply to notions of increasing forces of mastication and tooth abrasion. More accurately, there are changes in behavior and tooth use which occur with advancing age, and these play an increasingly important role in changes in cranial growth. Such changes can be observed for both sexes, although sexually dimorphic differences produce greater alterations in the male.

Although there are general tendencies indicated for changes in the cranial growth in the orang-utan, one cannot help but be impressed by the at times tremendous ranges of variation that exist for the various characteristics measured. Such degrees of individual variation are comparable to what one finds in modern *Homo* (i.e., the "intravital plasticity" of Kurth [1962]) (Röhrer-Ertl, 1984a). Perhaps now, against the data base for the cranium, more accurate results from study of the postcranium will be forthcoming (cf. Schultz, 1941).

The methods used here had earlier been applied to *Homo* (Röhrer-Ertl, 1984a) and point out the importance of taking into consideration the composition of the samples used, e.g., with regard to sex, age, and locality. The consistently smaller samples of Sumatran orang-utans for which such information is available reflect the unsystematic manner in which the material was collected.

Neurocranium

OI and OTI values remain fairly stable whereas LBI, LHI, GI/K, TFTI, and TCFI indicate the following: In both sexes, relative growth from birth to full maturity in cranial width exceeds height and length, but it is relatively more pervasive in males than in females. In general, though, this widening of the neurocranium results from influence by the splanchnocranium on the growth of the external lamina of the neurocranium. The emphasis on cranial widening creates a fully mature neurocranium that appears relatively flat when viewed in norma cranialis and which is notably different from the more globular cranium of younger individuals. It was because of this difference that, well into the 19th century, different-aged individuals were identified as different taxa.

Male neurocrania become relatively longer than those of females, but relative height in females surpasses that in males. The development in males of sagittal and nuchal crests is no doubt in part responsible for these differences.

Splanchnocranium

LBI/Mand, LHI/Mand, ZBI, PBI, PBIpr, JMI, PI, PHI, NI, and GBI all illustrate the tendency in both sexes for greater relative growth in width than in length or height. In mature individuals, sexual dimorphism is reflected in females having crania that appear distinctly higher than in males.

Range of Variation

The region of the orang-utan cranium that exhibits the least variability (i.e., is less influenced by "intravital plasticity" [Kurth, 1962]) is the middle face. The measurements involved include basinasal length (basion–nasion), basibregmatic height (basion–bregma), biorbital width (ektoconchion–ektoconchion), bimalar (bizygomatic) width (zygion–zygion), orbital height, width, and depth, and height of the nasal aperture (rhinion–nasion); these correspond to Martin's (1928) numbers 5, 17, 44, 51, 52, 53, and 55(1). These results corroborate earlier nonmetrical studies which found the midfacial region of the orang-utan to be relatively "stable" (varying little) morphologically (Röhrer-Ertl, 1982a,b, 1984a).

All other measurements reflect a wide range of cranial variation, which increases with age and in concert with the developing masticatory apparatus, and which is analogous to what one finds in modern *Homo* (Röhrer-Ertl, 1984b). But changes through adulthood, as well as variation, are due less to a simple increase in the forces of mastication than to changes in tooth use which result from changes in behavior (as a possible example, see Rodman's [this volume] discussion of bark-eating in the orang-utan).

CRANIAL CAPACITY

Cranial capacity was measured within 5 cm^3 of accuracy using summer vetch seeds, which are round and fall between peas and millet in diameter. The results are presented in Fig. 15–15 and Tables 15–1 to 15–4.

If we look just at the beginning phases of the plots for Skalau females and males and Sumatran males, we see that brain growth (as reflected by cranial capacity) proceeds on average somewhat gradually. In the Skalau males and Sumatran females, especially, and to some degree in the Skalau females, there is a more severe increase in cranial capacity in the later—up to and into the mature—phase of growth.

In all four plots there are peaks of cranial capacity at younger ages that are larger than those at older ages. Aside from simple sampling error, there might be another possible cause for this re-

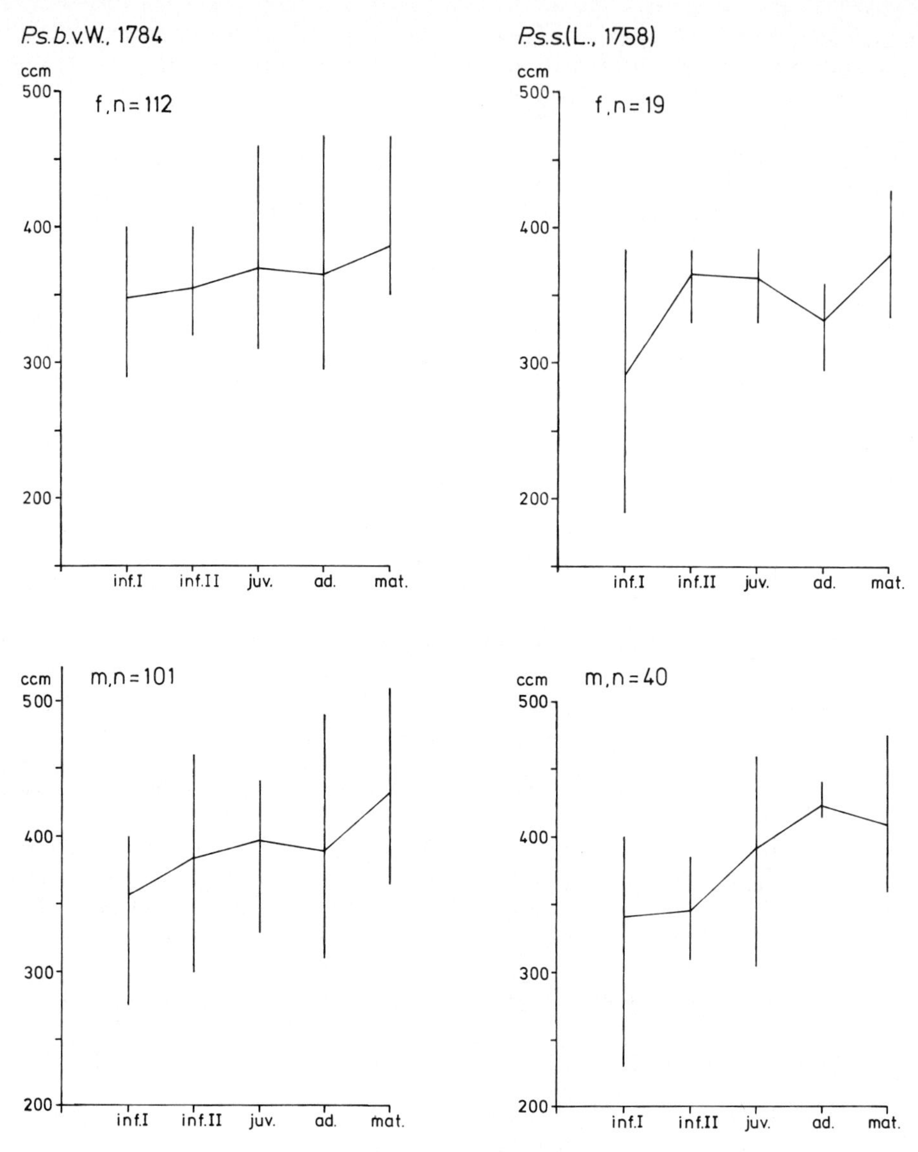

FIG. 15–15. Cranial capacity (CC). The ordinate is in cm³. Otherwise see Fig. 15–1 for explanation.

sult, one that reflects differences in mortality rates among the age groups. For example, in the Skalau males, the average cranial capacity of juveniles (399 cm³) is higher than that for adults (381 cm³), rather than being intermediate between the latter and the infant II (365 cm³) groups. This peak for juveniles reflects a higher accident and, thus, mortality rate. There is a juvenile peak in the Skalau female sample, but it is not as high as in the male sample. Since zoo research (e.g., Becker, 1984) demonstrates that, at least in these conditions, young males are far more active than similar-aged females, one would expect a higher accident, and possibly mortality, rate for subadult males. But since the samples of both sexes show a bias toward juveniles with higher than expected cranial capacities, we might suggest that there is a correlation between brain growth and hyperactivity and that higher mortality rates may reflect something along the lines of "too much, too soon." The peaks in cranial capacity of not yet fully mature Sumatran females and males can be interpreted similarly.

CONCLUSIONS

1. The present study confirms previous assessments of the broad range of variation possible in

orang-utan morphology. All ranges of variation are extreme and can be compared favorably with values obtained for modern *Homo*. The middle face, however, seems to be relatively stable.

2. The sample from Sumatra exhibits a greater range of variation than that from Skalau, probably because only one population is represented in the latter sample, whereas the Sumatran assemblage includes individuals from Atjeh and Deli. Thus, the extremes of variation, especially as reflected in the facial width index, are most likely due to overlapping values for the two Sumatran populations. Nevertheless, the samples of each subspecies do differ from one another in the distribution of single characteristics as well as indices.

3. Cranial growth is correlated with total body growth but is also influenced by the developing, erupting, and functioning dentition as well as by the development and function of the muscles of mastication. The latter, especially, affects the cranium in the final phases of development.

4. Evaluation of cranial capacities determined for the samples studied indicates not only that brain volume increases after closure of the sutures, but that there is an unexpected minor peak in brain size increase during the juvenile stage. This slight apparent increase may actually be reflective of sampling of hyperactive young individuals who would be most prone to accident and thus come from an age group with a higher mortality rate than the others.

5. All criteria demonstrate the existence in the orang-utan of a high degree of sexual dimorphism. The crania of females are generally higher and shorter than those of males of equivalent ages. In both sexes, however, growth in the width dimension of all parts of the cranium dominates during all ages but becomes particularly marked during the adult stage.

REFERENCES

Becker, C. 1984. Orang-Utans und Bonobos im Spiel, Untersuchungen zum Spielverhalten von Menschenaffen. Profil, Munich.

Kurth, G. 1962. Stellung und Aussagewert der gegenwärtig bekannten mittelpleistozänen Hominidae: Zuwachsrate und Wanderungsbeweglichkeit im Zeitmasstab sowie humane Leistungsfähigkeit als zusätzliche Kriterien für die Einstufung. *In* Evolution and Hominization, pp. 70–190, ed. G. Kurth. Fischer, Stuttgart.

Martin, R. 1928. Lehrbuch der Anthropologie. Fischer, Jena.

Oppenheim, S. 1927. Methoden zur Untersuchung der Morphologie der Primaten: Kraniologie, Osteometri und am Lebenden. *In* Handbuch der Biologischen Arbeitsmethoden, pp. 531–608, ed. E. Abderhalden. Urban and Schwarzenberg, Berlin and Vienna.

Röhrer-Ertl, O. 1982a. Über Subspecies bei *Pongo pygmaeus* Linnaeus, 1760. *Spixiana, 5:* 317–321.

Rħrer, Ertl, O. 1982b. Hinweise auf ein soziales Isolat des 18. /19. Jh. (St. Katherinenfriedhof in Braunsweig). *Homo, 33*:139–148.

Röhrer-Ertl, O. 1983. Zur Erforschungsgeschichte und Namengebung beim Orang-Utan. *Pongo satyrus* (Linnaeus, 1758), Synon. *Pongo pygmaeus* (Hoppius, 1763) (mit Kurzbibliographie). *Spixiana, 6:* 301–332.

Röhrer-Ertl, O. 1984a. Orang-Utan-Studien. Hieronymus, Neuried.

Röhrer-Ertl, O. 1984b. Ein Beitrag zur historischen Vertiefung der Akzelerationsforschung: Längenwachstumsänderungen in einer Population vom Tell es Sultan/Jericho aus dem Präkeramischen Neolithikum (ca. 9000–6000 v. Chr.). *Ärztliche Jugendkunde, 75:* 292–302.

Röhrer-Ertl, O. and Frey, K.-W. 1987. On secondary hyperparathyroidism in orang-utans, *Pongo satyrus* (Linnaeus, 1758).*Geganbaurs Morphologisches Jahrbuch, Leipzig, 133*: 361–383.

Schultz, A. H. 1941. Growth and development of the orang-utan. *Contributions to Embryology, 29:* 41–110.

16

Sexual Dimorphism in Exocranial and Endocranial Dimensions

LINDA A. WINKLER, GLENN C. CONROY, AND MICHAEL W. VANNIER

Sexual dimorphism in adult orang-utan crania produces pronounced differences in appearance between the sexes. Descriptions and measurements of the outer body, cranium, and teeth (Ashton and Zuckerman, 1956; Biegert, 1957; Hrdlička, 1907; Schultz 1941, 1962, 1969; Selenka, 1898) indicate that substantial sexual dimorphism is present in both exocranial and endocranial features of the skull. Sexually dimorphic differences in cranial morphology may influence phylogenetic comparisons between *Pongo* and other taxa, particularly if the specimens being compared are of different sexes. Therefore, a clear understanding of the degree and range of this sexual dimorphism is necessary in order to fully assess the taxonomic relationships between *Pongo* and the other hominoids, particularly the fossil "ramamorphs" (*Sivapithecus, Gigantopithecus, Ramapithecus,* etc.). Here, we will attempt to reassess sexual dimorphism in orang-utan exocranial and endocranial morphology utilizing statistical analyses of a series of 37 craniometric measurements and three-dimensional CAT-scanning computed tomography.

METHODS

A total of 130 orang-utan crania of known sex from the collections of the American Museum of Natural History, British Museum (Natural History), Cleveland Museum of Natural History, National Museum of Natural History (Smithsonian Institution), Rijksmuseum Van Natuurlijke Historie (Leiden), University of Zürich (Anthropological Institute), and Washington University (St. Louis) were measured for a series of 37 craniometric exocranial dimensions. All measurements were taken with a Helios needle point dial caliper or with a spreading caliper and recorded in centimeters to an accuracy of 0.01. In addition, data regarding states of tooth eruption were obtained from each specimen.

All craniometric points and measurements, as listed in Tables 16–1 and 16–2, were defined according to standard anthropological technique and procedure (Krogman, 1978; Schultz, 1962; Vallois, 1965); exceptions are discussed below. Vault length was measured from glabella to lambda, whereas total skull length was measured from prosthion to lambda. Maxillary width was measured from right zygomaxillare to left zygomaxillare and maxillary height from dacryon to the anterior alveolar edge of the upper first molar, when this was present. If, in a particular specimen, the upper first molar was unerupted, the anterior edge of the upper deciduous molar or the crypt of this tooth was utilized instead for this point. Coronoid process height was measured from coronion to gonion.

All specimens were measured for as many points as possible. However, damaged or missing cranial elements prevented obtaining all measurements from all specimens. No attempt was made to estimate or compensate for these missing values.

After all specimens of a given collection were measured, 10% of the specimens were remeasured at random to determine the degree to which measurement error occurred. Correlations between first and second measurements were found to be 95% or higher for all dimensions.

BMDP7D (Dixon, 1985) was used for the univariate statistics as well as for an analysis of variance (ANOVA) between the sexes for each dimension. Initially, males and females of all ages were compared (Table 16–1). Then, only the adults were compared (Table 16–2), an adult being defined by its possessing a fully erupted set of teeth.

In both ANOVA series, Levene's test for equal variances (Brown and Forsythe, 1974b) was computed to test for variance equality in the groups. Welch and Brown-Forsythe (1974a) statistics, which do not assume within-group variances to be equal, were computed and used if the Levene's test indicated variance inequality.

The heads of one male and one female orang-utan cadaver, as well as the head of a male gorilla, were subjected to a complete three-dimensional CT-scan analysis to examine intracranial structures. All specimens were from the collections of the Department of Anatomy and Neurobiology, Washington University. Both specimens of *Pongo* had been embalmed and retained skin, subcutaneous tissue, muscle, bone, and brain tissue. They had recently been partially dissected to expose aspects of facial and mandibular musculoskeletal morphology (e.g. Brown, 1986; Winkler, 1986). The computer-imaging techniques used have been described fully elsewhere (e.g., Conroy and Vannier, 1984, 1985; Vannier et al., 1983a–c, 1984, 1985).

The CT data were collected in the following manner for the gorilla and both orang-utan specimens. Sequential high-resolution, narrowly collimated (2 mm) CT scans were produced for the entire skull (using Siemens Somatom DR CT scanners at the Mallinckrodt Institute of Radiology, Washington University). All the CT scans (134 for the male skull and 106 for the female skull) were stored off-line on 8-inch floppy disks or magnetic tape. Disks and tapes containing the original scan data were then copied into a CT scan evaluation console (Siemens Evaluscope RC) for three-dimensional surface reconstruction. A 256×256 reconstruction matrix was used. Pixel edge length in the plane of section was approximately 0.5 mm and the pixel area was less than 1 mm^2 in the plane of section. The three-dimensional images were then copied to blank floppy disks, manipulated using window level and width controls, and photographed like any ordinary CT scan (for further discussion on three-dimensional surface reconstruction methods, see Vannier et al., 1985).

Two of us (Conroy and Vannier, 1985) recently devised a noninvasive, three-dimensional CT-scanning technique through which endocranial volume and shape can be determined with a high degree of accuracy. A three-dimensional "endocast" was generated by the computer and "reinserted" into the computer-generated three-dimensional skull image so that the precise geometric relationships between the two could be evaluated.

Endocranial volume was calculated for the two specimens. As each contiguous 2 mm CT slice was displayed on the Evaluscope console, window width and level controls were adjusted so that the endocranial borders were clearly demarcated. A marker stylus on a resistor pad x-y digitizer built into the Evaluscope was placed in the middle of the endocranial cavity and a "region highlighting" program was run which automatically "found" the endocranial boundary and computed the volume of all the pixels within the boundary for that CT slice. Total endocranial volume was then simply computed as the sum of all the "endocranial pixels" for each CT slice.

RESULTS

Exocranial Analysis

Table 16–1 presents the results of the univariate statistics and ANOVA comparing the males and females of all ages. With the exceptions of the staphylion-to-hormion dimension, as well as mandibular height, mandibular length (right side), maxillary width, and interorbital breadth, all traits exhibit significant differences ($p < .01$) between the sexes. Interorbital breadth, maxillary width, and mandibular length (right side) are also significantly different in this comparison of males and females of all ages but to a lesser degree ($p < .05$). Mandibular height and the staphylion-to-hormion dimension are not significantly different ($p > .05$) when the specimens of all ages are compared.

The results of the univariate statistics and ANOVA comparing adult male and female specimens demonstrate a substantial degree of sexual dimorphism. All 37 craniometric dimensions are significantly different ($p \leq .001$) between adult males and females, with the males being larger than the females in all dimensions. However, there is overlap of the ranges of the measurements between the sexes: i.e., the largest females approach the size of the smallest males.

Endocranial Analysis

Figures 16–1 to 16–3 illustrate computer-generated three-dimensional sagittal sections of the cranium of a male gorilla, a male orang-utan, and a female orang-utan. Figure 16–1 shows the large frontal sinus which separates the outer and inner tables of the frontal bone in the gorilla, the rostral limit of the inner table of the frontal bone, and the large sphenoid air sinus. Male and female orang-utans both lack the large browridges and true frontal sinuses that typify gorillas (Cave and Haines, 1940), in spite of the fact that the sexes are extremely sexually dimorphic in adult body weight and exocranial dimensions. Figures 16–4 and 16–5, which illus-

TABLE 16–1. ANOVA of Orang-utan Sexual Dimorphism in Specimens of All Ages, Infant to Adult, Measured in Centimeters

	♂			♀				
	N	$\overline{X} \pm SD$	SEM	N	$\overline{X} \pm SD$	SEM	F-Value	p
Vault length	52	12.413 ± 1.281	0.178	70	11.222 ± 0.971	0.116	31.54*	0.000
Head breadth[1]	49	11.380 ± 1.440	0.206	71	9.832 ± 0.799	0.095	46.75*	0.000
Bizygomatic breadth[1]	50	13.490 ± 3.593	0.508	69	11.632 ± 2.266	0.273	10.37*	0.002
Interorbitalbreadth[1]	53	1.493 ± 0.5	0.069	71	1.284 ± 0.349	0.041	6.75*	0.011
Biorbital breadth[1]	52	10.023 ± 2.03	0.282	72	8.917 ± 1.39	0.164	11.53*	0.001
Head height[2]	46	9.671 ± 0.795	0.117	62	9.102 ± 0.735	0.093	14.74	0.000
Skull length	52	20.526 ± 3.905	0.541	70	17.713 ± 2.733	0.327	19.79*	0.000
Maxillary width	53	10.034 ± 2.378	0.327	71	9.072 ± 1.913	0.227	5.85*	0.018
Maxillary height								
Right side	53	7.026 ± 1.668	0.229	70	6.097 ± 1.497	0.179	10.22*	0.002
Left side	54	7.111 ± 1.693	0.230	72	6.064 ± 1.514	0.178	13.32	0.000
Upper facial height[2]	55	8.977 ± 2.491	0.336	75	7.583 ± 1.758	0.203	12.61*	0.001
Maximum palatal length[1]	52	7.763 ± 1.893	0.262	68	6.789 ± 1.471	0.178	9.43*	0.003
Maximum mandibular breadth[2]	43	11.462 ± 2.620	0.400	60	10.047 ± 1.878	0.242	9.17*	0.003
Mandibular length[1]								
Right side	42	14.205 ± 3.812	0.588	62	12.508 ± 2.85	0.362	6.04*	0.017
Left side	45	14.549 ± 3.829	0.571	66	12.645 ± 2.873	0.354	8.04*	0.006
Mandibular height[2]								
Right side	40	6.485 ± 1.77	0.280	61	5.987 ± 1.519	0.194	2.28	0.135
Left side	44	6.646 ± 1.821	0.275	65	5.992 ± 1.489	0.185	3.91*	0.052
Coronoid process height								
Right side	42	7.124 ± 2.001	0.309	61	6.290 ± 1.546	0.198	5.17*	0.026
Left side	46	7.227 ± 2.050	0.302	64	6.293 ± 1.558	0.195	6.75*	0.011
Porion to nasion								
Right side	54	9.491 ± 1.654	0.225	74	8.430 ± 1.123	0.131	16.61*	0.000
Left side	55	9.484 ± 1.642	0.221	74	8.502 ± 1.099	0.128	14.78*	0.000
Porion to orbitale								
Right side	54	8.146 ± 1.642	0.223	73	7.183 ± 1.105	0.129	13.92*	0.000
Left side	55	8.217 ± 1.667	0.225	75	7.241 ± 1.152	0.133	13.98*	0.000
Porion to prosthion								
Right side	54	14.677 ± 3.774	0.514	73	12.813 ± 2.563	0.300	9.81*	0.002
Left side	55	14.906 ± 3.740	0.504	75	12.837 ± 2.656	0.307	12.29*	0.001
Staphylion to hormion	49	3.196 ± 1.031	0.147	59	2.902 ± 1.179	0.153	1.87	0.175
Sphenobasion to basion	48	2.626 ± 0.665	0.096	65	2.316 ± 0.415	0.052	8.07*	0.006
Staphylion to basion	47	6.674 ± 1.822	0.266	66	5.748 ± 1.148	0.141	9.46*	0.003
Staphylion to sphenobasion	50	4.348 ± 1.148	0.162	67	3.632 ± 0.789	0.096	14.35*	0.000
Hormion to basion	46	3.900 ± 0.857	0.126	58	3.442 ± 0.552	0.072	9.89*	0.002
Basion to prosthion	48	14.172 ± 3.629	0.524	66	12.138 ± 2.722	0.335	10.70*	0.002
Basion to nasion	48	9.360 ± 1.446	0.209	64	8.623 ± 1.162	0.145	8.39*	0.005
Condylion to prosthion								
Right side	53	14.158 ± 3.732	0.513	73	12.178 ± 2.531	0.296	11.19*	0.001
Left side	54	14.345 ± 3.796	0.517	74	12.280 ± 2.550	0.296	12.02*	0.001
Condylion to nasion								
Right side	53	9.413 ± 1.840	0.253	73	8.381 ± 1.229	0.144	12.59*	0.001
Left side	54	9.468 ± 1.889	0.257	74	8.400 ± 1.237	0.144	13.14*	0.001
Nasion to lambda	52	12.675 ± 1.395	0.194	71	11.333 ± 1.230	0.146	31.86	0.000

*Computed by Welch, Brown-Forsythe statistics; within-group variances not assumed to be equal.

[1]Measurement definitions taken from Schultz (1962).

[2]Measurement definitions taken from Vallois (1965).

TABLE 16–2. ANOVA of Orang-utan Sexual Dimorphism in Adult Specimens, Measured in Centimeters

	♂			♀				
	N	$\overline{X} \pm$ SD	SEM	N	$\overline{X} \pm$ SD	SEM	F-Value	p
Vault length	32	13.095 ± 0.931	0.165	38	11.778 ± 0.404	0.065	55.27*	0.000
Head breadth[1]	32	12.019 ± 1.198	0.212	39	10.137 ± 0.568	0.091	66.69*	0.000
Bizygomatic breadth[1]	28	16.285 ± 1.367	0.258	36	13.247 ± 0.997	0.166	97.80*	0.000
Interorbital breadth[1]	32	1.806 ± 0.314	0.056	40	1.485 ± 0.247	0.039	23.63	0.000
Biorbital breadth[1]	30	11.429 ± 0.897	0.164	40	9.822 ± 0.739	0.117	67.47	0.000
Head height[2]	28	10.114 ± 0.479	0.091	37	9.395 ± 0.467	0.077	36.98	0.000
Skull length	32	23.075 ± 1.642	0.290	37	19.585 ± 0.907	0.149	114.39*	0.000
Maxillary width	31	11.666 ± 1.082	0.194	39	10.268 ± 0.739	0.118	37.75*	0.000
Maxillary height								
Right side	32	8.177 ± 0.730	0.129	40	6.816 ± 0.447	0.071	85.58*	0.000
Left side	33	8.257 ± 0.773	0.135	40	6.787 ± 0.599	0.095	83.75	0.000
Upper facial height[2]	33	10.687 ± 1.094	0.191	41	8.780 ± 0.670	0.105	76.96*	0.000
Maximum palatal length[1]	31	9.117 ± 0.641	0.115	40	7.738 ± 0.597	0.094	87.45	0.000
Maximum mandibular breadth[2]	23	13.579 ± 1.040	0.217	32	11.413 ± 0.926	0.164	66.04	0.000
Mandibular length[1]								
Right side	22	17.370 ± 1.374	0.293	33	14.642 ± 0.873	0.152	68.34*	0.000
Left side	25	17.548 ± 1.319	0.264	35	14.735 ± 0.821	0.139	89.07*	0.000
Mandibular height[2]								
Right side	21	7.875 ± 0.982	0.214	32	7.109 ± 0.626	0.111	12.05	0.001
Left side	25	7.958 ± 1.031	0.206	34	7.055 ± 0.672	0.115	14.63*	0.000
Coronoid process height								
Right side	23	8.707 ± 0.911	0.190	31	7.453 ± 0.607	0.109	36.87	0.000
Left side	27	8.701 ± 1.047	0.201	33	7.346 ± 0.631	0.110	34.89*	0.000
Porion to nasion								
Right side	32	10.620 ± 0.729	0.129	41	9.125 ± 0.541	0.084	94.10*	0.000
Left side	33	10.581 ± 0.716	0.125	41	9.171 ± 0.516	0.081	90.16*	0.000
Porion to orbitale								
Right side	32	9.279 ± 0.694	0.123	41	7.851 ± 0.539	0.084	92.10*	0.000
Left side	33	9.347 ± 0.742	0.129	41	7.973 ± 0.516	0.081	81.44*	0.000
Porion to prosthion								
Right side	32	17.306 ± 1.723	0.305	41	14.537 ± 0.918	0.143	67.60	0.000
Left side	33	17.566 ± 1.254	0.218	41	14.659 ± 0.886	0.138	126.45*	0.000
Staphylion to hormion	29	3.931 ± 0.525	0.097	33	3.097 ± 0.298	0.052	57.09*	0.000
Sphenobasion to basion	28	3.065 ± 0.354	0.067	39	2.565 ± 0.234	0.038	42.42*	0.000
Staphylion to basion	27	8.066 ± 0.654	0.126	39	6.481 ± 0.465	0.074	132.96	0.000
Staphylion to sphenobasion	31	5.137 ± 0.517	0.093	39	4.120 ± 0.356	0.057	87.16*	0.000
Hormion to basion	26	4.509 ± 0.464	0.091	33	3.768 ± 0.345	0.060	49.55	0.000
Basion to prosthion	28	16.893 ± 1.177	0.223	39	13.755 ± 1.667	0.267	72.94	0.000
Basion to nasion	28	10.395 ± 0.589	0.111	38	9.308 ± 0.535	0.087	61.08	0.000
Condylion to prosthion								
Right side	32	16.833 ± 1.248	0.221	41	13.913 ± 0.876	0.137	126.66*	0.000
Left side	33	17.004 ± 1.347	0.234	41	14.024 ± 0.881	0.138	120.21*	0.000
Condylion to nasion								
Right side	32	10.675 ± 0.754	0.133	41	9.159 ± 0.564	0.088	90.13*	0.000
Left side	33	10.719 ± 0.899	0.157	41	9.182 ± 0.551	0.086	73.97*	0.000
Nasion to lambda	32	13.439 ± 0.981	0.173	38	11.977 ± 0.426	0.069	59.59*	0.000

*Computed by Welch, Brown-Forsythe statistics; within-group variances not assumed to be equal.

[1]Measurement definitions taken from Schultz (1962).

[2]Measurement definitions taken from Vallois (1965).

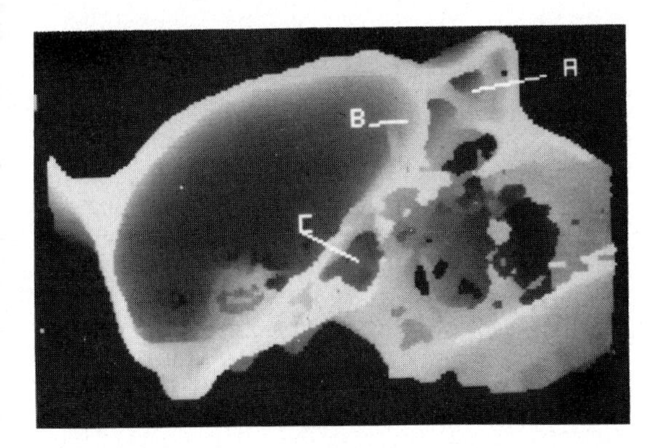

FIG. 16–1. 1. Three-dimensional computer-generated "sagittal" section of a male gorilla skull produced from contiguous 2-mm CT slices. (A) Frontal air sinus complex. (B) Inner table of frontal bone at rostral limit of frontal cortex. (C) Sphenoidal air sinus.

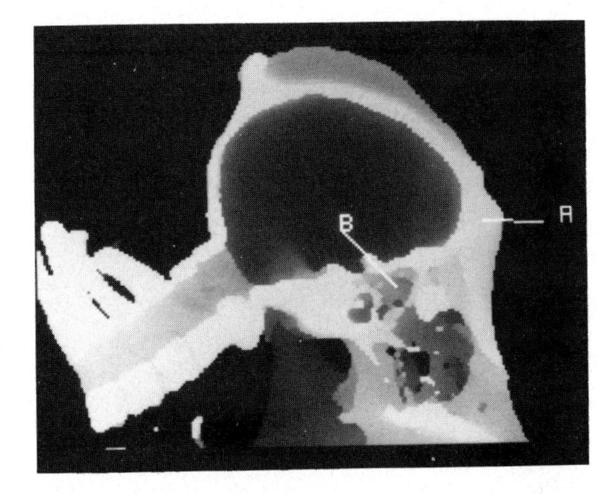

FIG. 16–2. Three-dimensional computer-generated "sagittal" section of a male orang-utan head produced from contiguous 2-mm CT slices. (A) Frontal bone (note absence of frontal sinus formation). (B) Sphenoidal air sinus.

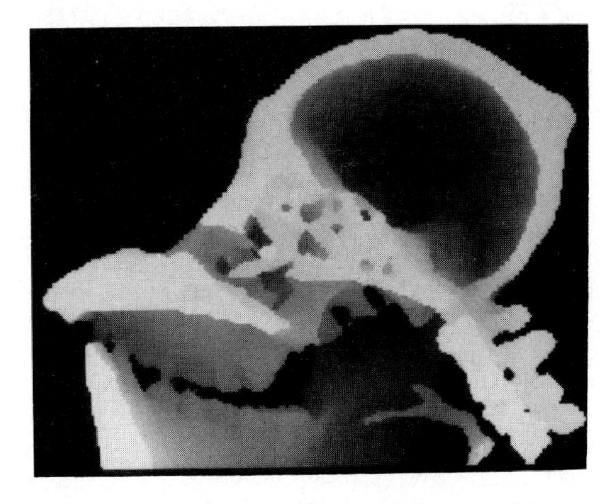

FIG. 16–3. Three-dimensional computer-generated "sagittal" section of a female orang-utan head produced from contiguous 2-mm CT slices.

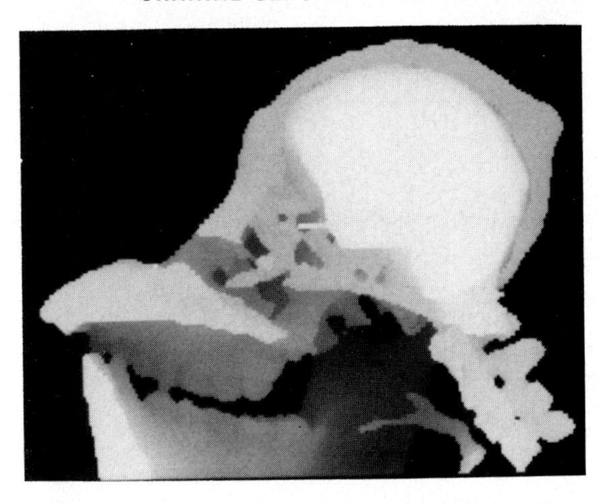

FIG. 16–4. Same view as in Fig. 16–3, including the opacified "endocast."

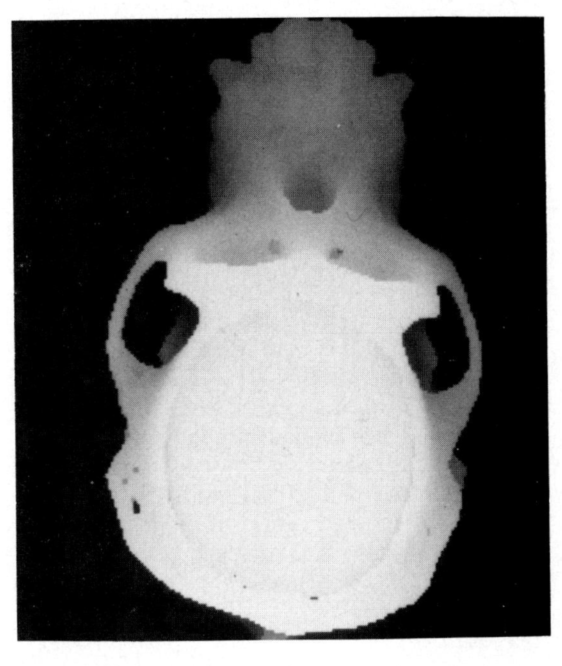

FIG 16–5. Same specimen as in Figs. 16–3 and 16–4 with opacified "endocast" seen from a "bird's-eye" view.

trate a three-dimensional "endocast" reinserted by computer into the female orang-utan skull, demonstrate that the outer and inner tables of the frontal bone (orbital roof and anterior cranial fossa floor) are more closely approximated than in *Gorilla.* It should be noted that skull and endocast are in perfect registration in these figures because they are both derived from the same CT scans. Although there is a significant amount of dimorphism in brain size, body size, and exocranial dimensions, the floor of the anterior cranial fossa nevertheless contributes to a large degree to the formation of the roof of the orbit in both orang-utan sexes. This configuration is very different from that seen in gorillas, where the floor of the frontal sinus often forms a significant portion of the orbital roof.

Endocranial volumes for the two orang-utan specimens were calculated to be 460 cm^3 for the male and 387 cm^3 for the female. The male gorilla specimen had an endocranial volume of abut 560 cm^3.

Another aspect of intracranial anatomy revealed in Fig. 16–1 to 16–3 is the difference in expression of airorhynchy not only between the gorilla and the orang-utans, but between the male and female orang-utans. Many investigators have remarked on the relative degree of "dorsiflexion" (airorhynchy) of the palate in relation to the cranial base in orang-utans and various other primates (see Shea, 1985, and references therein). Noninvasive imaging techniques allow some of these rather qualitative statements to be converted into more quantitative measures. For example, the male gorilla (Fig. 16–1) averages 44° for the endocranial base–hard palate angle, 143° for the angle between the plane of the foramen magnum and the endocranial base, 103° for the angle between the plane of the foramen magnum and the endocranial base, and approximately 16° for the angle between the basiocciput and the hard palate. Measurements of the same angles in the male orang-utan are 38°, 144°, 109°, and 1° respectively, and in the female orang-utan are 19°, 134°, 106°, and 2°, also respectively. Thus, it appears that airorhynchy is more pronounced in female than in male orang-utans (19° compared to 38° in endocranial base-hard palate angle). The male gorilla has a larger endocranial base-hard palate angle (44°), i.e., less airorhynchy, than orang-utans of either sex.

Discussion

As our results indicate, orang-utans are markedly sexually dimorphic cranially: differences between individual adult specimens are substantial, but statistical comparisons of the exocranial dimensions of all ages nevertheless demonstrate a high degree of sexual dimorphism. An attempt was not made in this study to compare immature specimens. However, Schultz (1962) has reported that measurements of infantile skulls vary nearly as much as adult skulls. Further research needs to be done in this area.

Sexually dimorphic differences among adults are well known (e.g., Ashton and Zuckerman, 1956; Biegert, 1957; Hrdlička, 1907; Schultz, 1941, 1962; Selenka, 1898; Winkler, 1986). Males and females differ greatly in absolute size of their cranial dimensions (Hrdlička, 1907; Schultz, 1962). In particular, males possess larger canines, become more prognathic, and have more developed cranial crests, higher endocranial volumes, and, in general, larger dimensions of the neurocranium, face, and jaw (see also chapter 15, this volume). The present research reaffirms these findings. Selenka (1898) suggested that the larger canines as well as masticatory and nuchal-occipital musculature of the male orang-utan are responsible for the pronounced sexual dimorphism in jaw size, prognathism, and cranial dimensions (see also chapter 15, this volume). Hrdlička (1907), as well, felt that facial growth was controlled by the development of the teeth and facial muscles which, in turn, contribute to the development of sexual dimorphism in the orang-utan cranium.

Other specialized structures, such as the fatty cheek pads of adult males, influence the dimorphic modeling of the orang-utan face (Winkler, 1986). In addition, Biegert (1957) has suggested that the development of the laryngeal sacs in the male orang-utan contributes to the expression of cranial sexual dimorphism, especially in regard to the lower jaw, cranial base, and the degree of airorhynchy. Our CT scans of the orang-utan and the gorilla (Figs. 16–1 to 16–3) do demonstrate some degree of sexual dimorphism in airorhynchy in the orang-utan. The gorilla exhibits an even greater endocranial base–hard palate angle, and, therefore, less airorhynchy than does the orang-utan of either sex.

The endocranial volumes computed for our specimens (460 cm^3 for the male; 387 cm^3 for the female) are within the range of previously published figures (Hrdlička, 1907; Schultz, 1969; Selenka, 1898), which indicate that endocranial volume is dimorphic in the orang-utan but with overlapping ranges between the sexes. Selenka (1898) has suggested a total range of 300–534 cm^3, with the male skull being on average 70 cm^3 larger than the female. Hrdlička (1907) proposed ranges of 355–540 cm^3 for male and 300–490 cm^3 for female orang-utans.

In addition to assessing sexual dimorphism in the orang-utan, we were particularly interested in the morphological relationships of browridge size, frontal sinus formation, and frontal lobe–bony orbit expansion (e.g., Moss and Young, 1960; Shea, 1985). Moss and Young noted that two functional demands are simultaneously met by the frontal bone: i.e., it contributes to the cerebral capsule of the frontal lobe (floor and wall of the anterior cranial fossa) and it also forms part of the orbital roof. If these two cranial components separate during growth, the orbital roof may become thick and pneumaticized. The separation of these functional units is seen when a small cerebral volume is associated with normal (or larger) orbital contents (as for example, in microcephaly). In such cases, a large supraorbital torus often results (Moss and Young, 1960).

It is also known that the topography of the ecto- and endocranial surfaces of the frontal bone is the result of functionally different sources: that is, the inner bone table molds itself to the size of the frontal lobes of the cerebrum, whereas the configuration of the outer bone table is largely influenced by masticatory or muscular stresses and restricted by its role in protecting the orbital contents (see Moss and Young, 1960; Russell, 1982). The average ratio of orbital to brain volume is 6.6% for male orang-utans, 5.9% for female orang-utans, 7.7% for male gorillas, and 7.7% for female gorillas (data from Schultz, 1940, 1969). Thus, we would predict that gorillas, the great ape with the lowest encephalization quotient and the largest orbital volume, would show a greater morphological disparity between the inner and outer tables of the frontal bone than *Pongo,* in spite of the fact that gorillas have absolutely larger endocranial cavities. This morphological disparity should, and does, manifest itself in greater browridge and frontal sinus formation in the large African ape than in *Pongo* (see discussion by Moss and Young, 1960, for further details). This study also clearly indicates that, as predicted, the outer and inner tables of the frontal bone are in closer proximity in *Pongo* than in *Gorilla.*

CONCLUSIONS

Our research is in agreement with previous descriptions of sexual dimorphism in the orang-utan cranium (Ashton and Zuckerman, 1956; Biegert, 1957; Hrdlička, 1907; Schultz, 1941, 1962, 1969; Selenka, 1898). Univariate statistics and ANOVA for 37 exocranial dimensions indicate substantial sexual dimorphism, particularly in adults. Results from CT-scanning confirm the presence of sexual dimorphism in endocranial volume and suggest that airorhynchy is slightly more pronounced in female orang-utans than in males. CT-scanning further indicates that the anterior cranial fossa contributes to the formation of the roof of the orbit in both male and female orang-utans, despite differences in brain size. The functional demands of the nuchal-occipital musculature, muscles of mastication, tooth size, laryngeal sacs and other specialized structures (such as the fatty cheek pads), when present, as well as orbital and cerebral volume, may explain the development of cranial sexual dimorphism in the orang-utan. Comparisons of the three-dimensional computed images of *Pongo* and *Gorilla* reveal differences in frontal bone morphology and the degree to which each is airorhynchous.

ACKNOWLEDGMENTS

We thank Dr. Jeffrey Schwartz for inviting our participation in this volume; Robert Knapp for technical assistance; Mrs. Cynthia Strickland for typing the manuscript; and the American Museum of Natural History, National Museum of Natural History (Smithsonian Institution), Cleveland Museum of Natural History, British Museum (Natural History), Rijksmuseum Van Natuurlijke Historie (Leiden), University of Zürich (Anthropological Institute), Washington University (St. Louis), and the St. Louis Zoological Park for allowing the use of their collections. This research was supported by the National Science Foundation (L.A.W. Grant BNS-831252), Yerkes Regional Primate Center, Emory University (NIH RR-00165), the L.S.B. Leakey Foundation, and Sigma Xi.

REFERENCES

Ashton, E. H. and Zuckerman, S. 1956. Cranial crests in the Anthropoidea. *Proceedings of the Zoological Society of London, 126:* 581–634.

Biegert, J. 1957. Der Formwandel des Primatenschädels und seine Beziehungen zur ontogenetischen (Entwicklung und den phylogenetischen Spezialisationen der Kopforgane.) *Gegenbaurs Morphologisches Jahrbuch, 98:* 77–199.

Brown, B. 1986. Mandibular symphyseal morphology and the evolution of the orang-utan. *American Journal of Physical Anthropology, 69:* 182.

Brown, M. B. and Forsythe, A. B. 1974a. The small sample behavior of some statistics which test the equality of several means. *Technometrics, 16:* 129–132.

Brown, M. B. and Forsythe, A. B. 1974b. Robust tests for the equality of variances. *Journal of the American Statistical Association, 69:* 364–367.

Cave, A. and Haines, R. 1940. The paranasal sinuses of the anthropoid apes. *Journal of Anatomy, 74:*493–523.

Conroy, G. and Vannier, M. 1984. Non-invasive three-dimensional computer imaging of matrix filled fossil skulls by high resolution computed tomography. *Science, 226:* 456–458.

Conroy, G. and Vannier, M. 1985. Endocranial volume determination of matrix filled fossil skulls using high resolution computed tomography. *In* Hominid Evolution: Past, Present and Future, pp. 419–426, ed. P. Tobias. Alan R. Liss, New York.

Dixon, W. J. 1985. BMDP Statistical Software. University of California Press, Berkeley.

Hrdlička, A. 1907. Anatomical observations of a collection of orang skulls in western Borneo, with a bibliography. *Proceedings of the United States National Museum, 31:* 539–568.

Krogman, W. M. 1978. The Human Skeleton in Forensic Medicine. Charles C Thomas Publishers, Springfield, IL.

Moss, M. and Young, R. 1960. A functional approach to craniology. *American Journal of Physical Anthropology, 18:* 281–292.

Russell, M. 1982. Tooth eruption and browridge formation. *American Journal of Physical Anthropology, 58:* 59–65.

Schultz, A. 1940. The size of the orbit of the eye in Primates. *American Journal of Physical Anthropology, 26:* 389–408.

Schultz, A. 1941. Growth and development of the orangutan. *Contributions to Embryology, Carnegie Institute, 29:* 57–110.

Schultz, A. 1962. Metric age changes and sex differences in primate skulls. *Zeitschrift für Morphologie und Anthropologie, 52:* 239–255.

Schultz, A. 1969. The Life of Primates. Universe Books, New York.

Selenka, E. 1898. Rassen, Schädel und Bezahnung des Orangutan. *Studien über Entwicklung und Schädelbau, 1:* 1–99.

Shea, B. 1985. On aspects of skull form in African apes and orangutans, with implications for hominoid evolution. *American Journal of Physical Anthropology, 68:* 329–342.

Vallois, H. V. 1965. Anthropometric techniques. *Current Anthropology, 6:* 127–145.

Vannier, M., Conroy, G., Marsh, J., and Knapp. 1985. Three dimensional cranial surface reconstructions using high resolution computed tomography. *American Journal of Physical Anthropology, 67:* 299–311.

Vannier, M., Marsh, J., and Warren, J. 1983a. Three dimensional computer graphics for craniofacial surgical planning and evaluation. *Computer Graphics, 17:* 263–273.

Vannier, M., Marsh, J., Warren, J., and Barbier, J. 1983b. Three dimensional computer aided design of craniofacial surgical procedures. *Diagnostic Imaging, 5:* 36–43.

Vannier, M., Marsh, J., Warren, J., and Barbier, J. 1983c. Three dimensional CAD for craniofacial surgery. *Electronic Imaging, 2:* 48–54.

Vannier, M., Marsh, J., and Warren, J. 1984. Three dimensional CT reconstruction images for craniofacial surgical planning and evaluation. *Radiology, 150:* 179–184.

Winkler, L. 1986. Relationships between the facial cheek pads and facial anatomy in the orangutan. *American Journal of Physical Anthropology, 69:* 280.

17

Phylogeny and Skull Form in the Hominoid Primates

BRIAN T. SHEA

There are a number of important reasons for the renewed attention being focused on the morphology, adaptations, and evolution of the orang-utan, *Pongo pygmaeus.* A primary one is the recent discovery of well-preserved Miocene hominoids that have rekindled debates concerning the relationships among the extant and fossil apes (Andrews and Tekkaya, 1980; de Bonis, 1983; Pilbeam, 1982; Wu et al., 1981, 1982, 1983). A second reason directly related to these newly described fossils is a series of recent morphological studies of hominoid primates analyzing the differences in craniodental form between the Asian and African great apes (e.g. Andrews and Cronin, 1982; Ciochon, 1983; Kay and Simons, 1983; Shea, 1985; Ward and Kimbel, 1983; Ward and Pilbeam, 1983; Wolpoff, 1982).

A third area of current interest involves the assessment of character polarity—of primitive and derived features—and the importance of such analyses for the reconstruction of ancestral morphotypes and phylogenetic relationships. Relevant papers on the orang-utan include those cited immediately above, in addition to several recent essays questioning the long-held implicit assumption that *Pongo* is on balance morphologically extremely derived, or autapomorphic (e.g., Hammond, 1983; Herbert, 1984; Lewin, 1983). Finally, we have the recent contributions by Schwartz (1983, 1984), who posited a specific phyletic connection (i.e., sister group relationship) between *Pongo* and *Homo* and their extinct relatives. This proposed link is, of course, contrary to the received wisdom advocating the African apes as the sister group for humans (e.g., Goodman et al., 1983). One ramification of Schwartz's controversial suggestion is that it has focused attention on an important and relatively dormant issue: exactly what derived *morphological features* do link the African great apes and humans?

In this chapter I discuss aspects of skull form in the extant and fossil hominoids. Detailed consideration is given to neurocranial/splanchnocranial positional relationships and the development of the supraorbital torus, since these features are among those which most clearly distinguish the orang-utan from the African apes. The phylogenetic implications of these differences are emphasized in a discussion of the evolution of hominoid skull form.

FACIAL POSITION IN EXTANT HOMINOIDS

Elsewhere I have reviewed the problem of neurocranial/splanchnocranial relationships in the Asian and African great apes (Shea, 1985). In this section I wish to extend this discussion to include the hylobatids or lesser apes. An extensive German literature on skull form provides the basis for these comparisons (see Angst, 1967; Biegert, 1957; Hofer, 1952, 1965; Kummer, 1952; Starck, 1953; Thenius, 1970). Hofer (1952) classified crania in terms of the angular relationship between the neurocranium and splanchnocranium and defined a variety of configurations. Two of these are of relevance to the present discussion: *klinorhynchy,* where the splanchnocranium is directed ventrally (or "downward") in relation to the cranial base; and *airorhynchy,* where the splanchnocranium is directed dorsally (or "upward") in relation to the cranial base. These particular configurations are problematic in at least two respects. First, there is the problem of using any particular plane as a reference axis for registration (Moyers and Bookstein, 1979); second, these are obviously *relative* assessments of neurocranial/splanchnocranial relations (in other words, a skull that is more airorhynchous than another may be more klinorhynchous than a third). In spite of these and other difficulties, this classification focuses attention and analysis on the important structural interrelationships among the neurocranium, basicranium, and splanchnocranium.

As noted by all of the German authors listed above, as well as various other craniologists (e.g., Delattre and Fenart, 1956; Vogel, 1968), the orang-utan provides an excellent example of airorhynchy. The position of the face, as measured by the alveolar plane, occlusal plane, or plane of the hard palate, is angled dorsally with respect to the cranial base (see Fig. 17–1). *Pongo*'s airorhynchy is especially marked when compared to the condition seen in the other extant large-bodied hominoids. *Homo, Pan,* and *Gorilla* all have a more ventral, or klinorhynchous, positioning of the face with respect to the basicranium (Delattre and Fenart, 1956; see Shea, 1985, Figs. 3 and 8).

Given this marked difference between the Asian and African great apes in facial position, it is of considerable interest to assess these relationships in the Asian lesser apes. The hylobatids have received much less attention in this regard, although Delattre and Fenart (1956) do note that they resemble *Pongo* in exhibiting the presence, if not degree of development, of airorhynchy. Based on the Lille School's technique of orientation along the horizontal vestibular axis, the siamang appears more airorhynchous than the gibbon, but both lesser apes differ from the klinorhynchous African apes in the direction of *Pongo* (Delattre and Fenart, 1956). In Fig. 17–1, I present skulls of some of the hominoids aligned along the posterior cranial base (sella-basion). The marked ventral deflection of the face in *Pan* is obvious. Although the gibbon skull appears much less airorhynchous than that of the siamang, an overlay of these two hylobatids and a further comparison with *Pan* demonstrates the structural uniformity of the lesser apes (Fig. 17–2). In fact, the degree of airorhynchy in the Asian hominoids appears roughly correlated with overall skull size, but whether this is due to a true growth-based or a biomechanically based causal relationship is at present uncertain.

SUPRAORBITAL TORUS FORM IN EXTANT HOMINOIDS

Many authors have described aspects of circumorbital and supraorbital morphology in the hominoids (see, e.g., references in Ehara, 1972, and Russell, 1985). Because I wish to discuss the possible relationships between facial position and supraorbital morphology, a brief description of the variation in supraorbital morphology seen in the extant hominoids is in order. Here I rely primarily on the recent descriptions given by Kimbel et al. (1984) in their work on the reconstruction of the *Australopithecus afarensis* skull.

African apes exhibit a strong, barlike supraorbital torus separated from the frontal squama by

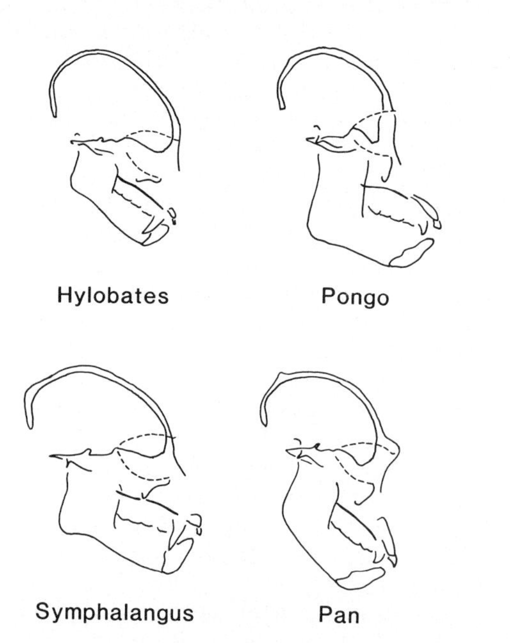

FIG. 17–1. A comparison of ape skulls oriented along the posterior cranial base and illustrating the varying degrees of angulation between the neurocranium and splanchnocranium. Note the strong dorsal flexion of the midface in *Pongo* and *Symphalangus* in particular. (Redrawn from Biegert, 1957).

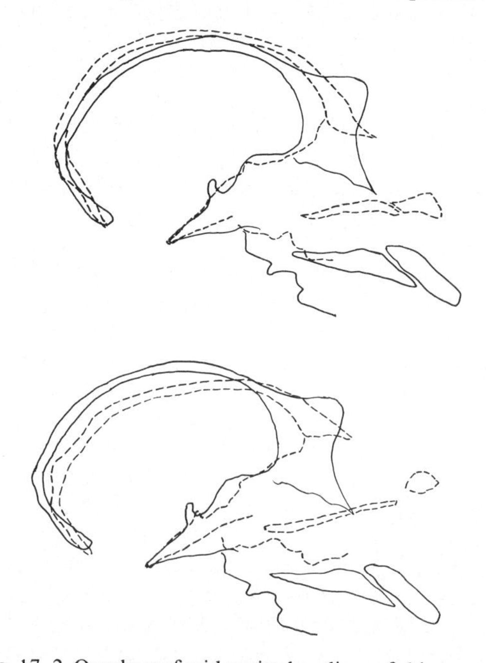

FIG. 17–2. Overlays of mid-sagittal outlines of chimpanzee crania (solid outlines) compared to those of a gibbon (dashed outline, above) and a siamang (dashed outline, below). The skulls have been reduced to the same overall length. Note that both lesser apes are considerably more airorhynchous than the chimpanzee.

a transverse depression identified as the sulcus supratoralis. Although the frontal bone is more vertical in the subadult skull, distinct supraorbital torus and sulcus supratoralis are nonetheless evident by the time the first permanent molar is erupting (Kimbel et al., 1984). In terms of interspecific variation, adult pygmy chimpanzees resemble juvenile common chimpanzees in the overall development of their supraorbital region (Cramer, 1977; Kimbel et al., 1984; Latimer et al., 1981; Shea, 1983; Weidenreich, 1941), whereas gorillas have a very well-developed and barlike torus above the orbits.

In *Pongo,* the supraorbital torus is very weakly developed and the frontal is relatively vertical. The weak tori arc over each orbit, are not continuous across the midline, and a sulcus supratoralis is absent (Kimbel et al., 1984). Although *Pongo* and the lesser apes differ in many aspects of circumorbital morphology (e.g., interorbital breadth, orbital shape), the latter hominoids are nonetheless also characterized by having a very weakly developed supraorbital tori that arc over each orbit and lack a sulcus supratoralis.

The supraorbital morphology of early hominids has been reviewed in detail by Kimbel et al. (1984). They stress the features in which the supraorbital tori of early *Homo habilis* and *Homo erectus* resemble those of the African apes (i.e., the presence of a distinct sulcus supratoralis and a supraorbital torus that is expressed independently of the position of the inferior temporal lines), as well the ways in which those of *Australopithecus (A. afarensis, A. africanus, A. robustus/boisei)* present a cohesive pattern somewhat distinct from that of the African great apes (e.g., no sulcus supratoralis). These interesting variations among the species of early hominids are not our concern here. What I would like to stress is that the hominids as a group resemble the African apes, and not the orang-utan and hylobatids, in their development of a marked supraorbital torus.

THE RELATIONSHIP BETWEEN FACIAL POSITION AND SUPRAORBITAL TORUS DEVELOPMENT

The preceding two sections have dealt with a phenetic description of neurocranial/splanchnocranial relations and the development of the supraorbital torus in the extant hominoids. In this section I will argue that these two aspects of craniofacial form are not unrelated in developmental, functional, and evolutionary terms.

The idea that the airorhynchous face and the relatively weakly developed supraorbital tori in *Pongo* are functionally interrelated is not new (e.g., Biegert, 1957; Delattre and Fenart, 1956; Moss and Young, 1960; Vogel, 1968). The mechanical basis of this argument is the claim that *one* variable affecting the degree of torus development is the spatial relationship among the neurocranial, orbital, and nasomaxillary regions. Smith and Ranyard (1980) have utilized these arguments in their analysis of browridge development in Neandertals and other hominids. I (Shea, 1986) have used the extant hominoids to illustrate the significance of these structural interrelationships in a response to a recent review of the function of the supraorbital torus (Russell, 1985). Russell rejected the claim that variation in facial, orbital, and neurocranial relationships is correlated with the degree of browridge development, arguing in the process that these spatial relationships do not vary during the ontogeny of *Pan, Gorilla,* or *Papio,* nor in a comparison of adult hylobatids, orang-utans, and African apes. Shea (1985, 1986) presents evidence to counter both of these assertions. Bony responses to stresses generated during mastication presumably play a role in the degree of supraorbital torus formation as well, as Russell (1985) and others have argued (although we badly need *in vivo* experimental data to confirm this). But these stresses interact with the structural or positional factors (see Shea, 1986).

Delattre and Fenart (1956) and Shea (1985, 1986) discuss these principles in the context of contrasting the orang-utan with the African great apes, and I will not repeat these arguments here. But it is worth stressing that the relatively weak development of the supraorbital tori in the hylobatids, although usually assumed to be the result of a very gracile skull form, may also be related to the dorsal position of their faces.

Arguments as to why the orbital region is "repositioned" in the presence of altered neurocranial/splanchnocranial relationships draw on concepts in craniofacial biology such as facial form or structure (e.g., Delattre and Fenart, 1956; Enlow 1982), pattern invariance (Moyers et al., 1979), and morphological integration (Olson and Miller, 1958; see Zingeser, 1973). This is an aspect of craniofacial biology that has received relatively little attention over the past several decades (see my discussion of the differences in skull form between the Asian and African great apes [Shea, 1985], the details of which will not be repeated here).

In order for these ideas to move beyond mere empirical statements about the maintenance of *form* in spite of changing *position* (Delattre and Fenart, 1956), we must explicate some of the functional interrelationships that are undoubtedly at the base of the observed patterns. In this regard, Zingeser's (1973) arguments about the maintenance of the angular relationship between the anterior face and the occlusal plane are of interest. In addition, Greaves's (1985) recent analyses of the postorbital bar and supraorbital morphology may also be very relevant and functionally informative here. Among other things, Greaves suggests that a consistent angular relationship (i.e., 45°) between a line fitted to the postorbital bar and one fitted from the jaw joint to the bite point in lateral perspective provides the most efficient resistance to the stresses generated during mastication. Thus, from a functional perspective, if a dorsal deflection of the face shifts the bite point anterosuperiorly, we might expect a corresponding rotation of the orbital region in order to maintain functional equivalence in stress-resisting capabilities of the postorbital bar. These and other such functional/structural relations are currently being investigated in a broad series of primates and nonprimates in order to address the question of structural invariance (M. Ravosa, work in progress).

Whatever the specific biomechanical bases of the observed correlation between supraorbital morphology and facial position in the extant hominoids, we can use it to make certain predictions and extrapolations for fragmentary fossils. For example, our knowledge of these structural relationships, combined with the presence of splanchnocranial, basicranial, and neurocranial remains, makes it possible to reasonably infer that *Australopithecus afarensis* must have had a supraorbital region more like that seen in the African apes than in *Pongo* (see Kimbel et al., 1984, for specific details concerning variation in supraorbital morphology *within* the African hominoid clade). These relationships also form the basis of the prediction (see below) that no Miocene or Plio-Pleistocene hominoid will be found to exhibit a markedly developed supraorbital torus in the presence of an airorhynchous face.

SOME ASPECTS OF SKULL FORM IN FOSSIL HOMINOIDS

The number of hominoid(?) fossil genera with reasonably well preserved cranial material can be counted on one hand. Excluding the hominids, we have two genera previously considered to be hominoids—*Aegyptopithecus* and *Pliopithecus*—as well as *Proconsul* and *Sivapithecus*. All of these genera are represented only by fragmentary and reconstructed skulls. What can be determined from them in regard to the features discussed above, i.e., facial position and supraorbital morphology?

Aegyptopithecus

Brief descriptions of cranial morphology in *Aegyptopithecus zeuxis,* based on the reconstruction of the 1966 skull, have been given by Simons (1972), Kay and Simons (1980), and Fleagle and Kay (1983). Simons (1984) presented a few comments and an illustration of three recently recovered and well-preserved faces of this genus. Figure 17–3, after Kay and Simons (1980), illustrates skull form in *Aegyptopithecus zeuxis* compared to that of a represen-

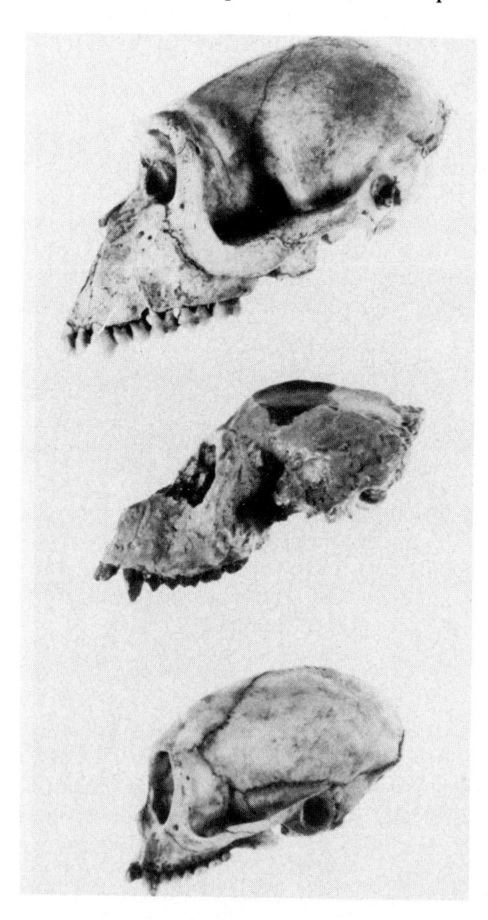

FIG. 17–3. A comparison of *Aegyptopithecus zeuxis* (cast of CGM40237) (center) with *Macaca sylvanus* (top) and *Cebus albifrons* (bottom). Note the dorsal inclination of the occlusal plane, midface, and orbits in *Aegyptopithecus* relative to the other skulls. (From Kay and Simons, 1980.).

tative cercopithecoid and ceboid. Note the airorhynchous face of *Aegyptopithecus:* the occlusal plane, nasomaxillary complex, and circumorbital region are all shifted dorsally relative to the condition observed in the other skulls. The dorsal inclination of the orbits has been noted by Fleagle and Kay (1983), and this is most clearly appreciated by fitting an imaginary vertical line to either the orbital opening or the lateral orbital bar. In terms of the supraorbital morphology, there is no distinct or marked torus and no sulcus supratoralis, even though the new faces exhibit a sagittal crest that originates quite far forward on the frontal (Simons, 1984). Although recent discoveries of other *Aegyptopithecus* material will require some reconsideration of the original reconstruction (i.e., they have more widely flared cheekbones and shorter snouts), they basically corroborate the above comments regarding facial position and supraorbital morphology (personal observation).

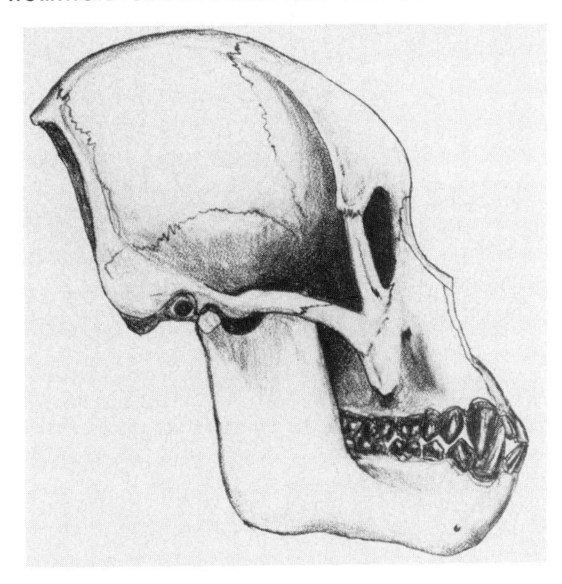

FIG. 17–4. A drawing of the reconstruction of the *Proconsul africanus* skull by Walker et al. (1983). See text for discussion. (Photo courtesy of A. Walker.)

Pliopithecus

The best-preserved skull of *Pliopithecus vindobonensis* was described by Zapfe (1958, 1960). It has been compared to those of *Aegyptopithecus* and *Proconsul* by Fleagle and Kay (1983) in an informative analysis. Some of the details of facial and orbital position in *P. vindobonensis* are difficult to ascertain due to incomplete preservation and the artificial material introduced during reconstruction (personal observation). Nevertheless, it does appear that the orbits are more vertically oriented relative to the Frankfort Horizontal than in *Aegyptopithecus* (Fleagle and Kay, 1983; see Fig. 17–4). The circumorbital region in *Pliopithecus* is quite similar to that of the hylobatids in that a raised lip-like projection of bone rings the orbits inferiorly (Fleagle and Kay, 1983) and a similarly small projection arches over them superiorly. A marked torus or associated sulcus is not present. Overall, the position of the face and development of the tori are best described as basically similar to those seen in the hylobatids or some of the extant colobines. It is not possible to determine whether the face of *P. vindobonensis* is strongly airorhynchous, as in the siamang, or moderately airorhynchous, as in the gibbon.

Proconsul

The skull of *Proconsul africanus* discovered by M. Leakey in 1948 remains one of the most completely preserved hominoid cranial specimens. Primary descriptions and reconstructions may be found in Clark and Leakey (1951), Robinson (1952), and Davis and Napier (1963). A more recent reconstruction, based on the addition of new basioccipital pieces, has been completed by Walker et al. (1983) (Fig. 17–4). The position of the face relative to the neurocranium is difficult to ascertain due to incomplete preservation. In addition, the degree of prognathism and possibly even facial position have varied significantly in the numerous reconstructions. The reconstructions of Walker et al. (1983), Clark and Leakey (1951), and Robinson (1952) are all in agreement in depicting a relatively strong acute angulation between the frontal and zygomatic processes of the malar, which may reflect a dorsal hafting of the face to the braincase (compare *Pongo* and *Pan* in Fig. 17–5). The morphology of the frontal suggests a relatively close approximation of the orbital capsules to the anterior neurocranium. As has been noted by many previous authors, the supraorbital region in the *P. africanus* skull is smooth, with no evidence of a torus. Subadult material of *Proconsul,* probably representing one of the larger species, i.e., *P. nyanzae* or *P. major,* also indicates little or no browridge development (Andrews et al., 1981).

Sivapithecus

New insights concerning craniofacial morphology in the genus *Sivapithecus* have emerged with

FIG. 17–5. Lateral views of the skulls of (from left to right) the orang-utan *(Pongo pygmaeus), Sivapithecus indicus,* and the chimpanzee *(Pan troglodytes).* Note the resemblances between *Sivapithecus* and *Pongo* in circumorbital and facial shape and position. See text for additional discussion. (Photo courtesy of D. Pilbeam.)

the discovery of a partial face (GSP 15000) from the late Miocene of Pakistan (Pilbeam, 1982). Finds from Turkey (Andrews and Tekkaya, 1980) and China (e.g., Wu et al., 1983) supplement the Pakistan material because they also preserve portions of the face and skull other than jaws and teeth. The *Sivapithecus (S. indicus)* face has been described (Pilbeam, 1982; Preuss, 1982) and analyzed (Andrews, 1982; Lipson and Pilbeam, 1982; Ward and Pilbeam, 1983), with most authors commenting on the phenetic and potential phylogenetic proximity to the extant orang-utan (but see Kay and Simons, 1983, for a dissenting view). GSP 15000 closely resembles the orang-utan, and clearly differs from the African apes, in the conformation of the subnasal region (Ward and Kimbel, 1983; Ward and Pilbeam, 1983), the dorsal inclination of the premaxilla and alveolar plane, the concave face, the zygomatic region, the shape of the orbits, the narrow interorbital distance, and the lack of a pronounced supraorbital torus with a sulcus supratoralis, among other features (Andrews and Cronin, 1982; Pilbeam, 1982; Preuss, 1982).

The morphology and orientation of the palate and circumorbital region in *Sivapithecus indicus* suggest a splanchnocranial/neurocranial positional relationship very much like that seen in extant *Pongo.* The angulation of the zygomatic arch also supports this interpretation (see Fig. 17–5), although fragments in the area of the zygomaticomaxillary suture are missing in GSP 15000 (Pilbeam, 1982). The more complete but crushed cranial material from Lufeng has only briefly been described and analyzed (Wu and Oxnard, 1983; Wu et al., 1981, 1982, 1983; Xu and Lu, 1980). Significantly, these skulls differ from *Sivapithecus indicus* in having a broad interorbital region and somewhat differently shaped orbits, if crushing has not distorted these features too much. Wu et al. (1983) described these skulls (one of which they identify as *Sivapithecus* and the other as *Ramapithecus*) as strongly resembling *Pongo* and not the African apes in their facial morphology. Specifically, they note the concave midfacial region, the dorsal inclination of the palate, and the supraorbital region as being similar to the orang-utan.

DISCUSSION

Historical Influences

The pervasive effect of our assumptions regarding the likelihood of primitive and derived states in analyses of phylogeny and adaptation cannot be overestimated. For example, several authors have recently considered some of the implications of past claims that Miocene hominoids could be fitted comfortably into the neontologically based dichotomy of "pongid" and "hominid" (Pilbeam 1979; Wolpoff, 1982). This organization resulted from a gradistic approach (Rosenberger, 1980) to the morphological variation in living and fossil hominoids, and it inculcated common conceptions that the "pongids" were largely primitive and the hominids derived. I think an interesting parallel case can be made in a comparison of the Asian great ape *(Pongo)* with the African great apes *(Pan/Gorilla),* though in this case the concept of different grades has not been involved, of course. Nevertheless, the traditional assumption has been that in terms of craniodental morphology, the African apes more closely approach the primitive condition for the large-bodied hominoids.

Probably the most significant reason for this belief is that Pilbeam's (1969) detailed study of the East African hominoid craniodental material from the early Miocene led him to suggest specific phyletic connections between the chimpanzee and gorilla and, respectively, the smaller and larger species of *Pronconsul.* Although Pilbeam (e.g., 1969:126) stressed that these early *Proconsul* fossils were by no means *identical* to the skulls and teeth of living African apes, the claim for direct ancestor-descendant relations dating back to the early Miocene implicitly suggested a strong resemblance between the fossil and extant groups. One interesting spinoff of particular relevance to the present discussion was the debate between Louis Leakey and Pilbeam over whether the *P. africanus* skull lacked a supraorbital torus simply because of its size and overall gracility (Pilbeam, 1969). Drawing on a comparison of the pygmy chimpanzee and the gorilla, Pilbeam (1969) argued that the early Miocene skull lacked a strong torus reminiscent of the African apes only because it was so small an animal, and not because it exhibited more fundamental differences in skull form. Leakey disagreed, but these types of arguments served to further strengthen general notions that the proconsuls and African great apes are similar in skull form.

These arguments on their own would not necessarily indicate that the African apes are generally primitive in skull form relative to *Pongo* in an analysis explicitly focusing on primitive and derived states. But such approaches were very rare in primatology in the 1960s (see Martin, 1968, for an exception), and phylogenetic relationships were based on general phenetic similarity combined with the fossil record. In addition, it is very significant that no specific ancestors for the orang-utan were posited, although it was generally maintained that *Pongo* had split off prior to the early Miocene hominoid radiation, perhaps in the late Oligocene (Pilbeam, 1972:46; Simons and Pilbeam, 1965:139). A further contributing factor is that specific comparisons between the Oligocene *Aegyptopithecus* and extant apes almost always involved the African great apes rather than the orang-utan. This is not surprising, since the comparisons of Oligocene to modern forms were invariably made via the proconsuls, which, as I have noted, were thought of as early members of the African ape lineages. There were no fossil "pre-orangs" to even consider in this view (with the exception of brief references to some of the *Sivapithecus* material by various authors). Perhaps the most explicit statement of the pervasive implicit ideas which I am trying to probe here was made by Simons (1972:217) when he wrote:

> It is also a remarkable coincidence that both *Aegyptopithecus zeuxis* and *Dryopithecus africanus* can be shown from cranial and dental studies to be plausible stages related to or in the ancestry of the modern chimpanzee. Inasmuch as the chimpanzee is possibly the closest relative of man among animals, the implications of these relationships are obvious. Somewhere along a line of ancestry in or near this one, the first hominids branched off. We cannot be certain when this was, but the oldest undoubted hominids are much younger than either of these two skulls. Thus the possibility exists that one or both of them represents a stage in human ancestry. Perhaps more important is the fact that even if neither is a direct human ancestor, they remain informative of general grades or stages of hominoid evolution at about 20 million and 30 million years ago, respectively.

Although most attention was focused on when and why the hominid lineage branched from this Miocene-Recent ape baseline, such scenarios also implied that, to the extent that the Asian and African great apes differed, it was *Pongo* that had changed the most, since *Aegyptopithecus* undoubtedly preceded the divergence of the orangutan lineage and then the African ape lineage.

Another prevalent idea contributing to notions that *Pongo* was fairly specialized in overall skull form derived from claims of similarity between humans and hylobatids. A relatively extreme version of this view was advanced by Gould (1975:284) when he favorably quoted Dubois (1896:10) to imply that simple isometric enlargment of the hylobatid skull would yield a distinctively hominid-like morphology:

> We have only to double the length and breadth, both of the thigh-bone and of the skull of a *Hylobates syndactylus,* to have dimensions corresponding to those of the Java form [*Pithecanthropus*]. By doubling all dimensions of a *Hylobates* we would obtain an imaginary product with a corresponding cranial capacity also.

A more recent example of such links is provided by Kluge (1983), in his phylogenetic analysis of hominoid relationships, in which he argues that *Homo* and the hylobatids share the primitive condition of nonprognathic faces. This argument derives from the notion that *Pongo,* as well as the African great apes, have diverged from the condition primitive for the large-bodied hominoids. In fact, a prognathic face is almost certainly primitive for the large-bodied hominoids, and the superficial resemblance between *Homo* and hylobatids in the face reflects nonhomologous, parallel development. That Kluge is almost certainly wrong in his claims is not essential here. Rather, the point is that very few links have been drawn between the hylobatids and *Pongo* in terms of skull form—usually noted are the similarities to *Homo* or the African apes. Phenetic connections between hylobatids and hominids, on the one hand, and proconsuls and African apes, on the other, have tended to practically insure the conclusion that *Pongo* is likely to be specialized in overall skull form.

Character Polarity

Having reviewed some of the predominant assumptions in analyses of hominoid skull form over the past several decades, it is now necessary to consider the phylogenetic implications of the morphological features discussed in this chapter. Bearing in mind that we are dealing here with relative degrees of facial angulation rather than a discrete character that is either present or absent, it is most likely that an airorhynchous face is primitive for Hominoidea. This would mean that a dorsally inclined splanchnocranium is also primitive for the large-bodied hominoids, the primitive condition being retained (though perhaps exaggerated—see below) in the orang-utan. A degree of airorhynchy comparable to, or greater than, that found in the gibbons is seen in *Pongo, Hylobates symphalangus,* and possibly all the fossil taxa discussed above (though the tentative nature of this conclusion must be granted, considering the fragmentary condition of the fossils). In the extant cercopithecoids, facial position also varies considerably, but Delattre and Fenart's (1956) and Vogel's (1968) results support the argument that a dorsal deflection is primitive for the group. It follows from this that African apes and hominids share a similarly derived condition of increased klinorhynchy, which may have evolved in parallel in some of the cercopithecines (this needs to be investigated in greater detail) and perhaps other taxa (e.g., *Daubentonia;* see Hofer, 1965).

Supraorbital morphology supports a similar polarity and set of linkages, regardless of whether or not the two features are related in the fashion I have suggested in this chapter. The review of fossil and extant hominoids and other taxa indicates that the relative lack of a marked torus is primitive for the Hominoidea as a whole, and thus for the large-bodied hominoids as well. This has also been suggested by Andrews (Andrews, 1982; Andrews and Cronin, 1982). It is more difficult to say whether the primitive condition precisely resembled the supraorbital morphology seen in *Aegyptopithecus, Proconsul,* the hylobatids, *Pongo, Sivapithecus,* or some other form, since there are some interesting variations among these taxa. The supraorbital morphology of the large form (RUD-44) from Rudabánya, Hungary, most closely resembles *Pongo* and not the African apes (Shea, personal observations; Martin and Andrews 1982; Wolpoff, 1980), providing some evidence for the lack of a well-developed torus in *Dryopithecus* as well. The data of facial position and supraorbital morphology clearly differentiate the extant African apes from the more primitive early Miocene proconsuls. As argued by McHenry et al. (1980) and Corruccini and Henderson (1978), there is little reason to posit specific links between these early Miocene apes and the African apes based on cranial morphology.

Other features of significance in deciphering hominoid phylogeny may be related to variation in facial position (Shea, 1985). The development of a true ethmofrontal sinus appears to be a derived trait linking the African apes and humans (Cave and Haines, 1940). This pneumatization may be related to bony expansion that is associated with klinorhynchy and the morphology of the supraorbital region in the African apes and hominids. I have also suggested that aspects of subnasal morphology, recently shown to be important in distinguishing among various hominoid groups (Ward and Kimbel, 1983; Ward and Pilbeam, 1983), may be related to facial position (Shea, 1985). Whatever the merits of this particular suggestion, it is important to stress in the present context that there are at least three different patterns of hominoid subnasal morphology, i.e., that characterizing (1) the hylobatids and dryopith hominoids; (2) the ramapith hominoids plus *Pongo;* and (3) the African apes and hominids (Ward and Pilbeam, 1983). Ward and Pilbeam suggest that pattern 3 is likely to be the primitive condition for the large-bodied hominoids and that pattern 2 is a derived feature linking the ramamorph hominoids and the orangutan. But this assessment is admittedly based on perceived "likelihood of transformation" of pattern 1→3 vs. pattern 2→3. I would submit that in our current state of knowledge of hominoid skull form this is a very difficult choice to make. Is this once again an example of the tendency to view *Pongo* as highly derived and African apes as resembling the primitive early Miocene forms? It is quite possible that the condition (or some slight variation thereof) seen in *Pongo* and the known ramapiths is primitive for the large-bodied hominoids. The alternative case as stated by Ward and Pilbeam (1983) (and Ward and Kimbel, 1983) would be much stronger if the subnasal morphology in African apes and early hominids were not derived and distinct from the pattern seen in *Proconsul, Rudapithecus,* and the hylobatids.

A number of recent popular articles (Hammond, 1983; Herbert, 1984; Lewin, 1983) have suggested that certain craniodental features of *Pongo* might indeed be primitive for the large-bodied hominoids, rather than derived as has usually been assumed. This introduces some interesting new perspectives in paleoanthropology. Nevertheless, the features discussed have involved predominanty the gnathic complex, especially the thick-enameled postcanine teeth and robust mandibular morphology. This renewed interest has been generated in part by the Miocene fossils from Buluk, western Kenya, which Leakey and Walker (1985) argue strongly resemble known *Sivapithecus* from Asia (but also see Delson, 1985). The present analysis of facial position, supraorbital morphology, and possibly related features (Shea, 1985) adds some other characters in which *Pongo* may exhibit or most closely approximate the primitive condition for the large-bodied hominoids. A broad skull which is relatively short anteroposteriorly and which bears a deep face may also prove to be primitive for this clade. Certainly the narrow elongate skull form of African apes (and hominids to a certain extent) is quite distinct from that of *Pongo* and the ramapiths (Wolpoff, 1982), and is also very different from that of the proconsuls. As new fossils are discovered and new analyses completed we will need to reassess these and other features more fully. It would be a great mistake to argue (as some recent popular articles have) that the orang-utan represents a "missing link" which preserves the primitive state *in toto.* Of course all creatures, living and extinct, represent a mixture of primitive and derived features. Latent assumptions that *Pongo* was on the balance extremely derived have misled interpretations of fossils and phylogeny. We do not want to replace these assumptions with a new, but potentially equally misleading, set.

Alternative Phylogenetic Hypotheses

One specific phylogenetic hypothesis which deserves mention is Schwartz's (1984) contention that *Pongo,* the ramapiths, and hominids comprise a clade to the exclusion of the African apes. Do the features reviewed in the present paper corroborate or falsify this phylogenetic hypothesis? I think a good argument can be made that, as the evidence stands at present, there is an important series of derived cranial characters shared by the human and African ape lineages. In addition to features of subnasal morphology (Ward and Pilbeam, 1983) and an ethmoidally derived frontal sinus (Cave and Haines, 1940), a splanchnocranium deflected ventrally (klinorhynchy) and a marked supraoribital torus appear to be important synapomorphies of a postulated human-African ape clade. The evidence discussed in this chapter corroborates the traditional linking of humans with the African apes, and demonstrates some of the important differences in skull form between these hominoids and either the lesser apes or the dryopiths.

There are currently three other primary alternative views regarding the phylogenetic placement of the ramapiths. These have been discussed by Kay and Simons (1983), who favor an interpretation of the ramapiths as the sister of hominids. This view requires that many of the craniodental similarities between *Pongo* and *Sivapithecus* (e.g., orbit shape, facial shape, nasal floor morphology) represent either (1) parallel developments, or (2) primitive retentions. The difficulties for the Kay and Simons scenario posed by the first choice are obvious, particularly since these features have not been shown to be as susceptible to parallel evolution as traits such as mandibular robusticity and enamel thickness. The second choice requires equally unlikely amounts of parallel evolution in aspects of the orbital and facial regions of the skulls of African apes and early undoubted hominids such as *Australopithecus.* Other parallel features of the skull would be the klinorhynchous faces and marked supraorbital tori of African apes and *Australopithecus,* since the lack of these features in *Pongo* and *Sivapithecus* is assumed to reflect the plesiomorphous condition for the large-bodied hominoid clade in this scenario. For the above reasons, I believe the features of facial position and supraorbital morphology can be taken as evidence against the phyletic linking of the ramapiths and hominids, whether in the scenario of Schwartz or that of Kay and Simons.

A second alternative posits phylogenetic links between *Pongo* and certain or all of the ramapiths (e.g. Andrews, 1982; Andrews and Cronin, 1982; Ciochon, 1983; Ward and Pilbeam, 1983). My argument that *Pongo* more closely approximates the primitive condition in several of the features which most clearly distinguish the African and Asian great apes from one another does not necessarily negate the links proposed between the orang-utan and *Sivapithecus.* Most of the derived features linking these two hominoids (Andrews and Cronin, 1982; Ciochon, 1983) remain as elsewhere strong evidence for a specific relationship (although I have argued here and in Shea, 1985, that nasal floor morphology and a few others may not be derived). Furthermore, although an airorhynchous face is probably primitive for Hominoidea, the *marked degree* of airorhynchy seen in *Pongo* may be a derived condition shared with at least some *Sivapithecus.*

The argument that *all* the ramapiths (i.e., *Ouranopithecus, Sivapithecus, Ramapithecus* [?], *Gigantopithecus,* etc.) are specifically related to *Pongo* is much harder to substantiate at this point. Just as the discovery of the upper face of *Sivapithecus indicus* from Pakistan dramatically clarified certain aspects of hominoid evolution, the recovery of similar parts of other Miocene hominoids will undoubtedly result in more surprises and reorientation. Taking a broader perspective on the evidence, however, and considering our lack of knowledge of facial anatomy in this very diverse middle to late Miocene radiation, I am inclined at present to follow Greenfield (1980) and Wolpoff (1982, 1983) in arguing that one or another of the ramapiths gave rise to all of the extant large-bodied hominoids. This scenario explains the marked similarities in the teeth of the ramapiths and early hominids as the result of common ancestry rather than parallel development.

The interesting aspect of this scheme in terms of the features of relevance here is that specific predictions can be made about future discoveries of ramapith fossils (or perhaps future clarification of those already recovered). If this phylogenetic scenario is correct, then we would expect to find ramapith species with relatively nonairorhynchous faces and marked supraorbital tori, relatively elongate skulls with an African ape-like nasal floor morphology, robust jaws, and thick-enameled megadont teeth, in middle to late Miocene deposits of Africa. This is an elaboration of Wolpoff's (1982) prediction of the recovery of a "dental ramapith" with an African ape-like skull. If future paleontological work fails to yield such an African ramapith, this would strengthen the linking of all the ramapith specifically with the orang-utan lineage. Of course, given the morphology of known Miocene apes from Africa, the development of marked supraorbital tori and a ventrally deflected face would still presumably be synapomorphic for African apes and early hominids. The recovery of large proconsuls or dryopiths with these features would likely falsify their status as shared derived in the African hominoids, and increase the likelihood of parallel evolution in the position of the face and the development of the supraorbital region among the hominoids.

Recently discovered finds announced as this paper goes to press provide an interesting test of the above predictions. The partial cranium of the early Miocene hominoid *Afropithecus turkanensis* from northern Kenya (Leakey and Leakey, 1986a) appears to exhibit a marked dorsal angulation of the face, with slight or absent supraorbital tori and orbits rotated under the anterior neurocranium. A second and smaller skull of *Turkanapithecus kalakolensis* (Leakey and Leakey, 1986b) appears less airorhynchous than that of *Afropithecus,* and it also exhibits a slightly more robust supraorbital torus. Nevertheless, facial position and supraorbital morphology roughly resemble aspects of that seen in *Proconsul africanus* and *Pliopithecus vindobonensis,* and perhaps the hylobatids among the extant apes. Thus, this new material strengthens the claim that airorhynchous faces and relatively weakly developed supraorbital tori are primitive for the Hominoidea.

CONCLUSIONS

It is clear that the elucidation of the major differences in skull form between the orang-utan and the African great apes will contribute directly to our understanding of hominoid phylogeny. Two such important morphological distinctions between the Asian and African apes are facial position and supraorbital morphology. These craniofacial features are probably interrelated in developmental, functional, and therefore evolutionary terms.

It is significant that the hylobatids resemble the orang-utan, and not the African apes, in facial position and supraorbital morphology. Moreover, a review of Oligocene-Miocene fossil skulls reveals that all lack the strongly developed supraorbital torus and marked ventral deflection of the face characteristic of African apes and early hominids.

A consideration of some previous work on hominoid evolution reveals a strong implicit assumption that in general skull form the African apes are quite primitive, while the orang-utan is markedly specialized. This assumption resulted in part from (1) purported phyletic links uniting the early Miocene proconsuls and the African apes; (2) comparisons of Oligocene and recent hominoids always being made via the proconsuls; and (3) comparisons between hylobatids and other hominoids usually involving humans or the African apes. Recent morphological investigations and the discovery of new fossil hominoids have resulted in a serious questioning of this assumption. Based on the present evidence, I conclude that *Pongo* is primitive in its weak development of a supraorbital torus and dorsally angled face. The claim that the relative lack of a torus and an airorhynchous face is indeed primitive for the Hominoidea does not rule out the possibility that *Pongo* and its relatives might exhibit derived states for specific components of this morphological matrix, e.g., the morphology of the glabella region or a hyperflexed splanchnocranial position.

Combining these features with others analyzed elsewhere has several implications for hominoid phylogeny. First, there are a number of derived cranial features linking the African apes and early hominids, and these therefore provide evidence against Schwartz's (1984) hypothesis that the hominids, ramapiths, and orang-utan form a clade. Second, it also appears unlikely that the ramapiths are the sister group of hominids, as has been suggested by Kay and Simons (1983), because of the required parallelism in circumorbital and facial morphology. The features I analyzed in this contribution, in addition to other characters, do not argue against the linking of the ramapiths with the orang-utan. For several reasons, however, I currently favor the view argued by Greenfield (1980), Wolpoff (1982, 1983), and Leakey and Walker (1985) that all of the extant large-bodied hominoids derive from this ramapith group. This hypothesis allows specific predictions about the skull morphology of unknown or incompletely known hominoids. Hopefully, these predictions will soon be tested by the discovery of more fossil material from the African middle to late Miocene.

REFERENCES

Andrews, P. J. 1982. Hominoid evolution. *Nature, 295:*185–186.

Andrews, P. J. and Cronin, J. E. 1982. The relationship of *Sivapithecus* and *Ramapithecus* and the evolution of the orangutan. *Nature, 297:*541–546.

Andrews, P., Harrison, T., Martin, L. and Pickford, M. 1981. Hominoid primates from a new Miocene locality named Meswa Bridge in Kenya. *Journal of Human Evolution, 10:*123–128.

Andrews, P. G. and Tekkaya, I. 1980. A revision of the Turkish Miocene hominoid *Sivapithecus meteai. Paleontology, 23:*85–95.

Angst, R. 1967. Beitrag zum Formwandel des Craniums der Ponginen. *Zeitschrift für Morphologie und Anthropologie, 58:*109–151.

Biegert, J. 1957. Der Formwandel des Primatenschadels und seine Beziehungen zur ontogenetischen Entwicklung und den phylogenetischen Spezialisationen der Kopforgane. *Gegenbaurs Morphologische Jahrbuch, 98:*77–199.

Cave, A. J. E. and Haines, R. W. 1940. The paranasal sinuses of the anthropoid apes. *Journal of Anatomy, 72:*493–523.

Ciochon, R. L. 1983. Hominoid cladistics and the ancestry of modern apes. *In* New Interpretations of Ape and Human Ancestry, pp. 783–843, ed. R. L. Ciochon and R. S. Corruccini. Plenus Press, New York.

Clark, W. E. Le Gros and Leakey, L. S. B. 1951. Fossil mammals of Africa. *British Museum (Natural History), 1:*1–117.

Corruccini, R. S. and Henderson, A. M. 1978. Palato-facial comparison of *Dryopithecus (Proconsul)* with extant catarrhines. *Primates, 19:*35–44.

Cramer, D. L. 1977. Craniofacial morphology of *Pan paniscus. Contributions to Primatology, 10:*1–64.

Davis, P. R. and Napier, J. 1963. A reconstruction of the skull of *Proconsul africanus* (R.S. 51). *Folia Primatologica, 1:*20–28.

De Bonis, L. 1983. Phyletic relationships of Miocene hominoids and higher primate classification. *In* New Interpretations of Ape and Human Ancestry, pp. 625–649, ed. R. L. Ciochon and R. S. Corruccini. Plenum Press, New York.

Delattre, A. and Fenart, R. 1956. Analyse morphologique du splanchnocrane chez les primates et ses rapports avec le prognathisme. *Mammalia, 20:*169–323.

Delson, E. 1985. The earliest *Sivapithecus? Nature, 318:* 107–108.

Dubois, E. 1896. On *Pithecanthropus erectus:* a transitional form between man and the apes. *Scientific Transactions of the Royal Society of Dublin, 6:*1–18.

Ehara, A. 1972. Morphologische Analyse uber Variabilitat und Funktionell Bedeutung der Jochbugen Form bei Katarrhinen Primaten. *Zeitschrift für Morphologie und Anthropologie, 66:*83–94.

Enlow, D. H. 1982. Handbook of Facial Growth, 2nd edition. Saunders, Philadelphia.

Fleagle, J. G. and Kay, R. F. 1983. New interpretations of the phyletic position of the Oligocene hominoids. *In* New Interpretations of Ape and Human Ancestry, pp. 181–210, ed. R. L. Ciochon and R. S. Corruccini. Plenum Press, New York.

Goodman, M., Baba, M. L. and Darga, L. L. 1983. The bearing of molecular data on the cladogenesis and times of divergence of hominoid lineages. *In* New Interpretations of Ape and Human Ancestry, pp. 67–86, ed. R. L. Ciochon and R. S. Corruccini. Plenum Press, New York.

Gould, S. J. 1975. Allometry in primates, with emphasis on scaling and the evolution of the brain. *In* Approaches to Primate Paleobiology, pp. 244–292 (vol. 5, Contributions to Primatology), ed. F. S. Szalay. Karger, Basel.

Greaves, W. S. 1985. The mammalian postorbital bar as a torsion-resisting helical strut. *Journal of Zoology, London (A), 297:*125–136.

Greenfield, L. O. 1980. A late divergence hyopthesis. *American Journal of Physical Anthropology, 52:*351–365.

Hammond, A. L. (1983) Tales of an elusive ancestor. *Science '83, 4*(9)*:*36–43.

Herbert, W. 1984. The living link. *Science News, 125:*41.

Hofer, H. O. 1952. Der Gestaltwandel des Schädels der

Säugetiere und Vögel mit besonderer Berücksichtigung der Knickungstypen der Schädelbasis. *Verhandlungen Anatomischen Gesellschaft, 50:*102–113.

Hofer, H. O. 1965. Die morphologische Analyse des Schädels des Menschen. *In* Menschliche Abstammungslehre, pp. 145–226, ed. G. Heberer. G. Fischer, Stuttgart.

Kay, R. F. and Simons, E. L. 1980. The ecology of Oligocene African Anthropoidea. *International Journal of Primatology, 1:*21–37.

Kay, R. F. and Simons, E. L. 1983. A reassessment of the relationship between later Miocene and subsequent Hominoidea. *In* New Interpretations of Ape and Human Ancestry, pp. 577–624, ed. R. L. Ciochon and R. S. Corruccini. Plenum Press, New York.

Kimbel, W. H., White, T. D. and Johanson, D. C. 1984. Cranial morphology of *Australopithecus afarensis:* a comparative study based on a composite reconstruction of the adult skull. *American Journal of Physical Anthropology, 64:*337–388.

Kluge, A. G. 1983. Cladistics and the classification of the Great Apes. *In* New Interpretations of Ape and Human Ancestry, pp. 151–177, ed. R. L. Ciochon and R. S. Corruccini. Plenum Press, New York.

Kummer, B. 1952. Untersuchungen uber die Entstehung der Schädelbasis Form bei Mensch und Primaten. *Verhandlungen Anatomischen Gesellschaft, 50:*122–126.

Latimer, B. M., White, T. D., Kimbel, W. H. and Johanson, D. C. 1981. The pygmy chimpanzee is not a living missing link in human evolution. *Journal of Human Evolution, 10:*475–488.

Leakey, R. E. F. and Leakey, M. G. 1986a. A new Miocene hominoid from Kenya. *Nature, 324:*143–146.

Leakey, R. E. F. and Leakey, M. G. 1986b. A second new Miocene hominoid from Kenya. *Nature, 324:*146–148.

Leakey, R. E. F. and Walker, A. C. 1985. New higher primates from the early Miocene of Buluk, Kenya. *Nature, 318:*173–175.

Lewin, R. 1983. Is the orang-utan a living fossil? *Science, 222:*1222–1223.

Lipson, S. and Pilbeam, D. 1982. *Ramapithecus* and hominoid evolution. *Journal of Human Evolution, 11:*545–548.

Martin, L. and Andrews, P. J. 1982. New ideas on the relationships of the Miocene hominoids. *Primate Eye, 18:*4–7.

Martin, R. D. 1968. Towards a new definition of primates. *Man (N.S.), 3:*377–401.

McHenry, H. M., Andrews, P. and Corruccini, R. S. 1980. Miocene hominoid palatofacial morphology. *Folia Primatologica, 33:*241–252.

Moss, M. L. and Young, R. W. 1960. A functional approach to craniology. *American Journal of Physical Anthropology, 18:*281–292.

Moyers, R. E. and Bookstein, F. L. 1979. The inappropriateness of conventional cephalometrics. *American Journal of Orthodontics, 75:*599–617.

Moyers, R. E., Bookstein, F. L. and Guire, K. E. 1979. The concept of pattern in craniofacial growth. *American Journal of Orthodontics, 76:*136–148.

Olson, E. C. and Miller, R. L. 1958. Morphological Integration. University of Chicago Press, Chicago.

Pilbeam, D. R. 1969. Tertiary Pongidae of East Africa: Evolutionary Relationships and Taxonomy. *Peabody Museum of Natural History, Bulletin 31,* 185 pp.

Pilbeam, D. R. 1972. The Ascent of Man. Macmillan, New York.

Pilbeam, D. 1979. Recent finds and interpretations of Miocene hominoids. *Annual Reviews of Anthropology, 8:*333–352.

Pilbeam, D. 1982. New hominoid skull material from the Miocene of Pakistan. *Nature, 295:*232–234.

Preuss, T. M. 1982. The face of *Sivapithecus indicus:* description of a new relatively complete specimen from the Siwaliks of Pakistan. *Folia Primatologica, 38:*141–157.

Robinson, J. T. 1952. Note on the skull of *Proconsul africanus. American Journal of Physical Anthropology, 10:*7–12.

Rosenberger, A. L. 1980. Gradistic views and adaptive radiation of platyrrhine primates. *Zeitschrift für Morphologie und Anthropologie, 71:*157–163.

Russell, M. D. 1985. The supraorbital torus: "a most remarkable peculiarity." *Current Anthropology, 26:*337–360.

Schwartz, J. H. 1983. Palatine fenestrae, the orangutan and hominoid evolution. *Primates, 24:*231–240.

Schwartz, J. H. 1984. The evolutionary relationships of man and orangutans. *Nature, 308:*501–505.

Shea, B. T. 1983. Size and diet in the evolution of African ape craniodental form. *Folia Primatologica, 40:*32–68.

Shea, B. T. 1985. On aspects of skull form in African apes and orangutans, with implications for hominoid evolution. *American Journal of Physical Anthropology, 68:*329–342.

Shea, B. T. 1986. Skull form and the supraorbital torus in primates. *Current Anthropology, 27:*257–260.

Simons, E. L. 1972. Primate Evolution. Macmillan, New York.

Simons, E. L. 1984. Dawn ape of the Fayum. *Natural History, No. 5:*18–20.

Simons, E. L. and Pilbeam, D. R. 1965. Preliminary revision of the Dryopithecinae (Pongidae, Anthropoidea). *Folia Primatologica, 3:*81–152.

Smith, F. H. and Ranyard, G. C. 1980. Evolution of the supraorbital region in Upper Pleistocene fossil hominids from South-Central Europe. *American Journal of Physical Anthropology, 53:*589–609.

Starck, D. 1953. Morphologische Untersuchungen am Kopf der Saugetiere besonders Prosimier, ein Beitrag zum Problem des Formwandels des Saugerschadels. *Zeischrift für Wissenschaftliche Biologie, 157:*169–219.

Thenius, E. 1970. Zum Problem der Airorynchie des Saugetierschadels: ein Deutungsversuch. *Zoologischer Anzeiger, 185:*159–172.

Vogel, C. 1968. The phylogenetical evaluations of some characters and some morpological trends in the evolution of the skull in catarrhine primates. *In* Taxonomy and Phylogeny of Old World Primates with References to the Origin of Man, pp. 21–55, ed. B. Chiarelli. Rosenberg and Sellier, Turin.

Walker, A., Falk, D., Smith, R. and Pickford, M. 1983. The skull of *Proconsul africanus:* reconstruction and cranial capacity. *Nature, 305:*525–527.

Ward, S. C. and Kimbel, W. H. 1983. Subnasal alveolar morphology and the systematic position of *Sivapithecus*. *American Journal of Physical Anthropology, 61*:157–171.

Ward, S. C. and Pilbeam, W. H. 1983. Subnasal alveolar morphology of Miocene hominoids from Africa and Indo-Pakistan. *In* New Interpretations of Ape and Human Ancestry, pp. 211–238, ed. R. L. Ciochon and R. S. Corruccini. Plenum Press, New York.

Weidenreich, F. 1941. The brain and its role in the phylogenetic transformation of the human skull. *Transactions of the American Philosophical Society, 31*:321–442.

Wolpoff, M. H. 1980. Paleoanthropology. Knopf, New York.

Wolpoff, M. H. 1982. *Ramapithecus* and hominid origins. *Current Anthropology, 23*:501–522.

Wolpoff, M. H. 1983. *Ramapithecus* and human origins: an anthropologist's perspective of changing interpretations. *In* New Perspectives on Ape and Human Ancestry, pp. 651–676, ed. R. L. Ciochon and R. S. Corruccini. Plenum Press, New York.

Wu, R., Han, D., Xu, Q., Lu, Q., Pan, Y., Zhang, X., Zheng, L. and Xiao, M. 1981. *Ramapithecus* skulls found first time in the world. *Kexue Tongbao, 26*:1018–1021.

Wu, R., Han, D., Xu, Q., Qui, G., Lu, Q., Pan, Y. and Chen, W. 1982. More *Ramapithecus* skulls from Lufeng, Yunnan—Report on the excavation of the site in 1981. *Acta Anthropological Sinica, 1*:101–108.

Wu, R. and Oxnard, C. E. 1983. *Ramapithecus* and *Sivapithecus* from China: some implications for higher primate evolution. *American Journal of Primatology, 5*:303–344.

Wu, R., Xu, Q., and Lu, Q. 1983. Morphological features of *Ramapithecus* and *Sivapithecus* and their phylogenetic relationships. *Acta Anthropological Sinica, 2*:1–14.

Xu, Q. and Lu, Q. 1980. The Lufeng skull and its significance. *China Reconstructs, 29*:56–57.

Zapfe, H. 1958. The skeleton of *Pliopithecus (Epipliopithecus vindobonensis)* Zapfe and Hürzeler. *American Journal of Physical Anthropology, 16*:441–458.

Zapfe, H. 1960. Die Primatenfunde aus der Miozanen Spaltenfullung von Neudorf an der March (Devinska Nova Ves), Tschechos lowakei. Mit Anhang: Der Primatenfund aus dem Miozan von Klein Hadersdorf in Niederosterreich. *Schweizerische Paleontogische Abhandlung, 78*:1–293.

Zingeser, M. R. 1973. Dentition of *Brachyteles arachnoides* with reference to alouattine and ateline affinities. *Folia Primatologica, 20*:351–390.

18

Basicranial and Facial Topography in Pongo *and* Sivapithecus

BARBARA BROWN AND STEVEN C. WARD*

The evolutionary history of the orang-utan is presently far better documented in the fossil record than is that of the chimpanzee and gorilla. Although there is a recently described maxilla from the later Miocene of Kenya that seems to show affinities to the gorilla (Ishida et al., 1984), fossil evidence for the ancestry of the African apes is, in reality, nonexistent. In contrast, fossil and subfossil teeth attributable to extinct species of *Pongo* have been collected in Asia for over 100 years. Recently, a consensus has emerged within the paleoanthropological community that the later Miocene hominoid genus *Sivapithecus* is likely to be a cladistic relative of the orang-utan. This view is by no means universally held (Kay, 1982; Kay and Simons, 1983), but a series of recent reports have marshaled considerable evidence in its support.

If *Sivapithecus* does prove to be cladistically related to *Pongo,* it would indicate that the orang-utan clade is at least 12 million years old. This is the oldest known appearance of *Sivapithecus* in the Siwaliks of Indo-Pakistan (Raza et al., 1983). At the present time there is no unequivocal evidence of *Sivapithecus* in the lower Miocene of East Africa or Asia Minor, despite the attribution to this taxon of part of the Buluk sample from northern Kenya (Leakey and Walker, 1985) and the Pasalar dental collection from Turkey (Andrews and Tobien, 1977). However, these samples do not preserve what are viewed as the most reliable synapomorphies linking *Sivapithecus* and *Pongo:* the configuration of the premaxilla and hard palate in the vicinity of the incisive canals, the width of the interorbital septum, and the shape of the orbits. These features have received considerable attention in recent attempts to reconstruct hominoid phylogeny (Andrews, 1985; Lipson, 1985; Pickford, 1985; Pilbeam, 1985, 1986; Ward and Kimbel, 1983). Other characters, especially enamel thickness and prism structure, once thought to be fairly unambiguous traits, are now known to represent the culmination of complex patterns of development. Accordingly, there are complexities in interpreting their polarities (Martin, 1983, 1985).

Another character, upward flexion of the face with respect to the cranial base, or airorhynchy, is peculiar to *Pongo* among living hominoids. Until recently it has not been possible to determine facial-basicranial relationships in Miocene hominoids due to the lack of requisite fossils. This deficiency has been alleviated to some extent by the addition of newly found parts of the *Proconsul africanus* cranium (KNM RU 7290) from Rusinga Island in Kenya, a series of hominoid crania from the Lufeng locality in Yunnan Province, China, and a relatively complete face (GSP 15000) from a locality near Kaulial Village, on the Potwar Plateau in northern Pakistan. These specimens are few in number, span vast distances in time and space, and all exhibit varying degrees of postmortem distortion and/or missing anatomy. Nevertheless, they represent the only direct evidence of basicranial topography in Miocene hominoids. In a recent review of the facial anatomy of *Sivapithecus indicus* (Ward and Brown, 1986), we discussed the possibility of identifying cranial base and facial relationships in the GSP 15000 specimen following reconstruction of the upper face, zygomaticomaxillary region, and repositioning of the temporomandibular joint. We have since completed a preliminary reconstruction of the specimen with emphasis on these regions and will here discuss the basicranial and facial relationships that can be inferred from its morphology as well as other relevant fossils from Pakistan and India. We will also compare these relationships with patterns present in extant Bornean and Sumatran orang-utans. Finally, we will conclude with an assessment of the phylogenetic relationships of later Miocene hominoids from localities in Greece, Hungary, Turkey, Indo-

*The sequence of authors is alphabetical.

Pakistan and China, with the intention of contributing to a better understanding of the early evolutionary history of *Pongo.*

BASICRANIAL AND FACIAL TOPOGRAPHY IN *PONGO*

Upward flexion (airorhynchy) of the face in *Pongo* has been the subject of extensive analysis and comment (Angst, 1967; Biegert, 1983; Krogman, 1931; Shea, 1985). Among primates, only orang-utans, howler monkeys, and the subfossil Malagasy prosimian *Megaladapis* are known to exhibit pronounced airorhynchy (Tattersall, 1972). It has been claimed by Delattre and Fenart (1956) that living hylobatids are also airorhynchous, but this has been neither supported nor refuted by subsequent observations. The subject of hominoid skull form, with special emphasis on facial position relative to the cranial base, has benefited from a recent detailed review by Shea (1985), who concluded that the unique elements of *Pongo* cranial form (e.g., attenuated supraorbital contours, tall and narrow orbits with narrow interorbital distance, concave midface, absence of true ethmofrontal sinuses, small sphenoid sinuses, and nontruncated premaxilla associated with minimal subpremaxillary deflection of the hard palate and small incisive canals) are the result of changes in facial position rather than a fundamental rearrangement of the large-bodied hominoid cranial *Bauplan.* It follows from Shea's argument that airorhynchy, at least as it is manifested in *Pongo,* is the proximate mechanism underlying these features in both the orang-utan and its cladistic relatives. Of particular interest is Shea's prediction that upward flexion of the face with respect to the cranial base should be associated with reduction or loss of ethmoidally derived frontal sinuses. The presence of these structures in the African apes and all known hominids, which accompanies ventral facial flexion in these primates, would seem to support Shea's conclusions. Finally, given the distribution of airorhynchy and its presumed suite of morphological correlates, Shea concluded that dorsal flexion of the face is quite likely to be primitive for hominoids and that the arrangement presented by *Pan, Gorilla,* and hominids represents a derived configuration. This view is based largely on a presence/absence comparison of fossil forms. It applies the operational criterion that lack of supraorbital tori in hominoids should be associated with some form of airorhynchy. We think that this interpretation is unlikely on several counts, which we discuss below.

Correlates of airorhynchy in *Pongo* are well documented. They include position of the temporal lobes of the brain, arrangement and size of the paranasal sinuses, and orientation of the nasal and oral cavities, pharynx, suprahyoid region, and possibly the posterior position of the foramen magnum. Comparative basicranial geometries in *Pan, Pongo,* and two early hominids (AL 58-22, a partial face and cranium of *Australopithecus afarensis* from the Hadar Formation, and KNM CH-1, a partial cranium of *Australopithecus boisei* from Chesowanja in north central Kenya) are shown in Figure 18–1. Each specimen is aligned along the Frankfort Horizontal. In this orientation, upward rotation of the palate and alveolar plane is quite apparent in *Pongo,* with adult males displaying a greater degree of airorhynchy than females. Both male and female chimpanzees show the expected downward rotation of the facial region. Both fossil hominids show the same pattern, with downwardly rotated faces and large frontal sinuses.

Lateral radiographs also show that the position of the cribriform plate of the ethmoid, which is the most anterior part of the chondrocranial portion of the cranial base, varies considerably among hominoids. In orang-utans, the cribriform plate is located in a lower vertical position than in both chimpanzees and early hominids. By using the Frankfort Horizontal as a reference plane, it is apparent that the cribriform plate in orang-utans is situated either at the level of the Frankfort Horizontal or just slightly above it. It is always well above the Frankfort Horizontal in chimpanzees and fossil hominids. Both Shea (1985) and Montagu (1943) have discussed variations in the anterior cranial base of hominoids. Montagu described the presphenoid region in orang-utans as being "stenosed." It is reasonable to impute the constriction of the ethmoid complex, as well as the ventrally positioned cribriform plate, to the airorhynchous condition, especially with respect to its effect on orbital shape and position (see below).

In coronal section, differences in facial-cranial relationships are also evident. Orang-utans have a vertically deep oral cavity. Its vertical dimension is due primarily to the deep mandibular corpus, which is characteristic of both male and female orang-utans. The nasal cavity is tall, but it is also transversely broad. In chimpanzees, as well as in gorillas, the nasal and paranasal spaces are relatively taller than in orang-utans, but these spaces are transversely narrow. Also in the African apes, the lateral walls of the maxilla are

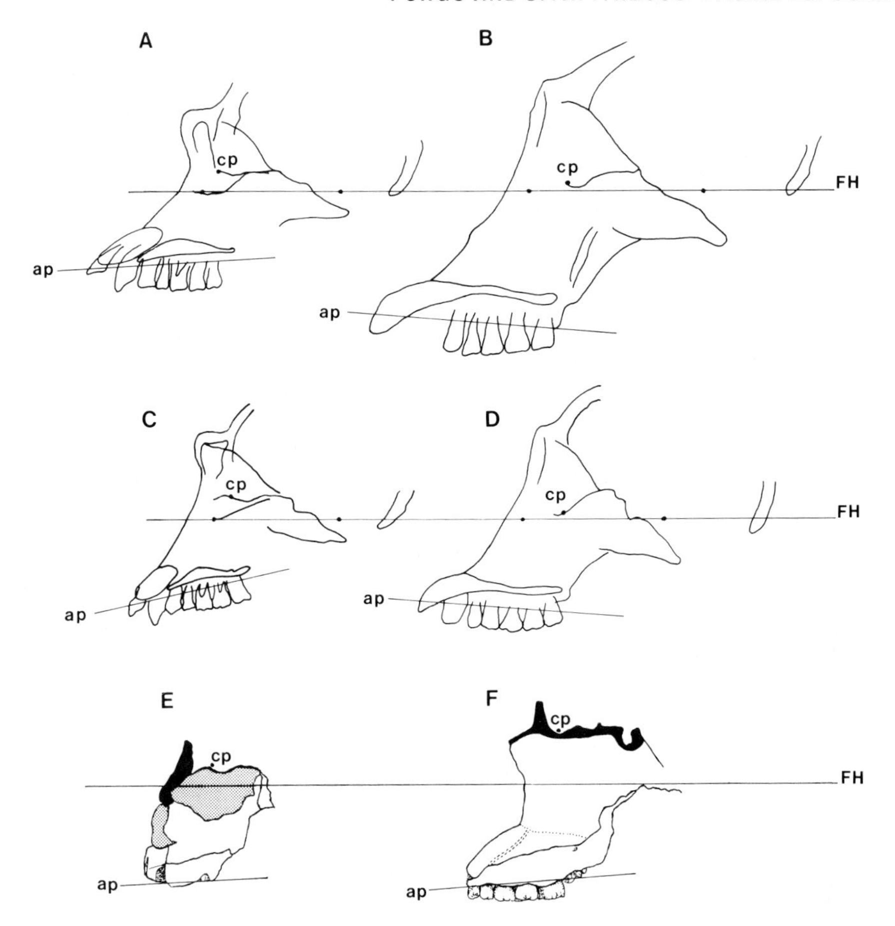

FIG. 18–1. Basicranial and facial relationships. (A) Male *Pan troglodytes;* (B) male *Pongo pygmaeus;* (C) female *Pan troglodytes;* (D) female *Pongo pygmaeus;* (E) *Australopithecus afarensis* (AL 58-22); (F) *A. boisei* (KNM BC-1). All specimens are oriented in the Frankfort Horizontal (FH). Other abbreviations: cp, cribriform plate; ap, alveolar plane. In the chimpanzee, the cribriform plate is situated well above the FH, while in orang-utans, it is located on, or quite close to the level of the FH. The two early hominids have a cribriform plate placed well above the FH. Note also the upward flexion of the alveolar plane in *Pongo* and its caudal deflection in the chimpanzee and early hominids.

vertically disposed, and the maxillary sinuses are narrow when compared to *Pongo* (Fig. 18–2).

There are also several soft tissue correlates of the airorhynchous condition as it is manifested in *Pongo.* These involve the arrangement of the muscles in the oral cavity floor, muscle attachment patterns on the hyoid bone, and the disposition of the laryngeal air sacs.

The orang-utan possesses a strongly developed platysma muscle. It extends laterally over much of the intrinsic facial musculature and posteriorly over the nuchal region. The platysma maintains a very strong attachment along the mandibular base, beginning at the mandibular symphysis and extending posteriorly to the vicinity of the masseter. This pattern is also present in chimpanzees and gorillas. In large male orang-utans, a massive basal swelling on the mandibular corpus is the site of an extensive attachment for the platysma. Other rugosities and platysmatic striae are often associated with this muscle. Immediately beneath the submandibular portion of the platysma, where the muscle is thickest, is the anterior extension of the laryngeal air sac. These diverticulae emerge beneath the greater hyoidal horns and converge in the midline (Fig. 18–3). All great apes possess these structures, as do siamangs, howler monkeys, and several other New World primates. Male orang-utans are reported to have the most extensive air sac systems, followed by male gorillas. These structures can be extremely invasive, extending into the axillae below, and upwards and dorsally onto the nuchal musculature (Brandes, 1932; Miller, 1941; Sonntag, 1924a,b).

We dissected the laryngeal air sacs of three orang-utans: an adult male, an adult female, and a juvenile male whose first molar was completely

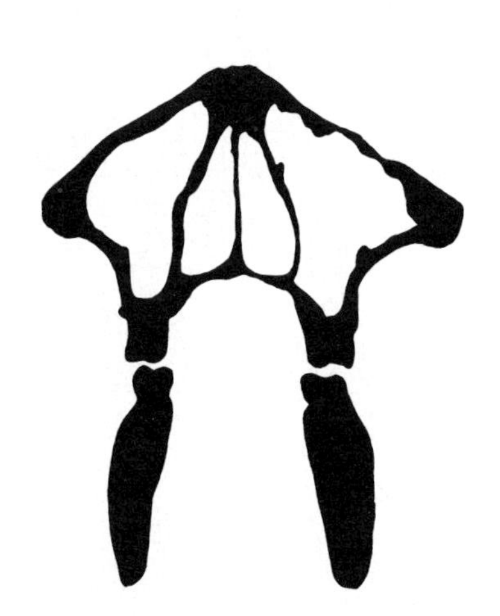

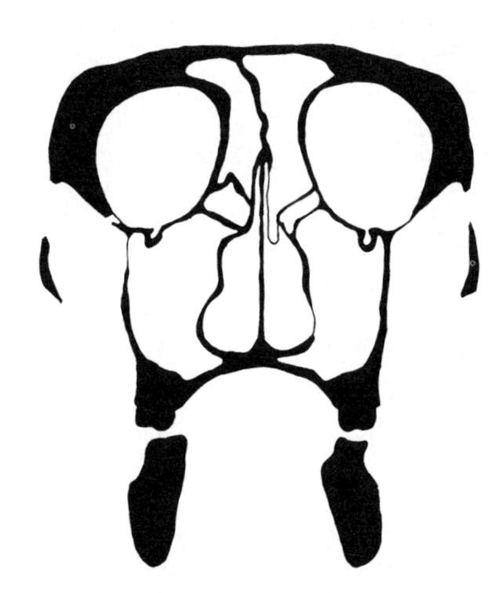

FIG. 18–2. Coronal sections through the second molars in an adult male orang-utan (A) and an adult male chimpanzee (B).

erupted. In both adult specimens, the submandibular extension of the air sac system was in contact with the base of the mandibular symphysis, while in the juvenile, it had grown anteriorly to a point approximately three quarters of the distance between the hyoid body and the inferior transverse torus of the mandible. Although there is no direct evidence to support a functional assessment, it may be that the massive and strongly attached platysma in *Pongo* in some way helps to regulate air sac volume and intra-sac pressures during vocalization and even, perhaps, during locomotion.

Beneath the submental portion of the laryngeal air sac lie the syprahyoid muscles as well as the muscular floor of the oral cavity. The most striking difference between *Pongo* and other hominoids in this region is the complete lack of anterior digastric muscles (Chapman, 1880; Sonntag, 1924a,b). Loss of the anterior diagastric in orang-utans has been accompanied by a complete separation of the posterior component of the muscle from both the hyoid bone and the stylohyoid muscle. The posterior diagastric inserts into the inner aspect of the mandibular angle and would be expected to depress the mandible upon contraction.

The first muscle observed upon removal of the laryngeal air sac in *Pongo* is the mylohyoid muscle (Fig. 18–4). It is astonishingly robust compared to the mylohyoid of African apes. The muscle retains the anteriorly downward origin on the internal surface of the mandibular corpus typical of hominoids, with both mylohyoid muscles meeting in the midline. There is no obvious raphe as the fibers from both sides decussate. Anteriorly, the mylohyoid does not attach to the mandibular symphysis but terminates a centimeter or so short of the midline. As a result, a diamond-shaped defect is present immediately posterior to the inferior transverse torus of the

FIG. 18–3. Submandibular topography of the laryngeal air sac system of an adult male orang-utan. Abbreviations: Sm, mental symphysis; Py, platysma muscle; Itt, inferior transverse torus of the mandible; smd, submental diverticulum of the laryngeal sac; Ms, masseter muscle; sms, submandibular portion of the laryngeal sac; lsms, lateral extension of the submandibular sac. Arrows indicate the position of the additus of the laryngeal ventricle on each side.

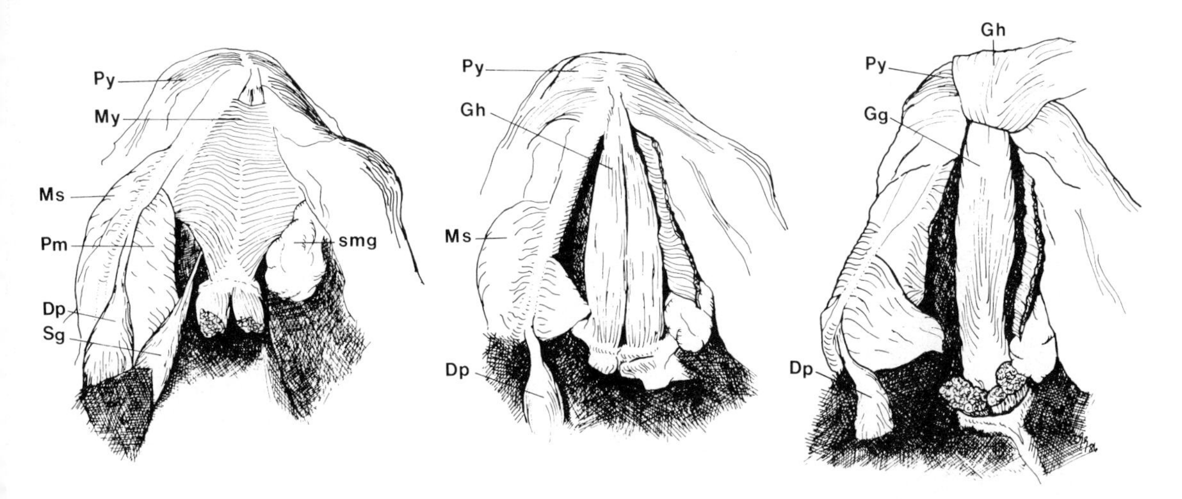

FIG. 18–4. Muscles in the floor of the oral cavity of an adult female orang-utan. Abbreviations: Py, platysma; My, mylohyoid; Ms, masseter; Pm, medial pterygoid; Dp, posterior digastric; Sg, styloglossus; smg, submandibular gland; Gh, geniohyoid; Gg, genioglossus. There is complete absence of the anterior belly of the digastric muscle in orang-utans.

symphysis. In our three specimens, there were variable connections between fibers of the mylohyoid and the underlying geniohyoid muscle.

The geniohyoids are uncomplicated, albeit very strongly developed in the orang-utan. They are attached to the inferior tranverse torus of the mandible by means of thick tendons. In some specimens, a thin bony septum develops in the midline between them. The geniohyoids insert by means of fleshy fibers onto the anterior surface of the hyoid body. The genioglossus muscles arise on either side of the midline from the symphyseal fossa and develop a series of posterior and superior curves which pass into the substance of the tongue. The length of the genioglossus exceeds the length of the oral cavity, so that the posterior extremity of the muscle curves sharply over the hyoid body, forming a pair of bilateral bulbs. A pair of reciprocal depressions for these bulbs occurs on the cranial surface of the hyoid body. A similar arrangement characterizes chimpanzees and gorillas.

Biegert (1963) has suggested that hypertrophy of the laryngeal sac system in orang-utans is a functional determinant of their pronounced airorhynchy. In his review of comparative skull form in hominoids, however, Shea (1985) is more cautious in linking air sacs with upward rotation of the face. We agree with Shea that it is difficult to identify a specific ontogenetic connection between the two. The howler monkey is often cited as exhibiting parallel patterns of hyolaryngeal enlargement and airorhynchy. Usually overlooked in this comparison is the fact that *Alouatta* possesses a highly modified hyoid bone. The hyoid body has been transformed into a bell-like resonating chamber, a modification that has not occurred in *Pongo. Alouatta* has a topographic frontal sinus, unlike the orang-utan, and it also possesses an anterior digastric muscle with a concomitant linkage to the hyoid bone (Schon, 1968). However, in howler monkeys, the anterior diagastric insertion is in a posterior position on the mandibular base, rather than in the symphyseal region as in most other anthropoids. Whether or not this configuration is a response to hyolaryngeal enlargement is not known. We must therefore conclude that the issue of laryngeal sac enlargment and airorhynchy is, in essence, fundamentally teleological. Does an enlarging sac system encroach upon the floor of the oral cavity, "forcing" the face upwards during development? Or does an upwardly growing face mandate the presence of a space-occupying structure beneath it as it retreats upwards? There are probably other alternatives as well, and the question is not likely to be resolved without longitudinal growth studies on a diverse array of primates with laryngeal sac systems.

Despite variations among living hominoids in basicranial flexion, facial hafting patterns, and laryngeal air sac development, there is little concomitant variation in the position of the hyoid bone. In our orang-utan cadaver series, the hyoid body was situated at a level slightly above the mandibular base and in the region of the gonial angle. There are no enlargements or other specializations in the orang-utan hyoid body that reflect the presence of expanded air sacs. In fact, the hyoid of female orang-utans is commonly more gracile than that of female chimpanzees. From this it may be concluded that the functional analogy drawn for *Pongo* and *Alouatta* is not an analogy at all. Although both forms are airorhynchous primates, their basic hyolaryngeal patterns are quite different. While Biegert's (1963) contentions concerning the general effects of hyolaryngeal enlargement and facial and cranial base relationships may well be correct, it is more than likely that different patterns of postnatal cranial growth and development are operative in *Alouatta* and *Pongo.*

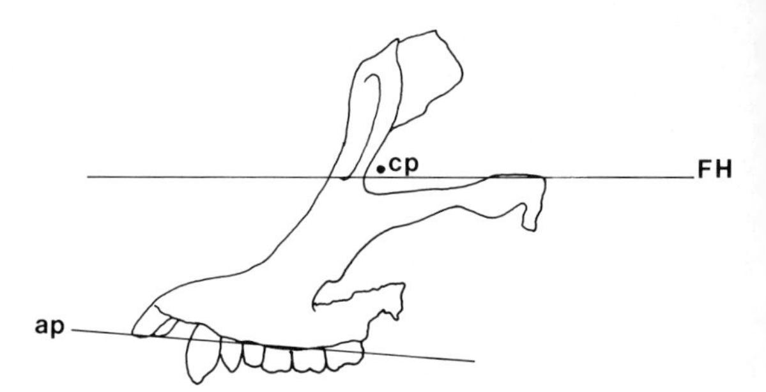

FIG. 18–5. GSP 15000 in lateral projection, and oriented in the Frankfort Horizontal (FH). The estimated position of the cribriform plate (cp) lies close to the FH, and the alveolar plane (ap) is deflected superiorly. Both topographic patterns are homologous to those described for *Pongo.*

CEPHALOFACIAL TOPOGRAPHY IN *SIVAPITHECUS*

WIth the recovery and analysis of a relatively complete face of *Sivapithecus indicus* near Kaulial Village on the Potwar Plateau in northern Pakistan, it is now possible to identify some important elements of cephalofacial topography in this later Miocene hominoid (Pilbeam, 1982; Preuss, 1982; Ward and Brown, 1986). This specimen (GSP 15000) consists of most of the face, minus the right zygoma and zygomatic arch. Since our recently completed analysis of basic facial anatomy (Ward and Brown, 1986), we have continued reconstructing the badly distorted left zygomatic root area and the position of the left temporomandibular joint. Two parts of the cranial base, the articular part of the temporal bone and the vicinity of the cribriform plate, are now situated in approximately correct positions, as estimated by the position of the frontal crest on the endocranial surface of the frontal bone. The temporal fragment is sufficiently well preserved to estimate the position of porion, and the orbit is sufficiently well preserved to identify orbitale. It is therefore possible to identify with reasonable accuracy the Frankfort Horizontal (Fig. 18–5). While we would prefer to have additional midline cranial base landmarks, application of the Frankfort Horizontal allows us to make direct comparisons with living great apes as well as a number of other fossil hominoids.

In frontal view, the reconstructed face of GSP 15000 appears somewhat narrower than the face of a comparably sized female orang-utan (Fig. 18–6). It is in some respects similar to the full-face reconstruction of *Proconsul major* (based on UMP 62-11) published by Pilbeam in 1969. The upper face of GSP 15000 has a narrow interorbital septum and tall, narrow orbits otherwise typical of *Pongo.* Other apparently unique similarities that the Kaulial face shares with orang-utans are the pattern of temporal line relationships to the supraorbital rims, the form of the anterior part of the temporal fossa, a downward and medially inclined frontozygomatic suture, supraorbital rims that are confluent with the superior orbital margins, lack of any supraorbital tori, absence of a posttoral sulcus, and absence of ethmofrontal sinuses (Kelley and Pilbeam, 1986; Pilbeam, 1982; Preuss, 1982; Ward and Brown, 1986). GSP 15000 differs from *Pongo* primarily in the degree of general facial robustic-

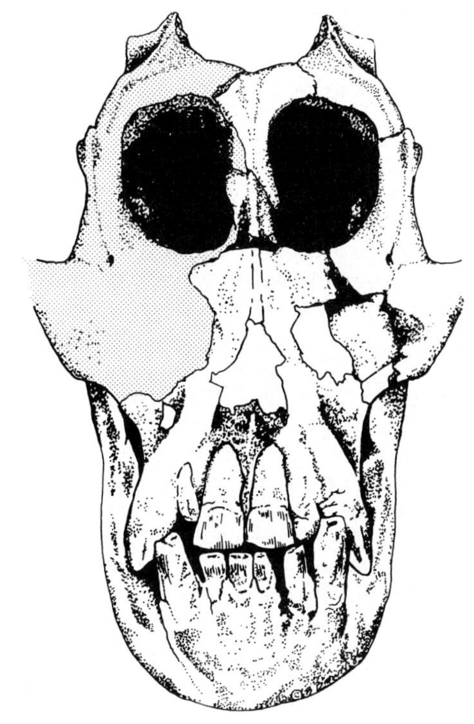

FIG. 18–6. GSP 15000, facial view. Reconstructed area is shaded.

ity. In most comparisons, the individual facial bones of the Kaulial face are more lightly constructed than in adult female orang-utans of comparative size.

In lateral view, the upper face of GSP 15000 is dominated by a marked midfacial concavity. This is associated with upward flexion of the face, as determined by the angle of the hard palate and alveolar planes with respect to the Frankfort Horizontal. Probable topographic correlates of airorhynchy in GSP 15000 include an upwardly and posteriorly oblique zygomatic arch and the position of the cribriform plate. One can conclude from the nature of the fractured medial orbital wall, the position of the frontal crest, and the configuration of the nasal attachment for the perpendicular plate of the ethmoid that the cribriform plate of GSP 15000 was situated in a low position between the orbits. This, in turn, suggests the development of presphenoidal pinching or stenosis. When projected onto the Frankfort Horizontal GSP 15000 displays the pattern present in *Pongo* in which the level of the cribriform plate falls approximately on the Frankfort Horizontal. As noted previously, the position of the cribriform plate in *Pan, Gorilla,* and the hominids *A. afarensis* and *A. boisei* is well above the level of the Frankfort Horizontal.

It is clear from these observations that *Sivapithecus indicus* shares with *Pongo* a pattern of craniofacial union that is unique among living large-bodied hominoids. However, there are also some notable differences between the two forms. These differences are primarily confined to the lower parts of the face, particularly the oral cavity. From the large sample of *Sivapithecus* mandibles recovered from India and Pakistan since the early part of this century, it can be determined that all species of this genus possessed anterior digastric muscles. Distinct digastric impressions are present on the basal surface of the inferior transverse tori of all specimens where the symphysis is preserved (Fig. 18–7). From the state of these impressions, it appears that the anterior digastric was a thin fan-like sheet of muscle, perhaps similar to that of the chimpanzee. This is the same pattern present in all species of *Proconsul,* regardless of size. We can therefore reasonably assume that, in these hominoids, the anterior digastric maintained the typical complex linkage with the hyoid bone that is characteristic of all other hominoids with the exception of *Pongo.* Interestingly, the earliest undoubted hominid, *A. afarensis,* possessed a robust anterior digastric quite different from the pattern present in *Sivapithecus* and other large-bodied apes. In the Hadar and Laetoli hominid mandibles, the anterior digastrics leave either deeply excavated slots or ovoid foveae on each side of the symphyseal midline. This is quite similar to the pattern of anterior digastric insertion in *Homo* and suggests changes in the functional role of this muscle.

Confusion about the presence or absence of the anterior digastric in *Sivapithecus* has resulted in some misinterpretation of symphyseal morphology. Brown et al. (1924) incorrectly identified the digastric fossae in AMNH 19411, presently recognized by Kay (1982) as *S. sivalensis.* What they labeled as the digastric fossae are actually facets for the geniohyoid muscles. This mandible does, however, have flat, crescentic digastric impressions in a more basal position, lateral to the geniohyoid facets.

Loss of the anterior digastric muscle in *Pongo,* with associated changes in the suprahyoid muscle complex and attachment of the posterior digastric to the mandibular angle, must be viewed as a significant change in craniofacial structure. The functional and topographic mechanisms underlying this change are at present not clear. If a major restructuring of the digastric complex is associated only with the development of airorhynchy, then we would expect that *Sivapithecus* and *Alouatta* would show evidence of similar changes. Another possibility is that development of a large submandibular diverticulum of the laryngeal sac system renders the standard suprahyoid muscle pattern topographically untenable. Male gorillas, however, have very large air sacs, yet they also have anterior digastrics linked to the hyoid. The key here may involve the development and size of the submandibular extension of the sac. Gorillas, as well as chimpanzees, have submandibular diverticulae that displace the submandibular salivary glands and stretch the overlying platysma (Miller, 1941) but neither African ape appears to develop as extensive a sac between the base of the mandibular symphysis and the hyoid body as does the orang-utan.

Another structural feature that may account for differences between *Sivapithecus* and *Pongo* in their suprahyoid spaces and submandibular regions is the shape of the mandibular symphysis. The inferior transverse torus in *Pongo* is vertically thin, and extends posteriorly much farther than does the vertically thick and bulbous torus in *Sivapithecus.* In addition, the symphysis is more obliquely aligned in orang-utans. The mandibular corpus is deeper and narrower in *Pongo* than it is in *Sivapithecus* and it is not as strongly buttressed externally.

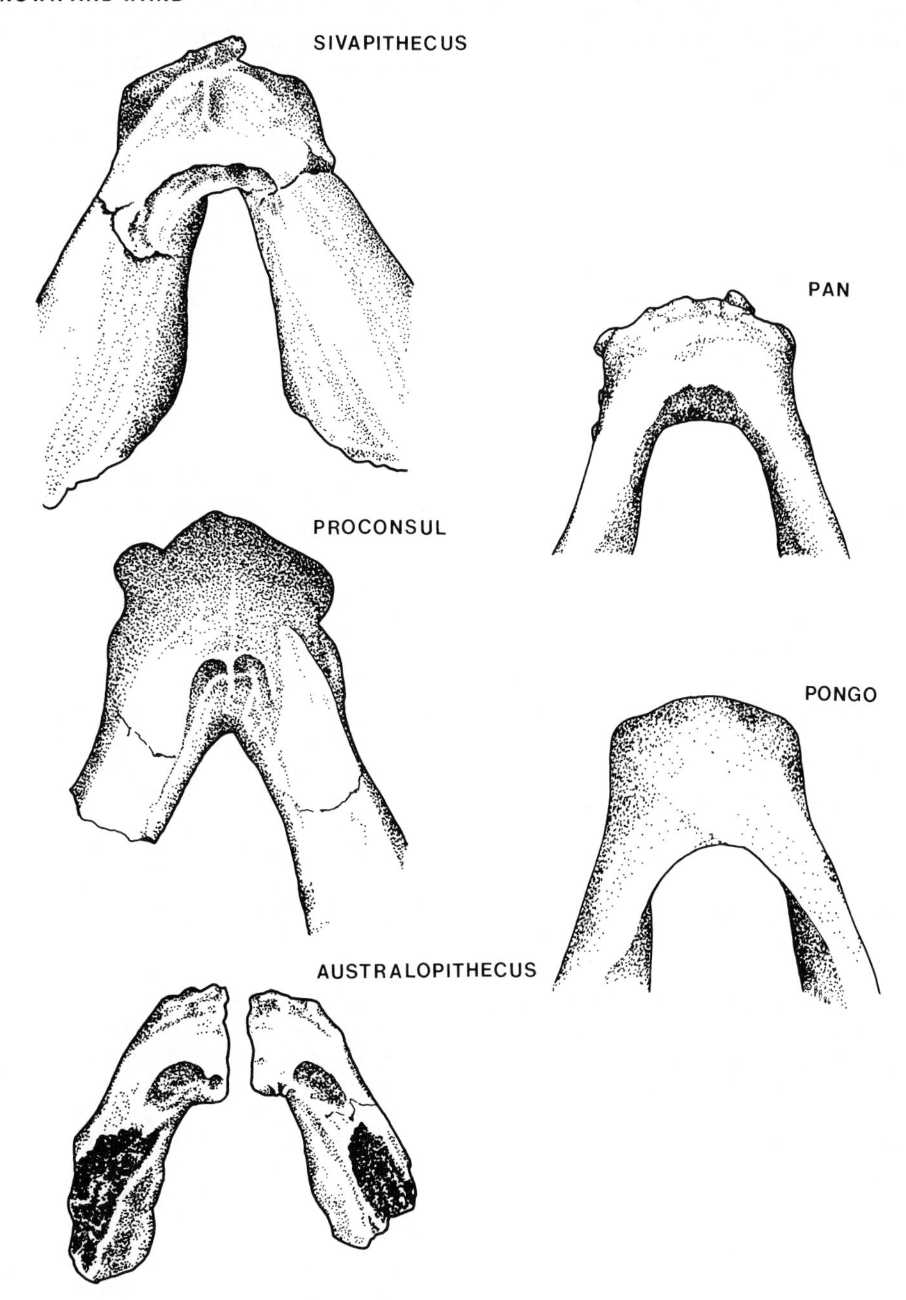

FIG. 18–7. Hominoid basal mandibular anatomy. *Sivapithecus* (GSP 9564), *Proconsul* (KNM-SO 396), *Australopithecus* (AL 400-1 on the left and AL 277-1 on the right), *Pan*, and *Pongo*. With the exception of the orang-utan, all show evidence of digastric impressions or foveae.

Other differences between the two genera extend beyond the floor of the oral region to the dentition. Such morphological differences extend to all tooth classes (Kelley and Pilbeam, 1986) but are particularly pronounced in the molars. The distinctive wrinkling of occlusal enamel in orang-utans, coupled with attenuation of cusp relief, is not expressed to the same extent in *Sivapithecus*. Unworn permanent molars of *S. indicus* do have crenulated occlusal surfaces, but the pattern of crenulation, as well as the reduction of cusp relief, appears to differ from the condition present in orang-utans.

From these observations it is clear that the most compelling similarities between *Pongo* and *Sivapithecus* are to be found in the upper and middle parts of the face and in the mode of craniofacial union. Below the level of the hard palate, the two forms are rather different. Despite a complex of anatomical features that are reasonably interpreted as synapomorphies which identify orang-utans and later Miocene ramamorph hominoids as members of a unified clade, *Sivapithecus* was not a Miocene orang-utan. The development of airorhynchy in the *Sivapithecus-Pongo* clade is probably a derived condition, and many of the other features in the upper and middle parts of the face discussed above are probably structural correlates of airorhynchy. Both of these conclusions have implications in hominoid as well as hominid phylogeny reconstruction.

BIOGEOGRAPHY AND TAXONOMIC STATUS OF MIOCENE HOMINOIDS

An important issue now confronting hominoid paleoprimatology is identifying the last common ancestor of the *Sivapithecus-Pongo* clade and the large-bodied "western" hominoids. A major step in that direction would be the discovery of an early species of *Sivapithecus* in east African deposits prior to early to middle Chinji times in the Siwaliks. Leakey and Walker (1985) claim to have made this important discovery at the 17-my (million years) Buluk locality in northern Kenya. The more prominent specimens are KNM WS 124, a right mandibular corpus, KNM WS 11599, the anterior portion of an associated mandible and left maxilla, and KNM WS 12606, an isolated lower left third molar. There are numerous features preserved that conclusively preclude assignment of the large Buluk hominoids to *Proconsul.* However, Delson (1985) has criticized assignment of the specimens to *Sivapithecus* on the grounds that sufficient diagnostic features are also not preserved, particularly the subnasal region, the orbits, and the interorbital septum. Delson further noted that greater consideration could have been given to assigning the Buluk hominoids to *Kenyapithecus,* but since not all specimens attributable to that taxon have been described, this course of action would seem to be precluded.

Assignment of KNM WS 124 to *Sivapithecus* would be supported on the basis of its lateral corpus morphology which is similar to the pattern described for *S. indicus* mandibles from the Potwar Plateau. We have recently completed detailed stereophotogrammetry of lateral corpus surfaces. Our preliminary analyses suggest differences in the ways the corpus and ramus portions of the mandible are joined. At the present time, we prefer to consider the Buluk hominoids as either an early form of *Kenyapithecus* or a new genus entirely. This issue will only be clarified when a complete premaxilla and upper face are recovered from Buluk or from another locality of equivalent age and community structure.

The taxonomic problems posed by the Buluk sample may well prove to have been at least partially resolved by the recent discovery of new early Miocene hominoid cranial material from sediments west of Lake Turkana (Leakey and Leakey, 1986a,b). Two new genera have been identified, *Afropithecus turkanensis* and *Turkanapithecus kalakolensis.* The sediments from which they were recovered are presently reported to be between 16 and 18 my old. *Afropithecus,* which is the larger of the two hominoids, shows striking similarities in its mandibular corpus and maxillary morphology to the larger Buluk specimens, prompting Leakey and Leakey (1986a) to suggest that the Buluk material previously attributed to *Sivapithecus* (Leakey and Walker, 1985) should now be attributed to *Afropithecus.* The hypodigm of *Afropithecus* includes a relatively complete cranium that shows evidence of upward facial flexion. Leakey and Leakey (1986a) report that the specimen, KNM WK 16999, has a long face, wide interorbital distance, and a steeply inclined frontal. They also note the presence of a frontal sinus. From outward appearances, it seems possible that KNM WK 16999 has an upwardly flexed face, although further work is required to determine the configuration of the anterior cranial base and its topographic relationships to the elongated face. However, if *Afropithecus* ultimately proves to have an airorhynchous craniofacial pattern, then Shea's proposal that dorsal facial flexion is a primitive hominoid character is strengthened.

Another important series of hominoid fossils destined to have a major impact on hominoid phylogeny reconstruction is the recently expanded *Kenyapithecus* collection. Pickford (1985, 1986) believes that two species are represented in current collections. Specimens have been recovered from a variety of localities in western Kenya and more recently in the region around Nachola in north central Kenya. From descriptions provided by Pickford (1982, 1985, 1986) and Ishida and his colleagues (1984), it seems evident that *Kenyapithecus* is not likely to be a common ancestor to both the *Sivapithecus-Pongo* clade and the African pongid-hominid lineage. It appears that *Kenyapithecus* has a subnasal pattern intermediate between that of *Proconsul* and *Australopithecus afarensis:* the palate is closer to the posterior pole of the nasoalveolar clivus than in *Proconsul* and the premaxilla is obliquely oriented in section, much as in *A. afarensis.* The canines are size dimorphic and externally rotated. Therefore, *Kenyapithecus* is the earliest African hominoid with this canine implantation pattern. The molars appear to be capped with thick enamel. A canine fossa is prominent behind the canine pillar and the zygomatic root is in a high position on the maxilla. In the authors' current view of hominoid phylogeny, this pattern of maxillofacial anatomy is not inconsistent with the expected cephalofacial organization of the last common ancestor of the African ape clade and the earliest hominid species.

Outside of Africa, middle to later Miocene hominoid systematics presents a set of complex problems. Some issues are closer to resolution, other remain intractable, and new problems are present on the horizon.

Kelley and Pilbeam (1986) have recently reviewed the morphology and geographic distribution of large-bodied Miocene hominoids and have done much to clarify an unnecessarily complicated taxonomic structure. They recognize two species of *Dryopithecus* in Europe, *D. laetanus* and *D. fontani* from Spain, France, Austria, several other localities in western Europe, and Hsialungtan in China. The important late Miocene locality at Rudabánya has produced a large hominoid sample, including postcranials. Kelley and Pilbeam prefer to retain Kretzoi's genus *Rudapithecus* for this material, due largely to the form of the mandibular corpus. For the present, we prefer to view the Rudabánya hominoids as one dimorphic species of *Dryopithecus.*

Sivapithecus, in the restrictive sense in which we recognize it, is known from Pasalar, Yassorien, and Candir in Turkey, as well as Pakistan, Nepal, and India. The earliest undoubted *Sivapithecus* specimens are from the middle of the Chinji Formation in Pakistan (Raza et al., 1983) and consist of two partial mandibles, a damaged maxilla, some isolated teeth, and some postcranial elements. The maxilla is sufficiently preserved anteriorly to confirm that it has the Asian premaxillary/maxillary pattern shared by both *Sivapithecus* and *Pongo.* We can therefore document the *Sivapithecus-Pongo* clade in Asia to just under 12 my B.P. (before present).

Two other hominoid samples, one in Europe and one in Asia, have been the subject of much discussion and uncertainty. The apparently large-bodied hominoid *Ouranopithecus* from Macedonia has been suggested by some workers (Andrews and Tekkaya, 1980; Kay and Simons, 1982; Szalay and Delson, 1979) to be a junior synonym of *Sivapithecus,* while its describers recognized elements of molar and mandibular anatomy that justified the new nomen (de Bonis and Melentis, 1977, 1984). *Ouranopithecus* was a dimorphic hominoid, with large molars and deep mandibular corpora. Recent cleaning of the one maxilla in the Macedonian sample revealed that the subnasal configuration is quite similar to that of the gorilla (de Bonis and Melentis, 1984). This, in conjunction with an assessment of canine morphology, prompted Kelley and Pilbeam (1986) to remove *Ouranopithecus* from the *Sivapithecus-Pongo* clade. Along with de Bonis, they identify it as belonging to the "western" hominoids in Europe and Africa.

The Lufeng hominoids have been assigned to *Ramapithecus* and *Sivapithecus* by Wu and his colleagues at the Institute of Vertebrate Paleontology and Paleoanthropology in Beijing (Lu et al., 1981; Wu, 1984; Wu et al., 1982, 1983, 1985). In an analysis of over one thousand isolated teeth from Lufeng, Wu and Oxnard (1983) suggest that at least two taxa are represented in the Lufeng sample, but more recently, Wu et al. (1985) proposed two taxonomic alternatives for the Lufeng hominoids. In the first scenario, all morphological and metric variation in the sample can be accounted for by invoking sexual dimorphism and recognizing the presence of one taxon. A suggested corollary of this model is that the Lufeng hominoid is closely related to the orang-utan. The second scenario recognizes two taxa: a large species of *Sivapithecus* that is closely related to *Pongo* and a species of *Ramapithecus* that is ancestral to *Australopithecus.* Wu et al. (1985) do not express a preference for either model.

We have recently had the opportunity to examine the Lufeng cranial material as well as some of the isolated teeth. In our opinion, the subnasal anatomy, interorbital region, orbits, and contours of the lateral mandibular corpus preclude assigning these specimens to *Sivapithecus.* The premaxillary pattern of the smaller cranium, PA 677, is the "western"/"African" type with an anteriorly placed incisive opening on the palate and a drop from the posterior pole of the nasoalveolar clivus into the incisive fossa. The interorbital distance is extremely wide, and its facial surface is formed into a vertically disposed concave trough. The medial ends of the supraorbital rims terminate as prongs that extend medially over the concave glabellar region. We are initially prepared to attribute this bizarre interorbital topography to the extensive postdepositional distortion suffered by all the Lufeng cranial material. Examination of both PA 644, the largest cranium, and PA 677, a smaller relatively complete cranium, clearly shows the same morphology, and we are confident that both specimens preserve a reasonably accurate pattern of interorbital contours.

Further examination of the large Lufeng cranium, PA 644, indicates that, like *Sivapithecus* and *Pongo,* it lacked an ethmofrontal sinus. A large plug of lignitic matrix, which was visible on the right side of the face, formed a cast of the maxillary sinus. This mass of matrix could be observed extending up into the interorbital region as an extension of the maxillary sinus. No ethmoid cells were discernible in the fractured interorbital region.

Kelley and Pilbeam (1986) compared the craniodental anatomy of the Lufeng sample with the hominoid material from Rudabánya, Hungary, and noted some similarities between the two assemblages. These similarities are most prevalent in the configuration of the upper and middle parts of the face. Of particular interest is the absence of a true supraorbital torus and posttoral sulcus in the Rudabánya specimens. It can be determined from RUD-44, a fragment of the interorbital region, that the specimen lacked an ethmofrontal sinus. The specimen is from a large adult individual and has a pair of asymmetric pneumatic cavities on its internal surface. These chambers end in the inferior broken surface of the specimen and do not extend above the level of the supraorbital rims. The position, size, and asymmetry of these spaces are consistent with the pattern present in *Pongo* and are likely to be frontal extensions of the maxillary sinuses. Unfortunately, either because of deformation or lack of preservation, cranial base relationships are not available for analysis in the Lufeng and Rudabánya hominoids.

Although *Pongo* is currently restricted to the islands of Borneo and Sumatra, its prehistoric distribution was much wider, extending throughout southeast Asia. Pleistocene and post-Pleistocene hominoid fossil dentitions attributed to *Pongo pygmaeus* (Hooijer, 1948; von Koenigswald, 1939) have been recovered from caves and excavations on Borneo (Kalimantan and Sarawan), Sumatra, and Java as well as on the mainland in North Vietnam, Laos, and the Yunnan, Kweichow, Kwangsi, and Kwangtung Provinces of southern China (Kahlke, 1972). The prehistoric hominoid dentition from Sumatra is 12.9% larger in maxillary occlusion surface area than that of extant *Pongo,* and the hominoid dentition from the middle Pleistocene of China is even larger (Hooijer, 1948; Smith and Pilbeam, 1980). Unfortunately, little other useful phenetic information can be gleaned from these samples.

From this brief review of fossil hominoid biogeography and anatomy we note that, rather than showing patterns of similarity, middle to later Miocene hominoid localities yield samples that are singular in their diversity. The more we know about Miocene hominoids, the more distinctive each sample becomes. We share with Kelley and Pilbeam (1986) and Pilbeam (1985, 1986) the view that only hominoids from Candir, Yassorien, the Siwaliks, and probably Pasalar are attributable to the genus *Sivapithecus.* This is a restrictive interpretation of the hypodigm but we believe it justified. At present, the cladistic relationships of the Lufeng hominoids are unclear. They differ profoundly from *Sivapithecus* in certain aspects of cephalofacial organizations, yet they have the most orang-like molars of any Miocene hominoid. We are currently inclined toward the view that the Lufeng specimens will have to be assigned to a new genus. Likewise, we believe that there are, *at present,* no East African Miocene hominoids attributable to *Sivapithecus.* It therefore would seem that *Sivapithecus* first appears shortly after closure of the eastern Tethyian basin, sometime after 18 my and perhaps as late as 13 my B.P. This would make the Buluk material an ideal candidate as an early species of *Sivapithecus,* but we are disinclined to support such an assignment on the grounds discussed above.

CONCLUSIONS

Airorhynchy in the skull of *Pongo* is unique among living large-bodied apes. The growth mechanisms that produce this pattern are not well understood, but they evidently involve prolonged activity of the facial sutures (Krogman, 1931). It is clear, however, that a complex of anatomical features and relationships are biologically integrated with airorhynchy in hominoids. As we have seen, this involves the orbits, the paranasal sinuses, the premaxilla, and the soft tissue environment of the oral and suprahyoid regions. We are unaware of the causes and effects of these associations, but we agree with Shea (1985) that the associations are real and significant.

Applying airorhynchy and its topographic correlates to the ancestry of *Pongo* is more problematic, but it is promising. We have been able to demonstrate that *Sivapithecus* is characterized by airorhynchy and that the expected facial correlates of hominoid airorhynchy involving the frontal bone, anterior cranial base, interorbital septum, palate, and premaxilla are all present in this hominoid as well. The oral region of *Sivapithecus,* however, shows a more generalized hominoid configuration, with retention of the anterior digastrics, a vertical mandibular symphysis, a stout inferior symphyseal torus, and low, broad mandibular corpora.

In order to proceed with a reconstruction of hominoid phylogenetic relationships, we must first address the issue of character weights, or valence. While it is necessary and appropriate to identify polarities and morphoclines dictated by the relevant characters, it is certain that some characters, for a variety of reasons, will prove to be better indicators of lineage relationships than others. To this point, we have emphasized characters that, in our view, support a cladistic relationship between *Pongo* and *Sivapithecus.* This is after making a case for each character as representing synapomorphies in the *Pongo* clade. There is, however, a complementary, and perhaps more useful approach to character assessment that has recently been developed by Lipson (1985) concerning the cladistic relationships of *Sivapithecus.* Lipson applied elements of Bayesian probability theory to the problem of character weighting in hominoid phylogeny reconstruction. She assigned probabilities to the likelihood that a set of characters were true synapomorphies in the postulated *Sivapithecus-Pongo* clade, rather than retained primitive features (symplesiomorphies) or parallel developments (homoplasies). She concluded that two characters—tall, oval orbits and a narrow interorbital distance—are reliable synapomorphies that support the suggestion that *Sivapithecus* and *Pongo* are members of an Asian hominoid lineage. The elegance of Lipson's approach is compelling because it emphasizes the quality of characters applied to phylogeny reconstruction and minimizes the pitfalls of unwittingly employing symplesiomorphies and homoplasies in the identification of cladistic relationships.

With the recognition of *Sivapithecus* and *Pongo* as members of one Asian hominoid lineage, we can now turn our attention to other pressing issues. The cladistic relationships of the Lufeng sample in China are uncertain, and hominoid evolution in Africa is a morass of complexity. We share the view of most workers that humans are more closely related to the African apes than to the orang-utan. Indeed, the molecular evidence for such a cladistic relationship is steadily accumulating (Koop et al., 1986). We therefore differ with Schwartz (1984a,b) on this issue and would expect the last common ancestor of *Pan, Gorilla,* and *Homo* to postdate the cladistic ancestor of the *Sivapithecus-Pongo* clade. We predict that the hominid-African ape ancestor would have retained a downwardly flexed face. On this point we differ with Shea (1985), who predicted that airorhynchy is primitive for hominoids. Other elements of facial anatomy in our hypothetical hominid-African ape ancestor would include a relatively high cribriform plate, orbits generally as broad as they are tall, supraorbital torus and posttoral sulcus, wide interorbital distance, ethmoid air cells and an ethmoidally derived frontal sinus, wide incisive fossa on the nasal surface of the hard palate, anteriorly placed double incisive foramina on the oral surface of the palate, thick molar enamel, and externally rotated canines. We interpret most of these features, with the possible exception of rotated canines, as primitive for post-early Miocene large-bodied hominoids. Rotated canines first appear in *Kenyapithecus* between 14 and 16 my ago and are also found in *Sivapithecus* and all known hominid species. We believe that we will be considerably better informed about what the common ancestor of African apes and early hominids might have looked like craniofacially when detailed comparative observations on the complete *Kenyapithecus* sample from the Nachola region are reported. As far as the ancestry of *Pongo* is concerned, new discoveries of *Sivapithecus* material from the lower Chinji faunal complex will in all likelihood provide new insights into the development of the derived cephalofacial topographic pattern that is so peculiar to *Sivapithecus* and the orang-utan.

REFERENCES

Andrews, P. 1985. Family group systematics and evolution among catarrhine Primates. *In* Ancestors: The Hard Evidence, pp. 14-22, ed. E. Delson. Alan R. Liss, New York.

Andrews, P. and Tekkaya, I. 1980. A revision of the Turkish Miocene hominoid *Sivapithecus meteai. Paleontology, 23:*85–95.

Andrews, P. and Tobien, H. 1977. A new Miocene locality in Turkey with evidence on the origin of *Ramapithecus* and *Sivapithecus. Nature, 268:*699–701.

Angst, R. 1967. Beitrag zum Formwandel des Craniums der Ponginen. *Zeitschrift für Morphologie und Anthropologie, 58:*109–151.

Biegert, J. 1963. The evaluation of characteristics of the skull, hands, and feet for primate taxonomy. *In* Classification and Human Evolution, pp. 116–145, ed. S. L. Washburn. Aldine, Chicago.

Bonis, L. de and Melentis, J. 1977. Les primates hominoides du Vallesien de Macedoine (Grèce). Étude de la manchoire inférieure. *Geobios, 10:*849–885.

Bonis, L. de and Melentis, J. 1984. La position phyletique d'*Ouranopithecus. In* The Early Evolution of Man with Special Emphasis on Southeast Asia and Africa, pp. 13–23, ed. P. Andrews and J. L. Franzen. Courier Forschungsinstitut Senckenberg, Frankfurt am Main.

Brandes, R. von. 1932. Über den Kehlkopf des Orangutan in verschiedenen Altersstadien mit besonderer Berücksichtigung der Kehlsackfrage *Morphologie Jahrbuch, 69:*1–61.

Brown, B., Gregory, W. K. and Hellman, M. 1924. On three incomplete anthropoid jaws from the Siwaliks, India. *American Museum Novitates, 130:*1–9.

Chapman, H. C. 1880. On the structure of the orang outang. *Proceeding of the Academy of Sciences of Philadelphia, 1880:*160–176.

Delattre, A. and Fenart, R. 1956. Analyse morphologique du splanchnocrâne chez les Primates et ses rapports avec le prognathisme. *Mammalia, 20:*169–323.

Delson, E. 1985. The earliest *Sivapithecus? Nature, 318:*107–108.

Hooijer, D. A. 1948. Prehistoric teeth of man and the orang-utan from central Sumatra, with notes on the fossil orang-utan from Java and southern China. *Zoologische Mededeelingen, 29:*175–301.

Ishida, H., Pickford, M., Nakaya, H. and Nakano, Y. 1984. Fossil anthropoids from Nachola and Samburu Hills, Samburu District, Kenya. *African Study Monographs Supplementary Issue, 2:*73–85.

Kahlke, H. D. 1972. A review of the Pleistocene history of the orang utan (*Pongo* Lacepede 1799). *Asian Perspectives, 15:*5–14.

Kay, R. F. 1982. *Sivapithecus simonsi,* a new species of Miocene hominoid, with comments on the phylogenetic status of the Ramapithecinae. *International Journal of Primatology, 3:*113–173.

Kay, R. F. and Simons, E. L. 1983. A reassessment of the relationship between later Miocene and subsequent Hominoidea. *In* New Interpretations of Ape and Human Ancestry, pp. 577–624, ed. R. L. Ciochon and R. S. Corruccini. Plenum Press, New York.

Kelley, J. and Pilbeam, D. 1986. The Dryopithecines: Taxonomy, comparative anatomy and phylogeny of Miocene large hominoids. *In* Comparative Primate Biology, vol. I: Systematics, Evolution, and Anatomy, pp. 361–411, ed. D. Swindler and J. Erwin. Alan R. Liss, New York.

Koeningswald, G. H. R. von. 1939. The relationship between the fossil mammalian faunae of Java and China, with special reference to Early Man. *Peking Natural History Bulletin, 13:*293–298.

Koop, B. F., Goodman, M., Xu, P., Chan, K. and Slightom, J. L. 1986. Primate γ-globin DNA sequences and man's place among the great apes. *Nature, 319:*234–238.

Krogman, W. M. 1931. Studies in growth changes in the skull and face of anthropoids. V. Growth changes in the skull and face of the orang-utan. *American Journal of Anatomy, 47:*343–365.

Leakey, R. E. and Leakey, M. G. 1986a. A new Miocene hominoid from Kenya. *Nature, 324:* 143–146.

Leakey, R. E., and Leakey, M. G. 1986b. A second new Miocene hominoid from Kenya. *Nature 324:*146–148.

Leakey, R. E. F. and Walker, A. 1985. New higher Primates from the early Miocene of Buluk, Kenya. *Nature, 318:*173–175.

Lipson, S. F. 1985. A phylogenetic analysis of the palatofacial morphology of *Sivapithecus indicus.* Ph.D. Dissertation. Harvard University, Cambridge.

Lu, Q., Xu, Q. and Zheng, L. 1981. Preliminary research on the cranium of *Sivapithecus yunnanensis. Vertebrata Palasiatica, 12:*101–106.

Miller, R. A. 1941. The laryngeal sacs of an infant and an adult gorilla. *American Journal of Anatomy, 69:*1–17.

Montagu, M. F. A. 1943. The mesethmoid-presphenoid relationships in the Primates. *American Journal of Physical Anthropology, 1:*129–141.

Pickford, M. H. L. 1982. New higher primate fossils from the middle Miocene deposits at Majiwa and Kaloma, Western Kenya. *American Journal of Physical Anthropology, 25:*1–6.

Pickford, M. H. L. 1985. A new look at *Kenyapithecus* based on recent discoveries in Western Kenya. *Journal of Human Evolution, 14:*113–143.

Pickford, M. H. L. 1986. Hominoids from the Miocene of East Africa and the phyletic position of *Kenyapithecus. Zeitschrift für Morphologie und Anthropologie, 76:*117–130.

Pilbeam, D. 1969. Tertiary Pongidae of East Africa: evolutionary relationships and taxonomy. *Bulletin of the Peabody Museum of Natural History, 31:*1–185.

Pilbeam, D. 1982. New hominoid skull material from the Miocene of Pakistan. *Nature, 295:*232–234.

Pilbeam, D. 1985. Patterns of hominoid evolution. *In* Ancestors: The Hard Evidence, pp. 51–59, ed. E. Delson. Alan R. Liss, New York.

Pilbeam, D. 1986. Distinguished lecture: hominoid evolution and hominoid origins. *American Anthropologist, 88:*295–312.

Preuss, T. M. 1982. The face of *Sivapithecus indicus:* description of a new, relatively complete specimen from the Siwaliks of Pakistan. *Folia Primatologica, 38:*141–157.

Raza, S. M., Barry, J. C., Pilbeam, D., Rose, M. D., Shah, S. M. I. and Ward, S. C. 1983. New hominoid Primates from the middle Miocene Chinji Formation, Potwar Plateau, Pakistan. *Nature, 306:*52–54.

Schon, M. A. 1968. The muscular system of the red howling monkey. *United States National Museum Bulletin, 273:*1–185.

Schwartz, J. H. 1984a. The evolutionary relationships of man and orang-utans. *Nature, 308:*501–505.

Schwartz, J. H. 1984b. Hominoid evolution: a review and a reassessment. *Current Anthropology, 25:*655–672.

Shea, B. T. 1985. On aspects of skull form in African apes and orangutans, with implications for hominoid evolution. *American Journal of Physical Anthropology, 68:*329–342.

Smith, R. and Pilbeam, D. 1980. Evolution of the orang-utan. *Nature, 284:*447–448.

Sonntag, C. F. 1924a. On the anatomy, physiology, and

pathology of the orang-outan. *Proceedings of the Zoological Society, 24:*349–450.

Sonntag, C. F. 1924b. The Morphology and Evolution of the Apes and Man. John Bale, Sons, and Danielsson Limited, London.

Szalay, F. S. and Delson, E. 1979. Evolutionary History of the Primates. Academic Press, New York.

Tattersall, I. 1972. The functional significance of airorhynchy in *Megaladapis. Folia Primatologica, 18:*20–26.

Ward, S. C. and Brown, B. 1986. The facial skeleton of *Sivapithecus indicus. In* Comparative Primate Biology, vol. I. Systematics, Evolution, and Anatomy, in press, edited by D. Swindler and J. Erwin. Alan R. Liss, New York.

Ward, S. C. and Kimbel, W. H. 1983. Subnasal alveolar morphology and the systematic position of *Sivapithecus. American Journal of Physical Anthropology, 61:*157–172.

Wu, R. 1984. The crania of *Ramapithecus* and *Sivapithecus* from Lufeng, China. *Courier Forschungsinstitut Senckenberg, 69:*41–48.

Wu, R., Han, D., Xu, Q., Qi, G., Lu, Q., Pan, Y., Chen, W., Zhang, X. and Xiao, M. 1982. More *Ramapithecus* skulls found from Lufeng, Yunnan. Report on the excavation of the site in 1981. *Acta Anthropologica Sinica, 1:*101–108.

Wu, R. and Oxnard, C. E. 1983. Ramapithecines from China: evidence from tooth dimensions. *Nature, 306:*258–260.

Wu, R., Xu, Q. and Lu, Q. 1983. Morphological features of *Ramapithecus* and *Sivapithecus* and their phylogenetic relationship—morphology and comparison of the crania. *Acta Anthropologica Sinica, 2:*1–10.

Wu, R., Xu, Q. and Lu, Q. 1985. Morphological features of *Ramapithecus* and *Sivapithecus* and their phylogenetic relationships—morphology and comparison of the teeth. *Acta Anthropologica Sinica, 4:*197–204.

V

DENTAL MORPHOLOGY

Although there is now a reasonable number of publications dealing in one way or another with the dental remains of the extant apes and various supposedly related Miocene taxa, basic descriptions of the teeth of especially the extant large hominoids are essentially lacking, and those that do exist are largely framed within the context of comparison with potential fossil relatives. The contributions here by J. Douglas Swarts, and by Daris Swindler and Andrew Olshan, will certainly go far toward filling in the lacuna in essential dental anatomy of the extant orang-utan.

The chapter by Swarts is one of the few studies on any primate that focuses on the details of the deciduous dentition. Since the deciduous premolars (deciduous molars) are the most complex of the primary teeth that are eventually shed, Swarts concentrated on their morphology. The restriction of the analysis at this time to the posterior upper and lower deciduous teeth is consistent with the typical state of preservation of the fossilized jaws and contained teeth of juvenile individuals and is of potential relevance to the interpretation of the "permanent" molars; the latter are part and parcel of the same primary set of teeth as the "deciduous" teeth that precede them in time and space. After describing and compiling an extensive catalog of features of the deciduous premolars of the orang-utan, Swarts collected similar data on the other extant hominoids and then more broadly on the New World anthropoid primates in order to establish a base from which to reconstruct a potential ancestral morphotype for the larger clade. It is against this larger background that Swarts first made phenetic comparisons, concluding from this endeavor that the orang-utan is most similar to *Homo* in overall dental morphology. Swarts then used the larger data base to determine character state polarity (i.e., the primitiveness or derivedness of the features delineated). Although Swarts could isolate a potential synapomorphy uniting the orang-utan with *Homo* and the gibbon and another derived feature aligning the orang-utan with *Homo* and the gorilla, he found that the greatest number of apomorphies were shared by humans and orang-utans.

Swindler and Olshan provide a description of the adult dentition of the orang-utan, both morphologically and metrically, and present not only a comparative overview of dental morphology among the extant hominoids, but the first broad comparison between the orang-utan and its apparent close relative *Sivapithecus.* For example, Swindler and Olshan document the heteromorphy of the upper incisors of *Pongo* and *Sivapithecus* in contrast to those of the other hominoids; this should quell disagreements that have arisen around the "correctness" of this interpretation. They also find that, metrically, upper and lower canines are sexually dimorphic among the great apes, and, specifically, that the canines of *Sivapithecus* are most similar to the orang-utan's. The upper premolars of the great apes are morphologically quite similar, with *Pongo* and *Pan* developing cingula much less frequently than the gorilla; lack of buccal cingulids on the lower premolars is a derived feature of the orang-utan relative to other apes. The lower anterior premolar of the extant great apes and *Sivapithecus,* as well, is sectorial and metaconid development is variable. Swindler and Olshan also suggest that the sometime occurrence of an upper molar protoconule and the frequent lack of upper molar cingula are derived features of the orang-utan's dentition and that, overall, dental morphology is indicative of the relatedness of *Sivapithecus* and the orang-utan. They conclude that dental morphology does not support the unity of a *Sivapithecus–Pongo–Homo* clade. On a broader evolutionary note, Swindler and Olshan argue that, at least with regard to the lower molars, the various groove and cusp patterns that we see are incidental rather than the direct results of selection.

19

Deciduous Dentition: Implications for Hominoid Phylogeny

J. DOUGLAS SWARTS

The renewed interest in orang-utan biology, reflected in this volume, is a direct consequence of newly acquired data from paleontological and neontological studies which bear on hominoid systematics. The recently discovered facial material of *Sivapithecus* (*Ramapithecus*) suggests this taxon's affinities to the orang-utan and thus seems to preclude a close phylogenetic relationship between *Sivapithecus* (*Ramapithecus*) and hominids (Andrews and Cronion, 1982; Andrews and Tekkaya, 1980; Kay, 1982; Pilbeam, 1982). Data from molecular (Sibley and Ahlquist, 1984) and chromosomal (Yunis and Prakash, 1982) studies of extant taxa have been interpreted to suggest a closer phylogenetic affinity between *Pan* and *Homo* than between *Pan* and *Gorilla.* Both of these hypotheses conflict with traditionally accepted interpretations of the phylogenetic position of *Sivapithecus (Ramapithecus)* relative to hominids (Kennedy, 1980) and of *Pan* relative to *Gorilla.* On the basis of these interpretations, three potentially falsifiable hypotheses of phylogenetic affinities within Hominoidea can be proposed: (1) *Pongo* is a member of a great ape clade which includes *Pan* and *Gorilla;* (2) *Pongo* is divergent from the other large hominoids, thereby implying an African hominoid group including *Homo, Pan,* and *Gorilla;* or (3) *Pongo* has affinities with *Sivapithecus* (*Ramapithecus*), which allies it with hominids (Schwartz, 1984a,b). Testing these hypotheses must be based on one of two approaches: a reexamination of the data presented when a phylogenetic hypothesis is proposed or the use of a new data set. This study is of the latter type, and develops a hypothesis of the phylogenetic affinities of *Pongo* based on the morphology of the posterior deciduous dentition.

Deciduous dental morphology has been studied only sporadically since Flower's (1871) lecture to the Odontological Society of Great Britain on the deciduous dentition of mammals. Recent work on deciduous teeth has concentrated either on size differences (Ashton and Zuckerman, 1950; Butler, 1968; Corrucini, 1977; Lavelle, 1977) or on sequences of crown development (Butler, 1967; Oka and Kraus, 1969; Swindler and McCoy, 1964; Swindler et al., 1968). Studies of the morphological structure of the deciduous dentition are few, and concentrate on either a restricted aspect of the dental morphology (Butler, 1979; von Koenigswald, 1967, 1979) or relatively few taxa (Hershkovitz, 1981, 1984). Remane's (1960) treatment of the primate deciduous dentition, in his treatise on primate dental anatomy, remains the most comprehensive investigation of this subject. The purpose of this study is to present for the first time a detailed description of the morphology of the deciduous premolar dentition of *Pongo,* make similar morphological comparisons of these teeth in extant hominoids, and, on the basis of these data, attempt to delineate the phylogenetic affinities of *Pongo* within Hominoidea.

METHODOLOGY

The sample analyzed came from the osteological collections of the Carnegie Museum of Natural History (CMNH), the American Museum of Natural History (AMNH), and the United States National Museum (USNM). The taxa examined and the taxonomic distribution of the sample are listed in Table 19–1. The modern human sample consists of juvenile material excavated from archeological sites in Carthage, Tunisia, and the Foley Site (36GR52), a prehistoric Amerindian village in southwestern Pennsylvania.

This study focused on the deciduous premolar rather than the anterior teeth because of their greater morphological complexity. This greater complexity yields more characters for analysis and makes the assessment of parallelism more reliable. Furthermore, since these are the deciduous teeth most frequently preserved in the fossil record, the information derived from this study will be of use in paleontological studies.

TABLE 19–1. Distribution of Specimens

Taxon	Number of Genera	Number of Species	Number of Individuals
Pongo	1	1	9
Pan	1	2	12
Gorilla	1	1	5
Homo	1	1	24
Hylobates	1	3	10
Cercopithecoidea	6	12	19
Ceboidea	12	17	42

Because ceboids were used as a potential outgroup for hominoids, the question of the homology of their deciduous premolars and those of hominoids arose. The deciduous premolars of hominoids are usually identified as dP3 and dP4. For the purposes of this chapter, it is assumed that the second deciduous molar of ceboids is homologous with the anterior or first deciduous premolar of catarrhines, that is dP3. Thus, ceboids possess the deciduous premolar complement dP2, dP3, and dP4, whereas catarrhines lose the dP2 and retain dP3 and dP4.

The deciduous premolar teeth of *Pongo* were examined, their morphological features identified, and the degree of expression of these characters noted. Using this character list, the character states for these features were determined for the other hominoids (*Homo, Pan, Gorilla,* and *Hylobates.*) The distribution of these character states among the hominoids yields the phenetic relationships of *Pongo.* However, to ascertain the phylogenetic affinities of *Pongo,* the polarities of the character states of the deciduous premolars were assessed using both the principles of commonality and outgroup comparison (Eldredge and Cracraft, 1980). The characters identified, their various character states, and their distribution among Hominoidea and the potential outgroups are presented in Table 19–2.

RESULTS

The Morphology of Orang-utan Deciduous Premolars

The length of the trigonid of the lower dP3 is approximately one-third the overall length of the tooth (Fig. 19–1a,b). The buccal face of the tooth is distended, giving it a broad shallow slope. Conversely, the lingual side is essentially vertical. The protoconid is the tallest cusp and its apex is located buccally, i.e., approximately one-third the distance posterior to the mesial end of the tooth. The metaconid is approximately three-fourths the height of the protoconid; its apex lies lingually and diverges somewhat from the apex of the protoconid. The protocristid is well developed and, consequently, the trigonid basin is isolated from the talonid basin. A strong paracristid courses anteriorly, down the face of the protoconid, to the mesial edge of the crown where it turns lingually toward the metaconid. The hypoconid is approximately one-half the height of the protoconid and its apex is connected to that of the protoconid by a strong, buccally emplaced cristid obliqua. The hypoconulid and entoconid, when present, are indistinct and poorly differentiated from the strong enamel ridge enclosing the shallow talonid basin.

The lower dP4 is larger than the lower dP3 and smaller than the lower first molar, but it is morphologically almost identical to the latter (Fig. 19–1a,b). The trigonid of the lower dP4 forms the mesial one-third of the tooth. As in the lower dP3, the buccal face of this tooth is well developed, forming a long buccal slope. The lower dP4 protoconid and metaconid are of approximately equal size and height and are connected by an arcuate anterior paracristid. A protocristid is variably present, and thus the trigonid and talonid basins may or may not be in contact. The hypoconid is about three-fourths the height of the protoconid and is connected to it by a strong, buccally coursing cristid obliqua. The entoconid varies in its height above the talonid basin, while the hypoconulid is generally weakly expressed in a distobuccal position. The talonid is closed lingually and distally by an enamel ridge, as in the lower dP3.

The upper dP3 is essentially square (Fig. 19–1c,d) as a result of the mesial position of the protocone apex and the infilling of the mesiolingual portion of the tooth. The cusp apices have a more centralized position due to the relatively long buccal and lingual slopes of the crown. The protocone is approximately equal in height to

TABLE 19–2. Deciduous Dentition–Character Distribution

LOWER dP3	1 Protoconid Apex (M-D)	2 Protoconid Apex (B-L)	3 Metaconid	4 Hypoconid	5 Entoconid	6 Hypoconulid	7 Cristid Obliqua	8 Para-cristid	9 Talonid Basin
Taxon									
Pongo	⅓	B	+	+	AB	AB	B	SA	CL
Homo	⅓	B	+	+	+	+	B	SA	CL
Pan	½	C	AB	+	WK	AB	B	SS	O
Gorilla	½	C	AB	+	WK	AB	B	SS	O
Hylobates	½	C	+	+	WK	AB	B	SS	O
Commonality	d	d	p	p	a	p	p	d	d
OWM	⅓	C	+	+	+	AB	B	SS	O
NWM	½	B	+	+	WK	AB	B	SS	O
Outgroup	d	p	p	p	d	p	p	d	d

LOWER dP4	10 Trig:Tal Length	11 Meta:Proto Size	12 Hypoconulid Size	13 Hypoconulid Position	14 CO Course	15 CO Termination	16 Talonid Basin	17 Proto-cristid
Taxon								
Pongo	⅓	=	WK	B	B	PR apex	CL	WK
Homo	⅓	=	STG	B	B	PR apex	CL	WK
Pan	½	GT	MOD	B	B	PR base	O	STG
Gorilla	½	GT	MOD	B	B	PR base	O	STG
Hylobates	½	=	WK	C-B	B	PR apex	CL	WK
Commonality	d	p	d	p	p	p	p	p
OWM	½	=	AB	-	B	PR base	0	WK
NWM	½	=	WK	C	B	PR base	CL	WK
Outgroup	d	p	p	d	p	d	p	p

TABLE 19–2 (*Continued*)

UPPER dP3	18 Proto:Para Size	19 Proto:Para Position	20 Pre:Post Paracrista Length	21 Distal Cingulum	22 Cusp Apices	23 Hypocone	24 Preproto-crista	25 Postproto-crista
Taxon								
Pongo	=	L	=	WK	C	AB	STG	AB
Homo	=	L	=	STG	C	AB	STG	AB
Pan	LT	D	SH	WK	P	WK	WK	WK-AB
Gorilla	LT	D	SH	WK	P	MOD	WK	WK
Hylobates	LT	L	SH	WK	P	AB	WK	AB
Commonality	d	p	d	p	d	p	d	p
OWM	LT	L	—	WK	P	VAR	WK	AB
NWM	LT	L	=	WK	P	AB	WK	AB
Outgroup	d	p	p	p	d	p	d	p

UPPER dP4	26 Proto:Para Size	27 Meta:Para Size	28 Hypo:Proto Size	29: Preproto-crista	30 Postproto-crista	31 Cusp Apices	32 Mesial Cingulum	33 Distal Cingulum	34 Lingual Cingulum	35 Pre-hypocrista
Taxon										
Pongo	=	LT	¾	STG	WK	C	AB	STG	AB	AB-WK
Homo	=	=	¾	STG	STG	C	AB	STG	AB	AB
Pan	=	=	¾	MOD	MOD	P	AB	WK	WK	WK
Gorilla	=	GT	¾	STG	MOD	P	AB	WK	STG	WK
Hylobates	=	=	½	WK	WK	P	AB	MOD	MOD	WK-MOD
Commonality	p	a	p	p	d	d	p	d	d	d
OWM	=	LT	¾	WK	AB	P	AB	MOD	MOD	WK
NWM	=	=	½	MOD	MOD	P	AB	MOD	MOD	WK-MOD
Outgroup	p	d	d	d	d	d	p	d	d	d

Abbreviations:
dP3, deciduous anterior premolar; dP4, deciduous posterior premolar
OWM, Old World monkeys, Cercopithecoidea; NWM, New World monkeys, Ceboidea
d, derived; a, autapomorphic; p, primitive
+, present; AB, absent
B, buccal; L, lingual
M, mesial; D, distal
P, peripheral; C, central
=, equal; GT, greater than; LT, less than
WK, weak; MOD, moderate; STG, strong
CL, closed; O, open
SS, short and straight; SA, short and angled
PR, protoconid
SH, short
VAR, variable

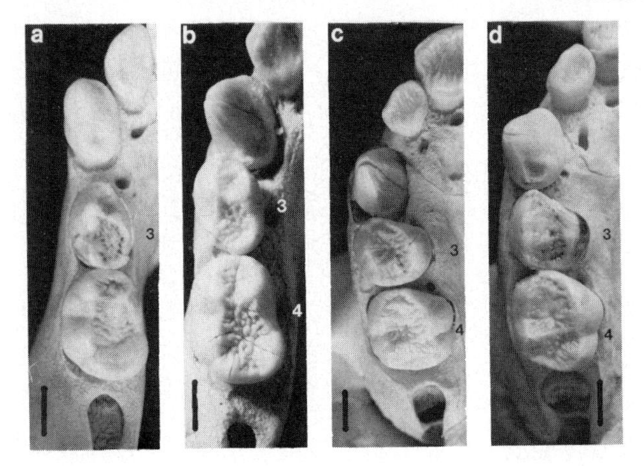

FIG. 19–1. Deciduous premolars (3 = dP3, 4 = dP4) of *Pongo*. Lower deciduous premolars: (a) USNM 292562 and (b) USNM 396920; upper deciduous molars: (c) USNM 396920 and (d) USNM 292562. Scale = 0.5 cm.

the paracone. The pre- and postparacristae are subequal in length and, as a consequence, the paracone's apex is mesiodistally centered. The preprotocrista is very strong, extending from the apex of the protocone to the mesial end of the preparacrista. The postprotocrista, on the other hand, is relatively weak. The distal cingulum, which runs from the apex of the protocone to the distal end of the postparacrista, is moderately developed and closes off the talon basin distally.

The upper dP4 possesses four well-developed cusps, the paracone, metacone, protocone, and hypocone (Fig. 19–1c,d). The trigon cusps are of approximately equal height, whereas the hypocone is about three-fourths the height of the protocone. The buccal and lingual flanks of the tooth are broad; therefore, the cusp apices are relatively centralized. The preprotocrista is very strong and runs from the apex of the protocone to the mesial terminus of the preparacrista. The postprotocrista is weaker than the preprotocrista and originates on the distal flank of the protocone. A strong distal cingulum runs from the apex of the hypocone to the distal end of the postmetacrista. Typically, there is no prehypocrista traversing the distance between the hypocone and the protocone apices.

The Distribution of Characters within Hominoidea

Initially, *Pongo* can be grouped phenetically with other hominoid genera on the basis of the overall similarity of their deciduous dentition (Figs. 19–2 and 19–3). All hominoids share the same character state for 6 of the 35 characters (4, 7, 13, 14,

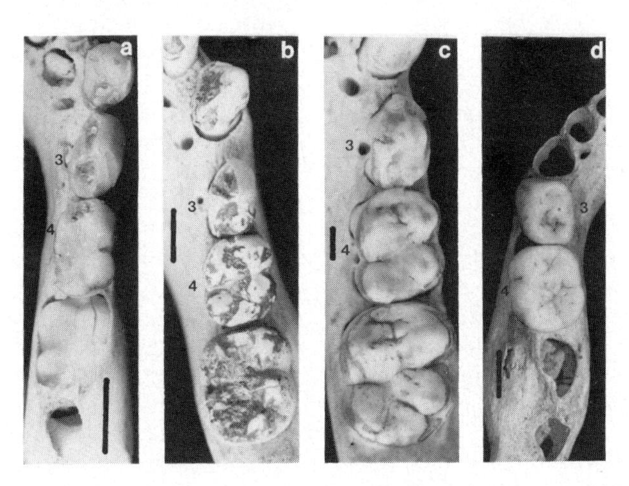

FIG. 19–2. Lower deciduous premolars of the other hominoid genera (3 = dP3, 4 = dP4): (a) *Hylobates leucogenys* (USNM 240493), (b) *Pan troglodytes* (USNM 220068), (c) *Gorilla gorilla* (USNM 176214), (d) *Homo sapiens* (Monongahela Indian). Scale = 0.5 cm.

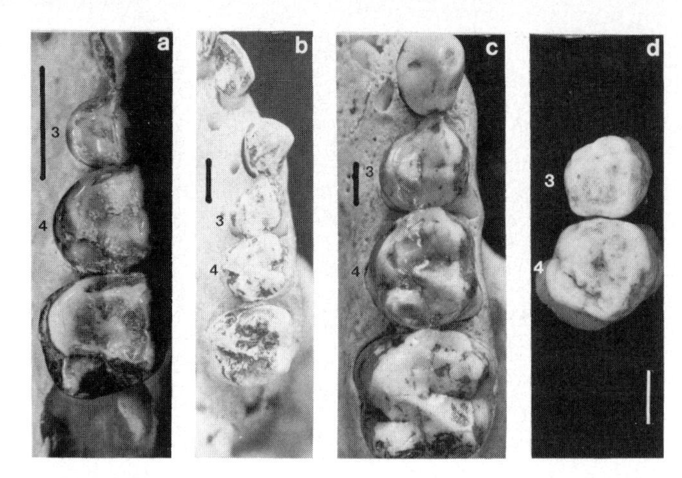

FIG. 19–3. Upper deciduous premolars of the other hominoid genera as in Fig. 19–2. Scale = 0.5 cm.

26, 32) of the deciduous dentition (Table 19–2). *Pongo* is unique among hominoids because it lacks an entoconid on the lower dP3 and the metacone of the upper dP4 is smaller than the paracone. Of the remaining 27 characters, *Pongo* and *Homo* share the same character state for 12 of them (1, 2, 8, 9, 10, 18, 20, 22, 24, 31, 33, 34), and both of these hominoids share with *Hylobates* the same character state for another 7 features (3, 11, 15, 16, 17, 19, 23). The remaining 7 characters produce 5 other subgroups within Hominoidea. On the basis of a simple preponderance of common deciduous premolar character states, it appears that *Pongo* and *Homo* are the most similar among hominoids. However, general similarity is not sufficient to delineate the phylogenetic affinities of a taxon. To do so the polarity of the character states must be established.

Character State Polarity

The polarities of the *Pongo* deciduous premolar character states were determined using the principle of commonality and outgroup comparison. The principle of commonality states that a character state shared by a majority of taxa within a larger, more inclusive taxon should be considered primitive within the group. On the basis of this criterion, character states common to three or more genera within Hominoidea should be considered primitive and therefore incapable of defining phylogenetically meaningful groups within the superfamily.

An analysis of the 33 characters, other than the two possessed uniquely by *Pongo* (5, 27), demonstrates that in 19 instances *Pongo* and two or more hominoid genera share the same deciduous premolar character state, and thus these characters cannot distinguish among potential taxonomic groups within Hominoidea (i.e., characters 3, 4, 6, 7, 11, 13, 14, 15, 16, 17, 19, 21, 23, 25, 26, 28, 29, 32, 35). The orang-utan character state for the remaining 14 characters (1, 2, 8, 9, 10, 12, 18, 20, 22, 24, 30, 31, 33, 34) is shared with only one other hominoid; these characters are therefore derived within the superfamily.

Outgroup comparison can also be used to determine the polarity of character states. A character state common in an outgroup (usually of a larger more inclusive taxonomic group), such as the rest of Catarrhini or perhaps Anthropoidea, is considered to be primitive. Extant Cercopithecoidea would appear to be the logical choice as an outgroup for Hominoidea, but the deciduous and permanent dentitions of cercopithecoids are highly specialized in comparison to other anthropoids. Their molars and deciduous premolars are bilophodont, a condition that probably developed after the divergence of cercopithecoids from their last common ancestor with hominoids. Therefore, the appropriate outgroup must be sought among the more inclusive taxon, Anthropoidea, as represented by the extant ceboids (Fig. 19–4). A hypothetical ancestral anthropoid condition for the deciduous premolars can be constructed by analyzing the ceboid deciduous premolar morphology.

The middle of three premolars, the dP3 of ceboids, as stated previously, is assumed to be homologous with the dP3 (the anterior deciduous premolar) of hominoids. Thus, the morphology of the ancestral anthropoid deciduous premolar teeth could be characterized as follows. The lower dP3 would have possessed a talonid and trigonid of approximately equal length, with the

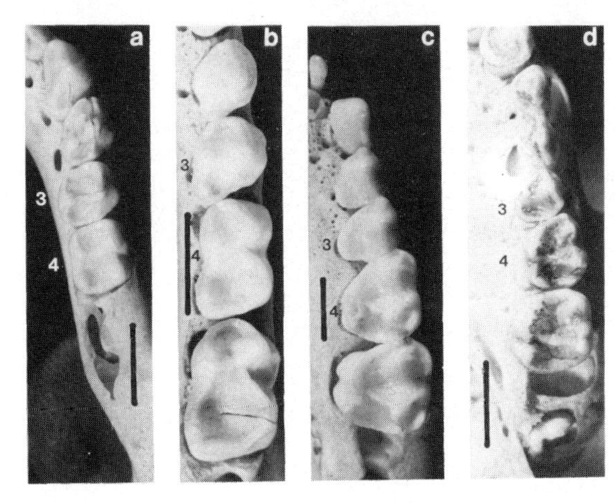

FIG. 19–4. Deciduous premolars (3 = dP3, 4 = dP4) of (a) *Cebus capucinus* (USNM 171072) and (b) *Alouatta fusca* (USNM 518250) (lowers); (c) *Alouatta fusca* (USNM 518250) and (d) *Ateles geoffroyi* (USNM 337685) (uppers). Scale = 0.5 cm.

talonid being narrow mesiodistally, distinctly basined, and open distally. The apices of the protoconid and metaconid would have been divergent. The paracristid would have coursed mesiodistally down the face of the protoconid and terminated without turning lingually. The lower dP3 would have possessed a moderately developed hypoconulid. The talonid and trigonid of the lower dP4 would have been subequal in length. The apices of all the cusps would have been situated near the margins of the crown. The apex of the hypoconid would have been connected to the base of the protoconid by a buccally coursing cristid obliqua. The talonid of the lower dP4 would have been more elaborated in detail, relative to the condition of the lower dP3, by the addition of a weak medial hypoconulid.

The upper dP3 of the hypothetical anthropoid ancestor would have been rectangular in outline, with the buccolingual dimension being the greater. The paracone would have been the largest cusp and it would have been situated medially on the buccal side of the tooth. The protocone would have been approximately three-fourths the height of the paracone and situated directly lingual to the latter cusp. The metacone and hypocone would have been lacking. The tooth would have been bounded mesially and distally by weakly developed cingula. The upper dP4 would have been more molariform, possessing a metacone that was only slightly smaller than the paracone. The hypocone would have been three-fourths the height of the protocone. A moderately developed distal cingulum would have originated from its apex. This cingulum would have coursed distally before turning buccally and ending distal to the metacone. A weak prehypocrista would have run from the apex of the hypocone to the distal flank of the protocone. The preprotocrista would have been moderately developed, connecting the protocone apex to the mesial terminus of the preparacrista. The weak postprotocrista would have originated on the distal flank of the protocone.

The polarity of the orang-utan character states can be assessed on the basis of this ancestral morphotype. In this context, *Pongo* possesses the primitive character state for 12 of the 35 deciduous premolar characters (2, 3, 6, 11, 12, 16, 17, 19, 20, 21, 23, 25, 28). It is autapomorphic (i.e., possesses a morphology unique to itself) because it lacks an entoconid on the lower dP3 and because the metacone is smaller than the paracone on the upper dP4. For the remaining 16 characters, *Pongo* possesses a derived character state (1, 8, 9, 10, 13, 15, 18, 22, 24, 28, 29, 30, 31, 33, 34, 35). The pattern of distribution of these derived character states among the extant hominoids forms the basis of an analysis of the phylogenetic affinities of *Pongo*.

CONCLUSIONS

Phenetically, *Pongo* is allied with *Homo* by the sheer number of common character states. A more discriminating analysis of the data yields the polarities of these character states, and delineates those of the orang-utan deciduous premolars which are derived. The character states of the deciduous premolars of *Pongo* which are derived according to both methods include 1, 8, 9, 10, 18, 22, 24, 30, 31, 33, 34, and 35. Among these characters, several could be strongly influenced by enamel thickness and thus might be expected to covary with it. These include the relative strengths of the cristae and cingula and the degree of peripheralization of the cusp apices. After these features are removed from further consideration, several characters nevertheless remain, especially in the lower deciduous premolars, which represent morphologies presumably unaffected by enamel thickness and which should be especially useful for phylogenetic reconstruction; these are characters 1, 8, 9, 10, and 18. The positions and heights of cusps, the relative sizes of the trigonid and talonid basins, and the morphology (course) of the lower dP3 paracristid may all be significant in an analysis of the phylogenetic relationships of *Pongo*.

Given that these character states appear to represent derived conditions, it follows that extant hominoids possessing these character states should have close phylogenetic affinities with *Pongo*. Examination of Table 19-2 reveals that the anterior shift of the protoconid apex (character 1), the length and course of the paracristid (8), and the enclosed talonid basin (9) of the lower dP3, in addition to the short trigonid (10) of the lower dP4 and the tall protocone relative to the paracone (18) of the upper dP3, are shared exclusively by the orang-utan and *Homo* (Fig. 19-4). The distribution of the other derived character states of *Pongo* within Hominoidea indicates further that this taxon (a) is autapomorphic with regard to the lack of an entoconid on the lower dP3 (5) and in the possession of an upper dP4 metacone which is small relative to the paracone (27); (b) could be allied with both *Homo* and *Hylobates* by its possession of a cristid obliqua that terminates at the protoconid apex on the lower dP4 (15); (c) shares with *Homo* and *Gorilla* a strongly developed preprotocrista on its upper dP4 (29); and (d) shares with all hominoids a buccally emplaced hypoconulid on its lower dP4 (13). All other similarities between *Pongo* and any other hominoid, or group of hominoids, are best interpreted as retentions from a common ancestor which predates the origin of Hominoidea.

Thus, based on the morphology of the deciduous premolars, it seems that among extant hominoids the phylogenetic affinities of *Pongo* lie with *Homo*. By extension, any extinct taxon with strong phylogenetic affinities to *Pongo* (e.g., *Sivapithecus*) must also have shared a more recent common ancestor with hominids than with other extant hominoids.

REFERENCES

Andrews, P. and Cronin, J. E. 1982. The relationships of *Sivapithecus* and *Ramapithecus* and the evolution of the orang-utan. *Nature, 297:*541–546.

Andrews, P. and Tekkaya, I. 1980. A revision of the Turkish Miocene hominoid *Sivapithecus meteai. Palaeontology, 23:*85–95.

Ashton, E. H. and Zuckerman, S. 1950. Some quantitative dental characteristics of the chimpanzee, gorilla and orang-outang. *Philosophical Transactions of the Royal Society of London, 234:*471–485.

Butler, P. M. 1967. Comparison of the development of the second deciduous molar and first permanent molar in man. *Archives of Oral Biology, 12:*1245–1260.

Butler, P. M. 1968. Growth of the human second lower deciduous molar. *Archives of Oral Biology, 13:*671–682.

Butler, P. M. 1979. Some morphological observations on unerupted human deciduous molars. *Ossa, 6:*23–38.

Corruccini, R. S. 1977. Cartesian coordinate analysis of the hominoid second lower deciduous molar. *Journal of Dental Research, 56:*699.

Eldredge, N. and Cracraft, J. 1980. Phylogenetic Patterns and the Evolutionary Process. Columbia University Press, New York.

Flower, W. H. 1871. Notes on the first or milk dentition of the Mammalia. *Transactions of the Odontological Society, 3:*211–33.

Hershkovitz, P. 1981. Comparative anatomy of Platyrrhine mandibular cheek teeth dpm4, pm4, m1 with particular reference to those of *Homunculus* (Cebidae), and comments on Platyrrhine origins. *Folia Primatologica, 35:*179–217.

Hershkovitz, P. 1984. More on the *Homunculus* Dpm4 and M1 and comparisons with *Alouatta* and *Stirtonia* (Primates, Platyrrhini, Cebidae), *American Journal of Primatology, 7:*261–283.

Kay, R. F. 1982. *Sivapithecus simonsi,* a new species of Miocene hominoid, with comments on the phylogenetic status of the Ramapithecinae. *International Journal of Primatology, 3:*113–173.

Kennedy, G. E. 1980. Paleoanthropology. McGraw-Hill, New York.

Lavelle, C. L. B. 1977. Ape tooth correlations. *Journal of Dental Research, 56:*702.

Oka, S. W. and Kraus, B. S. 1969. The circumnatal status of molar crown maturation among the Hominoidea. *Archives of Oral Biology, 14:*639–659.

Pilbeam, D. R. 1982. New hominoid skull material from the Miocene of Pakistan. *Nature, 295:*232–234.

Remane, A. 1960. Zähne und Gebiss. *Primatologia, 3:*637–846.

Schwartz, J. H. 1984a. The evolutionary relationships of man and orang-utans. *Nature, 308:*501–595.

Schwartz, J. H. 1984b. Hominoid evolution: a review and a reassessment. *Current Anthropology, 25:*655–672.

Sibley, C. G. and Ahlquist, J. E. 1984. The phylogeny of the hominoid primates, as indicated by DNA-DNA hybridization. *Journal of Molecular Evolution, 20:*2–15.

Swindler, D. R. and McCoy, H. A. 1964. Calcification of deciduous teeth in rhesus monkeys. *Science, 144:*1243–1244.

Swindler, D. R., Orlosky, F. J. and Hendrickx, A. G. 1968. Calcification of the deciduous molars in baboons *(Papio anubis)* and other primates. *Journal of Dental Research, 47:*167–170.

von Koenigswald, G. H. R. 1967. Evolutionary trends in the deciduous molars of the Hominoidea. *Journal of Dental Research Supplement, 46:*779–786.

von Koenigswald, G. H. R. 1979. Observations on the trigonid of the last lower deciduous molar (m2) of man and some higher primates. *Ossa, 6:*157–162.

Yunis, J. J. and Prakash, O. 1982. The origin of man: a chromosomal pictorial legacy. *Science, 215:*1525–1530.

20

Comparative and Evolutionary Aspects of the Permanent Dentition

DARIS R. SWINDLER AND ANDREW F. OLSHAN

In this chapter we are concerned primarily with the dentition of *Pongo pygmaeus pygmaeus* (Hoppius) and how it may help sort out and clarify some of the problems posed in the introduction to this book. The orang-utan is generally considered to be the most specialized of living pongids, whereas the chimpanzee and gorilla are more primitive. In an evolutionary context, this means that the latter taxa should more closely resemble the last common ancestor of great apes and hominids. However, as discussed in other chapters in this volume, this thesis has recently been challenged by several students, even to the point of asking the question: Is the orang-utan a living fossil? (Lewin, 1983). The arguments have involved many biological systems, including the dentition. Therefore, it is the purpose of this chapter to present a detailed description and analysis of the dentition of the orang-utan as a basis for comparison with other extant and extinct hominoids. In our opinion, the magnitude of dental variability, both morphological and metrical, must be established before reliable conclusions can be forthcoming regarding this hominoid's horizontal or vertical relationships.

DESCRIPTION AND COMPARISON

Incisors

Upper

The upper incisors are heteromorphic in *Pongo,* i.e., the central one (I^1) is larger and more robust than the lateral one (I^1). The incisal border of I^1 is particularly elongated mesiodistally in comparison with I^2; indeed, I^2 is frequently pointed (Fig. 20–1). The labial surfaces of I^{1-2} are convex vertically and transversely and may, on occasion, present shallow, vertical grooves. The mesial border of I^1 is higher and more vertical than the lateral border, which tends to be convex. On I^2, both of these borders are more rounded (convex) than they are on I^1, which accounts for the more pointed morphology of the tooth. Marginal ridges are present on I^{1-2}, albeit more pronounced on I^1 where they accentuate the concavity of the lingual surface. On both I^1 and I^2 the mesial and distal marginal ridges become confluent with the lingual cingulum, which varies in size on both incisors, but which is always better developed on I^1 and may come to occupy much of the lingual surface. The lingual surface of I^1 is frequently quite concave above the cingulum where a system of vertical ridges and grooves (tuberculum dentale) is observable (Fig. 20–1); the median ridge is usually more pronounced than the others and may extend to the incisal border. The incisor size sequence is always $I^1 > I^2$ (Figs. 20–2 and 20–3).

Lower

I_1 has a high, narrow crown with a nearly horizontal incisal border. The labial surface is flat to convex, while the labial surface is slighly concave. I_2 is similar to I_1, except that the labial surface is more convex mesiodistally and the distal border is rounded rather than vertical. The lingual surfaces of both teeth possess mesial and distal marginal ridges and lingual cingula. Median lingual ridges are also present on both teeth and may be quite prominent, extending as cone-like ridges as far as the incisal border; however, lingual tubercles are never as salient as they are on upper incisors. The incisor size sequence is either $I_1 = I_2$ or $I_1 > I_2$ (Figs. 20–2 and 20–3).

Comparison

The upper incisors are much more heteromorphic in *Pongo* than in other living pongids; indeed, I^2 is more narrow and frequently quite pointed compared to I^1 (Fig. 20–1). This condition obtains in the Lufeng *Sivapithecus* specimen (PA 644), where I^2 is only about half the size of I^1 (Etler, 1984), as well as in *Sivapithecus* GSP

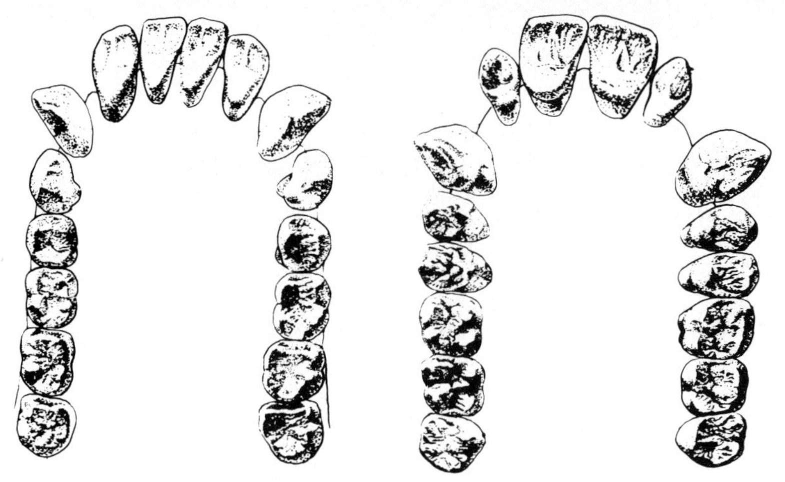

FIG. 20–1. Occlusal view of the upper and lower dentition of *Pongo pygmaeus.*

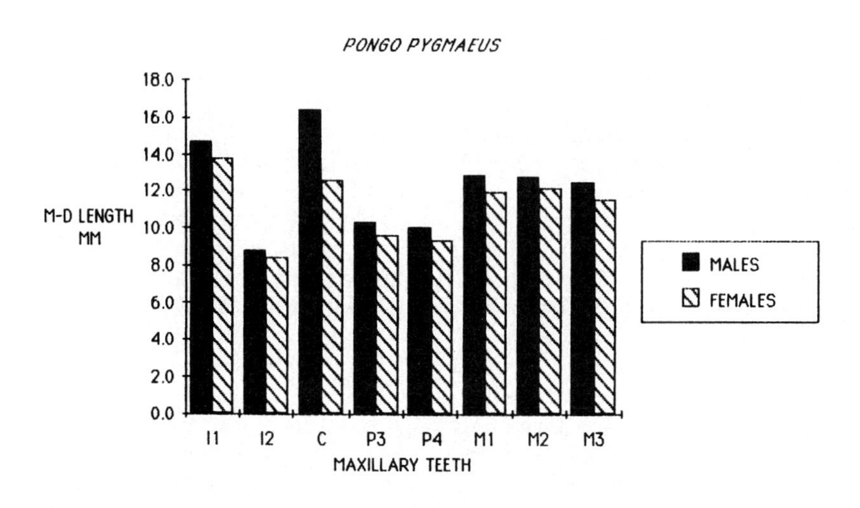

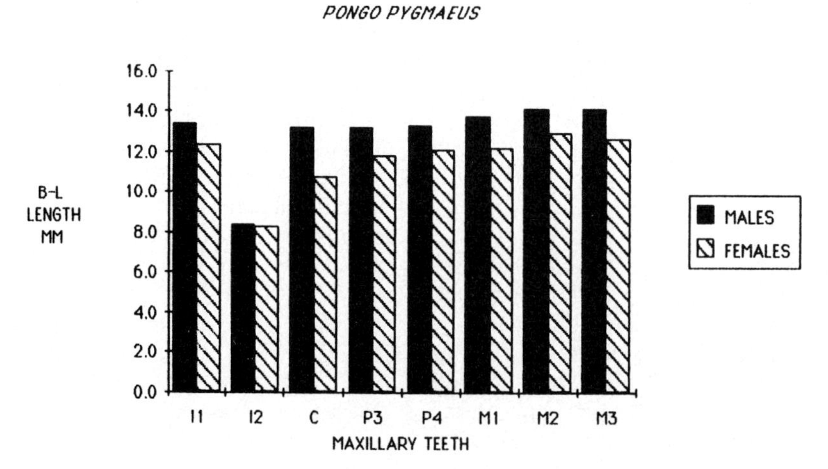

FIG. 20–2. Mesiodistal *(top)* and buccolingual *(bottom)* dimensions of the maxillary teeth of *Pongo pygmaeus.*

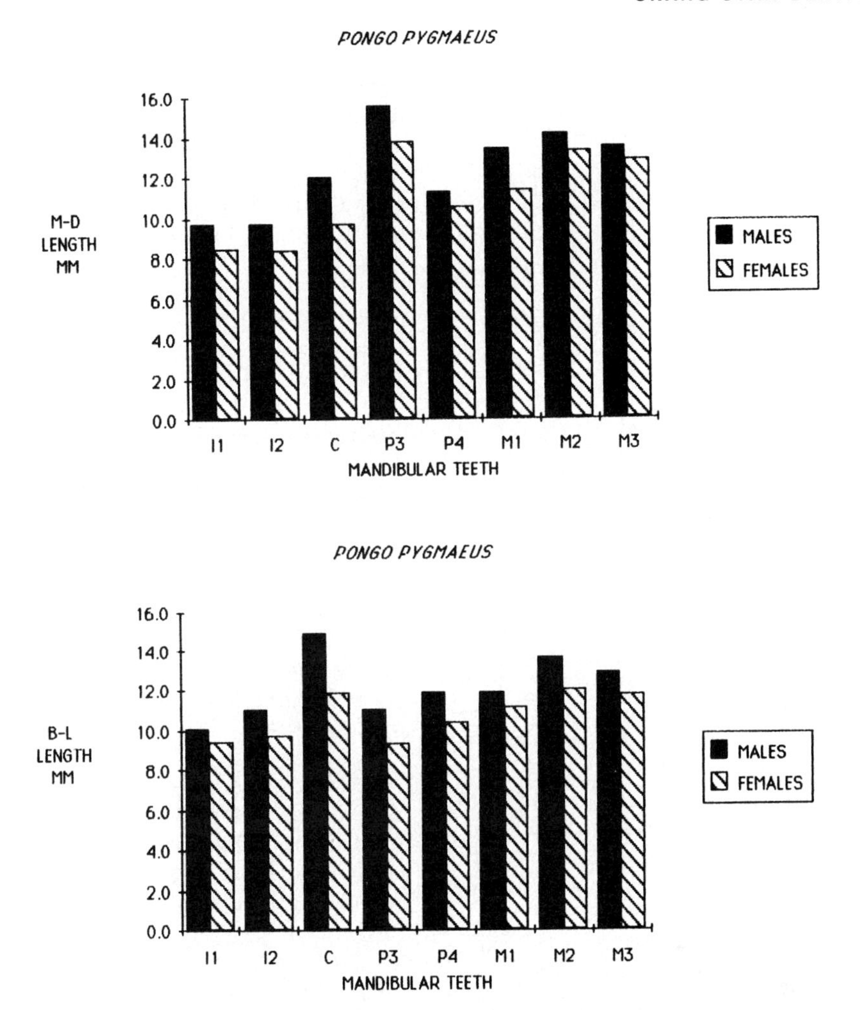

FIG. 20–3. Mesiodistal (*top*) and buccolingual (*bottom*) dimensions of the mandibular teeth of *Pongo pygmaeus.*

15000 from Pakistan. Thus, in this character, the Miocene hominoid is more comparable to the orang-utan than to the African pongids. It is also interesting to note that the Lufeng *Ramapithecus* incisors are morphologically similar to *Sivapithecus,* differing only in their smaller size (Etler, 1984).

Swindler (1976) calculated an I^1/I^2 ratio of 1.65 for extant orang-utans. Hooijer (1948) reported a mean of 1.62. There is obviously variation in this ratio from one orang-utan sample to another, but, in all cases, this ratio is always much larger in this hominoid than it is in the African apes. Although Andrews and Tekkaya (1980) argued that this I^1/I^2 disparity represents a derived character uniting *Pongo* with *Sivapithecus* phylogenetically, Kay (1982) suggested that the high ratio in both taxa resulted from different and independent evolutionary factors linked to molar size; i.e., similar I^{1-2} / M^1 ratios were brought about by the enlargement of I^1 in *Pongo* but, in *Sivapithecus indicus,* by the reduction of I^2; therefore, such similarity is the consequence of evolutionary parallelism.

The lingual surfaces of the upper incisors in these fossils are so worn that it is difficult to ascertain the original condition of the cingulum and tuberculum dentale. Weidenreich (1937) considered the latter structure a primitive trait in hominoids that undergoes a gradual reduction during the course of hominid evolution. In extant pongids, it is generally better developed and more complicated in gorillas than in chimpanzees or orang-utans. There is an isolated left I^1 (GSP 9898) from the Potwar Plateau of Pakistan that is relatively unworn. This is a small tooth (M-D length = 10mm), displaying a slight lingual cingulum from which minuscule vertical ridges (crenulations) ascend toward the incisal border. The lingual surface is concave, being circumscribed by the lingual cingulum mentioned above as well as by marginal ridges. There is no indication of a tuberculum dentale.

The I_1/I_2 ratio is higher in *Pongo* than in other pongids, and the data presented by Kay (1982) for *Sivapithecus* fall within the ranges of any of the three extant pongids.

Canines

Upper

The upper canines are sexually dimorphic in extant *Pongo*. Kay (1982) calculated the male-to-female mesiodistal ratio to be 1.36, while Swindler (1976) found both mesiodistal and buccolingual dimensions significantly different between the sexes (M-D, $p < 0.02$; B-L, $p < 0.03$) (see Figs. 20–2 and 20–3). In general, the upper canine is a large, conical tooth that is elongated mesiodistally. The mesial surface is convex and rounded; the distal surface may be either straight or convex. The male upper canine displays a vertical groove on the mesiolingual surface which is usually wanting in the female but which, if present, is weakly developed. A well formed vertical ridge lies distal to the lingual groove that frequently forms the mesial border of a second lingual groove. On occasion, the region of this second lingual groove may be occupied by a number of minor grooves and ridges. A small, narrow lingual cingulum is more limited to the mesial portion of the male upper canine than to its distal section, while in the female upper canine the cingulum passes along the entire lingual surface, forming a well-developed ridge which frequently possesses a central tubercle.

Lower

The lower canine is also sexually dimorphic. Kay (1982) gave the male-to-female mesiodistal ratio as 1.26, and Swindler (1976) detected a significant sexual dimorphism in the buccolingual dimension ($p < 0.005$) (see Figs. 20–2 and 20–3). The lower canine is conical in shape with its greatest convexity along the mesiobuccal surface of the tooth. The crown tip is curved slightly distolingually and a distinct distal ridge courses to the base of the crown. The tooth bears a shallow mesiolingual groove which is often better developed in males. The lingual cingulum is more pronounced in females and represents the major nonmetric sexually dimorphic difference between these teeth.

Comparison

Among extant hominoids canine sexual dimorphism is always greatest in *Gorilla,* followed by *Pongo, Pan, Hylobates,* and *Homo.* The upper canine of *Sivapithecus* from Lufeng (PA 644) is in size and morphology comparable to the orang-utan and most similar to Siwalik *Sivapithecus* (GSP 15000) (Etler, 1984). A structural difference between GSP 15000 and extant *Pongo* is the presence in the former species of a deep triangular groove on the mesial surface of the upper canine in this position. A groove is not present in *Pongo*; as noted above, a groove or grooves are located on the lingual surface of the orang-utan upper canines. There is a lingual cingulum on the Siwalik fossil but it is badly worn. All Miocene hominoids display canine sexual dimorphism but, according to Kay (1982), the difference is quite small compared to that in extant apes. Kay also found the greatest occlusal dimension of the upper canine to be "approximately buccolingual" in the Siwalik hominoids but to be oriented more mesiodistally in extant apes.

It is also interesting to note that several Chinese scholars maintain that the maxillary canines of the Lufeng *Ramapithecus* (PA 677) are quite different morphologically from those of *Sivapithecus* (Etler, 1984). Our tentative suggestion is that the upper canines of *Sivapithecus indicus* (GSP 15000) are different in size and morphology from those of the ramapithecines but are similar to those of the Lufeng *Sivapithecus* (PA 644).

The lower canine in extant *Pongo* lacks a buccal cingulum, but a weak buccal cingulum is reported for *Sivapithecus yunnanensis* (PA 568) (Etler, 1984). Hooijer (1948) also found a distinct mesiobuccal cingulum on a few of the lower canines of the orang-utans he described from the prehistoric caves of Sumatra.

Premolars

Upper

P^3 and P^4 (the anterior and posterior premolars, respectively) are bicuspid; the paracone is the larger of the two cusps and the smaller protocone lies opposite to it (Figs. 20–1 to 20–3). The protocone is lower and blunter than the paracone and the two cusps are separated by a deep mesiodistal groove. The mesiobuccal border of the paracone on P^3 is steeper than the distal border and projects further mesially than on P^4. Well-developed marginal ridges define the mesial and distal borders of the occlusal surfaces, which are usually better developed than the others. Between these two major cristae, several minor cristae descend from the tips of the paracone and protocone into the mesiodistal groove. These supplementary crenulations tend to obliterate the basic crown pattern and may be quite complex in some teeth. There is obviously much variation in the pattern of enamel wrinkles on the occlusal surfaces of these teeth in extant orang-utans. Buccal cingula are rare on P^3 and, when present, appear only as a slight enlargement along the mesial surface of the paracone. We have not observed these structures on P^4. Buccal cingula are also infrequent in the other extant pongids. A lingual cingulum, however, is common on P^{3-4} (85–90%) in *G. gorilla* (Swindler, 1976).

Lower

The lower premolars are heteromorphic, i.e., P_3 is essentially unicuspid, while P_4 is multicuspid (Figs. 20–1, 20–4, and 20–5). Thus, P_3 is generally referred to as sectorial, possessing as it does a large, mesiodistally elongated protoconid projecting to the buccal side of the canine, which creates a surface that is well adapted for honing with the distolingual surface of the upper canine. A small metaconid may be present distolingually to the apex of the protoconid and, if present, the two cusps are connected by the protocristid. Hooijer (1948) reported an entoconid and hypoconid in one specimen of orang-utan, but these cusps were not represented in our sample. In our opinion, the presence of these small cusps does not vitiate the honing ability of P_3, as long as the protoconid is mesiodistally elongated. A marginal ridge (paracristid) passes mesially down the protoconid to become confluent with the lingual cingulum, while a distal marginal ridge from the protoconid merges with the distal portion of the lingual cingulum. In this manner, the lingual surface of P_3 is separated into mesial (trigonid) and distal (talonid) depressions. The former is triangular and frequently bears a vertical groove, while the latter feature is more quadrangular. The wrinkle pattern is limited to the fossae, particularly the talonid fossa, where the wrinkles often radiate from the middle in herringbone fashion.

P_4 always has a large protoconid and metaconid; the metaconid may be larger than the protoconid. The two cusps are connected by a V-shaped protocristid that is created by a mesiodistal developmental groove that passes through it. The protocristid forms the distal border of the trigonid basin, which varies in shape from narrow-to-wide and shallow-to-deep, but which is always wider than the trigonid basin. The marginal ridges outlining the talonid basin occasionally possess an entoconid, or hypoconid, or both. Swindler (1976) found 85% entoconids and 18% hypoconids in his study of orang-utan teeth. Enamel wrinkles are present in both P_4 basins and tend to radiate out from the center. Buccal cingula are absent on P_3 and P_4, as well.

Comparison

The lower premolars are quite variable in extant pongids (Frisch, 1963; Hooijer, 1948; Remane, 1960; Swindler, 1976). For example, Hooijer (1948) described two morphological varieties of P_3 in *Pongo,* as did Frisch (1963) for gibbons. The major structural difference among P_3's in both extant and extinct pongids relates to the development of the metaconid and the morphology of the crown. In truly sectorial teeth, the metaconid is either absent or small when compared to the large buccolingually compressed protoconid, which is oriented in a mesiodistal direction. The resultant extensive buccal surface is functionally adapted for shearing against the upper canine. In extant *Pongo,* the metaconid is present 60% of the time and ranges in size from being practically absent to large, although we have never observed one as large as the protoconid (Swindler, 1976). Such a magnitude of variation in the metaconid of *Pongo* has been noted previously by Selenka (1898) and Hooijer (1948) and is also characteristic of Siwalik *Sivapithecus* (Kay, 1982; Swindler and Ward, in press). For example, Kay (1982) described two extreme conditions: PUA 1047-69 from India *(S. sivalensis)* has an "extremely small" metaconid and the occlusal axis of the tooth is mesiodistally oriented; GSP 9563 *(S. sivalensis)* from Pakistan also has a "very large" metaconid and the tooth is elongated buccolingually. Interestingly, P_3 of *Sivapithecus yunnanensis* (PA 548) from China is a sectorial, unicuspid tooth with no metaconid development (Etler, 1984). This tooth also possesses a buccal cingulid which is otherwise not present in the Miocene hominoids from India and Pakistan or in extant *Pongo.*

Such variable expression of dental polymorphisms within a single species implies either polygenic control, strong epigenetic influences during ontogeny, or, more probably, the combined effects of both. Whatever the ultimate conclusion regarding the mode of inheritance of such characters, one should be cautious of attaching too much taxonomic importance to them. Indeed, Palomino et al. (1977) concluded that nonmetric characters are not sensitive indicators of genetic relations because of the great variation existing between neighboring populations. Perhaps many quantitative features of the dentition that show differences within species (subspecies), as well as distinguishing between related species, are due to drift rather than being the result of specific adaptations (Butler, 1983; Swindler and Orlosky, 1972).

Molars

Upper

M^1 possesses four well-formed cusps; in the unworn condition, the buccal ones are more pointed than the lingual ones. The protocone is the largest cusp, followed by the paracone, metacone, and hypocone. The paracone lies mesial to the protocone, while the hypocone is situated distal to the metacone. A similar cusp arrangement is found on M^2, although the hypocone is usually more distal than it is on M^1. On M^3 the distal cusps, especially the hypocone, are frequently reduced in size even to the point of being absent (Fig. 20–1). A postprotocrista connects the metacone and protocone and is the most prominent crest on the occlusal surface of the upper molars. This crest is often bisected by a mesiodistally directed developmental groove. There are usually three occlusal fossae present, although the mesial one may on occasion be absent. The central or talon basin is always the most spacious of the three. In the unworn molar, the occlusal surface is covered with enamel wrinkles that radiate from the center of each fossa. Oftentimes, these crenulations are the most dominant features of a molar's occlusal surface. A distolingual groove (sulcus obliquus) separates the hypocone from the other cusps (Fig. 20–1). Remnants of lingual cingula may be present on all three molars: M^1 (50%), M^2 (51%), and M^3 (20%) (Swindler, 1976). Buccal cingula are very rare in *Pongo* (<35% [Swindler, 1976]). A small protoconule, lying just mesial to the protocone, is also variably expressed. Male and female molar size relations are shown in Figs. 20–2 and 20–3. The mesiodistal molar relations are $M^1 < M^2 > M^3$.

Lower

The lower molars possess five cusps which are generally arranged in the well known *Dryopithecus* (Y-5) pattern (Fig. 20–1). This pattern was originally defined by Gregory (1916) on the basis of the following characteristics: (1) the presence of five primary cusps; (2) when observed from the lingual side of the tooth, the developmental grooves form a Y configuration; and (3) there is contact between the metaconid and hypoconid. In agreement with others, we believe that the contact between the metaconid and hypoconid (e.g., Erdbrink, 1965; Frisch, 1965; Johanson, 1979; Jørgensen, 1955) is one of the principal characteristics of the pattern and agree with Johanson's interpretation that "if it is not accompanied by a point contact or a clear contact between the protoconid and entoconid it must be scored a 'Y.'"

The three buccal cusps are lower and less pointed than the two lingual cusps. The metaconid is frequently emplaced slightly distally to the protoconid on M_1, whereas on M_2 and M_3 this cusp is usually opposite the protoconid. The hypoconulid is always the smallest cusp and is located either central or buccal to the midline of the tooth. There are three occlusal fossae on each of the three molars. The anterior fossa (between the mesial marginal ridge and the protocristid) is a transverse slit representing a vestigial trigonid basin. This fossa is usually wider on M_1 than on M_2 or M_3; indeed, it is often absent on the latter tooth. The central part of the crown is rather spacious and represents the vestige of a talonid basin. A crest between the entoconid and hypoconulid demarcates the talonid basin from a narrow, obliquely oriented posterior fossa. This posttalonid fossa is generally no more than a narrow depression and is frequently absent on M_3. The wrinkle system is complex on all lower molars. For the most part, the wrinkles begin along the marginal ridges and pass downward toward the basins. One of these crenulations is more consistent in position and structure than the others and is known as the deflecting wrinkle; it will be discussed in more detail later. Supernumerary cusps, e.g., the tuberculum sextum and tuberculum intermedium, are present. Buccal cingula are rare in extant *Pongo*. The molar size relations are presented in Figs. 20–4 and 20–5. The mesiodistal molar length relations are $M_1 < M_2 > M_3$. In the orang-utan, M_2 may be longer mesiodistally than M_3, as presented here, but there is obvious variability in this relationship (see below).

Comparison

The molars of extant orang-utans conform to the general great ape morphotype, as noted above; however, certain details of their anatomy require more discussion. The upper molars of the orang-utan are more oval in occlusal outline than in other great apes and are thus more similar to the upper molars of *Homo* (Korenhof, 1960; Swindler, 1976). To the best of our knowledge, a metaconule has not been reported on the upper molars of orang-utans, nor was it present in the reconstructed ancestral morphotype of Catarrhini (Delson and Andrews, 1975). However, according to the latter authors, a small protoconule, lying just mesial to the protocone, was probably present (at least on M^1) in the early catarrhines. Delson and Andrews considered the condition in *Pongo* to represent a primitive retention and in *Pan* and *Gorilla* to reflect a more derived configuration. Indeed, a protoconule does occur on the upper molars of extant *Pongo* over 80% of the time (Hooijer, 1948; Korenhof, 1960; Remane, 1921, 1960; Selenka, 1898; Swindler, 1976). In addition, Hooijer reported a similar incidence of protoconule development on the molars of prehistoric *Pongo.* Unfortunately, the protoconule has not been mentioned in studies of Miocene hominoids, and the upper molars from Pakistan *(Sivapithecus)* are so worn that it is impossible to see if the cuspule was present (personal observation, D.R.S.). The protoconule occurs less frequently in *Pan* and *Gorilla* than in *Pongo* and, particularly in the former two genera, this feature is barely distinguished (Korenhof, 1960; Swindler, 1976). If Delson and Andrews's reconstruction of the molar features of the ancestral catarrhine is correct, then *Gorilla* might better reflect the primitive condition (at least in regard to the protoconule) while *Pan* and *Pongo* represent differently derived (apomorphic) conditions.

The cingulum on the upper molars of extant pongids is quite variable in its development. When present, it is always better developed on the lingual side of the tooth. The lingual cingulum may be large in gorillas, somewhat smaller in chimpanzees and, as Frisch (1965) noted, it is always smallest in Pongo, although it is still present. For example, Hooijer (1948) reported cingula on M^{1-3} in his sample of prehistoric and recent *Pongo,* and Swindler (1976) found remnants of it present in *Pongo* on M^1 (50%), M^2 (51%), and M^3 (20%). Korenhof (1960) even reported traces of what he considered a Carabelli cusp in *Pongo.* Regrettably, the upper molars of *Sivapithecus* from Pakistan are so worn they reveal nothing regarding this feature. Moreover, there is no mention of a lingual cingulum in *Sivapithecus* from China (Etler, 1984). The lingual cingulum is well developed in *Proconsul,* but it is much reduced in *Sivapithecus africanus* (BMNH 16649) (Frisch, 1965). Obviously, cingular reduction has proceeded at different rates among the hominoids since the Miocene, reaching its greatest extent in pongids (orang-utans) and in hominids (both *Australopithecus* and *Homo*). In our opinion, the condition of lingual cingulum in *Pongo* and in hominids represents a derived character state, whereas the two African pongids have retained a more ancestral appearance.

The upper molar size sequences are always more variable than those of the lowers in all extant pongids (Mahler, 1980). In the gorilla and orang-utan, Mahler found the most common size sequences to be $M^2 > M^1 > M^3$ (41.5% of orang-utans) and $M^2 > M^3 > M^1$ (22.6% of orang-utans), while in the chimpanzee, the most frequent progression was $M^1 > M^2 > M^3$, which is also the assumed human sequence. Even the $M^3 > M^2 > M^1$ progression was found in 12.8% of the orang-utan sample. The points to be made here are the extreme variability in these sequences and the fact that no sequence can be considered typical for a species in either the upper or lower molars (see below).

The lower molars of all Hominoidea display the Y-5 pattern or one of several modifications of it, i.e., Y-6, Y-4, +5, +6, +4, X-5, X-6, X-4. The Y-5 pattern itself probably had its roots in the Oligocene, as is suggested by its presence in *Aegyptopithecus* (Skaryd, 1971). Certainly *Aegyptopithecus* had the metaconid-hypoconid contact. The *Dryopithecus* (Y-5) pattern is a character state of all three lower molars of *Dryopithecus,* as well as of all other hominoid genera of the Miocene and Pliocene. One of us (D.R.S.) recently studied the *Sivapithecus* material from Pakistan and found that all lower molars had the Y-5 pattern except for one M_2, which had a Y-6 arrangement. In addition, the fossils from Rudabánya, Hungary, and the material from Pasalar, Turkey, have well-developed Y-5 patterns, as do all *Proconsul* specimens examined.

The incidence of the different lower molar groove patterns and cusp numbers in extant pongids is presented in Table 20–1. It is immediately apparent that the Y-5 pattern is the dominant one in these taxa, particularly if the Y-6 pattern is included in the Y-5 category, as it is by some authors. Only in the chimpanzee is there a

TABLE 20–1. Distribution of Mandibular Molar Groove Patterns and Cusp Number in Recent Pongids (Percentages-Sexes Combined)

		Number of teeth	Y-6	Y-5	Y-4	+6	+5	+4	Y	+
Pan troglodytes verus[1]	M_1	157	11.5	88.5					100.0	
	M_2	143	11.9	77.6		2.8	7.7		89.5	10.5
	M_3	109	2.8	44.0	0.9	1.8	33.9	16.5	47.7	52.2
Pan troglodytes schweinfurthi[2]	M_1	124		100.0					100.0	
	M_2	81		100.0					100.0	
	M_3	62		82.3			14.5	3.2	82.3	17.7
Pan troglodytes troglodytes[2]	M_1	211		97.6					97.6	
	M_2	175		98.3	0.6				98.9	
	M_3	129		71.3	14.7		4.7	3.1	86.0	7.8
Pan paniscus[2]	M_1	103		100.0					100.0	
	M_2	60		96.6					96.6	
	M_3	32		75.0	6.2				81.2	
Pongo pygmaeus pygmaeus[1]	M_1	300	2.0	98.0					100.0	
	M_2	293	7.2	91.5			1.4		98.7	1.4
	M_3	208	9.1	86.5	0.5		3.4	0.5	96.1	3.9
Gorilla gorilla gorilla[1]	M_1	88		100.0					100.0	
	M_2	94	14.9	85.1					100.0	
	M_3	80	8.7	91.3					100.0	
Gorilla gorilla beringei[1]	M_1	71		100.0					100.0	
	M_2	69	43.5	56.5					100.0	
	M_3	65	33.8	66.2					100.0	

[1]Present study.

[2]Johanson, 1979 (did not record 6th cusp).

noticeable increase in the other lower molar configurations, and even in this taxon, it is limited mainly to the Liberian subspecies *Pan troglodytes verus.* The independence of cusp number and groove pattern was demonstrated by Jørgensen (1955) and is obvious from the data in Table 20–1. But what of the role of selection on these different cusp and sulci patterns? Although little direct information is available for analysis, Johanson (1979) was led to speculate that, since groove patterns are more variable than alterations of cusp number, selection has been stronger for five-cusped lower molars and reduced for retention of the "Y" configuration. This may be true for hominids, but the data presented in Table 20–1 for pongids do not support this thesis. To the contrary, when one compares the gorilla and orang-utan data for cusp number and groove pattern, it appears that selection has been equally strong for both characters. The chimpanzee data are more variable, but even here, it is difficult to ascertain any definite pattern except perhaps in the Liberian subspecies, in which groove configuration is more variable than cusp number. Skaryd (1971) found a pattern similar to the one presented here for the chimpanzee and gorilla. In extant *Homo,* five-cusped lower molars are fairly common on M_1, but are less frequently developed on M_2, and are quite rare on M_3. It would appear that, of the different groove patterns and cusp numbers observable today among hominoids, the most derived patterns are in *Homo* and the least derived (primitive) ones are in gorillas and orang-utans. In fact, the findings for the latter two taxa indicate few significant temporal modifications of the lower molar patterns for millions of years, thus implying a long period of stability or stasis. The chimpanzee, on the other hand, particularly the Liberian subspecies *P. t. verus,* appear to be the most derived extant hominoid with regard to lower molar patterns. It is thus interesting to recall that Frisch (1965:103) anticipated this finding regarding the gorilla and orang-utans when he wrote that the chimpanzee "probably stands alone among the great apes for this particular *(progressive)* trait" (italics added).

It would appear that these morphological features (groove patterns and cusp numbers) of the lower molars of hominoids have been under different selective forces since the various lineages separated, resulting in the frequencies given in Table 20–1. This is the traditional view and the one profferred by Johanson (1979) in his attempt to explain the adaptive significance of the *Dryopithecus* pattern (see above). But what were or are these selective mechanisms? To date, there has not been a satisfactory explanation of this conundrum. Therefore, we ask, is it necessary to invoke selection or can these different character states be explained by the "effect hypothesis"?

The "effect hypothesis" predicts that long-term directional trends in evolution may be the result of unselected effects of characters within species rather than, as traditionally viewed, the result of selection for specific adaptations (Vrba, 1983; Williams, 1966). According to Williams, the adaptations of organisms that are shaped by selection to perform certain functions may also have incidental effects that are not the direct result of natural selection. In other words, a particular effect may be an incidental consequence of some more fundamental function. Thus, the lower molar patterns of hominoids (Y-5, +4, etc.) may be incidental effects rather than the direct result of selection. In this view, the molars are shaped by natural selection to perform certain basic mechanical functions, i.e., reduce particles of food for swallowing and digestion, while the various occlusal relations among the cusps and grooves are interpreted as incidental effects representing the by-products of the more basic adaptations of the dentition. This interpretation helps to explain the persistence of the *Dryopithecus* pattern in various hominoid lineages as well as its modifications through time, since a phenotypic trend does not have to be more adaptive or successful than alternatives to be maintained in a lineage.

Among extant great apes, lower molar buccal cingulids are more frequent and better developed in gorillas and least frequent and least developed in orang-utans. The chimpanzee is intermediate in such features. When present in *Pongo,* this cingulid is usually a small ridge-like structure on the buccal surface of the protoconid or in the groove or hypoflexid notch between the protoconid and hypoconid. Swindler (1976) found such a buccal cingulid present in orang-utans about 20% of the time on M_{1-2} and over 25% of the time on M_3; in any case, it was always very small. Hooijer (1948) described a similar condition for the Pleistocene orang-utans from Sumatra, which suggests that there has been little if any reduction since then. Interestingly, the fossil hominoid *Sivapithecus yunnanensis* (PA 548) from the late Miocene (ca. 8 my B.P.) of China displays a small buccal cingulid between the protoconid and hypoconid on M_{1-2}; M_3 had not erupted. Also, in two other sivapithecines, *Sivapithecus sp.* (GSP 15255) and *Sivapithecus indicus* (GSP 11536) from Pakistan (8–12 my B.P.) a slight buccal cingulid passes from the developmental groove onto the protoconid on M_1. The size and position of these cingulids are quite similar in all of these taxa.

A sixth cusp may develop between the hypoconulid and entoconid on all lower molars in pongids and is usually referred to as the tuberculum sextum (see Table 20–2). In a recent study by Wood and Abbott (1983) the incidence of the tuberculum sextum was used as one of several morphological criteria in separating "robust" from "gracile" australopithecines. Among extant great apes, the greatest frequency of this cusp is in gorillas, where it is found only on M_1 and M_2; especially, in the mountain gorilla, *Gorilla gorilla beringei,* the frequency of this cusp is significantly different ($p < 0.01$) from all other extant hominoids. The orang-utan in general has the lowest incidence of a tuberculum sextum, except on its M_3 where the incidence of this cusp is higher ($p < 0.05$) than in *Pan troglodytes verus.* It is also present in 12% of lower molars (M_{1-2} combined) of eight *Sivapithecus* fossils from Pakistan (personal observation, D.R.S.). This structure has not been described in the material from China. According to our findings there has been a decrease in the incidence of the tuberculum sextum in extant compared to Pleistocene *Pongo* (Table 20–2).

Another variable cusp, the tuberculum intermedium (seventh cusp), when present, is located between the metaconid and entoconid on the molars of hominoids (Table 20–3). The seventh cusp (C7) is also one of the discrete morphological features used by Wood and Abbott (1983) in their study of Plio-Pleistocene hominids. They

TABLE 20–2 Tuberculum Sextum in the Hominoidea in %

	M_1	M_2	M_3
P. troglodytes verus	(157) 11.5	(143) 11.9	(109) 2.8
P. pygmaeus pygmaeus	(300) 2.0	(293) 7.2	208) 9.1
P. pygmaeus (subfossils)[1]	(88) 19.3	(93) 9.7	(86) 19.0
G. gorilla gorilla	(88) 0	(94) 14.9	(80) 8.7
G. gorilla beringei	(71) 0	(69) 43.5	(65) 33.8
Sivapithecus (Pakistan)[2]	(8) 0	(8) 12.5	(5) 0
Proconsul (casts)[3]	(35) 0?		(11) 0

[1]Hooijer, 1948.

[2]Harvard Collection.

[3]Harvard Collection.

?M_1 or M_2.

TABLE 20–3. Tuberculum Intermedium in the Hominoidea in %

	M_1	M_2	M_3
P. troglodytes	(84) 0	(75) 9.3	(58) 0
P. pygmaeus pygmaeus	(126) 10.3	(124) 36.3	(96) 24.0
P. pygmaeus (subfossils)[1]	(88) 25.0	(93) 22.6	(86) 27.0
G. gorilla gorilla	(39) 8.3	(38) 31.6	(33) 51.5
G. gorilla beringei	(34) 2.7	(37) 48.6	(34) 76.5
Sivapithecus (Pakistan)[2]	(8) 37.5	(8) 37.5	(5) 80.0
Proconsul (casts)[3]	(35) 20.0?		(11) 36.4

[1]Hooijer, 1948.
[2]Harvard Collection.
[3]Harvard Collection.
?M_1 or M_2.

found C7 to be practically absent in robust hominoids, while in the gracile forms, it was present in over 50% of M_{2-3}. Among extant great apes, *Pan* has the lowest frequency of a tuberculum intermedium, while the mountain gorilla possesses the highest incidence of this trait. Indeed, the major significant difference in the expression of this character state is between gorillas and the other extant great apes (M_2, $p < 0.01$, and M_3 $p < 0.05$). *Pongo* is somewhat intermediate in the frequency with which a tuberculum intermedium is developed and, in this regard, has apparently changed little since the Pleistocene (Table 20–3). A tuberculum intermedium occurs rather frequently in the *Sivapithecus* from Pakistan and is also present in the African *Proconsul* material. As with the sixth cusp, the presence or absence of a tuberculum intermedium has not been mentioned for the Chinese material.

It has been stated that M_2 is larger than M_3 in *Pongo* and *Homo,* whereas M_3 is the largest lower molar in the other living apes (Delson and Andrews, 1975), and thus, this suggests a closer affinity of *Pongo* and *Homo* (Schwartz, 1984). However, lower molar size relationships are so notoriously variable in extant hominoids that "no sequence can be said to be typical of all great apes or even of any particular species" (Mahler, 1980:751). Such variation therefore makes one suspicious of the usefulness of this trait for comparative taxonomic purposes.

It is well known that the occlusal surfaces of the premolars and molars of hominoids possess ridges and furrows known as wrinkles. In extant hominoids, wrinkles are best developed on the low-cusped cheek teeth of *Pongo* and least developed on the low-cusped cheek teeth of *Homo.* According to Weidenreich (1937), it was Schlosser (1887) who first noted that the molars of *Dryopithecus* had wrinkles similar to those of the chimpanzee, the orang-utan, and recent *Homo.* In unworn orang-utan molars, the entire surface is covered with wrinkles, which tends to obliterate the primary groove system (Fig. 20–1). In fact, it is easier to identify the groove pattern if there has been some wear. The wrinkle patterns are complicated, but they often begin perpendicular to the marginal ridges around the crown and course toward the central basin. The wrinkles are variable in expression, ranging from fine, narrow wrinkles to coarse wrinkles. Both of these varieties are found in all fossil and extant orang-utans and should not be used for taxonomic purposes, a point thoroughly discussed by Hooijer (1948). The cheek teeth of *Sivapithecus* from Lufeng have rather complex occlusal surface crenulations with moderately high cusps (Etler, 1984). The crenulations on the unworn molars of *Sivapithecus* from Pakistan look very much like the patterns in extant orang-utans (personal observation, D.R.S.). We believe these crenulations may define a larger orang-utan clade consisting of *Sivapithecus* and Pleistocene and Recent *Pongo,* a condition first suggested by Schwartz (personal communication).

CONCLUSIONS

We have discussed the dentition of the orang-utan in some detail and compared it with the dentition of other extant and extinct hominoids (but especially *Sivapithecus*) in an attempt to discover any unique dental features shared by these taxa. This seemed appropriate, since there has been so much written recently about the closeness of the affinities among hominids, *Pongo,* and *Sivapithecus* (see Schwartz, 1984, and this volume for references). The results of this study suggest the following:

1. The upper incisors are more heteromorphic (I^2 is much more narrow than I^1) in *Pongo* than in other extant great apes. This condition obtains in *Sivapithecus.* However, whether this represents a derived state or evolutionary parallelism in these forms is difficult to decide due to insufficient fossil material. Certainly, there has been a trend to reduce I^2 in the orang-utan since the Pleistocene (Hooijer, 1948). The tuberculum

dentale is considered an ancestral state in hominoids.

2. Both upper and lower canines are sexually dimorphic in *Pongo*, a condition usually considered ancestral for hominoids (Delson and Andrews, 1975). The lack of sexual dimorphism in *Hylobates* and *Homo* therefore represents a derived condition. The canines attributable to *Sivapithecus* display similar degrees of sexual dimorphism as in *Pongo*.

3. The upper premolars of *Pongo* and *Pan* possess fewer cingular remnants than *Gorilla*, and thus, in this regard, *Pongo* and *Pan* are derived compared to *Gorilla*. When present in *Pongo*, the cingulum is limited to the buccal surface of the paracone of P^3 and appears as a slight swelling. Upper premolar cingula are not mentioned with regard to the *Sivapithecus* from China (Etler, 1984), but they are definitely not present in *Sivapithecus* from Pakistan (personal observation, D.R.S.). The lower premolars of great apes are heteromorphic: P_3 is typically unicusped and sectorial while P_4 is multicusped. Variability in P_3 morphology is produced by the development of the metaconid, which ranges in size from virtually absent to large in *Pongo*. A metaconid is also present but variable in size in *Sivapithecus*. Since the unicusped, sectorial P_3 seems to represent the ancestral condition, the development of P_3 metaconids in *Pongo* and *Sivapithecus* is of phylogenetic interest. Buccal cingula are absent on P_{3-4} of extant *Pongo; Sivapithecus yunnanensis* has one on P_3 (Etler, 1984). Retention of these buccal cingula is considered ancestral, while their absence represents a derived condition.

4. The upper molars of *Pongo* as well as of other great apes possess four cusps, three occlusal fossae, a postprotocrista, and a sulcus obliquus. Buccal cingula are extremely rare, but lingual cingula may be found on all three molars. Cingular development is the least pronounced in *Pongo*, and this is apparently the derived condition. A protoconule is variably expressed on these molars, and this is considered to represent a derived character state for *Pongo*. The lower molars of hominoids display the *Dryopithecus* Y-5 pattern or some modification of it. It is suggested that these different patterns are incidental effects, i.e., by-products of the more fundamental adaptations of the dentition. The Y-5 pattern is considered to represent the ancestral character state for hominoids. The importance of molar size sequences for taxonomic purposes is questioned because of the work of Mahler (1980), which demonstrated the great variability of molar sizes in extant great apes. Among hominoids, an occlusal wrinkle pattern is present in both upper and lower molars. It is always most complex in *Pongo* and may identify a taxonomic clade consisting of *Pongo* and *Sivapithecus*. The accessory cusps, C6 and C7, are variably present in extant and extinct hominoids, but they can still provide important data regarding taxonomic affinities (Swindler and Orlosky, 1972; Wood and Abbott, 1983). Finally, and although we did not study molar enamel thickness, we refer the reader to the recent work of Martin (1985), who has suggested that enamel thickness is a primitive trait for *Homo* and great apes and is thus of little value in identifying a hominid.

5. The information presented here supports the suggestion that sivapithecines share many dental character states with orang-utans and that these two taxa probably represent a single clade. Some of these common features are I^2 reduction; similar degree of canine sexual dimorphism; molar cusps moderately high; general lack of buccal cingula on the cheek teeth; enamel wrinkling comparatively complex on all cheek teeth. That these and other characters are held in common by these two groups and not found in this same constellation in other extant taxa has motivated many students to reevaluate the phylogenetic position of the orang-utan (Andrews and Cronin, 1982; Gantt, 1983; Pilbeam, 1982; Schwartz, 1984; Ward and Pilbeam, 1983). Whatever the ultimate outcome of this research, the present findings, in our opinion, do not appear to support a close dental relationship between *Pongo* and *Homo*.

ACKNOWLEDGMENTS

We wish to thank the following Museums and staff for the generous use of collections and facilities: National Museum of Natural History (Smithsonian Institution), Washington, D.C.; American Museum of Natural History, New York; Senckenberg Museum, Frankfurt am Main; Zoologische Staatssammlung, Munich; and the Musée Royal de l'Afrique Centrale, Tervuren. Our gratitude is expressed to the following universities for permitting use of their collections and facilities: Anthropologisches Institut der Universität Frankfurt, Frankfurt am Main, and Anthropologisches Institut der Universität Zurich, Zurich. The senior author wishes to extend special thanks to the Alexander von Humboldt-Stiftung for an award (1982–83), which supported the collection of much of the original data, and to the L. S. B. Leakey Foundation for a grant which permitted the study of original *Sivapithecus* material at Harvard University. We wish to thank

Linda Curtis for drawing Fig. 20–1. Finally, we wish to acknowledge the support of NIH grant DE-02955 from 1962 to 1968.

REFERENCES

Andrews, P. J. and Tekkaya, I. 1980. A revision of the Turkish Miocene hominoid *Sivapithecus meteai. Paleontology, 23:*85–95.

Andrews, P. J. and Cronin, J. 1982. The relationships of *Sivapithecus* and *Ramapithecus* and the evolution of the orangutan. *Nature, 297:*541–546.

Butler, P. M. 1983. Evolution and mammalian dental morphology. *Journale de Biologie Buccale, 11:*285–302.

Delson, E. and Andrews, P. J. 1975. Evolution and interrelationships of the catarrhine primates, *In* Phylogeny of the Primates, pp. 357–402, ed. W. P. Luckett and F. S. Szalay. Plenum Press, New York.

Erdbrink, D. P. 1965. A quantification of the *Dryopithecus* and other lower molar patterns in man and some of the apes. *Zeitschrift für Morphologie und Anthropologie, 57:*70–108.

Etler, D. A. 1984. The fossil hominoids of Lufeng, Yunnan Province, The People's Republic of China: a series of translations. *Yearbook of Physical Anthropology, 27:*1–56.

Frisch, J. E. 1963. Dental variability in a population of gibbons. *In* Dental Anthropology, vol. V, pp. 15–28, ed. D. R. Brothwell. Macmillan, New York.

Frisch, J. E. 1965. Trends in the Evolution of the Hominoid Dentition. Bibliotheca Primatologica Fasc. 3. S. Karger, Basel.

Gantt, D. 1983. The enamel of Neogene hominoids: structural and phyletic implications. *In* New Interpretations of Ape and Human Ancestry, pp. 125–161, ed. R. Ciochon and R. Corruccini. Plenum Press, New York.

Gregory, W. K. 1916. Studies on the evolution of the primates. *Bulletin of the American Museum of Natural History, 35:*239–355.

Hooijer, D. A. 1948. Prehistoric teeth of man and of the orangutan from central Sumatra, with notes on the fossil orangutan from Java and southern China. *Zoologisches Mededeelingen, 29:*173–301.

Johanson, D. C. 1979. A consideration of the "Dryopithecus pattern." *Ossa, 6:*125–138.

Jørgensen, K. D. 1955. The *Dryopithecus* pattern in recent Danes and Dutchman. *Journal of Dental Research, 34:*195–208.

Kay, R. F. 1982. *Sivapithecus simonsi,* a new species of Miocene hominoid, with comments on the phylogenetic status of the Ramapithecinae. *International Journal of Primatology, 3:*113–173.

Korenhof, C. A. W. 1960. Morphogenetical Aspects of the Human Upper Molar. Druk: Uitgevermaatschappij Neerlandia, Utrecht.

Lewin, R. 1983. Is the orangutan a living fossil? *Science, 222:*1222–1223.

Mahler, P. E. 1973. Metric variation in the pongid dentition. Ph.D. Dissertation, University of Michigan.

Mahler, P. E. 1980. Molar size sequence in the great apes: gorilla, orangutan, and chimpanzee. *Journal of Dental Research, 59:*749–752.

Martin, L. 1985. Significance of enamel thickness in hominoid evolution. *Nature, 314:*260–263.

Palomino, H. Chakraborty, R. and Rothhammer, F. 1977. Dental morphology and population diversity. *Human Biology, 49:*61–70.

Pilbeam, D. 1982. New hominoid skull material from the Miocene of Pakistan. *Nature, 295:*232–234.

Remane, A. 1921. Beiträge zur Morphologie des Anthropoidengebisses. *Weigmann Archives für Naturgeschichte, 87:*1–179.

Remane, A. 1960. Zähne und Gebiss. *Primatologia, 3:*637–846.

Schlosser, M. 1887. Die Affen, Lemuren, Chiropteren, Insectivoren, Marsupialier, Creodonten, und Carnivoren des europäischen Tertiärs und deren Beziehungen zu ihren lebenden und fossilen ausereuropäischen Verwandten. *Beiträge Paläontologie Oesterreich-Ungarns Orients, 6:*1–162.

Schwartz, J. H. 1984. The evolutionary relationships of man and orangutans. *Nature, 308:*501–505.

Selenka, E. 1898. Menschenaffen (Anthropomorphae). Studien über Entwicklung und Schädelbau. I. Rassen, Schädel und Bezahnung des Orangutan. Kriedel, Wiesbaden.

Skaryd, S. M. 1971. Trends in the evolution of the pongid dentition. *American Journal of Physical Anthropology, 35:*223–239.

Swindler, D. R. 1976. Dentition of Living Primates. Academic Press, London.

Swindler, D. R. and Orlosky, F. J. 1972. Metric and morphological variability in the dentition of colobine monkeys. *Journal of Human Evolution, 3:*135–160.

Swindler, D. R. and Ward, S. C. (in press). Morphology and evolution of the hominoid lower third premolar.

Vrba, E. S. 1983. Macroevolutionary trends: new perspectives on the roles of adaptation and incidental effect. *Science, 221:*387–389.

Ward, S. C. and Pilbeam, D. R. 1983. Maxillofacial morphology of Miocene hominoids from Africa and Indo-Pakistan. *In* New Interpretations of Ape and Human Ancestry, pp. 211–238, ed. R. L. Ciochon and R. S. Corruccini. Plenum Publishing Co., New York.

Weidenreich, F. 1937. The dentition of *Sinanthropus pekinensis:* a comparative odontography of the hominoids. *Palaeontologia Sinica, 101:*1–177.

Williams, G. C. 1966. Adaptation and Natural Selection, Princeton University Press, Princeton.

Wood, B. A. and Abbott, S. A. 1983. Analysis of the dental morphology of Plio-Pleistocene hominids. I. Mandibular molars: crown area measurements and morphological traits. *Journal of Anatomy, 136:*197–219.

VI

POSTCRANIAL MORPHOLOGY, ONTOGENY, AND FUNCTIONAL ANATOMY

Mary Ellen Morbeck and Adrienne Zihlman open this final section with a look at orang-utan adaptations and evolution from a novel perspective: using body segment, tissue composition, and joint surface area data to broaden the base for functional interpretations of locomotion and sexual dimorphism beyond the scope attainable from linear measurements. They conclude that, overall, orang-utan morphology emphasizes mobility, rather than resistance to compressive weight-bearing, and its association with slow, cautious arboreal climbing, bridging, and hanging allows females and much larger males to be both large-bodied and highly arboreal. Morbeck and Zihlman argue that "sexual dimorphism" should be viewed as a morphological mosaic to which body segment and tissue composition, surface areas, body weight, and canine size contribute differentially and to varying degrees. Male and female morphologies reflect their respective adaptations to feeding, social behavior, and reproduction. These features—reduced to the common denominators of arboreality and extreme sexual dimorphism—must be taken into account in reconstructing the selective forces that resulted evolutionarily in the orang-utan.

Michael Rose discusses the arboreal adaptations of the cheiridia, or hands and feet. Although the orang-utan is unusual for an animal of its size in having long forelimbs and short hindlimbs, it is still able to clamber, climb, and suspend itself effectively. The cheiridia act as hooks, allowing the orang-utan to take firm grips. Grasping function in the hand is localized in the lateral four fingers. The carpus, which acts as a universal joint to position the fingers for grasping, is constructed to withstand considerable compressive and tensile forces. Rose finds that the toes act in a similar way to the fingers and that, although the joints of the tarsus are morphologically very different from those of the carpus, mobility in the ankle also approximates that of a universal joint. Rose concludes that features related to stability are less emphasized in the orang-utan's tarsus than in its carpus, partly because of the primacy of the forelimb in positional activities.

Russell Tuttle and Gerald Cortright provide an invaluable synthesis of evidence relating to the positional behavior, adaptive complexes, and evolution of the orang-utan. Their review of the functional morphology of the fore- and hindlimb complements the preceding chapters and reinforces the inescapable conclusion that the orang-utan is a highly advanced, versatile, arboreal climbing and foraging machine. As Tuttle and Cortright point out, these features must be taken into serious consideration when choosing between the "always arboreal" and "ancestrally terrestrial" hypotheses that have been alternatively proposed in the interpretation of the orang-utan's evolution. Indeed, one is hard put at best to identify and delineate morphological features that are specifically related to the occasional terrestrial habits of some orang-utans. Russell and Cortright conclude that, although there is a strong possibility that some pre-Pleistocene and Pleistocene members of the orang-utan lineage were more terrestrial than modern orang-utans, the adaptive complexes of the extant form argue substantially for intense and perhaps long-term selection for arboreality in the direct ancestry of the now insular southeastern Asian *Pongo.*

With the chapter by John Anderton, we shift focus from functional anatomy to more descriptive and developmental anatomy. Although Anderton's account of the "normal" appendicular myology of the orang-utan may in some detail overlap and complement the presentations in the previous chapters, the major thrust of the contribution is not the typical muscular arrangements of the fore- and hindlimb of the orang-utan, but the variations and anomalies that he and others have found. Of particular interest is how these variations compare to those described in other hominoids and whether, in form, they resemble configurations otherwise typical of less closely related primates. In this regard, Anderton finds "reversions" or atavistic arrangements involving the brachialis, triceps brachii, and the peroneus digiti quinti. In terms of other variations, Anderton delineates a muscle he identifies as the "subgluteus," which has not been described in other primates; in the orang-utan, this muscle may be a derivative of the gluteus superficialis, with which it shares innervation, but it could also be related somehow to the piriformis. Variation in pedal flexor anatomy, not just in the orang-utan but in the chimpanzee and gorilla as well, which may superficially suggest the monophyly of a great ape group, Anderton interprets as independent solutions by the Asian and African apes to locomotory and behavioral demands. Indeed, peculiarities of pedal flexor anatomy not seen in the African apes may be added to the suite of characteristics already recognized as special to the hindlimb adaptations of the orang-utan.

William Jungers and Steve Hartman conclude this section with their study on relative growth of the locomotor skeleton in the orang-utan and, comparatively, among the hominoids. Using linear measurements on long bones and the pelvic and pectoral girdles, they pursue answers to the questions: (1) Is postnatal growth isometric (i.e., do shape relationships remain constant from birth into adulthood) and (2) Does growth exaggerate differences already evident at birth in the large-bodied hominoids, or do these species become more similar in body proportions as they mature? Perhaps not unexpectedly, Jungers and Hartman find that patterns of growth allometry separate humans from the great apes, particularly with regard to emphasis on the hindlimb. The African apes display some degree of negative allometry in upper and lower elements of the forelimb, but the orang-utan deviates from the trend in showing slight positive allometry in relative growth of the radius. In general, all five species of large-bodied hominoids depart significantly from multivariate isometry throughout growth. When Jungers and Hartman assessed overall phenetic similarity in multivariate growth among these five species, the common chimpanzee and gorilla clustered together, and then the orang-utan, followed last by the pygmy chimpanzee; the minimum spanning tree suggested similarity, but with relatively great distance, between humans and the common chimpanzee. Upon using a numerical approach to cladistic analysis of relative growth within species, Jungers and Hartman found that the shortest tree unites humans and orang-utans. Only the third "most" parsimonious tree unites the common and pygmy chimpanzee, in which case humans are joined next, then the gorilla, and last the orang-utan. Jungers and Hartman favor this particular arrangement because it is congruent with the branching sequence derived from DNA hybridization studies.

21

Body Composition and Limb Proportions

MARY ELLEN MORBECK AND ADRIENNE L. ZIHLMAN

The orang-utan *(Pongo pygmaeus),* the large-bodied arboreal "loner" of the Bornean and Sumatran rain forest, is an unusual mammal and an unusual primate. It is the largest arboreal species, and the largest mammalian frugivore. Its social life is limited; only mother-young groups are stable, and associations of adults including mating pairs are temporary. However, in other features, including relative brain size and cognitive abilities, orang-utans are similar to the other great apes.

Locomotor, postural, and manipulative skills within the hominoid radiation have been built upon vertical orientation of the trunk and enhanced shoulder mobility through positioning the large, well-developed clavicle on a broad chest, humeroulnar stability, forearm rotation, wrist flexibility, and grasping hands (Schultz, 1968, 1969b; Washburn, 1968). Hominoids present a spectrum of forelimb and hindlimb adaptations, with orang-utans at one end and humans at the other. Human upper limbs are adapted for fine-tuned manipulations rather than for locomotion or body support; those of African apes for arboreal climbing and terrestrial knuckle-walking; and those of orang-utans for slow climbing among the trees—elongated forelimbs with large hands and very mobile shoulders. In hindlimb morphology, at one end of the spectrum, humans have long lower limbs and reduced mobility of hip, knee, and ankle joints; at the other, orang-utans have short lower limbs, large feet, and pronounced hip, knee, ankle, and foot mobility. Among the hominoids, humans have the shortest upper and longest lower limbs, relative to trunk length, orang-utans the longest upper and shortest lower limbs.

Locomotion and posture integrate all aspects of an animal's activity within a particular environment and require an efficient system of joints, links, and segments. The distribution of weight to these segments and the relative proportion of tissues also reflect an animal's locomotor type and way of life (Grand, 1977). The body weight of an individual animal reflects many interrelated variables of physiological function and structural design: metabolism, stage of growth, health and diet, locomotor energetics, and mechanics. Total body weight, limb proportions, and tissue composition in adults, like other morphological and physical features, are a product of the species' evolutionary history, as well as an individual's genotype, an individual's sex, and the environmental influences acting on an individual during its life history. These factors affect the time of onset, speed, and duration of stages in growth and development.

In this chapter we analyze several parameters of morphology—linear measurements, joint surface area, tissue composition, and segment weight. Through these types of data it is possible to obtain a more comprehensive picture of the relationship between anatomy and locomotor abilities, and also of the components of sexual dimorphism, than is available from simple weight and linear dimensions. Measures of joint surface area, for example, reflect dual functions of weight transfer and mobility and so provide a functional sense of locomotor dynamics. Such information provides clues to the evolutionary history of the species.

MATERIALS AND METHODS

Dissections were carried out on adult male and female orang-utans. Also skeletons of nine wild-shot Bornean orang-utans were studied: six females and three males, collected by W. L. Abbott and housed in the Smithsonian Institution. These data form part of a larger study of fossil and extant catarrhines designed to interpret and explain morphological variation as it relates to locomotion, posture, and sexual dimorphism.

Both cadaver and skeletal data on individuals of known sex and body weight are used to quantify relationships among (1) total body weight; (2) composition and distribution of weight in segments determined from dissections; (3) size and shape of postcranial joint surface areas; (4) linear values taken from latex templates and di-

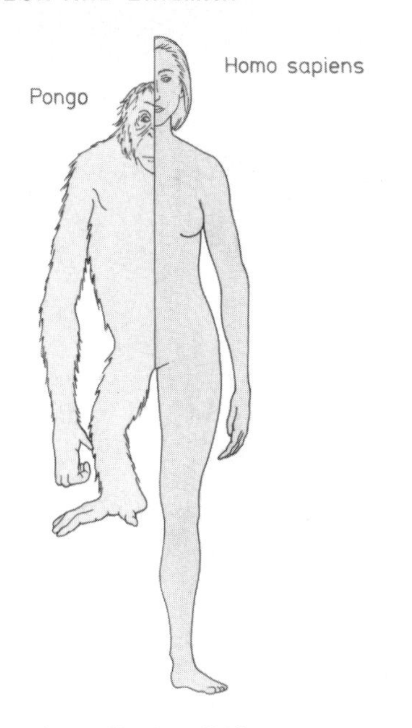

FIG. 21–1. Comparison of body build in *Pongo pygmaeus* and *Homo sapiens.* Drawn approximately to scale.

rectly from bones and teeth; and (5) weights of cleaned bones. Free-ranging adult males average 66 kg (Sumatra) and 73 kg (Borneo), whereas adult females average 37 kg (Eckhardt, 1975). These substantial differences between males and females raise questions about the kinds of differences of linear, surface area, and volume dimensions associated with weight differences.

Cadavers

Two orang-utans were dissected using body segment techniques (Grand, 1977; Zihlman, 1984) to determine body tissue composition and distribution of weight to segment. These captive individuals include (1) a 15–20-year-old adult male Bornean orang-utan and (2) a 9-year-old female orang-utan. The cadavers, frozen to preserve tissue weights, were then thawed, reweighed, and dissected following procedures outlined in Grand (1977) and modified in Zihlman (1984).

Total body weight (TBW) is taken as body weight at time of death prior to postmortem procedures. Some tissue weights were lost as a consequence of necropsy. For example, since the adult male brain was removed but not weighed, its weight was estimated.[1]

One side of each body was dissected segmentally. Head, trunk, pelvis, upper arm, forearm, hand, thigh, leg, and foot segments were separated at the relevant joints. Muscles crossing these joints were cut at the attachment sites. Skin, muscle, bone, fat, and "other" were dissected and weighed wet to the nearest gram. Here, "other" includes only tissue that cannot be classified in another category, unlike Grand (1977) who included trunk, neck, and masticatory muscles in the "other" category.

On the opposite side of the body, muscles and joints were dissected in detail. Muscles were weighed individually and muscle attachments noted, as were details of joint structure. Bones were weighed fresh and again after cleaning in order to be comparable to data derived from the skeletal series.

Within each segment, tissue weight is expressed as a percentage of segment weight. Combined tissue weight is expressed as a percentage of total body weight. Body segment weights also are expressed as a percentage of total body weight; forelimb and hindlimb segment weights are doubled to represent the whole body condition.

From the dissected animals it was possible to obtain information on (1) head, trunk, and upper and lower limb segments relative to total body weight; (2) relative proportions of tissues (muscle, bone, skin, fat, and other); and (3) bone weights, joint surface areas, and linear measurements.

Skeletons

Data on the nine wild-shot *Pongo pygmaeus pygmaeus* from the Smithsonian provide a sample for comparison with the dissected specimens. The requirement that individuals have known body weights and sex limited the potential sample. However, at least for long bone length and breadth measurements, we can compare data from a larger sample that includes both Bornean and Sumatran individuals ($N = 25$: 12 males, 13 females, Morbeck, unpublished data).

Bones of the cranium, mandible, vertebral column, forelimb, and hindlimb were weighed. Thirty-seven linear measurements (in millimeters) were taken directly on the skull, face, teeth, limb bones, and trunk, and 22 linear measurements were taken from latex templates of joint surfaces. Joint surface areas were measured via latex templates on cleaned, dry bones (Gomberg, 1981; Gomberg and Morbeck, 1983). Modifying the method described in Gomberg (1981), latex templates were mounted with tape on clear plastic and measured using a Zeiss MOP 3 Image Analyzer.

Joint surface area and shape reflect stability

and weight transfer, as well as the direction and range of motion. For instance, a low ratio of the area of the glenoid fossa to the humeral head area is characteristic of orang-utans and other hominoids and reflects shoulder mobility of a large humeral head. Further, the high ratio of the humeral trochlear area to capitular area, combined with the distinctive distal humeral joint shape, indicates a greater emphasis on weight transfer and stability in the medial aspect of the elbow joint complex.

Joint size and shape determined from bones alone, as with linear measurements, tell only part of the functional story. Soft tissue contributes to the functional joint surface and increases the articular area. In the orang-utan wrist joint, for example, the triangular disc of cartilage connects the radius and broad ulnar head; the shoulder and hip joints each have a labrum surrounding the proximal joint surface. Size, shape, and placement of ligaments can both facilitate and limit motion. However, the joint surface area data used here provide a better reflection of joint features than do linear measurements alone.

We recognize the problems of a small sample—two dissected animals and nine museum skeletons. However, the pattern of variation in the dissected specimens parallels the variation observed in the skeletons. Statistical analyses have been carried out where appropriate. Selected variables are compared trait by trait in an effort to delineate patterns of variation in body size, linear, and joint area size and proportions.

RESULTS

Our data confirm marked sexual dimorphism in weight and linear measures in orang-utans (Eckhardt, 1975; Schultz, 1941). The male-female differences are further expressed through segmental masses, tissue proportions and joint surface areas, which vary with body weight and secondary sexual characteristics.

Body Weight, Body Segments, and Tissue Composition

Body Weight

Ranges of male and female body weight in our Bornean sample do not overlap. The captive Bornean male (102 kg) exceeds the group range (83.9–90.7 kg) which represents the heavier males in Eckhardt's (1975) compilation using a larger sample (Bornean male range 34.0–90.7 kg, mean 72.8 kg). The captive Sumatran female (27.8 kg) falls below the group range of known body weights for free-ranging adult females (31.7–45.4 kg).[2]

Body Segments

Body proportions and features within segments differ in the adult pair (Figure 21–2; Table 21–1). The male's head is similar in relative weight to the female's but his trunk is relatively larger. The forelimbs are similar (16.7 vs. 16.3% of TBW), but the male's hindlimbs are considerably lighter (12.0 vs. 17.8% of TBW). Within the forelimb, the forearm of the male is relatively heavier, the upper arm and hand lighter. However, despite the marked difference in hindlimbs relative to TBW, within the hindlimb, the segmental proportions of the thigh, ankle, and foot are nearly identical in both individuals (Table 21–1). This unexpected finding suggests geometric similarity.

Tissue Composition

Differences in the contribution of body segments to TBW and variation in tissue composition within segments relate to overall body size and, in part, to sexual dimorphism, especially in muscle size and distribution of fat deposits. The male has a higher proportion of musculature, 35% of TBW compared to 27% in the female. Bone, skin, and fat (and other) are relatively greater in the female (Table 21–2A), even though the male has extensive laryngeal sacs with fat deposits and an additional fat deposit on its upper back.

Brain weight in the female (341 g) represents 1.2% of body weight and about 18% of head segment weight. In the male, brain weight, based on our estimate (434 g) is 0.4% of body weight and only 6% of the head segment.

In the forelimb and hindlimb, tissue composition differs slightly (Table 21–2B). In the forelimb, the amounts of muscle and bone are similar; in the hindlimb, however, the male has a greater percentage of muscle (57 vs. 50%). The most marked differences are in trunk/hip muscle, being one and one-half times greater in males than females (64 vs. 36%).

Sexual dimorphism is evident within the head. Although the heads are similar in relative size (7 vs. 6.9%, Fig. 21-1), the muscles of mastication in the male are more than twice that of the female (15.9 vs. 6.6%): the temporalis is about four times larger and the masseter is more than twice as large (Table 21–3).

FIG. 21–2. Major body segments expressed as a percentage of total body weight (TBW).

Adult males have extensive connective tissue and fat deposits in the cheek flanges. Cheek pad weight in this male accounts for about 13% of head segment weight. Detailed dissection reveals that skin contributes 28% and fat with connective tissue 72% of total cheek pad weight.

The relative weight of the segments, and tissue composition within them, reflect male-female differences in secondary sexual characteristics and body size. Sex differences are easily delineated. Body size relationships will become clearer in a larger sample of several adults which vary in body weight.

TABLE 21–1. Segments Within Forelimbs and Hindlimbs Expressed as a Percentage of Forelimb and Hindlimb Weight

	Adult Male	Adult Female
Forelimb		
Upper arm	39.3	44.9
Forearm	46.7	39.7
Hand	14.0	15.4
(Total = 100%)		
Hindlimb		
Thigh	52.1	53.1
Leg	28.9	26.8
Foot	19.0	20.1
(Total = 100%)		

Linear, Joint Surface, and Weight Measurements

In the Bornean skeletal sample, the extreme sexual dimorphism in body size is expressed in bone lengths, breadths, circumferences and

TABLE 21–2A. Tissue Composition Expressed as a Percentage of Total Body Weight (TBW)

	Adult Male	Adult Female
Muscle	35.0	26.9
Bone	10.7	14.7
Skin	13.2	14.1
Other	41.1	44.3
(Total = 100%)		

TABLE 21–2B. Tissue Composition Expressed as a Percentage of Trunk/Hip, Forelimb and Hindlimb Body Segments

	Adult Male	Adult Female
Trunk/hip		
Muscle	63.9	36.3
Bone	13.4	24.7
Skin	16.7	19.9
Other	6.0	19.1
(Total = 100%)		
Forelimb		
Muscle	53.0	52.2
Bone	15.7	16.5
Skin	27.2	24.2
Other	4.1	7.1
(Total = 100%)		
Hindlimb		
Muscle	56.7	49.8
Bone	17.6	18.0
Skin	21.3	22.1
Other	4.4	10.1
(Total = 100%)		

weights, joint size, and in some cranial and dental measurements. Male and female ranges tend not to overlap. However, the male-female differences in linear, area, and weight measurements do not show the same pattern. For example, in this sample, male and female ranges in long bone lengths overlap in the humerus, but not in the radius, femur, or tibia. But in long bone lengths from a larger *Pongo* sample (Morbeck, unpublished data) there is overlap in the ranges between males and females of each of these bones. Long bone lengths average 118% longer in males than in females (Bornean sample) and 114% in the larger Bornean-Sumatran sample. (This difference in ranges may reflect subspecific variation or error related to a small sample.)

On the other hand, with the exception of only the acetabular area, males have relatively larger joint surface areas than observed in linear values, ranging from 125 to 225% of female size. Although there is a wide range of variation, male postcranial bone weights also are relatively greater than linear values and average 210% of female weights.

TABLE 21–3A. Muscles of Mastication Expressed as a Percentage of Head Segment and Total Body Weight (TBW)

	Adult Male	Adult Female
% Head segment	15.9	6.6
% Total body weight	1.1	0.5

TABLE 21–3B. Muscles of Mastication Expressed as a Percentage of the Head Segment[1]

	Adult Male	Adult Female
Temporalis	7.9	2.2
Masseter	5.2	2.2
Lateral Pterygoid	0.8	0.6
Medial Pterygoid	2.0	1.4

[1]Percents are based on muscle weights for the left side only. Left-side weights are doubled to express to bilateral condition. These data differ slightly from summary data in Fig. 21–2 since there is some asymmetry between right and left sides and the summary data represent added values.

In the head, male/female ratios of cranial and mandibular bone weight from the museum skeletal collection mirror the cadaver results: compared to those of females, male mandibles are 196% larger, and their crania are 171% larger. The heavier mandible of the male parallels the heavier musculature and larger anterior teeth. The skulls and jaws comprise only a slightly larger percent of total body weight in females (1.5%) than in males (1.1%).

Four linear skull measurements (cranial length, cranial base length, palatal length, and palatal breadth) show some overlap in ranges. Male/female ratios are lower than those observed in bone weight variables, (e.g., 124% male/female ratio in cranial base length and 109% in palatal breadth). In contrast, the length *(L)* and breadth *(B)* of the lower second molar (M_2) is slightly larger in females than in males (female: M_2L = 13.9 mm; M_2B = 12.9 mm; male: M_2L = 13.6 mm, M_2B = 12.7 mm). This is particularly interesting because M_2 measurements have been used to predict total body weight in fossils.

Joint surface area of limb bones shows pronounced sexual dimorphism, although the lumbar and sacroiliac joint areas are more variable. Apparently, there is an increase in the relative area of the shoulder and a decrease in hip joint with increasing body weight. Males have absolutely larger scapular glenoid fossa, humeral head, acetabulum, and femoral head surface areas. Male glenoid fossa and humeral head areas are, respectively, 168 and 174% larger than those of females; male acetabulum and femoral head areas are, respectively, 152 and 158% larger than those of females. Thus, the areas of the acetabulum and femoral head in females are relatively larger, compared in the surface areas of the glenoid fossa and humeral head, than in males.

An alternative way of demonstrating this is by comparing the area of the glenoid fossa to the area of the acetabulum as well as humeral head to femoral head surface areas. Males exhibit a larger surface area in the upper limb joints: glenoid fossa/acetablum is 55% in males vs. 50% in females. Humeral to femoral head is 115% in males and 105% in females.

Male-female differences may be seen in comparisons of joint surface to total "limb bone joint area" (LBJA). The LBJA is the sum of each of the smaller components of limb bone joints in our data set (i.e., glenoid fossa, ulna trochlear notch, radial head, acetabulum, tibia proximal and distal facets). Glenoid fossa-to-limb bone-joint area is similar in males and females (14.4 vs. 14.2%), and humeral head-to-limb bone joint area is 57% in males vs. 54% in females. The humeral head is relatively larger in males.

The hip joint is somewhat larger in females. Acetabulum-to-limb bone joint area is 26% in males and 29% in females, whereas the femoral head is 49% in males and 52% in females. The acetabulum-to-femur head ratio also suggests that males have relatively smaller acetabula than females (53 vs. 56%, respectively). Analysis by least-squares regression suggests a strong relationship between increasing body weight and increasing humeral head surface area, but the plots of the other joint surfaces show a wide scatter, especially among the females.

The distal humerus appears to increase with larger size in males. The articular surfaces are large in males when one compares corresponding radial and ulnar joint surfaces. More skeletal and cadaver data will clarify the relationships of body weight to joint surface area.

DISCUSSION

Ever since Schultz's (1930, 1936, 1937, 1968) early systematic studies of hominoid morphology using linear dimensions of skeletons it has been clear that orang-utans are the most unusual of the large-bodied apes. We extend these observations and offer insights into the orang-utan's arboreality, sexual dimorphism, and evolutionary history.

Two aspects of posture and motion are of particular interest. First, all feeding and most movement occurs within the forest canopy. Second, orang-utans move quite differently from the other apes. Their morphological and behavioral adaptations to aboreality are well defined.

Adult females and young orang-utans live virtually independently of the forest floor (MacKinnon, 1974a). Habituated males may travel long distances on the ground (Galdikas, 1979), or may become increasingly terrestrial as age and size interfere with their ability to travel through the forest canopy (MacKinnon, 1974b).

Orang-utans are slow, cautious, quadrumanous, arboreal climbers (Cant, 1987; MacKinnon, 1971, 1974b; Sugardjito, 1982). They are limited in daily travel through the forest canopy, ranging from 305 m in one study location (Rodman, 1977) to 800 m in another (Galdikas, 1979). Since the tree canopy is irregular, with branches of different sizes and angles, no single locomotor technique suffices. In transferring from tree to tree, orang-utans intentionally coordinate all four limbs and use the mechanical properties of the trees themselves in order to grab and reach adjacent branches (Chevalier-Skolnikoff et al., 1982; MacKinnon, 1974b).

Some arboreal adaptations are long upper limbs, fairly equal distribution of body weight to both fore- and hindlimbs, mobile shoulder and hip joints, and large hands and feet for gripping small lianas, branches, or large trunks. And, indeed, climbing may be THE primary locomotor adaptation of the hominoids (Fleagle et al., 1981; Washburn, 1963). For orang-utans, as for the other apes, the long forelimbs and flexible hands are used in bridging gaps, in foraging for fruits and other food items, and in nest-building. In fact, Washburn (1963:194) defines hominoid "brachiation" in its broadest sense as "climbing and eating by reaching." Thus, orang-utans have solved the problems of moving a large body through the forest canopy by an extreme version of arboreal adaptation (Tuttle, 1975), with advanced cognitive abilities (Chevalier-Skolnikoff et al., 1982) and with selective use of the forest structure (Horr, 1977; MacKinnon, 1974b).

Anatomy and Movement Capabilities

The arboreal adaptation of orang-utans is reflected in the musculoskeletal system and the flexibility of the joints, which maximize stability and movement of the trunk and forelimb (Schultz, 1930, 1937, 1956, 1968, 1969b; Washburn, 1968). All of the hominoids, with the exception of humans, have relatively long forelimbs. The weight variable gives these comparisons added significance. The living apes have relatively heavy upper limbs compared to monkeys: e.g., among monkeys, relative upper limb weight ranges from 9% in owl monkeys, to 12% in macaques and *Cebus,* to 14% in spider monkeys (Grand, 1977); it is 16% in female chimpanzees and 20% in siamangs (Zihlman, 1984).

In female orang-utans, as shown here, the upper and lower limbs may be nearly equal in relative weight (16.3 and 17.8%). This pattern of muscle and mass distribution closely resembles that of other slow climbers: the lorises (forelimbs 12.4% and hindlimbs 14% of TBW) (Grand, 1967) and sloths (11 and 12.8% of TBW) (Grand, 1978). Similar to the orang-utan, the loris and sloth have mobile hip, knee, ankle, and foot joints, which allow the animals to reach in any direction.

Proportions shift throughout the postnatal period. Compared to adults, infant orang-utans have relatively longer hindlimbs (Schultz, 1941, 1956). In adults, forelimbs are emphasized. Orang-utan hindlimb bones, expressed as a percentage of total skeletal weight, are lighter and their forelimb bones are heavier than in chimpanzees and gorillas (Schultz, 1962). In fact, the orang-utan skeleton as a whole is relatively lighter than in the African apes.

The thorax is broad and shallow with a wide manubrium and long clavicle that positions the shoulder at its side and allows the scapula to ride on it posteriorly. Upper trunk and shoulder musculature emphasizes forelimb strength and mobility. Associated with the vertical orientation of the trunk, the vertebral column migrates during growth to a more central position within the thorax, as is indicated by the angulation of the dorsal aspect of the ribs.

The thoracic region of the vertebral column is long; the lumbar region is short. The iliac crests (and the position of the origin for latissimus dorsi and the gluteal muscles) rise above the lumbosacral articulation and, thus, limit lower trunk flexibility. The vertebrae reflect the distinct distribution of body weight and the weight-bearing pattern. Relative thickness of the middle lumbar vertebra is less in quadrumanous orang-utans than in bipedal humans with their smaller total body weight but compressive weight-bearing trunk (Schultz, 1953). The internal stucture of the orang-utan trabeculae also reflects their climbing, hanging, and bridging behaviors, which have little emphasis on compressive weight-bearing in the vertebral column. Lumbar vertebrae trabeculae in orang-utans show a complex "honeycomb pattern," in contrast to other large-bodied hominoids, whose trabeculae show a primarily vertical and horizontal alignment (Oxnard, 1984).

The elongated scapula and associated musculature promotes scapular rotation and wide-ranging glenohumeral joint movement, especially in elevated positions (Oxnard, 1984). Compared to African apes and humans, the orang-utan shoulder joint is characterized by (1) broader acromial and coracoid processes that roof the joint; (2) more cranial orientation of the glenoid fossa; (3) a large medially directed humeral head, with the articular surface rising well above the insertion sites on the tuberosities for rotator cuff muscles; (4) a narrow bicipital groove.

Orang-utans exhibit full extension of the elbow and stability of the humeroulnar joint. This ability is reflected in the short ulnar olecranon process, in the very broad trochlea on the humerus with its narrowed midportion, and in the corresponding ulnar trochlear notch which bears a prominent central ridge. Forearm mobility is reflected in the humerus in its prominent lateral trochlear edge and rounded capitulum and in the horizontally oriented, circular radial head. Distally, at the radioulnar joint, the radius rotates around a very expanded ulnar head.

The medial humeral epicondyle as well as the forearm rotators and wrist and hand flexors are large. The wrist and midcarpal joints emphasize radial and ulnar deviation, flexion, and dorsiflexion. As in chimpanzees, flexors of the orang-utan wrist and hand are twice as heavy as the extensors. Unlike the African apes, orang-utan wrist extensors and flexors are about the same size (Tuttle, 1969). The ulnar styloid process is very short and separated from a small triquetral and distally displaced pisiform, which permits increased ulnar deviation (Lewis, 1972).

Among the large hominoids, hand length is greatest in orang-utans, whose long, curved metacarpals and phalanges and reduced thumb are well adapted to suspensory grasping. Digital flexors are large and the double-locking mechanism (Napier, 1960) allows gripping of many-sized substrates.

Orang-utans are unique among the Hominoidea in their extreme mobility of hip, knee, ankle, and foot, which allows variable positioning of the limb. Weight-bearing and compression are minimal and are reflected in bone weights of the lower limb, joint size and shape, and muscle differentiation. The unusual gait of adult males in traveling on the ground strongly underscores that the hindlimb has *not* been selected for terrestrial locomotion.

The mobile hip joint has a shallow acetabulum (Schultz, 1969a), and a high, rounded femoral head lacking a ligamentum teres. Hip and thigh musculature in orang-utans have a different arrangement compared to other apes (Sigmon, 1974). The cranial portion of gluteus maximus is thicker and covers a more extensive area of the orang-utan hip than it does in other pongids. The distal, or ischiofemoral portion is a separate muscle, whereas it is connected to the cranial part of gluteus maximus in the other pongids. Gluteus minimus is two separable muscles in orang-utans, which may allow greater independent action and, thus, more hip mobility (Sigmon, 1974).

The knee joint in orang-utans is also more mobile than in chimpanzees and gorillas. The medial tibial joint surface extends posteriorly with increased rotation contributing to marked inversion of the ankle joint. The popliteus muscle, a major knee rotator, is very large and the lateral collateral ligaments are long and lax.

The distinctive ankle and foot emphasize posterior joint mobility, lateral digital flexion and reduction, and reorientation of the hallux (Gomberg, 1981). The calcaneal tuberosity is short and narrow. Extrinsic digital flexors weigh almost as much as plantar flexors (Tuttle, 1970). Tarsal bones are smaller than in African apes and humans, but metatarsals and phalanges are long and account for the orang-utan's pronounced foot length (Schultz, 1963a,b).

The greatest potential mobility (almost "wrist-like") lies in the orang-utan talocrural, subtalar, transverse tarsal, and tarsometatarsal joints (Gomberg, 1981). The combined motion of these joints allows variable positioning of the flexed digits when gripping. The long metatarsals and phalanges increase the leverage of the large digital flexors and facilitate the double-locking mechanism (Schultz, 1963a,b; Gomberg, 1981). The hallux is small, especially in females (Tuttle and Rogers, 1966); functionally, it opposes the sole and not the other digits, as in the African apes and humans. The orang-utan's foot, therefore, emphasizes mobility but retains a powerful pincer function of the hallux.

Sexual Dimorphism

Orang-utans are one of the most sexually dimorphic species of primate. This aspect of dimorphism is most often reduced in the literature to adult body weight or canine tooth size differences, but this is misleading because (1) other anatomical features exhibiting dimorphism are neglected; (2) different traits exhibit different kinds of variability; (3) dimorphic traits may vary, at least in part, independently; and (4) the developmental mechanisms and timing in growth to produce the adult pattern are not emphasized. Sexual dimorphism is a mosaic and the overall pattern of a species may be distinct (Zihlman, 1976, 1982, 1985). Our findings show that different features may be more or less dimorphic in orang-utans. Although males may be twice the weight of females, other features do not scale similarly in length or surface area in females.

For example, head weights are similar, but the masticatory muscles are more than twice as heavy in males. Within the dentition, the degree of dimorphism varies. The maxillary canine male-to-female ratio is 133% in length and 132% in breadth; similarly, the mandibular canine male-to-female ratio is 127% in length and 142% in breadth, based on data compiled by Oxnard et al. (1985). In our skeletal sample of lower second molars, molars of females are slightly larger than those of males. In the much larger sample of Oxnard et al., molars of males are 10% larger. This pattern is reported in other species. In gorillas, canines of males are 60% larger than those of females, whereas molars of males are only 6% larger than those of females (McCown, 1982). And among rhesus monkeys, adult females have much larger molar teeth for their body weight than do males (Cochard, 1985).

McCown (1982) has argued that the female expresses the basic anatomy of the species. Thus, male morphology is viewed as the "anatomy of aggression," which is added on to the female baseline. Male-female differences are presented as part of a total morphological pattern, rather than as a sum of single, unrelated characters. This approach, however, does not take into account the constraints of female anatomy for reproduction.

Most secondary sexual characteristics do not appear until puberty. Female orang-utans are on a "fast track" and reach reproductive maturity at 8 years (in the wild) compared to 14–15 years for males (MacKinnon, 1974b). In addition to differences in the onset of reproductive potential, rates of growth and duration of growth differ between the sexes (Schultz, 1941). Secondary sexual characters of males—for example, canine teeth and specifically cheek flanges, laryngeal apparatus, and large skull (Short, 1981)—are not fully expressed until adulthood.

What maintains large body size in males? Is it sexual selection and mating pattern, or niche utilization, some combination, or other factors? Rather than ask why males are so large, we might ask why females are smaller. Reproductive demands provide some clues. Lactation, especially, may be the primary factor in determining body size in females (Lancaster, 1984). That females eat as much as males is suggested by both anatomy and behavior. Females spend more time feeding than males (61 vs. 59%) in spite of the much larger male body size (Galdikas and Teleki, 1981). The proportions of food may also differ. Rodman (1977) reports that a male orang-utan consumed less fruit than females (58.6 vs. 67%) but four times as much bark (16.5 vs. 4.9%) and half as many insects (0.8% vs. 1.9%). Female requirements differ from the male due to the energetics of long pregnancy and lactation as well as carrying the young for several years. Perhaps female orang-utans cannot afford to increase too much in size and still meet reproductive demands.

Anatomically and behaviorally, adult males are set apart from other age/sex classes by their large body size, inflatable throat pouches, enlarged fatty cheek pouches, and fatty neck region. Adult males eat a greater proportion of less nutritious foods, which is reflected in their larger masticatory muscles. They also use the lower levels of the canopy more frequently, sometimes travel on the forest floor, have larger home ranges and greater mean and maximum day ranges, give long calls, and are the least social of all age/sex categories (Galdikas, 1979, 1985; Rodman, 1977, 1979). The adult male long calls may keep adult males apart and may attract receptive females (Galdikas, 1983; Mitani, 1985). These calls, combined with the visual signals of the face and large body, may be part of the "anatomy of aggression" that minimizes fighting.

EVOLUTIONARY HISTORY

The details of orang-utan evolution, as for the other large-bodied hominoids, are unknown. Several scenarios and suggestions about lineages have been proposed: the large body size of orang-utans indicates descent from some terrestrial ancestor (Smith and Pilbeam, 1980); or orang-utans are direct descendants of *Sivapithecus* (Andrews and Cronin, 1981). Whatever picture emerges, morphology, behavior, and ecology of living orang-utans, as well as the molecular, fossil, and paleoecological data, must all be evaluated.

The fossils (i.e., morphological data) provide the critical test for hypotheses about when, where, and under what circumstances each hominoid evolved. Molecular data place orang-utans within branching patterns and approximate times of divergence. Studies on DNA and proteins (Cronin, 1983; Cronin et al., 1984; Hasegawa and Yano, 1984; Hasegawa et al., 1985; Sibley and Ahlquist, 1984) show that orang-utans separated from other hominoids after the gibbon group and prior to the African hominoid group. Using the molecular clock estimates from nucleic acid, sequence, immunological, and electrophoretic data, the orang-utan lineage may have diverged about 10–11 million years before present, after the gibbons (13–15 my B.P.), and prior to African hominoids (5–6 my B.P.) (Cronin et al., 1984).

As discussed here, the postcranial morphology of orang-utans distinguishes them from other hominoids and emphasizes their arboreal adaptations. Specializations of the craniofacial and dental complexes are also distinctive. The skull, face, and jaws of *Pongo* differ from African apes and humans in their extreme airorhynchous condition. Many of the distinctive features of the orbits, midface, and jaws may relate to the particular positioning of the face relative to the cranium (Shea, 1985). Molar enamel structure is also distinctive with its "intermediate/thick enamel" (Martin, 1985) and occlusal surface wrinkling.

Fossils with these defining traits would provide direct information about the evolution of modern orang-utans. Pleistocene and recent deposits in China, North Vietnam, and islands of southeast Asia have yielded more than 5,000 subfossil and fossil *Pongo* teeth (von Koenigswald, 1982). These teeth are larger and perhaps had a greater degree of sexual dimorphism than in modern *Pongo* (Hooijer, 1948). However, as discussed above, larger teeth may or may not indicate larger body weight.

Recently, with the recovery of new specimens and reinterpretation of previously known pre-Pleistocene fossils from Eurasia and Africa, the evolution of the *Pongo* lineage and its relation to other hominoids, including humans, has provoked much discussion. Fossil taxa that may be close to the divergence of modern lineages are, of course, likely to have a mosaic of features. Currently, in part because of a lack of associated postcranial material, Miocene and Pliocene hominoids are not clearly linked to particular modern lineages, and controversy about descendant relationships continues.

Sivapithecus, in particular, has been promoted as a possible orang-utan ancestor and hominid relative (Andrews and Cronin, 1982; Pilbeam, 1982; Schwartz, 1984). This group, defined in different ways by different researchers, includes fragmentary skulls, faces, jaws, teeth, and unassociated limb bones from Eurasia and Africa.

The fragments range from more than 17 (at Buluk in Kenya [Leakey and Walker, 1985]) to 5½ my B.P. (in the Siwaliks [Sankhyan, 1985]). *Sivapithecus* facial and dental remains display some traits found in *Pongo* (Andrews and Cronin, 1982; Pilbeam, 1982; Shea, 1985). Associated postcranial remains are not yet known for any of the proposed taxa. And, although some forelimb fossils exhibit characteristic hominoid features (Morbeck, 1983), no specialized orang-utan features are present in any of the fossil postcranial remains.

The paleoecology of the middle and late Miocene must also be considered in orang-utan evolutionary scenarios. The shift from a closed forest habitat in the early Miocene of Africa to one of the open forest–woodland in the middle Miocene was fundamental to the argument that *Ramapithecus–Sivapithecus* hominoids were more ground-dwelling (Andrews, 1981; Kennedy, 1978). However, the middle to late Miocene environment in Indo-Pakistan was one of increasing seasonality with a cooler and drier climate (Laporte and Zihlman, 1983), and such conditions do not seem to be conducive for evolving the highly arboreal habits of modern orang-utans.

At the present time, there is no proposed scenario of orang-utan evolution that we find convincing. The large tooth size of extinct Pleistocene orang-utans is insufficient to argue in favor of a terrestrial stage of evolution. *Sivapithecus* without associated limb bones cannot be confirmed as an orang-utan ancestor. The decrease in forests further complicates hypotheses about orang-utan life style and lineage. Unequivocal data, in the form of associated postcranial fossils, will contribute to solutions (Temerin, 1980), and for any evolutionary proposal, all these lines of evidence must be included.

CONCLUSIONS

In this chapter, data and interpretations based on new methods and approaches provide a framework for explaining orang-utan adaptations and evolution. Body segment, tissue composition, and joint surface area data broaden the base for functional interpretations of locomotion and sexual dimorphism more than when only linear measurements are used.

Overall, orang-utan morphology emphasizes mobility rather than compressive weight-bearing. It is associated with slow, cautious arboreal climbing, bridging, and hanging. Specializations allow females and much larger males to be both highly arboreal and large-bodied.

Sexual dimorphism is viewed as a mosaic of morphological features with varying degrees of difference as measured by segments, tissue composition, and surface areas, as well as the usual measures of body weight and canine size. Male and female morphologies reflect their respective adaptations to feeding, social behavior, and reproduction. Reproductive demands on the female, combined with the "anatomy of aggression" of the male, provide a functional way of interpreting male-female differences.

The arboreality and extreme sexual dimorphism in extant orang-utans must be taken into account when reconstructing the selective forces that produced *Pongo pygmaeus.* The fragmentary fossil record, with only bits of the phenotype, comprises only part of the data required for interpreting the way of life of and the ancestral-descendant relationships in orang-utan evolution. Morphological, behavioral, and environmental information must be integrated at all levels in any discussion of orang-utan biology and evolution.

ACKNOWLEDGMENTS

We thank Mary Marzke for access to the adult male orang-utan and for assistance in dissection; Lynda Brunker for assistance in dissection; Ted Grand for technical guidance and helpful discussion; Alis Temerin for comments on the manuscript; C. Simmons for Fig. 1 and A. Hetrick for Fig. 2. We appreciate technical assistance from M. Fogelman, C. Payne and D. Sample, and access to materal from R. W. Thorington, Jr., Smithsonian Institution, Arizona State University, Phoenix Zoo, Yerkes Regional Primate Center *(NIH Base Grant RR-000165)* and Oregon Regional Primate Center. We acknowledge with appreciation grants from Wenner-Gren Foundation for Anthropological Research; University of Arizona, Department of Anthropology; and University of California, Santa Cruz, Faculty Research Committee.

REFERENCES

Andrews, P. 1981. Hominoid habitats of the Miocene. *Nature, 289:*749.

Andrews, P. and Cronin, J. E. 1982. The relationships of *Sivapithecus* and *Ramapithecus* and the evolution of the orang-utan. *Nature, 297:*541–546

Cant, J. G. H. 1987. Positional behavior of the female Bornean orang-utan *(Pongo pygmaeus). American Journal of Primatology, 12:*71–90.

Chevalier-Skolnikoff, S., Galdikas, B. M. F. and Skolnikoff, A. Z. 1982. The adaptive significance of higher intelligence in wild orang-utans: a preliminary report. *Journal of Human Evolution, 11:*639–652.

Cochard, L.R. 1985. Ontogenetic allometry of the skull and dentition of the rhesus monkey *(Macaca mulatta). In* Size and Scaling in Primate Biology, pp. 231–255, ed. W. L. Jungers. Plenum Press, New York.

Cronin, J. 1983. Apes, humans, and molecular clocks: a reappraisal. *In* New Interpretations of Ape and Human Ancestry, pp. 115–150, ed. R. L. Ciochon and R. S. Corruccini. Plenum Press, New York.

Cronin, J. E., Sarich, V. M. and Ryder, O. 1984. Molecular evolution and speciation in the lesser apes. *In* The Lesser Apes: Evolutionary and Behavioural Biology, pp. 467–485, ed. H. Preuschoft, D. J. Chivers, W. Y. Brockelman and N. Creel. Edinburgh University Press, Edinburgh.

Eckhardt, R. B. 1975. The relative weights of Bornean and Sumatran orang-utans. *American Journal of Physical Anthropology, 42:*349–350.

Fleagle, J. G., Stern, J. T., Jr., Jungers, W. L., Susman, R. L., Vangor, A. K. and Wells, J. P. 1981. Climbing: a biomechanical link with brachiation and with bipedalism. *Symposium, Zoological Society, London, 48:*359–375.

Galdikas, B. M. F. 1979. Orang-utan adaptation at Tanjung Puting Reserve: mating and ecology. *In* The Great Apes, pp. 194–233, ed. D. A. Hamburg and E. R. McCown. Benjamin/Cummings Publishing Company, Menlo Park, CA.

Galdikas, B. M. F. 1983. The orang-utan long call and snag crashing at Tanjung Puting Reserve. *Primates, 24:*371–384.

Galdikas, B. M. F. 1985. Orang-utan sociality at Tanjung Puting Reserve. *American Journal of Primatology, 9:*101–119.

Galdikas, B. M. F. and Teleki, G. 1981. Variations in subsistence activities of female and male pongids: new perspectives on the origins of hominoid labor division. *Current Anthropology, 22:*241–256.

Gomberg, N. 1981. Form and function of the hominoid foot. Ph.D. dissertation. University of Massachusetts, Amherst.

Gomberg, N. and Morbeck, M. E. 1983. The use of joint surface areas in primate morphology. *American Journal of Physical Anthropology, 60:*199.

Grand, T. I. 1967. The functional anatomy of the ankle and foot of the slow loris *(Nycticebus coucang). American Journal of Physical Anthropology, 26:*207–218.

Grand, T. I. 1977. Body weight: its relation to tissue composition, segment distribution, and motor function. I. Interspecific comparisons. *American Journal of Physical Anthropology, 47:*211–240.

Grand, T. I. 1978. Adaptations of tissue and limb segments to facilitate moving and feeding in arboreal folivores. *In* The Ecology of Arboreal Folivores, pp. 231–241, ed. G. G. Montgomery. Smithsonian Institution Press, Washington, D.C.

Hasegawa, M., Kishino, H. and Yano, T. 1985. Dating of the human-ape splitting by a molecular clock of mitochondrial DNA. *Journal of Molecular Evolution, 22:*160–174.

Hasegawa, M. and Yano, T. 1984. Phylogeny and classification of Hominoidea as inferred from DNA sequence data. *Proceedings of the Japan Academy, 60(b)*(10):389–392.

Hooijer, D. A. 1948. Prehistoric teeth of man and the orang-utan from central Sumatra with notes on the fossil orang-utan from Java and southern China. *Zoologische Mededelingen, 29:*175–301.

Horr, D. A. 1977. Orang-utan maturation: growing up in a female world. *In* Primate Bio-Social Development: Biological, Social, and Ecological Determinants, pp. 289–321, ed. S. Chevalier-Skolnikoff and F. E. Poirier. Garland, New York.

Kennedy, G. E. 1978. Hominoid habitat shifts in the Miocene. *Nature, 271:*11–12.

Lancaster, J. B. 1984. Evolutionary perspectives on sex differences in higher primates. *In* Gender and Life Course, pp. 3–27, ed. A. S. Rossi. Aldine, New York.

Laporte, L. F. and Zihlman, A. L. 1983. Plates, climate and hominoid evolution. *South African Journal of Science, 79:*96–110.

Leakey, R. E. F. and Walker, A. 1985. New higher primates from the early Miocene of Buluk, Kenya. *Nature, 318:*173–175.

Lewis, O. J. 1972. Osteological features characterizing the wrists of monkeys and apes, with a reconsideration of this region in *Dryopithecus (Proconsul) africanus. American Journal of Physical Anthropology, 36:*45–58.

MacKinnon, J. 1971. The orang-utan in Sabah today: a study of a wild population in the Ulu Segama Reserve. *Oryx, 11:*141–191.

MacKinnon, J. 1974a. In Search of the Red Ape. Collins, London.

MacKinnon, J. 1974b. The behaviour and ecology of wild orang-utans *(Pongo pygmaeus). Animal Behaviour, 22:*3–74.

Martin, L. 1985. Significance of enamel thickness in hominoid evolution. *Nature, 314:*260–263.

McCown, E. R. 1982. Sex differences: the female as baseline for species description. *In* Sexual Dimorphism in *Homo sapiens:* A Question of Size, pp. 37–83, ed. R. L. Hall. Praeger, New York.

Mitani, J. C. 1985. Sexual selection and adult orang-utan long calls. *Animal Behaviour, 33:*272–283.

Morbeck, M. E. 1983. Miocene hominoid discoveries from Rudabánya: implications from the postcranial skeleton. *In* New Interpretations of Ape and Human Ancestry, pp. 369–404, ed. R. L. Ciochon and R. S. Corruccini. Plenum, New York.

Napier, J. R. 1960. Studies of the hands of living primates. *Proceedings of the Zoological Society, London, 134:*647–657.

Oxnard, C. E. 1984. The Order of Man. A Biomathematical Anatomy of the Primates. Yale University Press, New Haven.

Oxnard, C. E., Lieberman, S. S. and Gelvin, B. R. 1985. Sexual dimorphisms in dental dimensions of higher primates. *American Journal of Primatology, 8:*127–152.

Pilbeam, D. 1982. New hominoid skull material from the Miocene of Pakistan. *Nature, 295:*232–234.

Rodman, P. S. 1977. Feeding behaviour of orang-utans of the Kutai Nature Reserve, East Kalimantan. *In* Pri-

mate Ecology: Studies of Feeding and Ranging Behaviour in Lemurs, Monkeys and Apes, pp. 383–413, ed. T. H. Clutton-Brock. Academic Press, New York.
Rodman, P. S. 1979. Individual activity patterns and the solitary nature of orang-utans. *In* The Great Apes, pp. 235–255, ed. D. A. Hamburg and E. R. McCown. Benjamin/Cummings Publishing Company, Menlo Park, CA.
Sankhyan, A. R. 1985. Late occurrence of *Sivapithecus* in Indian Siwaliks. *Journal of Human Evolution, 14:*573–578.
Schultz, A. H. 1930. The skeleton of the trunk and limbs of higher primates. *Human Biology, 2:*303–438.
Schultz, A. H. 1936. Characters common to higher primates and characters specific for man. *Quarterly Review of Biology, 11:* 259–283, 425–455.
Schultz, A. H. 1937. Proportions, variability and asymmetries of the long bones of the limbs and the clavicles in man and apes. *Human Biology, 9:*281–328.
Schultz, A. H. 1941. Growth and development of the orang-utan. Carnegie Institution of Washington Publication 525. *Contributions to Embryology, 29*(182):57–110.
Schultz, A. H. 1953. The relative thickness of the long bones and the vertebrae in primates. *American Journal of Physical Anthropology, 11:*277–311.
Schultz, A. H. 1956. Postembryonic age changes. *Primatologia, 1:*887–964.
Schultz, A. H. 1962. The relative weights of the skeletal parts in adult primates. *American Journal of Physical Anthropology, 20:*1–10.
Schultz, A. H. 1963a. Relations between the lengths of the main parts of the foot skeleton in primates. *Folia Primatologica, 1:*150–171.
Schultz, A. H. 1963b. The relative lengths of the foot skeleton and its main parts in primates. *Symposium Zoological Society, London, 10:*199–206.
Schultz, A. H. 1968. The recent hominoid primates. *In* Perspectives on Human Evolution, pp. 122–195, ed. S. L. Washburn and P. C. Jay. Holt, Rinehart and Winston, New York.
Schultz, A. H. 1969a. Observations on the acetabulum of primates. *Folia Primatologica, 11:*181–199.
Schultz, A. H. 1969b. The Life of Primates. Weidenfeld and Nicolson, New York.
Schwartz, J. H. 1984. The evolutionary relationships of man and orang-utans. *Nature, 308:*501–505.
Shea, B. T. 1985. On aspects of skull form in African apes and orang-utans, with implications for hominoid evolution. *American Journal of Physical Anthropology, 68:*329–342.
Short, R. V. 1981. Sexual selection in man and the great apes. *In* Reproductive Biology of the Great Apes: Comparative and Biomedical Perspectives, pp. 319–341, ed. C. E. Graham. Academic Press, New York.
Sibley, C. G. and Ahlquist, J. E. 1984. The phylogeny of the hominoid primates, as indicated by DNA-DNA hybridization. *Journal of Molecular Evolution, 20:*2–15.
Sigmon, B. A. 1974. A functional analysis of pongid hip and thigh musculature. *Journal of Human Evolution, 3:*161–185.
Smith, R. J. and Pilbeam, D. R. 1980. Evolution of the orang-utan. *Nature, 284:*447–448.
Sugardjito, J. 1982. Locomotor behaviour of the Sumatran orang-utan *(Pongo pygmaeus abelii)* at Ketambe, Gunung Leuser National Park. *Malayan Nature Journal, 35:*57–64.
Temerin, L. A. 1980. Evolution of the orang-utan. *Nature, 288:*301.
Tobias, P. V. 1975. Brain evolution in the Hominoidea. *In* Primate Functional Morphology and Evolution, pp. 353–392, ed. R. H. Tuttle. Mouton Publishers, The Hague.
Tuttle, R. H. 1969. Quantitative and functional studies on the hands of the Anthropoidea. I. The Hominoidea. *Journal of Morphology, 128:*309–363.
Tuttle, R. H. 1970. Postural, propulsive, and prehensile capabilities in the cheiridia of chimpanzees and other great apes. *In* The Chimpanzee, vol. 2, pp. 167–253, ed. G. H. Bourne. S. Karger, Basel.
Tuttle, R. H. 1975. Parallelism, brachiation, and hominoid phylogeny. *In* Phylogeny of the Primates. A Multidisciplinary Approach, pp. 447–480, ed. W. P. Luckett and F. S. Szalay. Plenum Press, New York.
Tuttle, R. H. and Rogers, C. M. 1966. Genetic and selective factors in reduction of the hallux in *Pongo pygmaeus. American Journal of Physical Anthropology, 24:*191–198.
von Koenigswald, G. H. R. 1982. Distribution and evolution of the orang-utan, *Pongo pygmaeus* (Hoppius). In The Orang-utan: Its Biology and Conservation, pp. 1–15, ed. L. E. M. DeBoer. W. Junk, The Hague.
Washburn, S. L. 1963. Behavior and human evolution. *In* Classification and Human Evolution, pp. 190–203, ed. S. L. Washburn. Aldine Publishing Company, Chicago.
Washburn, S. L. 1968. The study of human evolution. Condon Lectures. Oregon State System of Higher Education, Eugene.
Zihlman, A. L. 1976. Sexual dimorphism and its behavioral implications in early hominids. *In* Les Plus Anciens Hominides, pp. 268–293, ed. P. V. Tobias and Y. Coppens. CNRS, Paris.
Zihlman, A. L. 1982. Sexual dimorphism in *Homo erectus. In* L'*Homo erectus* et la Place de L'Homme de Tautavel Parmi les Hominides Fossiles. *Congres International de Paleontologie Humaine, Ier Congres, 2:*947–970.
Zihlman, A. L. 1984. Body build and tissue composition in *Pan paniscus* and *Pan troglodytes* with comparisons to other hominoids. *In* The Pygmy Chimpanzee, pp. 179–200, ed. R. L. Susman. Plenum Press, New York.
Zihlman, A. L. 1985. *Australopithecus afarensis:* two sexes or two species? *In* Hominid Evolution: Past, Present and Future, pp. 213–220, ed. P. V. Tobias. Alan R. Liss, New York.

NOTES

1. The adult male, "Ben," was acquired by M. W. Marzke from the Phoenix Zoo. He died of natural causes while in residence at the Los Angeles Zoo. Ben weighed 102.06 kg at death. He was dissected by L. Brunker, M.

W. Marzke, M. E. Morbeck, and A. L. Zihlman at Arizona State University. The brain was removed prior to dissection and no weight was provided. An estimate of 434 g was taken to represent average male orang-utan cranial capacity (Tobias, 1975). Little difference in percentage of TBW was found using estimates of 415 and 455 g.

2. The captive-born female, "Bunga," was obtained by T. I. Grand from Yerkes Regional Primate Research Center and dissected primarily by L. Brunker at the Oregon Regional Primate Research Center. She weighed 27.8 kg at death. Zihlman (1984) reported that this female lies within the adult female range published by Schultz (1941). Smithsonian Institution individual number 153822 listed at 27.22 kg in Schultz's data, however, apparently was gutted before being weighed (Smithsonian Institution records). Therefore, Bunga lies below the known range for free-ranging adult females.

22

Functional Anatomy of the Cheiridia

M. D. ROSE

The following account is based on a review of the recent literature, supplemented by personal observations derived from dissection of an orang-utan and other ape cadavers and by examination of skeletal material. All figures cited are mean values.

Pongo pygmaeus, the orang-utan, is one of the most specialized catarrhine primates in terms of positional behavior and positional anatomy. Cant (1985) has provided the most detailed study of orang-utan positional behavior in natural environments. In clambering, which accounts for 51% of its arboreal locomotion, an upright orang-utan uses a combination of fore- and hindlimb suspension and hindlimb support. Other arboreal activities include climbing up and down vertical supports (18%), quadrupedalism (12%), and forelimb suspension (11%). About half of the postural activities used during feeding include suspension, usually with the body held horizontally, and typically involving a combination of hand and foot grasping. As a result of marked sexual size dimorphism, females use more postural suspension than males, while males come to the ground more than females (MacKinnon, 1974a,b). Thus positional behavior is dominated by activities during which the fore- and hindlimbs are oriented at varying degrees to the trunk so that different combinations of hand and foot grips can be taken among supports of different diameters and orientations. Most activities involve some form of suspension. Many features of overall body proportions and of the detailed construction of the cheiridia provide the anatomical basis for these activities.

The orang-utan has an intermembral index of 144, which is exceeded only by the 149 of the siamang, a lesser ape (Schultz, 1973). Jungers (1984, 1985) has shown that, when the orang-utan's size is taken into consideration, this is due to a unique combination of forelimb elongation and, especially, of hindlimb shortening. These proportional features are advantageous for a number of arboreal activities. The advantages of long forelimbs for forelimb suspension have been extensively documented (see, e.g., Jungers and Stern, 1984; Preuschoft and Demes, 1984). The combination of long forelimbs with short hindlimbs may be of particular significance in the more frequently used types of suspension involving the use of three or four limbs. Because of great mobility at the shoulder and hip, the cheiridia can grasp almost anywhere around an animal. By using short-range (hindlimb) and long-range (forelimb) grasping, an animal can choose from among a relatively large number of supports and can select these grasping points so that its limbs are in positions where large turning moments at joints do not have to be resisted by muscular effort. In addition, orang-utan limb proportions are advantageous for climbing on vertical supports, enabling the animal to distribute its weight so that frictional forces are maintained and muscular effort is minimized (Sarmiento, 1985).

The orang-utan's hands and feet are both used in the same way during most types of arboreal locomotion: the lateral four digits are used as grasping hooks (Fig. 22–3). Limb proportions and mobility, including the use of what are essentially universal joints in the tarsus and carpus, enable the cheiridia to be positioned for grasping. Cheiridial use can be analyzed in terms of the three overlapping functions of action as a load-bearing platform, action as a propulsive lever, and action as a grasping mechanism (Conroy and Rose, 1983). Orang-utans are unique in the degree to which the last function is emphasized; the first two functions are deemphasized in favor of intra-cheiridial mobility. To the extent that the hands and feet are used in similar ways, the term *quadrumanualism* is quite accurate for describing orang-utan positional behavior. To the extent that specifically hook-like grasping takes place, the term *quadruhamulitism* might be used (Rose, 1985).

FUNCTIONAL ANATOMY OF THE HAND

Among catarrhines the orang-utan is outstanding neither in the length of its hand compared to forelimb length (27%, Fig. 22–1) nor in hand

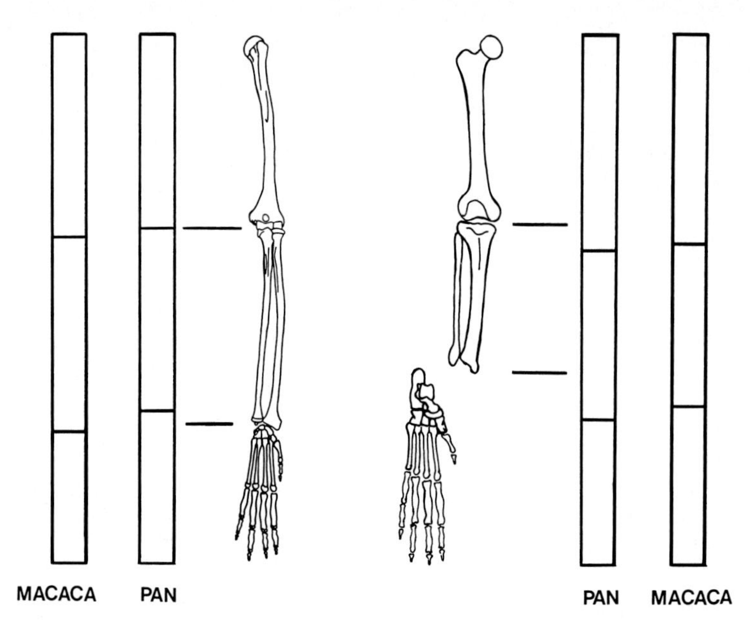

FIG. 22–1. The orang-utan forelimb (on left) and hindlimb (on right) drawn to the same length. Compared to other catarrhines the internal proportions of the forelimb are not remarkable, while in the hindlimb the foot accounts for a relatively large proportion of limb length. The foot is relatively and absolutely longer than the hand.

breadth compared to hand length (34%). As measured along the third ray (Fig. 22–2), the phalangeal part of the orang-utan hand is relatively long (53%) compared to the metacarpal (36%) and carpal (11%) parts of the hand (Erikson, 1963; Napier and Napier, 1967; Schultz, 1956). As in other apes, the formula of finger projection is most frequently III > IV > II > V, while the formula of metacarpal projection is most frequently II > III > IV > V. However, it should be noted that, in more than 40% of orang-utans, the length of the fourth proximal phalanx equals or exceeds that of the third (Susman, 1979). The middle and terminal phalanges of the third ray may also be longer than those of the fourth ray. The effect of emphasizing the fourth finger is to bring it more in line with the second and third fingers, which might be of particular importance

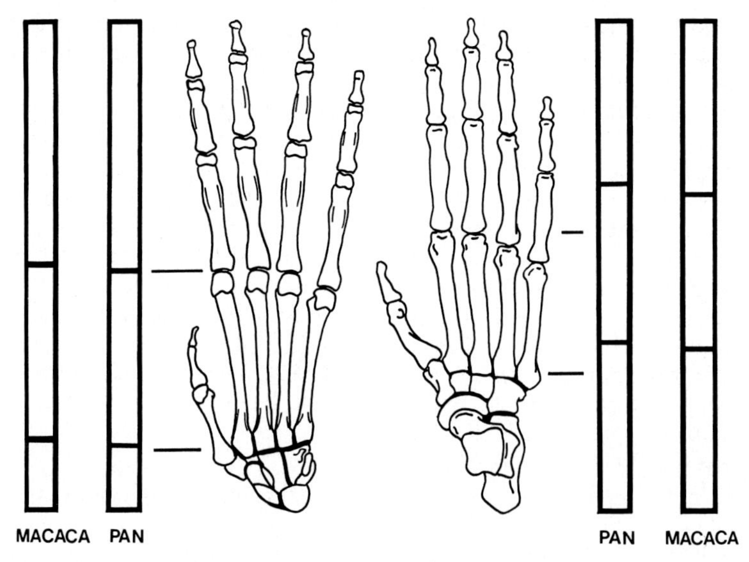

FIG. 22–2. The orang-utan hand (on left, palmar view) and foot (on right, dorsal view) drawn to the same length. Compared to other catarrhines the internal proportions of the hand are not remarkable, while in the foot the toes are relatively long and the tarsus is relatively short. The toes are relatively and absolutely shorter than the fingers.

for hook grips and double-locked grips (Fig. 22–4) taken on small-diameter branches.

In the lateral four fingers the middle and terminal phalanges receive the insertions of the long digital muscle tendons. These segments of the fingers provide the lock for finger grips. The proximal phalanges receive at least part of the insertion of the intrinsic finger muscles and are the main points of suspension for the rest of the body during suspensory grasping. The metacarpals provide not only the basis for movement between the grasping part of the hand and the rest of the body, but also the means by which the digits are hafted onto the carpus.

Orang-utan phalanges show a number of unique specializations. Although, as in other apes, the proximal phalanx of a ray is approximately equal in length to the combined lengths of the middle and distal phalanges (Napier, 1960), the proximal phalanx itself is long. Thus the orang-utan third proximal phalanx is 73% as long as the third metacarpal (Susman, 1979). Only lesser apes have proximal phalanges of comparable length. It is these intraphalangeal and metacarpophalangeal proportions that, together with the emphasis on flexion at interphalangeal and metacarpophalangeal joints (Napier, 1959), underlie the double-locked grip used by all great apes (Fig. 22–4), but especially by the orang-utan (Napier, 1956). Since double-locked grips are particularly advantageous for grasping small-diameter supports, they are an alternative to the thumb-assisted power grip used by many other primates (Napier, 1956). In the orang-utan the approximately transverse alignment of the proximal interphalangeal joints of the third, fourth, and fifth fingers, in conjuction with tapering finger tips, is another feature related to the efficacy of the double-locked grip.

The phalanges are gracile (Fig. 22–2). Thus the midshaft mediolateral diameter of the proximal phalanges in orang-utans (and lesser apes) is only 16% of the proximodistal length; the range for other catarrhines is 19–31%. The proximal and middle phalanges of the orang-utan are also more curved longitudinally than in other primates (Figs. 22–3 and 22–4). In terms of the size-independent measure of included angle of Susman et al. (1984), orang-utan phalangeal curvature is greater than 60°, which surpasses that of the lesser apes (53°) and chimpanzees (42°). The curvature is greatest in the distal third of the proximal phalanx, but is more uniform along the middle phalanx. Sarmiento (1985) has shown that, with phalangeal curvature of the degree found in the orang-utan, the muscular effort required to maintain frictional forces between the fingers and a support is minimized. Sarmiento also suggested that the particular patterns of curvature in the orang-utan are related to forces applied to the phalanges via the long flexor tendons when the interphalangeal joints are in their fairly flexed working positions (Figs. 22–3 and 22–4). The phalangeal shafts are oval in cross section

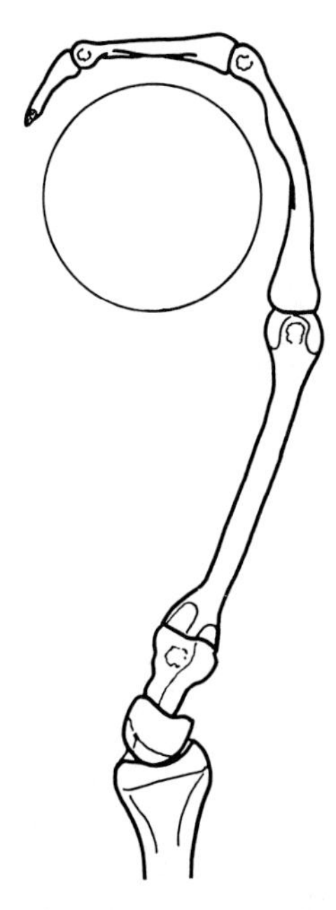

FIG. 22–3. An orang-utan suspensory hook grip. Only the left third finger is shown. The curvature of the phalanges and the flexion set of the finger joints result in a neutral position that fits a support of the diameter illustrated. In suspensory postures the wrist is extended at the midcarpal joint (capitatolunate joint shown) and radiocarpal joint (radiolunate joint shown).

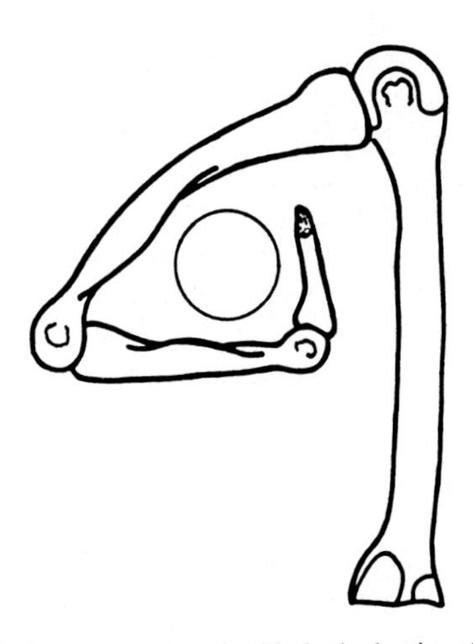

FIG. 22–4. An orang-utan double-locked grip of the left third finger on a small-diameter support. The metacarpophalangeal and interphalangeal joints are in their close-packed positions of full flexion.

and approach circularity proximally, which is in marked contrast to the situation in African apes, where the palmar surface of the shafts presents a concavity that is deepest between the ridges for the flexor tendon sheaths (Susman, 1979). Thus, although orang-utan phalanges are relatively gracile, they present features of longitudinal curvature and strategic shaft robusticity that are adequate to withstand the considerable forces that may act on them.

Orang-utan interphalangeal joints are basically uniaxial structures that are stabilized by the trochlear shape of the articular surfaces, which are particularly deep dorsopalmarly (Susman, 1979), and by strong collateral ligaments. They close-pack (ie., assume their most stable configuration) in full flexion (Fig. 22–4). As in other apes, the orang-utan has an extensive flexion range at its interphalangeal joints (Napier, 1960; Tuttle, 1969). As in all primates, finger grips are primarily dependent on flexion of the distal interphalangeal joints, produced by the action of *m. flexor digitorum profundus*, and at the proximal interphalangeal joints, produced by the action of *m. flexor digitorum superficialis*. However, this mechanism shows a number of specializations in orang-utans. Within the orang-utan forearm the finger flexor mass is characterized by separate origins of the two muscles and by separation of the muscle fascicules inserting into tendons for individual fingers. This allows for independent action of individual fingers and of individual joints within the fingers so that precise adjustments to grips can be made. Frequent reference will be made below to quantitative aspects of cheiridial muscle weight and joint excursions which, unless otherwise stated, are all taken from the invaluable studies of Tuttle (1969a, 1970, 1972).

As in other apes, orang-utan finger flexors account for a large proportion (over 40%) of the weight of the forearm musculature. The bulky flexor mass originates extensively from the fascias as well as from the interosseous membrane of the forearm. In both orang-utans and chimpanzees, the deep flexor is 50% heavier than the superficial flexor, whereas in lesser apes, the two muscles are more equally developed. This implies that, in great apes, distal interphalangeal flexion is particularly important for double-locked grips as well as for climbing on large-diameter supports, during which there is a premium on applying force as far as possible around the support.

Within the hand the long flexor tendons are tunnelled through the palm and along the fingers to prevent bowstringing. Orang-utans do not have a particularly deep carpal tunnel (Robertson, 1984; see also Fig. 22–6D). However, within the orang-utan wrist and palm, the tendons are tied down by a series of well-developed soft tissue structures that include the flexor retinaculum (which closes the carpal tunnel), the palmar aponeurosis, dorsopalmarly oriented septa separating the midpalmar space from the thenar and hypothenar eminences (to which the palmaris longus tendon largely contributes), and dorsopalmarly oriented septa connecting around individual tendons in the palm. On orang-utan phalanges the tendons are tunnelled by fibrous sheaths attached to ridges on the phalanges (Figs. 22–3 and 22–4). These sheaths are relatively long, extending for approximately 30% of the length of the bone. The sheaths are placed more distally than in African apes and are thickest distally, in the region where the phalanges are maximally curved. These sheaths are thus strategically developed and positioned so as to withstand forces tending to bowstring the long flexor tendons at points of maximal curvature of the phalanges.

In orang-utans the finger flexors weigh approximately four times as much as the finger extensors; the ratio of flexor to extensor weight is approximately 5:1 in the African apes and 6:1 in the lesser apes. This feature in orang-utans may reflect their more precise control of body movements, especially during suspensory grasping.

The extensor mass of orang-utans is not as well differentiated as the flexor mass. The long extensor tendons expand into hoods over the proximal phalanges and there receive a part of the insertions of the *m. lumbricalis* and *mm. interossei*. Well-developed retinacular (oblique) ligaments arise from the flexor tendons on the middle phalanges and insert into the extensor tendons on the middle phalanges. As in human fingers, these retinacular ligaments probably serve to stabilize the flexed distal phalanx (Bendz, 1985). Cutaneous ligaments, described for the human hand (Kaplan and Milford, 1984), are also well developed in the orang-utan and serve to control skin movements over the joints. Load-bearing subcutaneous fibrofatty cushions are situated superficially on the palmar aspects of the distal parts of the metacarpals, and proximal and middle phalanges.

Similar to the phalanges, the lateral four metacarpals of the orang-utan are relatively gracile (Figs. 22–2 to 22–4). These metacarpal shafts differ distinctively from those of other hominoids in being greater in dorsopalmar than in mediolateral diameter and in having thicker cortices (Susman, 1979) and can thus withstand considerable forces tending to flex them. Furthermore, in orang-utans, the metacarpal heads are broad, reaching their maximum mediolateral diameter palmarly. As in lesser apes, the orang-utan metacarpal heads point somewhat palmarly (Susman, 1979). The metacarpophalangeal joints are thus markedly flexed in the mid-position of the flexion-extension range (that is, they have a *flexion set*) and close-pack in full flexion (Fig. 22–4). The proximal articular surfaces of the orang-utan's proximal phalanges are also directed somewhat palmarly, being oriented at an angle of 108° to the long axis of the bone, which is beyond the range of 85–102° of other catarrhines. The combined effect of the flexion set to the finger joints and longitudinal curvatures to the bones is to give a markedly hook-like position of rest to the hand (Fig. 22–3). Flexion-extension at the lateral four metacarpophalangeal joints is extensive (170° [Napier, 1960]), but extension (19°), while greater than in lesser apes (0°), is restricted compared to African apes (50°). In apes, slight conjunct abduction-adduction and axial rotation accompany flexion at the second and fifth metacarpophalangeal joints (Lewis, 1977).

The proximal surface of the orang-utan proximal phalanx is circular in outline and equally curved across all diameters; in dorsal view, the metacarpal heads are also quite strongly curved. Thus abduction-adduction at the metacarpophalangeal joints is probably extensive. As in the African apes, the *mm. interossei* of the orang-utan account for a large proportion of its intrinsic hand muscle weight (over 60%). The dorsal *mm. interossei* weigh approximately three times as much as the palmar *mm. interossei* and may serve to stabilize the abducted fingers against forces tending to adduct them. However, because of their insertions into the proximal phalangeal bases, the dorsal *mm. interossei* may more importantly serve to flex the metacarpophalangeal joints. The lateral four orang-utan metacarpals articulate with the carpus as a transverse series of carpometacarpal joints that are strengthened by well-developed interosseous ligaments.

The orang-utan pollex is extremely short, being only 30% of the total hand length (Fig. 22–2). Relative pollical length in other catarrhines ranges between 36 and 46% (Napier and Napier, 1967). In the great apes, the pollical muscles account for about 25% of the mass of intrinsic hand muscles, whereas these muscles are more prominent in the lesser apes (35%). However, orang-utans differ markedly from the African apes in that *m. adductor pollicis* is less well developed, thus reflecting the reduced importance of pollex-assisted power gripping in orang-utans and supporting Tuttle's (1969a) suggestion that the orang-utan pollex is used primarily during manipulation rather than grasping. The particularly well-diffentiated intrinsic pollical muscles and the nature of their fasciculation in the orang-utan are consistent with this interpretation.

Within the wrist, overlapping sets of osteoarthrological features are representative of the set of joints, their ranges of movement, and their ability to withstand tensile and/or compressive forces. As in African apes, total muscle mass inserting around the wrist in the orang-utan is approximately half that inserting into the fingers. In orang-utans, wrist flexor mass is only slightly greater than wrist extensor mass (109%), whereas in other apes, flexor mass is considerably greater than extensor mass (chimpanzee, 141%; lesser apes, 156%). As suggested for the finger musculature, this wrist flexor-to-extensor ratio may reflect the orang-utan's control of body movements during suspensory grasping. Flexion-extension range at the orang-utan wrist is 216°, which is greater than in African apes (chimpanzee, 159°) but less than in lesser apes (233°). Extension range of the orang-utan wrist (85°) is greater than in other apes (42–76°).

Osteological features suggest that flexion-extension movements are partitioned between the radiocarpal and midcarpal joints. The scaphoid and lunate are relatively deep and strongly curved dorsopalmarly, but the distal surface of the radius is less strongly curved. In the midcarpal joint, the basically ball-and-socket articulation around the capitate and hamate heads provides a good basis for flexion-extension movements (Figs. 22–6B,D). In terms of axis alignment and the placement of the proximal articular surface of the lunate on the body of the bone, Robertson (1984) has shown that there is an extension set to the orang-utan radiocarpal joint. This contasts with the more flexed configuration in other apes. A stable position of wrist extension occurs during orang-utan suspensory grasping (Fig. 22–3) and may also occur during climbing.

In the great apes, wrist abductors (radial deviators) are about twice as massive as wrist adductors (ulnar deviators); wrist abductors are even more massive in lesser apes (230%). Abduction-adduction range is 110° in African apes, 120° in lesser apes, and 147° in orang-utans, in which adduction range (98°) is greater than abduction range (49°). In terms of the orientation of the proximal articular surface of the lunate with respect to the transverse plane, there is a slight adduction set to the orang-utan radiocarpal joint (Robertson, 1984; see also Fig. 22–6A). As indicated by the orientation of the articular surface of the hamate (in the distal row) relative to the triquetral (in the proximal row), there is an even more marked adduction set to the midcarpal joint (Robertson, 1984; see also Figs. 22–6A,B); this is more marked in orang-utans than in other higher primates. Jenkins and Fleagle (1975) and Sarmiento (1985) point to the long, even, and strong mediolateral curvature of the radiocarpal joint (Fig. 22–6A) as evidence that abduction-adduction movement is primarily localized there. The proximal surface of the relatively large lunate is a major contributor to this complex. However, the shapes of the bones concerned indicate that some abduction-adduction also takes place at the midcarpal joint (Fig. 22–6B). The presence of a soft-tissue-filled space between the distal ulna and the triquetrum (Fig. 22–5) may facilitate adduction in the wrist.

Other orang-utan carpal features related to both flexion-extension and abduction-adduction movements have been studied by Sarmiento (1985). He (also Robertson, 1984) points out that the pisiform is angulated to point as much distally as it does palmarly (Figs. 22–5 and 22–6A); it is also relatively short (Robertson, 1984). The pisiform is placed distally within the wrist because of the orientation of the joint between the hamate and the triquetrum (upon which the pisiform is carried). As a result, the pisiform provides a favorable lever arm for *m. flexor carpi ulnaris* acting as a flexor, despite its relative shortness (Sarmiento, 1985). This action is effective even when the wrist is already in a fairly flexed position, which, in effect, enables an animal to keep its wrist in a flexed position, against extending forces, during finger grasping occurring during climbing as well as suspensory activities.

The distal placement of the pisiform also provides *m. flexor carpi ulnaris* with a favorable lever arm for adduction at the radiocarpal joint (Sarimento, 1985). Approximation of the pisiform to the hamate, in conjunction with a strong pisohamate ligament, allows the flexor to produce as much adduction as occurs at the midcarpal joint. Because of the proportions of the carpus as a whole, these actions are equally effective throughout the abduction-adduction range. Thus, although abductor musculature is emphasized, the development of favorable lever arms and an extensive range of movement into adduction suggests that, in the orang-utan, the adducted position is an important one. During climbing the hand is placed in an adducted position prior to hoisting the body (Fig. 22–5). In addition, the adductor mechanism may serve to counteract net forces that otherwise would tend to abduct the wrist during the power phase of climbing (Sarmiento, 1985).

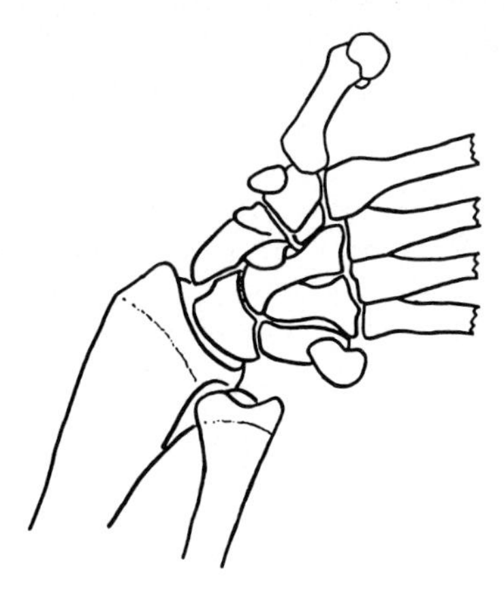

FIG. 22–5. Orang-utan left hand in an adducted position. Note the distal placement of the pisiform within the wrist. During loading in this position force is transmitted between the lateral three digits, the capitate and hamate in the distal carpal row, the relatively large lunate in the proximal row, and the radius. Note the medial extension of the distal radius, forming a shelf between the radiolunate and distal radiolunar joints. Figure redrawn with permission from Sarmiento (1985). Original drawing based on a radiograph.

The ability to effectively pronate-supinate the wrist is critical for suspensory activities and, in the orang-utan, a clearly defined complex of features within the midcarpal region suggests that these movements are primarily centered there (Fig. 22–6). One would expect these movements to result either from passive action of the suspended body or from rotating forces generated at the shoulder and/or radioulnar joints. Thus, the effect of muscles inserting into the hand should be minimal. In this regard, Robertson (1984) has shown that there is a supination set to the radiocarpal and, especially, to the midcarpal joint, which she relates to hand placement during climbing. Certainly, such a set would be particularly effective in bringing the ulnar side of the hand into contact with a vertical support, without necessitating rotation of the forearm beyond the midpoint of its range of pronation-supination. This would in turn allow for the most balanced action of the long digital flexors to each of the lateral four fingers.

The work of Jenkins and Fleagle (1975), Jenkins (1981), and Sarmiento (1985) provides a fairly clear picture of the mechanisms underlying these movements. In orang-utans, the midcarpal joint approximates a ball-and-socket. The ball is formed by the heads of the capitate and

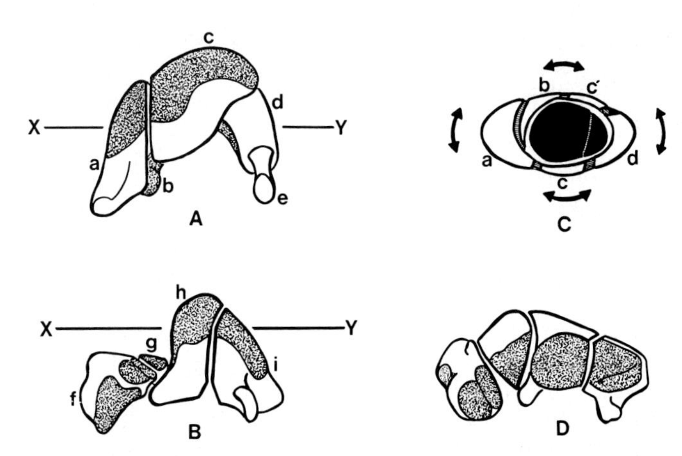

FIG. 22–6. Bones of the right wrist. The articular surfaces are stippled. (A) Palmar view of the bones of the proximal row. From the lateral (radial) to the medial (ulnar) side: (a) scaphoid, (b) part of centrale, (c) lunate, (d) triquetrum, (e) pisiform. Note the strong mediolateral curvature to the articular surfaces of the scaphoid and lunate. (B) Palmar view of the bones of the distal row: (f) trapezium, (g) trapezoid, (h) capitate, (i) hamate. Note that the proximal hamate and capitate form a ball that articulates with the socket formed by the distal surfaces of the bones of the proximal row. Note also the gutter formed between the trapezium and trapezoid laterally, and the capitate medially. (C) Diagrammatic transverse section in the plane of X–Y in A and B. The capitate and hamate heads are in black. The bones surrounding them are (a) scaphoid, (b) centrale, (c) palmar lip of lunate, (c′) dorsal lip of lunate, (d) triquetrum. Rotation of the bones of the proximal row around the bones of the distal row takes place during wrist pronation-supination movements. (D) Proximal view of the bones of the distal row.

hamate (Fig. 22–6), which are closely bound by a strong interosseous ligament; the capitate contributes to most of the articular surface. The socket is formed by the following structures: the relatively small triquetrum, which articulates with the hamate on the medial (ulnar) side of the wrist; the large lunate, which articulates with the head of the capitate; the scaphoid and centrale, which articulate with the head and lateral side of the capitate. It is thus probable that pronation and supination are equally extensive from the neutral position. During these movements (Fig. 22–6C), the lunate largely spins in place, while the scaphoid, carrying the centrale with it, moves along a path that approximates the arc of a circle, dorsally during supination, or palmarly during pronation; meanwhile the triquetrum follows a similar path, but in the opposite direction. These movements are guided in part by the movement of the distal centrale and scaphoid within a gutter formed between the radial side of the distal capitate and the proximal surfaces of the trapezium and trapezoid (Figs. 22–6B,D). Because this gutter does not narrow markedly along its length, pronation and supination in the orang-utan are unrestricted.

Sarmiento (1985) has also investigated a number of features related to the ability of the orang-utan carpus to withstand compressive and tensile forces. The orang-utan hand is compressively loaded while in an adducted position during climbing and during quadrupedal activities (Fig. 22–5). The lunate occupies a key position in transmitting force between the hand and forearm. The triquetrum is largely excluded from a load-bearing function because of its distomedial placement on the hamate (Figs. 22–6A,B). The extensive contact between the lunate and capitate and the close association of the capitate and hamate facilitate the transmission of force through the carpus via the lateral three digits. The radiolunate joint is extensive, due in large part to an ulnar extension of the distal radius (Fig. 22–5) that ends so as to align the distal radioulnar joint more towards a parallel than an orthogonal position relative to the radiolunate joint. While reducing shear between the bones concerned, this orientation also allows force to be transmitted to the distal ulna via the radial shelf, as well as directly to the radius.

The relatively large scaphoid (Fig. 22–6A) becomes more important when the orang-utan hand is loaded from the radial side. The scapho-centrale and radioscaphoid joints are aligned approximately parallel to each other and are orthogonal to compressive force when the hand is in the abducted position. Additional features related to the orang-utan hand's ability to withstand tensile stress include strong carpometacarpal ligaments and strong interosseous ligaments between the bones of the distal row; a palmar radiocarpal ligament that is well developed, especially with regard to its insertion into the lunate and capitate; and a well-developed palmar projection of the distal capitate (Fig. 22–6D) that serves as an attachment site for the palmar radiocarpal ligament as well as for a strong flexor retinaculum.

FUNCTIONAL ANATOMY OF THE FOOT

The orang-utan foot is unique in a number of its proportions (Figs. 22–1 and 22–2). It is remarkably long within the short hindlimb, accounting for 35% of total hindlimb length (Erikson, 1963; Lessertiseur and Jouffroy, 1973; Schultz, 1956), and being therefore relatively longer than the feet of all other catarrhines (19–31%). In contrast with the African apes, the orang-utan's foot is also considerably longer than its hand (142%, Erikson, 1963; 117%, Schultz, 1956); in lesser apes the hand is frequently longer than the foot (Schultz, 1956). Within the orang-utan foot, the phalangeal segment (43% of total length) is longer than the metatarsal (31%) and tarsal (26%) segments (Lessertiseur and Jouffroy, 1973). The phalanges are longer (36%) while the tarsus is shorter (35%) than those of the African apes. Although orang-utan toes are only about 80% as long as the fingers, the tarsus is longer than the carpus. Orang-utan foot proportions thus differ markedly from those of other higher primates, and, together with other features concerning the detailed structure and function of the foot, reflect the fact that the orang-utan's hands and feet act similarly during quadrumanous positional activities. However, differences in the absolute lengths of the fingers and toes, together with the fact that total forearm and hand musculature weighs 128% of total leg and foot musculature, point to the primacy of the forelimb. In higher primates other than orang-utan and lesser apes, leg and foot musculature outweighs forearm and hand musculature.

The digital projection formula of the orang-utan is most frequently III > IV > II > V while the metatarsal formula is typically II > III > IV > V (Fig. 22–2). The third and fourth toes are therefore the longest and most robust in orang-utans. In dorsal view, the metatarsophalangeal and proximal interphalangeal joints are aligned obliquely, distomedially-to-proximolaterally, rather than transversely, as is the case in the equivalent joints of the hand (Fig. 22–2). This suggests that there may be more obliquity to foot grips than to hand grips and that the feet may grasp vertical supports more frequently than horizontal supports.

Most of the features noted for the orang-utan's fingers concerning phalangeal morphology, the organization of the finger-moving musculature and its tendons, and the importance of the lateral digits in grasping, hold true for the toes as well. However, the toe phalanges are even more curved than the finger phalanges and are the most curved among hominoids (Gomberg, 1981). The proximal toe phalanges resemble finger phalanges and differ from other hominoids in the plantar set of their proximal articular surfaces (Gomberg, 1981; Tuttle, 1970). There is a similar flexion set to the orang-utan's metatarsophalangeal joints. As in the corresponding joints of the hand, stable positions of the joints are in flexion (Susman et al., 1984). Similar to the lesser apes, toe flexor musculature in the orang-utan accounts for 29% of the total weight of leg and foot musculature, which is much greater than in other catarrhines (14–25%). If it is present, the flexor hallucis longus tendon does not insert into the hallux, but instead, reinforces the long tendons to the third and fourth toes. The orang-utan foot has a well-developed transverse arch (Oxnard and Lisowski, 1980) that partly serves as a tunnel for the long flexor tendons (Conroy and Rose, 1983). This tunnel is closed superficially by the plantar aponeurosis and, at a deeper level, by transverse bands derived in large part from the insertions of the tibialis posterior tendon.

The tarsometatarsal joints of the orang-utan are in line with each other (Fig. 22–2) and are oriented transversely (Lewis, 1980a). This configuration, together with the shape of the articular surfaces, permits more flexion and extension than is possible in the corresponding joints of the hand. Mobility increases from the third to the fourth and fifth tarsometatarsal joints (Gomberg, 1981). Flexion here, together with intratarsal flexion, orients the foot for grasping and also serves as a passive mechanism during the load-bearing phase of quadrupedal activities.

The orang-utan hallux is extremely short, being only 26% of total foot length, and is thus even shorter relatively than the pollex (Fig. 22–2); relative hallucal length in other anthropoids ranges between 31 and 51% (Lessertiseur and Jouffroy, 1973; Schultz, 1963; Tuttle and Rodgers, 1966). Hallucal shortness in the orang-utan is due in part to the fact that there is usually only one hallucal phalanx. Hallucal musculature is correspondingly reduced. As in the hand, the adductor is particularly small. Thus, hallux-assisted power gripping is not an important aspect of foot use in the orang-utan.

The orang-utan talocrural, subtalar, and transverse tarsal joints are all very mobile. Dorsiflexion-plantarflexion is extensive at the talocrural joint, which is stabilized by a relatively deep trochlea on the talus (Langdon, 1984). The plantar flexors only account for 20% by weight of the leg musculature, compared to 43% in the chimpanzee (Gomberg, 1981). This is probably related to the fact that during the orang-utan's climbing and clambering the more powerful forelimbs do a larger proportion of the work of moving the body. Inversion-eversion is extensive at the subtalar joint (Fig. 22–7), with the inversion part of the range being emphasized.

Since the distal talocalcaneal joint is minimally developed in orang-utans, the proximal talocalcaneal joint is the major articulation for inversion-eversion movements at the subtalar joint. Although the relative articular area on the calcaneus is large compared to other hominoids, the corresponding area on the talus is relatively smaller than in other hominoids (Gomberg, 1981); this is a major factor influencing the mobility of the joint. The curvatures of the talar and calcaneal components of the proximal talocalcaneal joint are such that there is minimal contact (congruency) when the joint is in the neutral position (Langdon, 1984), which implies that, during load-bearing, the foot is used in positions of more or less full inversion or eversion, where congruency is maximal. These positions are considered below. The axis of the subtalar joint is aligned closer to the long axis of the foot in orang-utans than in other hominoids, indicating that a relatively large component of movement at the joint is in the same plane as movement at the transverse tarsal joint: this is the supination-pronation component of inversion-eversion. However, the proximal end of the subtalar joint axis is nevertheless placed laterally and plantarly with respect to its distal end. This results in the subtalar part of the foot becoming somewhat adducted and plantarflexed around the talus (Fig. 22–7) during inversion.

In the orang-utan, there is a marked supination set to the transverse tarsal joint (Fig. 22–8) at which pronation-supination is extensive, especially into the supination part of the range. Supination at the transverse tarsal joint combined with inversion at the subtalar joint results in the sole of the foot facing almost directly medially (Fig. 22–8); in conjunction with full dorsiflexion at the talocrural joint the foot becomes positioned so that the toes are ideally placed to grip vertical supports. This type of grasping is common during all types of arboreal positional activity.

The focus of orang-utan transverse tarsal movement is the pivot-like calcaneocuboid joint (Lewis, 1980a,b), which is formed by a peg-like process on the proximal cuboid and a corresponding pit on the distal calcaneus (Figs. 22–7 and 22–8). The axis of this pivot roughly parallels the long axis of the foot, and a plantar calcaneocuboid ligament attaches across the plantar aspect of this part of the joint. Movement between the cuboid and the calcaneus about the axis of the pivot and ligament is guided by a

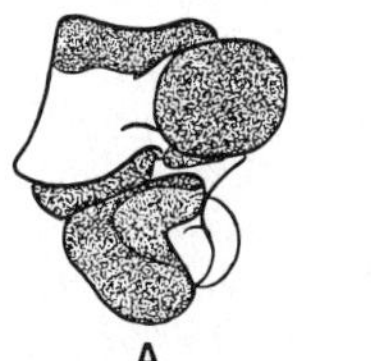

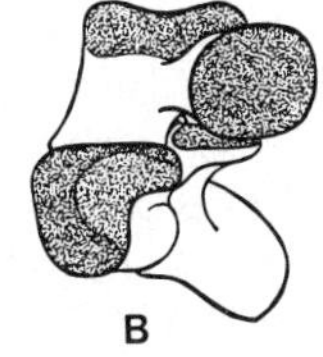

FIG. 22–7. Distal view of right talus (upper) and calcaneus (lower) in inversion (A) and eversion (B). The calcaneus is taken as being the moving element. Note that in the transition from inversion to eversion the calcaneus, carrying the rest of the foot with it, has conjunctly dorsiflexed and abducted.

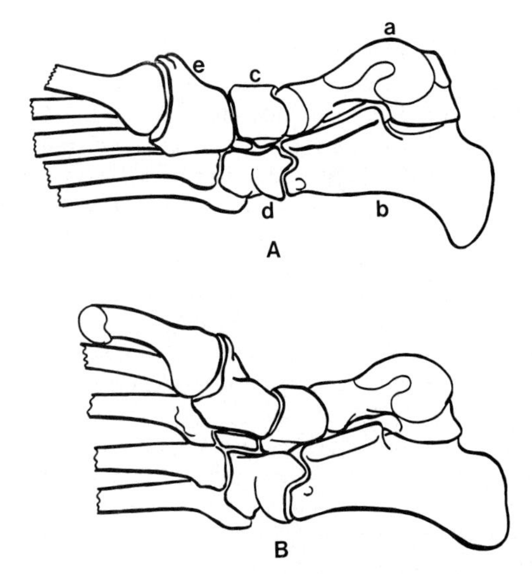

FIG. 22–8. Medial view of the right tarsus in eversion and pronation (A) and inversion and supination (B). Note that there is a supination set, even in the pronated position, and that in B the foot faces almost directly medially. (a) Talus, (b) calcaneus, (c) navicular, (d) cuboid, (e) medial cuneiform. The lateral cuneiform is seen in the interval between the navicular, cuboid, and medial cuneiform. The intermediate cuneiform is hidden by the medial cuneiform in both views.

transversely aligned, approximately planar area surrounding the pivot part of the joint. Since the orang-utan lacks a dorsal calcaneocuboid ligament, a possible constraint to this movement is eliminated (Gomberg, 1981). A planar naviculocuboid joint, which is relatively large compared to those of other hominoids (Gomberg, 1981), enables the cuboid and navicular to move together during transverse tarsal joint movements, and the absence of a plantar naviculocuboid ligament indicates differential movement of the two bones, especially during supination.

In keeping with the general emphasis on the lateral side of the orang-utan foot, the area of the calcaneocuboid joint is greater than that of the talonavicular joint (Gomberg, 1981). The talar head is thus relatively smaller than in other hominoids (Langdon, 1984) and it has a more highly curved surface (Gomberg, 1981). Considerable flexion-extension is possible at the transverse tarsal joint, especially when the joint is in the supinated position. As described above, this capability is probably expressed in conjunction with tarsometatarsal flexion. While there is considerable intratarsal mobility in African apes, their tarsal construction reflects more of a compromise between mobility and load-bearing functions than in the orang-utan.

Plantarflexed, everted, and pronated foot positions occur towards the end of the propulsive phase of orang-utan climbing. Plantarflexed foot positions also occur when the foot grips at shoulder level or when suspension is accomplished by the feet alone. Sarmiento (1983) has shown that a well-developed superficial head of *m. flexor digitorum brevis*, arising from an enlarged heel process on the calcaneus, contributes to effective toe flexion when the foot is plantarflexed. The pronounced heel process of orang-utan also permits an effective power grip, which results from the toes gripping in opposition to the heel part of the foot.

Because the orang-utan forefoot is emphasized and has a relatively large bulk of musculature inserting into it, its construction is adequate to withstand the largely tensile stresses it experiences during positional activities. The fact that the hindlimbs act in concert with the more powerful forelimbs serves to reduce the stress experienced by the foot. However, compressive forces must be experienced by the proximal tarsus during many acitvities. The relative reduction in size of the tarsus, extensive intratarsal mobility, and the reduction or absence of many ligaments (Langdon, 1984) imply that stabilizing mechanisms are only minimally adequate. Most of the intratarsal joints are close-packed in the everted, pronated position of the tarsus (Langdon, 1984), which does stabilize the foot during load-bearing in this position. There is a pronounced dorsolateral-to-plantarmedial set to both the talocrural and subtalar joints (Gomberg, 1981; Langdon, 1984), which, in a supinated, inverted, and dorsiflexed foot position, are brought into planes orthogonal to the forces acting across them, thus contributing to stability in this position. In the case of the talocrural joint, this configuration is most likely to occur if there is concomitant lateral rotation at the hip and flexion at the knee, as occurs during the power stroke of climbing.

QUADRUPEDALISM

Of all higher primates, the orang-utan has the least well-adapted cheiridia for quadrupedal progression on the ground or on large-diameter horizontal supports (Tuttle, 1970). This is indicated in part by the large variety of cheiridial positions the orang-utan assumes during quadrupedal locomotion (MacKinnon, 1974a,b; Sarmiento, 1985; Tuttle, 1967, 1969b). However, the most frequently used positions are instructive in that they illustrate how advantage is taken of functional complexes primarily used during non-quadrupedal arboreal activities. In the relatively infrequently used palmigrade hand position, the robust proximal carpus bears most of the load on the hand. In other hand positions, the fingers are flexed into their positions of maximum stability, and, in different positions, the load is shared by the fingers, the relatively robust ulnar side of the hand, and the proximal carpus. During the frequently used fist-walking position, the hand is loaded through its ulnar side, with the wrist in its stable adducted and extended position (Sarmiento, 1985). Foot positions are as varied as hand positions. As with the fingers, the toes are frequently held in their stable, flexed positions during load-bearing. The foot most frequently bears load in the supinated, inverted, and dorsiflexed position during bipedal as well as quadrupedal activities (Kimura, 1985; Napier, 1964; Sarmiento, 1985; Tuttle, 1969b). However, toward the end of the power stroke of quadrupedal walking, load may be transferred across the flexed toes as the foot moves into its plantarflexed, everted, and pronated configuration.

CONCLUSIONS

In terms of their cheiridial structure and function, and in comparison with other living and fossil primates, orang-utans are highly specialized animals. However, the timing of the acquisition of these specializations, and the environmental changes requiring them, are not clear. On the basis of facial similarities, it has been suggested that *Sivapithecus indicus* from the middle and late Miocene of Asia is a probable orang-utan ancestor (e.g., Andrews and Cronin, 1982; Pilbeam, 1982). However, postcranial specimens that probably belong to *S. indicus* show virtually none of the specializations seen in the modern orang-utan (Pilbeam et al., 1980; Rose, 1984). Instead, these postcranials seem to belong to a fairly generalized great ape, most similar overall to living African apes but lacking their knuckle-walking specializations. Whether or not *S. indicus* is an orang-utan ancestor, evidence of the evolution of the orang-utan's many unique specializations remains to be discovered.

ACKNOWLEDGMENTS

I would like to thank E. E. Sarmiento for his comments and suggestions. I am grateful to G. G. Musser, chairman of the Department of Mammalogy, American Museum of Natural History, for providing access to the collection in his care. Orang-utan and other primate cadavers used for dissection were acquired from the Yerkes Regional Primate Research Center. Results from research supported by NSF Grant BNS 84-18909 are included in this chapter.

REFERENCES

Andrews, P. and Cronin, J. E. 1982. The relationships of *Sivapithecus* and *Ramapithecus* and the evolution of the orang-utan. *Nature, 297:*147–157.

Bendz, P. 1985. The functional significance of the oblique retinacular ligament of Landsmeer. A review and new proposals. *Journal of Hand Surgery, 10*B:25–29.

Cant, J. G. H. 1985. Locomotor and postural behavior of orang-utan *(Pongo pygmaeus)* in Borneo and Sumatra. *American Journal of Physical Anthropology, 66:*153.

Conroy, G. C. and Rose, M. D. 1983. The evolution of the primate foot from the earliest primates to the Miocene hominoids. *Foot and Ankle, 3:*342–364.

Erikson, G. E. 1963. Brachiation in New World monkeys and in anthropoid apes. *Symposia of the Zoological Society of London, 10:*135–164.

Gomberg, D. N. 1981. Form and Function of the Hominoid Foot. Ph.D. Dissertation, University of Massachusetts.

Jenkins, F. A., Jr. 1981. Wrist rotation in primates: a critical adaptation for brachiators. *Symposia of the Zoological Society of London, 48:*429–451.

Jenkins, F. A., Jr. and Fleagle, J. G. 1975. Knuckle walking and the functional anatomy of the wrist in living apes. *In* Primate Functional Morphology and Evolution, pp. 213–228, ed. R. H. Tuttle. Mouton, The Hague.

Jungers, W. L. 1984. Aspects of size and scaling in primate biology with special reference to the locomotor system. *Yearbook of Physical Anthropology, 27:*73–97.

Jungers, W. L. 1985. Body size and scaling of limb proportions in primates. *In* Size and Scaling in Primate Biology, pp. 345–381, ed. W. L. Jungers. Plenum Press, New York.

Jungers, W. L. and Stern, J. T., Jr. 1984. Kinesiological aspects of brachiation in lar gibbons. *In* The Lesser Apes: Evolutionary and Behavioral Biology, pp. 119–134, ed. H. Preuschoft, D. J. Chivers, W. Y. Brockelman, and N. Creel. Edinburgh University Press, Edinburgh.

Kaplan, E. B. and Milford, L. W. 1984. The retinacular systems of the hand. *In* Kaplan's Functional and Surgical Anatomy of the Hand, pp. 245–281, ed. M. Spinner. Lippincott, Philadephia.

Kimura, T. 1985. Bipedal and quadrupedal walking of primates: comparative dynamics. *In* Primate Morphophysiology, Locomotor Analyses and Human Bipedalism, pp. 81–104, ed. S. Kondo. University of Tokyo Press, Tokyo.

Langdon, J. H. 1984. A Comparative Functional Study of the Miocene Hominoid Foot Remains. Ph.D. Dissertation, Yale University.

Lessertiseur, J. and Jouffroy, F. K. 1973. Tendances locomotrices des primates traduites par les proportions du pied. L'adaptation à la bipédie. *Folia Primatologica, 20:*125–160.

Lewis, O. J. 1977. Joint remodelling and the evolution of the human hand. *Journal of Anatomy, 123:*157–201.

Lewis, O. J. 1980a. The joints of the evolving foot. Part II, The intrinsic joints. *Journal of Anatomy, 130:*833–857.

Lewis, O. J. 1980b. The joints of the evolving foot. Part III, The fossil evidence. *Journal of Anatomy, 131:*275–298.

MacKinnon, J. 1974a. Behaviour and ecology of wild orang-utans *(Pongo pygmaeus). Animal Behaviour, 22:*3–74.

MacKinnon, J. 1974b. In Search of the Red Ape. Holt, Rinehart, and Winston, New York.

Napier, J. R. 1956. The prehensile movements of the human hand. *Journal of Bone and Joint Surgery, 38*B: 902–913.

Napier, J. R. 1959. Fossil metacarpals from Swartkrans. Fossil Mammals of Africa No. 17. British Museum (Natural History), London.

Napier, J. R. 1960. Studies of the hands of living primates. *Proceedings of the Zoological Society of London, 134:*647–657.

Napier, J. R. 1964. The evolution of bipedal walking in the hominids. *Archives de Biology (Liége), 75:*673–708.

Napier, J. R. and Napier, P. H. 1967. A Handbook of Living Primates. Academic Press, New York.

Oxnard, C. and Lisowski, F. 1980. Functional articulation of some hominoid foot bones: implications for the Olduvai (Hominid 8) foot. *American Journal of Physical Anthropology, 52:*107–117.

Pilbeam, D. 1982. New hominoid skull material from the Miocene of Pakistan. *Nature, 295:*232–234.

Pilbeam, D. R., Rose, M. D., Badgley, C. and Lipschutz, B. 1980. Miocene hominoids from Pakistan. *Postilla, 181:*1–94.

Preuschoft, H. and Demes, B. 1984. The biomechanics of brachiation. *In* The Lesser Apes: Evolutionary and Behavioral Biology, pp. 96–118, ed. H. Preuschoft, D. J. Chivers, W. Y. Brockelman, and N. Creel. Edinburgh University Press, Edinburgh.

Robertson, M. L. 1984. The Carpus of *Proconsul africanus:* Functional Analysis and Comparison with Selected Non-Human Primates. Ph.D. Dissertation, University of Michigan.

Rose, M. D. 1984. Hominoid postcranial specimens from the Middle Miocene Chinji Formation, Pakistan. *Journal of Human Evolution, 13:*503–516.

Rose, M. D. 1985. Functional anatomy of orang-utan hands and feet. *American Journal of Physical Anthropology, 66:*221–222.

Sarmiento, E. E. 1983. The significance of the heel process in anthropoids. *International Journal of Primatology, 4:*127–152.

Sarmiento, E. E. 1985. Functional Differences in the Skeleton of Wild and Captive Orang-Utans and Their Adaptive Significance. Ph.D. Dissertation, New York University.

Schultz, A. H. 1956. Postembryonic age changes. *Primatologia, 1:*887–964.

Schultz, A. H. 1963. The relative lengths of the foot skeleton and its main parts in primates. *Symposia of the Zoological Society of London, 10:*199–205.

Schultz, A. H. 1973. Age changes, variability and generic differences in body proportions of recent hominoids. *Folia Primatologica, 19:*338–359.

Susman, R. L. 1979. Comparative and functional morphology of hominoid fingers. *American Journal of Physical Anthropology, 50:*215–236.

Susman, R. L., Stern, J. T., Jr. and Jungers, W. J. 1984. Arboreality and bipedality in the Hadar hominids. *Folia Primatologica, 42:*113–156.

Tuttle, R. H. 1967. Knuckle-walking and the evolution of hominoid hands. *American Journal of Physical Anthropology, 26:*171–206.

Tuttle, R. H. 1969a. Quantitative and functional studies on the hands of the Anthropoidea. I. The Hominoidea. *Journal of Morphology, 128:*309–364.

Tuttle, R. H. 1969b. Knuckle-walking and the problem of human origins. *Science, 166:*953–961.

Tuttle, R. H. 1970. Postural, propulsive, and prehensile capabilities in the cheiridia of chimpanzees and other great apes. *In* The Chimpanzee, vol. 2, pp. 167–253, ed. G. Bourne. Karger, Basel.

Tuttle, R. H. 1972. Relative mass of cheiridial muscles in catarrhine primates. *In* The Functional and Evolutionary Biology of Primates, pp. 262–291, ed. R. H. Tuttle. Aldine-Atherton, Chicago.

Tuttle, R. H. and Rodgers, C. M. 1966. Genetic and selective factors in the reduction of the hallux in *Pongo pygmaeus. American Journal of Physical Anthropology, 24:*191–198.

23

Positional Behavior, Adaptive Complexes, and Evolution

RUSSELL H. TUTTLE AND GERALD W. CORTRIGHT

Look deeply into the face of a healthy orang-utan. The intelligence that emanates from its eyes gives cause to wonder how so bright a beast ended up with such a bizarre body. If we cannot explain *Pongo pygmaeus*, how can we expect to chronicle the morphological evolution of other evolutionary oddballs, including *Homo sapiens*?

Two theories on the evolution of *Pongo pygmaeus* are extant. Classically, the orang-utan is thought to have been derived from a stock of versatilely climbing, brachiating apes. Its remote and proximate ancestors were arboreal, and it remained largely treebound into historic times (Tuttle, 1974, 1975).

According to a newer theory, orang-utans are secondarily, though profoundly, adapted to life in the canopy because expanding populations of increasingly sophisticated hominid hunters forced them from the ground to relative safety in trees, where they not only developed peculiar locomotor structures but also became desocialized, presumably to enhance their obscurity (MacKinnon, 1971, 1974; Rijksen, 1978; Tuttle, 1986).

As with the vast majority of evolutionary biological puzzles, we do not have sufficient fossils and natural historical and experimental morphological data to resolve questions of orang-utan origins and transformations during the 10 million or more years (Cronin, 1983) that their lineage has graced the earth. However, during the past two decades orang-utans have been studied intensively at several localities in Kalimantan (Mitani 1985a,b; Galdikas, 1979, 1981a, 1982, 1983, 1984, 1985a,b; Rodman, 1973, 1977, 1979, 1984), East Malaysia (Davenport, 1967; Horr, 1972, 1975, 1977; MacKinnon 1971, 1974, 1979) and Sumatra (MacKinnon, 1974, 1979; Rijksen, 1978; Sugardjito, 1982, 1983). Further, comparative and functional morphological studies have been conducted on significant and small samples of captives (Cortright, 1983; Day and Napier, 1963; Tuttle, 1969a,b, 1970a,b, 1972, 1974; Tuttle and Basmajian 1974, 1977, 1978a,b; Tuttle et al., 1978; Tuttle, Basmajian and Ishida, 1979; Tuttle, Cortright and Buxhoeveden, 1979). And fossils that ultimately may prove to be ancestral to *Pongo* have been recovered (Andrews, 1983; Andrews and Cronin, 1982; Andrews and Tekkaya, 1980; Lipson and Pilbeam, 1982; Pilbeam, 1982, 1985; Pilbeam and Smith, 1981; Pilbeam et al., 1977, 1980; Preuss, 1982; Raza et al., 1983; Rose, 1984; von Koenigswald, 1982). A synthesis will be instructive now, modestly, to show that even large quantities of data can be inadequate vis-à-vis big evolutionary questions and, ambitiously, as a premise for informed speculations on the career of the red apes.

LOCOMOTION

Orang-utans are preeminently versatile climbers (Tuttle, 1975, 1977, 1986; Fig. 23–1). They traverse the canopy, sometimes on regular "highways," by clever, cautious bridging transfers, vertical ascents and descents on trunks and vines, hoisting, and occasional pedally assisted arm-swinging and quadrepedal suspensory movements beneath branches. If an orang-utan cannot reach a new support to cross a gap directly, it may sway the base support until the distance is closed and it can grab the new substrate. Sturdy horizontal and inclined boughs near the cores of large trees serve as substrates for quadrupedal walking, during which the subject holds on with hands and feet (Tuttle, 1986). If a bough is quite thick the subject may fist-walk or move palmigrade atop it (McKinnon, 1974).

During quadrupedal locomotion on branches and when climbing vines and apprehensible tree trunks, the subject protracts one forelimb and then the ipsilateral hindlimb, followed by the other forelimb and the second hindlimb. If a trunk is too thick for manual grips, the orang-

FIG. 23–1. Sumatran orang-utans climbing and hanging on vertical supports. (Photo courtesy of H. D. Rijksen).

utan hugs it and reaches upward concurrently with both forelimbs and then pushes upwards with its hindlimbs in a bear-like fashion. Descents are effected by walking or sliding down the substrate feet-first (MacKinnon, 1974; Rijksen, 1978; Sugardjito, 1982).

Brachiation, which is unassisted by pedal grasps, is secondary to versatile climbing as a mode of travel; the orang-utans generally employ it only for short distances (Tuttle, 1986). MacKinnon (1974) noted that the brachiation of orang-utans lacks the speed and flow of hylobatid ricochetal arm-swinging. Further, orang-utans swing their forelimbs overarm instead of underarm, which is the common swing stroke in gibbons and siamang (MacKinnon, 1974; Sugardjito, 1982). Rijksen (1978) reported that brachiation is also a part of certain displays by orang-utans.

Although fleeing subjects sometimes leap (Rijksen, 1978), it is uncommon for orang-utans to jump and drop vertically over notable distances (Tuttle, 1986). Davenport (1967) and Sugardjito (1982) observed no leaping or jumping in free flight by their subjects. MacKinnon (1971, 1974) described "tumble descents" in which the fugitive fell rapidly through the foliage while momentarily grasping and releasing supports en route with its hands and feet. Schaller (1961) saw one orang-utan jump a distance of 3 feet to a lower branch. Davenport (1967) and Rijksen (1978) witnessed dramatic display dives by Bornean and Sumatran orang-utans, respectively. The show-offs held onto branches pedally and lunged or fell forward so that ultimately they dangled headlong from toeholds alone (Tuttle, 1986).

Bipedalism is also rare in wild orang-utans. It does not form a significant part of their locomotor repertoire (Tuttle, 1986). Among recent observers, Davenport (1967) and Rijksen (1978) noted brief episodes of arboreal bipedalism. Bipedal stances were more commonly executed by females and subadult males than by adult males in Rijksen's (1978) study population.

Sugardjito (1982) quantified the locomotor behavior of the Sumatran orang-utans that he observed at Ketambe in the Gunung Leuser National Park, Indonesia. Overall, adults engaged in quadrumanous climbing, including dives, in 41% of locomotor bouts; brachiation, in 21% of bouts; tree swaying, in 15% of bouts; arboreal quadrupedal walking, in 13% of bouts; and vertical climbing, in 10% of bouts. Females engaged in more quadrumanous climbing and quadrupedal walking than males did. The bulky males were more inclined to sway trees in order to transfer between them and to brachiate than the females were. Females and males engaged equally in vertical climbing.

Bornean orang-utans, particularly adult males, sometimes walk quadrupedally on the ground, not only to cross artifactual breaks in the forest but also in more naturalistic contexts within the forest (Tuttle, 1986). Rodman (1979) found that one adult male in the Kutai Nature Reserve of Kalimantan Timur, Indonesia, spent 20% of his travel time on the ground. Galdikas (1979:223) reported that as large males became fully habituated to the presence of human observers in the Tanjung Puting Reserve, Kalimantan Tengah, Indonesia, "almost all of their long distance traveling was done on the ground." In the Ulu Segama Reserve of Sabah, East Malaysia, and the Ranun River region of eastern Sumatra, MacKinnon (1974) was chased by unhabituated male orang-utans, some of which climbed down from the trees to pursue him. Male Bornean orang-utans chase and flee from one another on the ground (Tuttle, 1986).

MacKinnon (1974) saw female and juvenile orang-utans traveling briefly on the ground, especially during wet weather. Galdikas (1979) noted that females in the Tanjung Puting Reserve forage on the ground much less frequently than adult males do. Rijksen (1978) and Sugardjito (1982) concluded that Ketambe orang-utans rarely descend to the ground. They do so in order to cross gaps in the forest, to feed, and to flee from other orang-utans.

The basic gait of grounded orang-utans is classed as diagonal sequence, diagonal-couplets (Fig. 23–2d). The subjects commonly place their feet to the same side of the ipsilateral hands (Hildebrand, 1967; MacKinnon, 1971, 1974). Occasionally they "crutch-walk," swinging the lower body through an arch formed by their firmly planted, extended and abducted forelimbs (Fig. 23–2c).

Terrestrial orang-utans variably place their

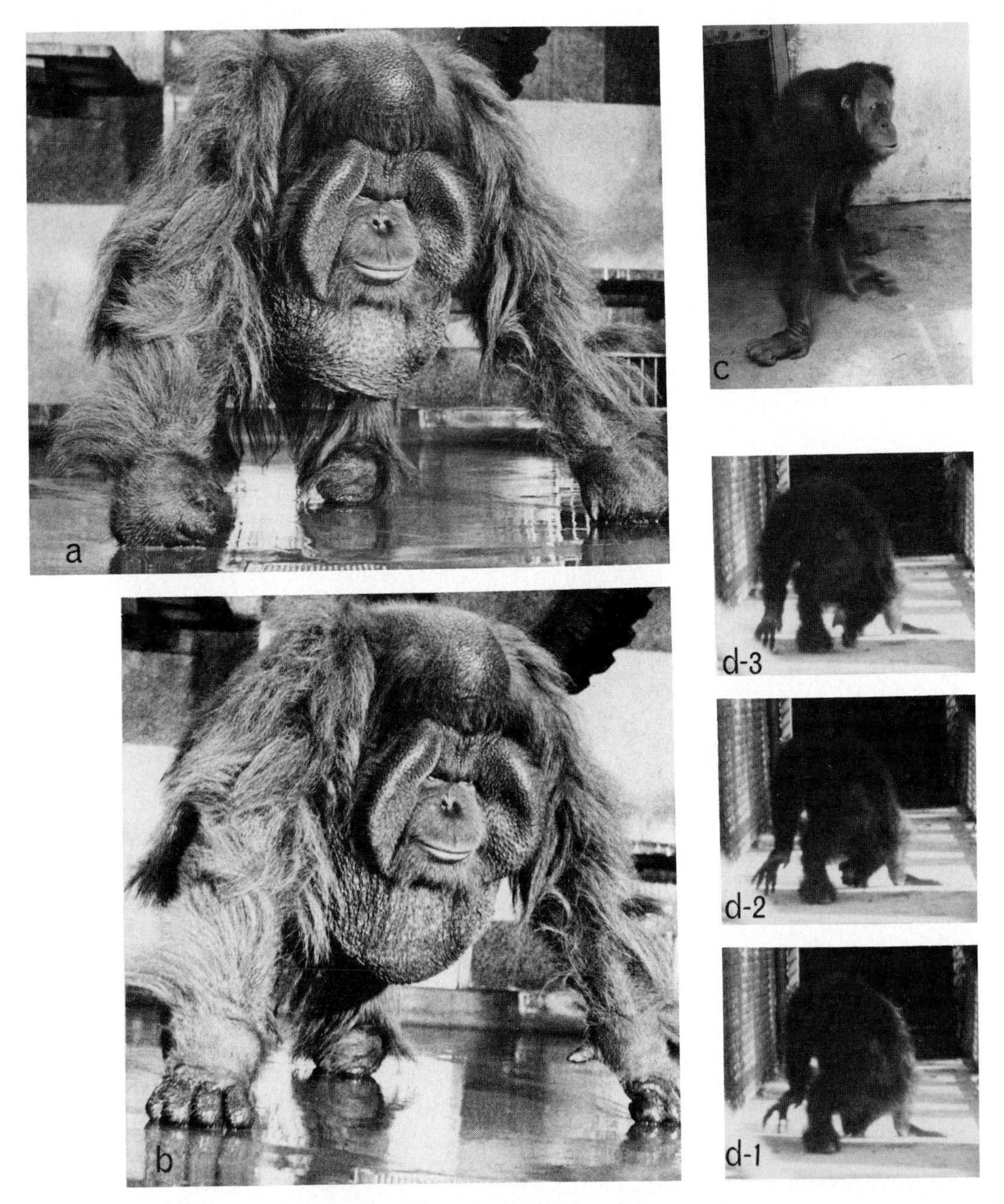

FIG. 23–2. Bornean orang-utan fist-walking (a) and knuckle-sliding (b) on wet cage floor. Sumatran orang-utans crutch-walking with hands in modified palmigrade posture (c) and walking quadrupedally with hands fully palmigrade (d). (Photos a and b courtesy of Brookfield Zoo).

hands in fisted, palmigrade, and modified palmigrade postures (Tuttle, 1967, 1969a,b, 1970a, 1975). Certain captives employ knuckled hand postures as they slide and walk on slick cage floors (Susman and Tuttle, 1976; Susman, 1974; Tuttle and Basmajian, 1974; Tuttle and Beck, 1972; Fig. 23–2b).

Some subjects walk on the lateral aspects of markedly inverted feet while others evert them to the extent that the halluces touch the ground; the toes are variably flexed or extended during progression (Figs. 23–2 and 23–3). MacKinnon (1974) noted that Sumatran orang-utans are more likely than Bornean orang-utans to approximate plantigrade foot postures. This was characteristic of new arrivals at Yerkes Regional Primate Research Center in 1963 (Tuttle, personal observation).

FEEDING AND REST POSTURES

Orang-utans are about as versatile in their feeding postures as they are in their locomotion. Although they usually sit on firm boughs or branches and unimanually hold supports overhead while gathering foods with a free hand, they sometimes adopt precarious suspensory postures. For instance, they may hang out from or beneath a branch while suspended from an ipsilateral hand and foot and use the other hand and foot to gather and hold food. And they occasionally hang by toeholds alone while collecting and opening fruits bimanually (MacKinnon, 1974; Rijksen, 1978). Both young and mature orang-utans have been seen hanging unimanually as they fed on fruit (Harrisson, 1962; MacKinnon, 1974). But they also collect large fruits and clusters of smaller foods and carry them to the security of large branches and nests (Harrisson, 1962; MacKinnon, 1971; Rijksen, 1978).

Orang-utans sleep at night in arboreal nests and commonly build nests for daytime rests. They repose supine, prone, and on their sides. Resting orang-utans also sit upright in nests or on boughs while holding overhead supports unimanually. MacKinnon (1974) saw subjects lying prone and supine along branches with their limbs dangling beneath them.

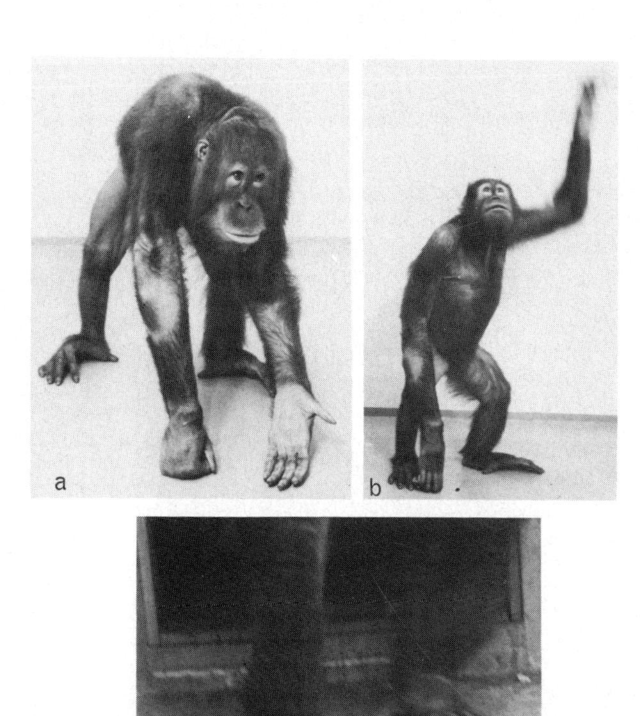

FIG. 23–3. Sumatran orang-utans whose halluces contact the ground during quadrupedal and bipedal postures.

MANIPULATION

Orang-utans are quite adept at manipulating a great variety of small and larger objects (Fig. 23–4). This is facilitated by independent control of the five fingers, which is most manifest for the index finger and the thumb (Tuttle, 1970a).

Orang-utans' hands and feet function importantly during their construction of arboreal nests (Tuttle, 1969a, 1970a, 1986). They hold crosspieces in place with their feet (or rumps) while drawing in additional supports. But the bits are not interlaced (Schaller, 1961).

Wild subjects have been observed plucking small fruits one-by-one with thumb-index grips and larger fruits with one or both hands (Schaller, 1961; Tuttle, 1970a, 1986). Their powerful jaws, flexible feet, and highly mobile, prehensile lips also come into play as they collect food.

Hungry rehabilitants sometimes use tools (e.g., sticks) to stab into durian fruits (Rijksen, 1978); but wild orang-utans have not demonstrated instrumental foraging (Galdikas, 1982; Tuttle, 1986).

Captive orang-utans appear to be somewhat less adept than chimpanzees at grasping tiny food objects and large spherical fruits (Tuttle, 1969a). But both hominoids employed Napierian (1960) power grips to grasp carrots, much as humans hold on to handles and poles.

ADAPTIVE COMPLEXES

Authors have used the term adaptive complex more commonly than they have defined it. We will follow the definition which states that an adaptive complex is composed of (1) the morphological complex—a limited set of coadapted features which act together to produce one or more functions characteristic of the species; (2) the genetic complex—structural genes upon which the morphological complex is premised and the mechanism whereby its components are developmentally coordinated; and (3) the selective complex—environmental features that acted with the genetic complex to produce the morphological complex (Tuttle, 1975:450–451; Tuttle and Watts, 1985:270).

Only the first of these three major components is readily accessible to the researcher and theoretician. The genetic complex and the selective complex are elusive, though the latter may sometimes be inferred with fair reliability from paleoenvironmental data. *Pace* modern biomedical equipment, it is still difficult to resolve questions of function for many morphological complexes in the apes (Tuttle, Cortright, and Buxhoeveden, 1979). This is especially true for wide-ranging arboreal species like *Pongo pygmaeus*, which, however much they might display their capacities in captivity, have not been challenged with the full array of situations to be en-

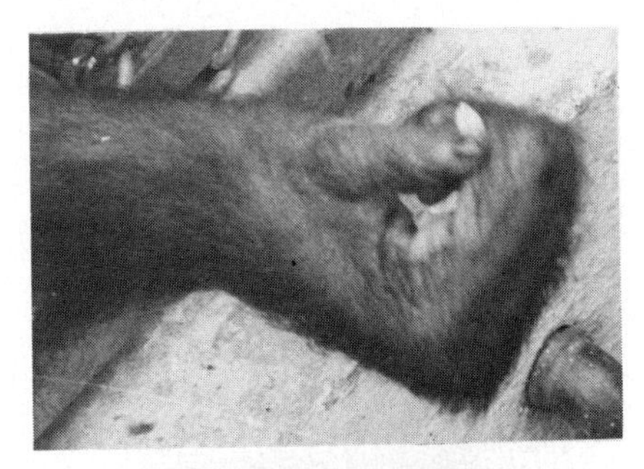

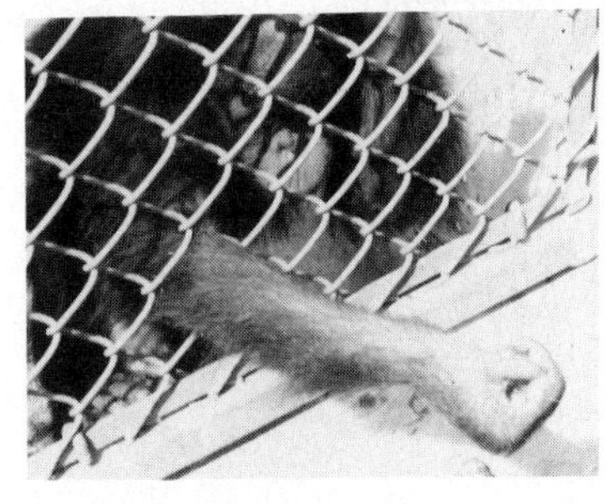

FIG. 23–4. Orang-utans holding small objects. (A) A kernel of maize between the thumb tip and side of the index finger. (B) A grain of rice between the thumb tip and side of the index finger. (C) An apple slice between the index and middle fingers.

countered in the forest. These caveats should be kept in mind as we review the locomotor anatomy of *Pongo pygmaeus* with special reference to its limbs.

The Forelimb

Orang-utan forelimbs are relatively immense. They are long (Fig. 23–5), highly mobile, and powerfully muscled (Erikson, 1963; Jungers, 1985; Schultz, 1936, 1956; Tuttle, 1975). Manual digits II–V are elongate and curved. The distal phalanges are conical, and concomitantly the fingertips taper.

The hook-like appearance of the medial four fingers results from ventral curvature of the respective metacarpal, proximal phalangeal, and middle phalangeal bones as well as the flexion set of the metacarpophalangeal and interphalangeal joints (Fig. 23–6a). Curvature is especially pronounced in the proximal and middle phalanges. On the palmar aspects of these phalanges, ridges at the points of maximum arc anchor strong flexor sheaths that prevent bowstringing by the long digital flexor tendons, which attach to the middle and distal phalanges (Susman, 1979; Tuttle, 1970a).

Orang-utans have relatively elongate phalanges in manual digit IV, which correlates with an emphasis on the ulnar aspect of the hand during climbing of vertical supports (Susman, 1979; Fig. 23–1). The interphalangeal proportions in digits II–V underpin a special double-locking mechanism whereby slender twigs and vines can be grasped securely (Napier, 1960; Susman, 1979; Tuttle and Rogers, 1966). Although extension of metacarpophalangeal joints II–V and the manual interphalangeal joints is markedly limited in *Pongo* (Fig. 23–6a), they can acutely flex their finger joints (Tuttle, 1969a,b, 1970a,b).

The intrinsic muscles of the hand, which power the medial four fingers, are quite robust, while the muscles of the thumb are dwarfed beside them. Nevertheless, the pollex contributes to a wide range of fine manipulation and some power gripping. The intrinsic muscles to the thumb are proliferated (Tuttle, 1969a); some of them send tendons to the distal phalanx, thereby replacing or augmenting the pollical tendon of the deep long digital flexor muscle, which is absent or, at best, vestigial in *Pongo* (Day and Napier, 1963; Straus, 1942; Tuttle, 1969a, 1970a; Fig. 23–7).

In *Pongo* the robust extrinsic musculature from the radial aspect of the forearm, which flexes the distal pollical phalanx in most nonpongid anthropoid primates (e.g., the flexor pollicis longus muscle of *Homo sapiens*), attaches distally via the tendon of the index finger (Fig. 23–7).

The orang-utan wrist is highly mobile, especially in comparison with those of gorillas and common chimpanzees. It can be dorsiflexed (Fig. 23–6a), volarflexed, abducted, and adducted (Fig. 23–6c) over wide arcs (Tuttle,

FIG. 23–5. A Sumatran orang-utan reaching overhead for food incentives.

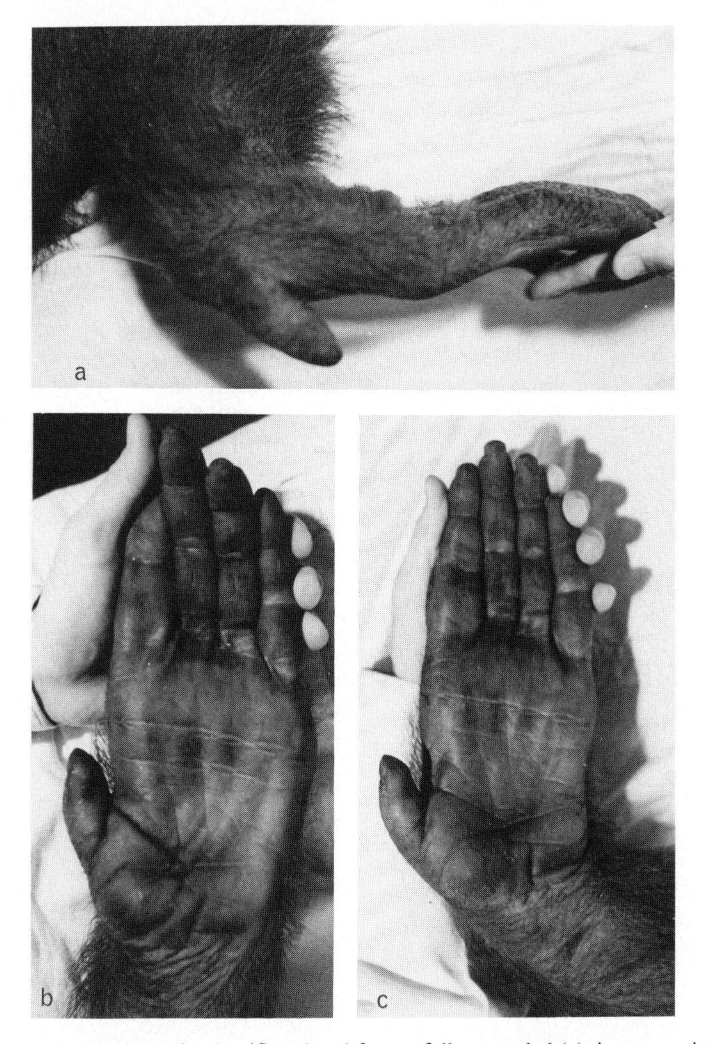

FIG. 23–6. An orang-utan hand with wrist dorsiflexed and fingers fully extended (a), in rest position (b), and maximally adducted (c).

1970a,b). Movement is facilitated by the fact that the distal ulna participates minimally in the wrist joint (Lewis, 1965, 1969, 1971, 1972, 1974) and certain carpal bones articulate as ball-and-socket joints in the radiocarpal and midcarpal regions (Corruccini, 1978; Corruccini et al., 1975; Jenkins and Fleagle, 1975; McHenry and Corruccini, 1983). Presumably also, the carpal ligaments are elastic and oriented so that wrist motion is facilitated, particularly dorsally and medially.

In keeping with the osteoligamentous potential for wrist movements, the muscles acting on the radiocarpal joint, including not only the proper flexors and extensors of the wrist but also the long digital flexors and extensors, are well developed (Tuttle, 1969a, 1970a,b), giving orang-utan forelimbs a Popeye appearance.

The elbow complex of *Pongo pygmaeus* allows full extension so that the forearm is aligned with the arm. The olecranon process of the ulna does not project notably above the proximal extremity of the trochlear notch, and thus it can enter the olecranon fossa when the humeroulnar joint is extended fully. The radioulnar joint permits a range of manual pronation and supination at least as wide as in healthy *Homo sapiens*. But quantification in respectable samples of *Pongo* is lacking.

Electromyographic studies on an adolescent female orang-utan have provided information on the activities of brachial muscles during a limited repertoire of spontaneous and incentive-induced positional and manipulatory behaviors (Tuttle et al., 1983).

Bimanual hoisting actions were accompanied by high activity in the brachialis muscle, but the biceps brachii muscle was inactive or modestly

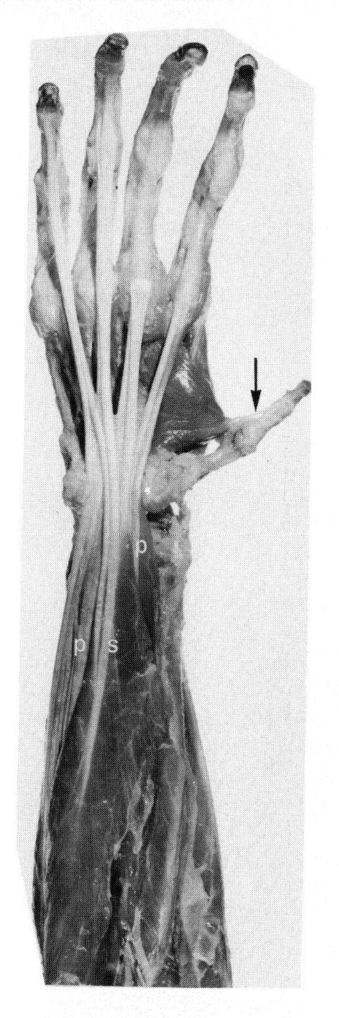

FIG. 23–7. Fresh dissection of the forearm and hand of *Pongo pygmaeus,* showing long digital flexor muscles (p, flexor digitorum profundus; s, flexor digitorum superficialis) and tendon from the adductor pollicis muscle to the tip of the thumb (arrow). Note that the radial component of the flexor digitorum profundus muscle does not have a tendon to the tip of the thumb.

active and the brachioradialis muscle acted at high levels only during the final segment of the hoist. *Per contra*, unimanual hoists elicited high activity in all three of these muscles (Fig. 23–8). Thus it may be inferred that the brachialis muscle is the prime flexor of the humeroulnar joint while the brachioradialis and biceps brachii muscles act as augmentary sources of power during more demanding activities by the orang-utan.

As expected, the triceps brachii muscle was silent during hoisting actions. However, it was surprising to find that not only the triceps brachii muscle but also the brachialis, biceps brachii, and brachioradialis muscles were silent during quiet pendant suspension with the humeroulnar joints fully extended. This indicates that osteoligamentous mechanisms in the joints, perhaps assisted by transarticular muscular force from the contracted long digital flexor muscles, is sufficient to support the suspended orang-utan's bulk. Hence, despite what might have been intuited, suspensory feeding is not necessarily excessively costly energetically.

During terrestrial quadrupedal progression, the three heads of the triceps brachii exhibited low levels of electromyographic (EMG) activity from midswing phase of the forelimb through the load-bearing segment of the stance phase. The anconeus muscle acted during a more limited part of the locomotor cycle, viz., the load-bearing segment of the stance phase.

During a brief bout of crutch-walking, the orang-utan's triceps brachii muscle was markedly active while the limb was load-bearing. During quiescent tripedal and quadrupedal stances, slight and moderate EMG potentials occurred in the medial head of the triceps brachii muscle.

Low EMGs were exhibited by the brachialis and biceps brachii muscles at prerelease of the hand during fist-walking and modified palmigrade progression; the brachioradialis muscle was generally silent. The biceps brachii muscle was also active at the outset of swing phases during crutch-walking.

While reaching for incentives overhead, the medial head of the orang-utan's triceps brachii muscle was initially active at low levels but later increased to moderate levels as the goal was reached. The anconeus muscle was even more highly active during elbow extension toward elevated incentives and while rubbing a towel on the floor.

Pulling against resistance and elbow flexion while waving the arm overhead were accompanied by high EMG potentials in the brachialis and biceps brachii muscles. High potentials also occurred in the brachialis and biceps brachii muscles when the orang-utan pressed objects against her head and trunk and hugged the investigator. The brachioradialis muscle was generally much less active during these movements, though high potentials did accompany one bout of pushing a ball against her mouth with a fully supinated hand.

During quiet feeding, elbow flexion characteristically elicited low EMG in the brachialis and biceps brachii muscles and the brachioradialis muscle was often silent. Some maneuvers evokes synergistic activity between the triceps

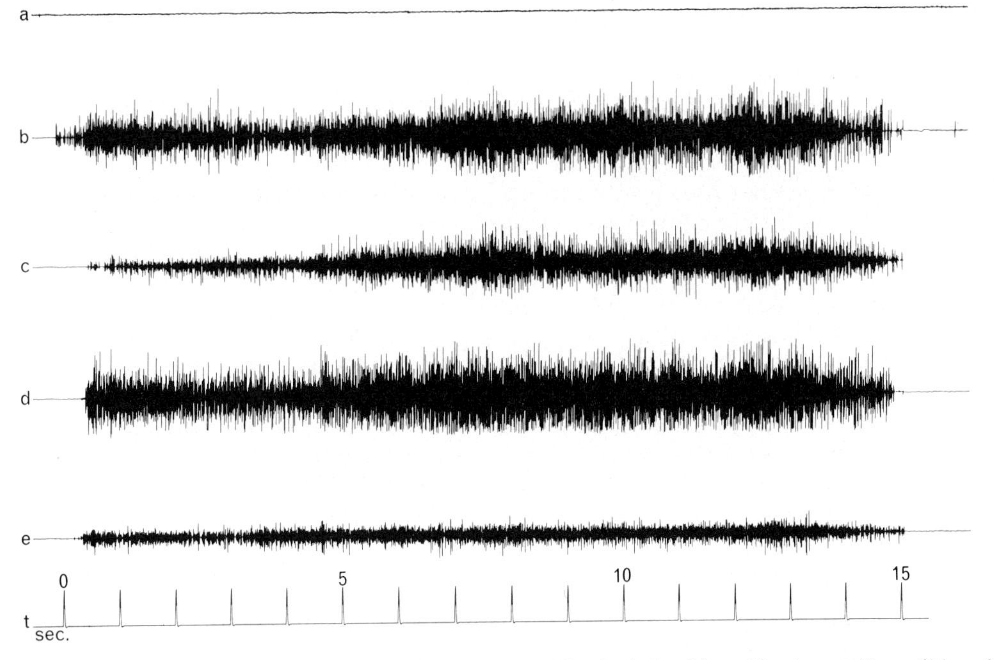

FIG. 23–8. EMG recording of brachial muscles in an orang-utan while she hoisted herself unimanually until her elbow was approximately orthogonal (first 5 sec) and then sustained the posture (sec 5–14) as she cautiously groped for fruit in a pan that was attached near the ceiling. Thereafter (sec 14–15) she descended to the floor. Symbols: (a) m. triceps brachii, caput longum; (b) m. brachialis; (c) m. biceps brachii, caput longum; (d) m. brachioradialis; (e) m. biceps brachii, caput breve. t, Time in seconds. The subject and experimental area are shown in Fig. 23–5.

brachii and certain brachial flexor muscles. For instance, as the suspended subject rapidly extended her elbow to reach for an object on the floor, the long head of the triceps brachii muscle acted simultaneously with the short head of the biceps brachii muscle. And it acted concurrently with brachial flexors when she flexed her upraised elbow. The two-joint muscles (long heads of the biceps brachii and triceps brachii) that were active on these occasions probably stabilized the shoulder joint in addition to acting on the elbow complex.

The shoulder complex of *Pongo pygmaeus* is particularly adapted for using the forelimb in upraised positions. The scapula is craniocaudally elongate and its glenoid cavity faces cranially. The acromion process, to which the deltoid and trapezius muscles attach, is large, and the coracoid process, which gives attachment to the short head of the biceps brachii, pectoralis minor, and coracobrachialis muscles, is robust. The long clavicle struts the scapula on the dorsoventrally flattened chest wall so that the glenoid cavity faces laterally as well as cranially (Ashton and Oxnard, 1963; Erikson, 1963: Oxnard, 1967; Roberts, 1974).

Ashton and Oxnard (1963) found that certain arm-raising muscles (deltoid, trapezius, serratus anterior pars caudalis) of apes (including an orang-utan) were, collectively, relatively larger than those of monkeys, and so were the latissimus dorsi and teres major muscles.

In the orang-utan, the supraspinatus muscle serves as an important initiator and accelerator of arm-raising in some contexts but it is generally less important than the deltoid as a maintainer of many upraised postures (Tuttle and Basmajian, 1978a). In the young female orang-utan, high EMG potentials characterized the intermediate part of the deltoid muscle during most abducent arm-raising episodes, including ones in which she was cautious and leisurely. High potentials persisted as she held her arm at 150–160°. When she raised her forelimb anteriorly, the anterior deltoid muscle was highly active; it was silent or only modestly active during lateral arm-raising. The posterior deltoid muscle was also variably active during arm-raising. However, during medial rotation of the upraised arm the posterior deltoid muscle was silent, which is to be expected because it is antagonistic to medial rotation of the humerus (Tuttle and Basmajian, 1978a).

High EMGs in the subscapularis muscle when the subject foraged overhead indicated that it is probably a major, if not the prime medial rotator of the orang-utan's arm (Tuttle and Basmajian, 1978a; Fig. 23–9).

In contrast with *Homo sapiens,* in *Pongo pygmaeus* the cranial trapezius and serratus anterior pars caudalis muscles do not act continuously as the arm is raised overhead. Instead there are brief bursts of EMG activity early in arm-raising, followed by diminution to silence at the zenith of the reach. Because the glenoid cavity is already oriented cranially, there is no need to rotate the pongid scapula extensively when reaching overhead (Tuttle and Basmajian, 1977).

Extensive attachment of the latissimus dorsi muscle to the lateral iliac crest is a unique feature of the great apes among primates. Bimanual and unimanual hoists by *Pongo* are accompanied by high EMGs in the iliac segment of the latissimus dorsi and teres major muscles. The vertebrocostal segment of the latissimus dorsi muscle generally exhibits low EMGs during hoisting (Tuttle and Basmajian, 1977). The teres minor muscle is also commonly active at high levels during hoists especially unimanual ones (Fig. 23–9). It most certainly acts as a retractor of the humerus, particularly when the humerus moves in a nearly parasagittal plane (Tuttle and Basmajian, 1978a). The cranial trapezius, caudal serratus anterior, pectoralis major, rhomboid, supraspinatus, infraspinatus, and subscapularis muscles were silent or exhibited low EMGs when the young female orang-utan hoisted herself bimanually and unimanually (Tuttle and Basmajian, 1977, 1978a).

As she hung bimanually and unimanually with the forelimb(s) extended and under tension, her latissimus dorsi, teres major, pectoralis major, cranial trapezius, serratus anterior pars caudalis, rhomboid, and deep scapulohumeral ("rotator cuff") muscles were virtually silent. Even the deep scapulohumeral muscles remained silent or exhibited only inconsequential single EMG potentials when she rotated passively while hanging by one hand (Tuttle and Basmajian, 1977, 1987a).

During all quiescent quadrupedal stances, the young female orang-utan's pectoralis major muscle was active at low levels; her supraspinatus and subscapularis muscles were commonly active also at low levels. Many other shoulder muscles were silent during quadrupedal stance (Tuttle and Basmajian, 1978b). When the subject stood tripedally, EMGs were basically similar to those that accompanied quadrupedal stances. But low potentials occurred in the iliac segment of the latissimus dorsi muscle and sometimes moderate potentials were exhibited by the supraspinatus muscle (Tuttle and Basmajian, 1978b).

Eccentric loadings of the forelimb during forward descents from elevated areas onto the floor elicited strikingly greater EMG activity in the shoulder muscles than occurred during stances. In the young female orang-utan the deltoid, supraspinatus, and, to a lesser extent, infraspinatus muscles were quite active while the forelimb was load-bearing. Further, the pectoralis major and serratus anterior pars caudalis muscles were commonly more active during forward descents onto the forelimb (Tuttle and Basmajian, 1978b).

During crutch-walking, the pectoralis major, rhomboid and iliac segment of the latissimus dorsi muscles variably exhibited low or moderate EMGs; the serratus anterior pars caudalis, teres major, cranial trapezius, and supraspinatus muscles were characteristically silent (Tuttle and Basmajian, 1978b).

Quadrupedal locomotion in *Pongo* differed from *Pan* during EMG experiments in that the orang-utan's shoulder did not pass over the supportive hand during midstance phases of the locomotive cycle. During stance phases, the pec-

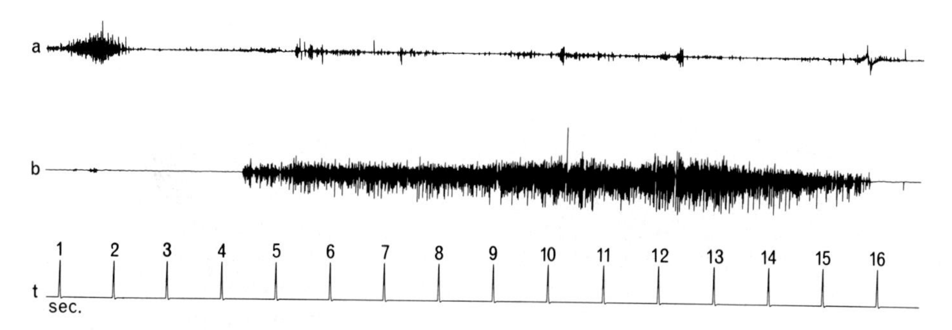

FIG. 23–9. EMG recording of right subscapularis (a) and teres minor (b) muscles in a bipedal orang-utan as she released the trapeze bar (Fig. 23–5) and reached further overhead with the right hand (sec 1–2) and then hoisted herself bimanually (sec 4–9) and unimanually (sec 9–16) on the bar until her right shoulder and elbow were at 100° as she foraged overhead with her left hand.

toralis major, intermediate and posterior deltoid, and subscapularis muscles of the orang-utan were active regularly, apparently providing support for the glenohumeral joint as well as propulsive force (Tuttle and Basmajian, 1978b). Additional propulsive force and support are severally provided by the rhomboid and supraspinatus muscles. The latter was also active early in swing phase of the forelimb. The intermediate and posterior deltoid and cranial trapezius muscles were typically active at release of the hand into swing phase. As the hand descended toward the substrate during late swing phase, the anterior deltoid muscle was active (Tuttle and Basmajian, 1978b).

The Hindlimb

Orang-utans have remarkably short hindlimbs, especially relative to the length of their forelimbs (Erikson, 1963; Jungers, 1985; Fig. 23–5). This is most manifest when they stand bipedally. Nevertheless, their feet are very long (Figs. 23–3 and 23–10) and powerful, and their thighs and legs are robustly muscled. The design of the orang-utan hindlimb clearly reflects the functional demands of frequent suspensory behavior in which a characteristic form of pedal prehension is employed.

Like their hands, orang-utan feet are preeminently built for gripping a wide variety of arboreal supports (Figs. 23–1 and 23–11). Pedal digits II–V are elongate and concavely curved ventrally (Schultz, 1956, 1963). The long digital flexors constitute a major portion of the leg musculature (Tuttle, 1970a; Fig. 23–12) and the intrinsic foot musculature is also very large relative to total hindlimb muscle mass (Ishida, 1972). The metatarsophalangeal and interphalangeal joints permit extensive flexion but only limited extension of digits II–V (Tuttle, 1970a, Fig. 23–10a). In contrast with the plantigrade feet of African apes and humans, orang-utan feet have small heels and their heel-raising (triceps surae) muscles constitute a smaller proportion of the leg muscles (Tuttle, 1970a,b, 1972).

The orang-utan hallux is greatly reduced skeletally and myologically (Schultz, 1936, 1941; Straus, 1942; Tuttle 1970a,b, 1972; Tuttle and Rogers, 1966). In approximately 60% of orang-utans, the hallucal distal phalanx and nail are lacking; occasionally, the proximal phalanx is also undeveloped. Moreover, absence of the distal phalanx is expressed about twice as often in females than in males (Tuttle and Rogers, 1966). Rijksen (1978) noted that among Sumatran orang-utans absence of the hallucal terminal phalanx and nail is characteristic of the "the light-haired, short-fingered type," while the "dark-haired, long-fingered type" have complete halluces; the former type may also lack thumbnails.

Nailed or not, the hallux of *Pongo* is characterized by absence of a functional tendon from the flexor digitorum fibularis muscle and a reduced mass of the intrinsic muscles that power pedal digit I. The terminal phalanges of complete halluces in *Pongo* receive partially compensatory tendons from certain preaxial intrinsic

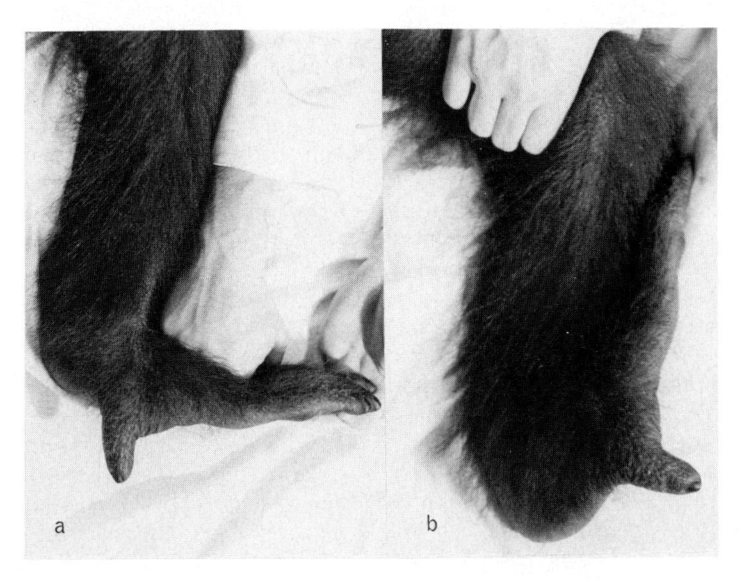

FIG. 23–10. Left foot of *Pongo pygmaeus* with toes maximally extended (a) and apposed to the anterior aspect of the leg (b). Note that the toe tips nearly reach the knee.

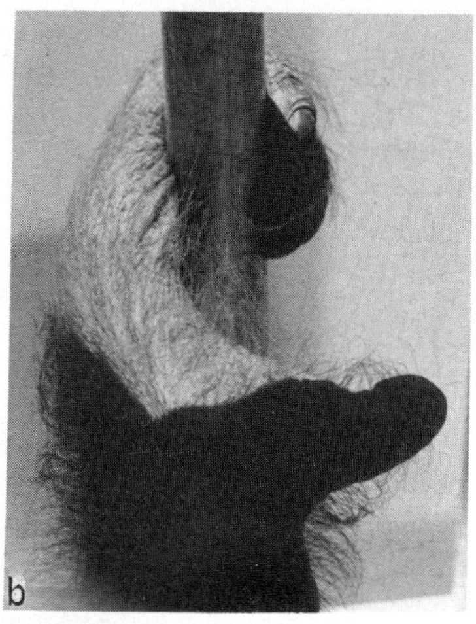

FIG. 23–11. Orang-utans hanging from bars in a cage roof (a, b) and cyclone fencing in a cage roof (c). Note the versatile posturing that the hip joints permit.

muscles of the foot (Tuttle, 1970a,b, 1972; Fig. 23–13). Even without terminal phalanges, orang-utans can grip powerfully with their halluces (Tuttle, personal bruises).

While the knee joint of *Pongo* does not differ greatly from the knees of the African apes (Tardieu, 1983), its hip joint is apparently adapted for greater mobility than in *Pan*. Asfaw (1985) noted that in *Pongo* ($N = 3$) the articular surface of the femoral head is more extensive than in *Pan* ($N = 4$). Indeed, orang-utans can circumduct their hips extensively in order to hang by footholds from superstrates and feed manually while their trunks are poised upright between the hindlimbs (Fig. 23–11). Once during a tantrum, a prone orang-utan repeatedly circumducted her hindlimbs simultaneously at the hip joints by abducting them, drawing them under her body, and then extending them. While prone she also widely adducted her hindlimbs at the hip joints and then rapidly adducted them so that her inverted soles clapped loudly (Tuttle, videotape of EMG experiment "Guchi #6").

In the hip and thigh musculature, *Pongo* is quantitatively most easily distinguished from other hominoids in the relative development of the major knee extensors (vasti) and flexors (hamstrings) (Fig. 23–14). The vasti permit the limb to act as an extensible strut during knee extension or controlled flexion under superincumbent body weight. Among anthropoid primates, the vasti are uniquely small in orang-utans and relatively largest in the most frequent and proficient bipeds—humans and hylobatids (Fig. 23–14A). Conversely, the hamstrings are significantly larger in *Pongo* than in other hominoids (Fig. 12–14B). Although functional interpretation of these differences is hampered by lack of kinesiological studies of orang-utan positional behavior, several inferences can be made that provide a satisfactory first-order explanation for several aspects of posture and locomotion peculiar to these apes: (1) the large hamstrings may reflect the need for powerful knee flexion during hindlimb suspension; and (2) the small size of the vasti may reflect a diminished capacity for

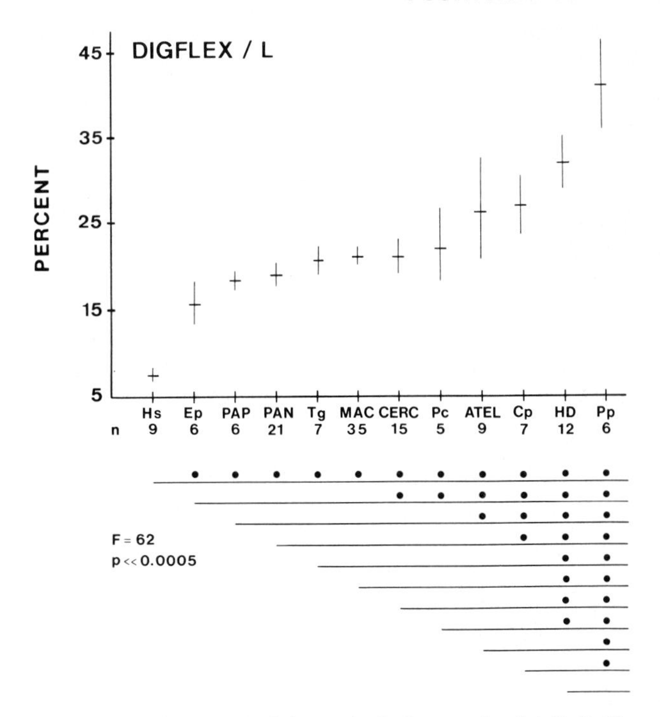

FIG. 23–12. The relative mass (expressed as percent) of the extrinsic flexors of pedal digits (flexor digitorum tibialis and flexor digitorum fibularis) compared to all other muscles of the leg. Here and in Fig. 23–14 the graphic display presents the mean and 95% confidence limits for each group. Statistical results presented are the *F*-ratio from a one-way ANOVA and the results of Scheffe's multiple comparison tests among taxa. A mark (+) on a line in any column indicates a significant ($p < 0.05$) difference between that taxon and the one under which the line begins at the left. Key to taxonomic abbreviations: ATEL *(Ateles);* CERC *(Cercopithecus);* Cp *(Colobus polykomos);* Ep *(Erythrocebus patas);* HD (hylobatid apes); Hs *(Homo sapiens);* MAC *(Macaca);* PAN (*Pan troglodytes* and *Pan gorilla*); PAP *(Papio);* Pc *(Presbytis cristatus);* Pp *(Pongo pygmaeus);* Tg *(Theropithecus gelada).*

the limb to act as a powerful strut, particularly in bipedal posture and locomotion, when the hindlimbs alone support body weight.

Studies of free-ranging and captive subjects provide a picture of characteristic orang-utan bipedal behavior that is consistent with this analysis. Orang-utan bipedalism is very infrequent; it consists largely of short postural episodes; and, it is usually performed only when additional support can be obtained by grasping available handholds (Cortright, 1975; Davenport, 1967; MacKinnon, 1971, 1974). When standing bipedally on the ground, orang-utans elevate the trunk by powerful hamstring action to extend the hip and thrust back their shoulders in an attempt to maintain a precarious balance of the trunk over fully extended hindlimbs (MacKinnon, 1971, 1974; Napier, 1964). By thus aligning the center of gravity vertically over the mechanical axis of the hindlimb, they avoid large flexor moments at the hip and knee that would necessitate powerful muscular action to stabilize these joints. In a flexed-hips, flexed-knees posture, the combined action of body weight and maximal hamstring activity may produce a knee flexor moment too great to be adequately balanced by sustained activity of the small vasti for a lengthy period of time. During bipedal walking, stabilization against passive knee flexion would be even more difficult to achieve. In the early stance phase of each step, the knee of the forwardly inclined limb would lie anterior to the vertical position of the center of gravity and would experience a very large flexor moment, especially in the single stance phase after the contralateral limb has begun to swing forward in preparation for the next step.

In further view of the foregoing analysis, it seems likely that the powerful forelimbs play a more dominant role than the hindlimbs during vertical climbing. Specific adaptations for arboreal suspension are not readily apparent in the orang-utan hip and thigh musculature. But the major contrasts in muscular development suggest that the influence of arboreality on orang-utan hindlimb design is more pervasive than

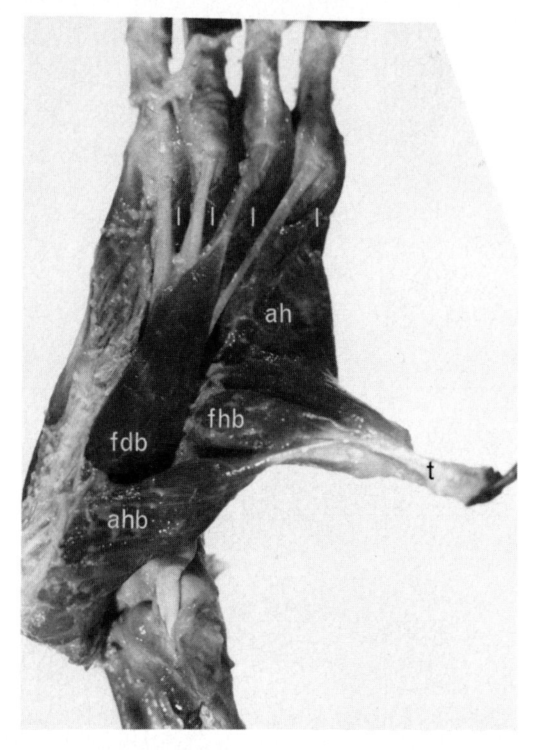

FIG. 23–13. Plantar view of a fresh dissection of a right foot of *Pongo pygmaeus.* The adductor hallucis (ah), flexor hallucis brevis (fhb), and abductor hallucis brevis (ahb) muscles contribute to a common tendon (t) which inserts to the distal phalanx. The pedal lumbrical muscles (l) and flexor digitorum brevis muscle (fdb) are also exposed.

might be inferred from considering details of structure and attachment alone and is not confined to the distal limb segments.

The hip musculature of *Pongo pygmaeus* represents a unique pattern among the Hominoidea. The lower portion of the gluteus maximus complex forms a discrete ischiofemoralis muscle (Fig. 23–15) and the lateral part of the gluteus minimus constitutes a discrete scansorius muscle (Sigmon, 1975; Tuttle et al., 1978). Electromyographic studies have shown that in *Pongo* the gluteus maximus proprius muscle is primarily an abductor and lateral rotator of the thigh at the hip joint while the ischiofemoralis, gluteus medius, and gluteus minimus proprius muscles are basically extensors of the hip joint. During bipedal stances, the gluteus maximus proprius muscle of the young female orang-utan was usually silent; but when the subject abducted her hip to shift sideways, moderate EMGs were exhibited. The ischiofemoralis, gluteus medius, gluteus minimus proprius, and rectus femoris muscles were variably active at high and low levels both when the subject stood bipedally with flexure of the hip and knee joints and when she rose to bipedal positions. EMGs ceased in these muscles as the hip and knee joints became aligned and overextended (Tuttle et al., 1978; Tuttle, Basmajian, and Ishida, 1979).

The broad, flat, elongate ilia of the orang-utan pelvis not only enhance the power of the hip extensors but also provide attachment for the latissimus dorsi muscles, which are major retractors of the arms. The reduced lumbar region of the vertebral column may further enhance the power of the latissimus dorsi muscles and also lower the center of mass in such a top-heavy animal.

In brief, then, the postcranial anatomy of *Pongo pygmaeus* provides dramatic evidence that this ape is highly advanced as a versatile arboreal climbing and foraging machine. *Per contra,* features specifically related to the terrestrial habits of some modern orang-utans are elusive and perhaps nonexistent. Intense selection over a relatively short time would have to be posited in order to transform large-bodied, ground-dwelling quadrupeds into the creatures we encounter in insular southeastern Asian forests.

EVOLUTION

At present we must await the discovery of postcranial fossils and refine our paleogeographic and paleoecological perspectives before championing either "always arboreal" or "ancestrally terrestrial" theories on orang-utan evolution. Evidence for the former view is rich, but predominantly of one kind, viz., morphological (Temerin, 1980). However, other factors suggest that orang-utans had part-time terrestrial ancestors or collaterals, or both, between the middle of the Miocene and the Pleistocene (Andrews, 1983; Galdikas, 1981b; Smith and Pilbeam, 1980a,b, 1981; von Koenigswald, 1982).

Craniodental similarities between *Pongo* and Eurasian (and perhaps African) Miocene "ramamorph" hominoids suggest that *Pongo* may have evolved from such an ancestor, perhaps, specifically, *Sivapithecus* (Andrews, 1983; Andrews and Tekkaya, 1980; de Bonis and Melentis, 1984; Ciochon, 1983; Greenfield, 1979; Pilbeam, 1985; Shea, 1985; Ward and Pilbeam, 1983; Wu et al., 1983; but see de Bonis, 1982 and Kay and Simons,1983).

Sivapithecus is found in deposits which suggest habitats that were more open than the tropical rain forests in which *Pongo pygmaeus* is now endemic. Woodlands were particularly favored

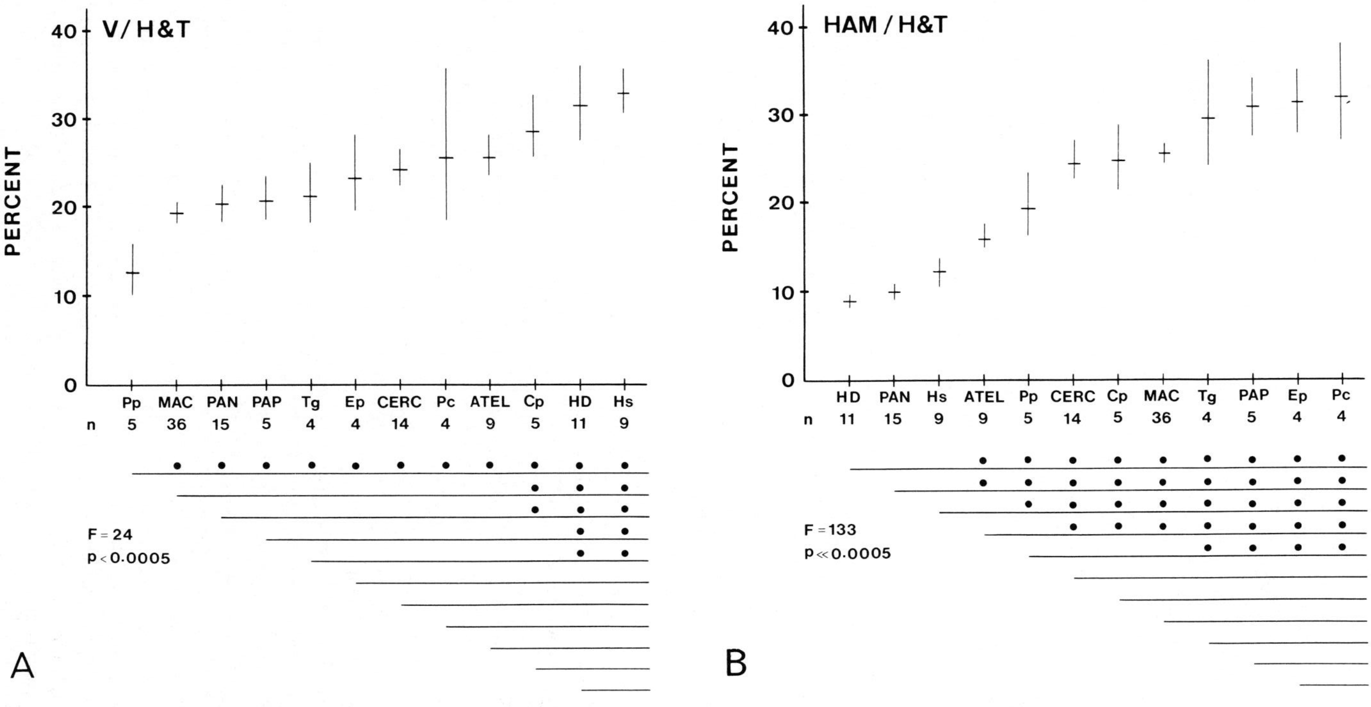

FIG. 23–14. The relative masses of the vasti (A) and the hamstrings (B) compared with all other muscles of the hip and thigh in anthropoid primates. See Fig. 23–12 for details.

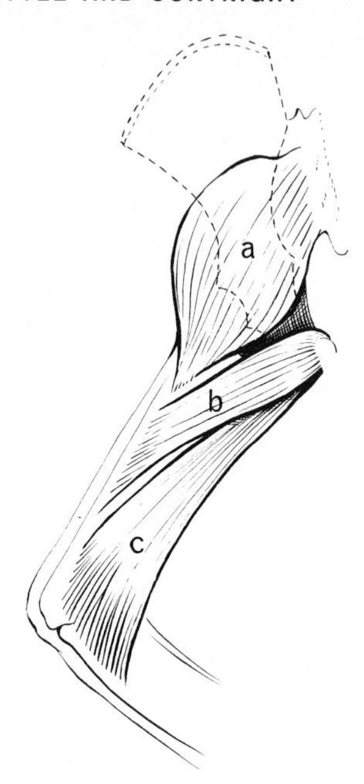

FIG. 23–15. Lateral hip and thigh muscles in *Pongo pygmaeus:* gluteus maximus proprius (a), ischiofemoralis (b), and biceps femoris caput longum (c). (Courtesy of B. Sigmon.)

by *Sivapithecus* (Andrews, 1983; Landau et al., 1982; Pilbeam et al., 1977, 1979). Since estimates of body sizes (based on tooth dimensions) for several species indicate that they were weighty mammals (Andrews, 1983; Gingerich et al., 1982), it is reasonable to assume that *Sivapithecus* was not strictly arboreal.

The thick enamel on their molars further suggests that they ate quantities of tough foods. Extending the model that Hatley and Kappelman (1980) applied to Plio-Pleistocene Hominidae, Andrews (1983) speculated that, like pigs and bears, *Sivapithecus* may have grubbed geophytes during seasonal food shortages above ground. But nuts, hard seeds, and tough-rinded fruits might have been important in addition to, or instead of, underground foodstuffs. If so, orang-utans (and several other arboreal anthropoid primates, which harvest notable quantities of tough food items in tropical rain forests) developed relatively thick molar enamel for reasons other than those that might be associated with terrestrial activities (Kay, 1981).

The few known postcranial remains of *Sivapithecus* are inadequate for elucidating just how arboreal this hominoid was. For instance, the well-preserved hallux (GSP 14046) from Pakistan shows clear evidence for powerful gripping (Conroy and Rose, 1983; Pilbeam et al., 1980; Tuttle, personal observations). But chimpanzees and gorillas, which also have hefty prehensile halluces, spend a good deal of time on the ground (Tuttle, 1970a, 1986). We can conclude only that at least one species of late Miocene *Sivapithecus* probably lacked the pedal versatility in climbing and suspensory behavior that is manifest in *Pongo pygmaeus.*

Despite a superabundance of orang-utan teeth ($N > 5{,}000$) from Pleistocene deposits in Java, Borneo, Sumatra, China, and Vietnam, no postcranial remains of Pleistocene orang-utans have been recovered (von Koenigswald, 1982). Thus we must rely solely on paleoenvironmental constructs to infer the arboreality or terrestriality of *Pongo.*The Pleistocene orang-utans of Java, Borneo, and Sumatra probably lived in humid, forested environments (von Koenigswald, 1982; de Vos, 1983, 1984). They are absent from Javan deposits that are inferred to represent open woodland habitats (de Vos, 1983).

According to von Koenigswald (1982), the oldest specimens of *Pongo* are from Javanese deposits that are about 2 million years old. De Vos (1984) argued that there is no evidence for the presence of *Pongo* in South China until the late middle or late Pleistocene. Could *Homo* have had a hand in their introduction into South China from the Indonesian Archipelago?

Based on "very large" Pleistocene orang-utan teeth from Chinese drugstores and their common association with teeth of pandas and mountain goats, von Koenigswald (1982:14) concluded that, in addition to there being conventional orang-utans in southeastern Asia, there had been mountain orang-utans in China, analogous to extant African mountain gorillas, and they, too, probably would have been at least partly terrestrial.

CONCLUSIONS

Due to the present dearth of postcranial evidence, we are left hanging anent the extent to which Pleistocene *Pongo* and their earlier Neogene ancestors of tropical southeastern Asia differed in terrestriality from extant *Pongo.* If swamp forests had been prevalent, it is likely that these earlier *Pongo* were very arboreal apes (Tuttle, 1969b, 1975).

The special adaptations of orang-utans for versatile climbing and processing tough arboreal foods allowed them to flourish in the vine-laden, fruitful tropical rain forests of Sumatra and Borneo. Before the evolution of technologically sophisticated *Homo*, there were few, if any, species that seriously threatened orang-utans as competitors for food or as predators, at least while they remained aloft (Tuttle, 1986). This may be reason enough for the special adaptive complexes of *Pongo*.

The recent catastrophic decline of *Pongo pygmaeus* through deforestation and human hunting is lamentable because, concurrently, natural scientists have reached a threshold of understanding these apes ecologically and paleontologically and, as such, may now finally look forward to comprehending them evolutionarily (Tuttle, 1986).

ACKNOWLEDGMENTS

This essay was written while R. Tuttle was a Guggenheim Fellow and was partially supported by NSF grant BNS-8504290 and the Marian and Adolph Lichtstern Fund of the University of Chicago. The EMG and comparative morphological studies were conducted with support from NIH Grant RR-00165 from the Division of Research Resources to the Yerkes Primate Research Center, PHS Research Career Development Award 1-K04-GM16347-01, and NSF Grants GS-834, GS-1888, GS-3209, and SOC 75-02478.

REFERENCES

Andrews, P. J. 1983. The natural history of *Sivapithecus*. *In* New Interpretations of Ape and Human Ancestry, pp. 441–463, ed. R. L. Ciochon and R. S. Corruccini. Plenum Press, New York.

Andrews, P. J. and Cronin, J. E. 1982. The relationships of *Sivapithecus* and the evolution of the orang-utan. *Nature, 297:*541–545.

Andrews, P. J. and Tekkaya, I. 1980. A revision of the Turkish Miocene hominoid *Sivapithecus meteai*. *Palaeontology, 23:*85–95.

Asfaw, B. 1985. Proximal femur articulation in Pliocene hominids. *American Journal of Physical Anthropology, 68:*535–538.

Ashton, E. H. and Oxnard, C. E. 1963. The musculature of the primate shoulder. *Transactions of the Zoological Society of London, 29:*553–650.

de Bonis, L. 1982. Reflexions sur la phylogenie et la classification des hominoides. *Geobios, Mémoire Special, 6:*305–319.

de Bonis, L. and Melentis, J. 1984. La position phyletique d'*Ouranopithecus*. *Courier Forschungsinstitut Senckenberg, 69:*13–23.

Ciochon, R. L. 1983. Hominoid cladistics and the ancestry of modern apes and humans: a summary statement. *In* New Interpretations of Ape and Human Ancestry, pp. 783–843, ed. R. L. Ciochon and R. S. Corruccini. Plenum Press, New York.

Conroy, G. C. and Rose, M. D. 1983. The evolution of the primate foot from the earliest primates to the Miocene hominoids. *Foot and Ankle, 3:*342–364.

Corruccini, R. S. 1978. Comparative osteometrics of the hominoid wrist joint, with special reference to knuckle-walking. *Journal of Human Evolution, 7:*307–321.

Corruccini, R. S., Ciochon, R. L. and McHenry, H. M. 1975. Osteometric shape relationships in the wrist of some anthropoids. *Folia Primatologica, 24:*250–274.

Cortright, G. W. 1975. A comparative and analytical study of gibbon bipedalism. M.A. thesis, The University of Chicago.

Cortright, G. W. 1983. The relative mass of hindlimb muscles in anthropoid primates: functional and evolutionary implications. Ph.D. dissertation, The University of Chicago.

Cronin, J. E. 1983. Apes, humans, and molecular clocks: a reappraisal. *In* New Interpretations of Ape and Human Ancestry, pp. 115–135, ed. R. L. Ciochon and R. S. Corruccini. Plenum Press, New York.

Davenport, R. K., Jr. 1967. The orang-utan in Sabah. *Folia Primatologica, 5:*247–263.

Day, M. H. and Napier, J. R. 1963. The functional significance of the deep head of the flexor pollicis brevis in primates. *Folia Primatologica, 1:*122–134.

Erikson, G. E. 1963. Brachiation in New World monkeys and in anthropoid apes. *Symposia of the Zoological Society of London, 10:*135–164.

Galdikas, B. M. F. 1979. Orangutan adaptation at Tanjung Puting Reserve: mating and ecology. *In* The Great Apes, pp. 194–233, ed. D. A. Hamburg and E. R. McCown. Benjamin/Cummings, Menlo Park, CA.

Galdikas, B. M. F. 1981a. Orangutan reproduction in the wild. *In* Reproductive Biology of the Great Apes, pp. 281–300, ed. C. E. Graham. Academic Press, New York.

Galdikas, B. M. F. 1981b. Modern adaptations in orangutans? *Nature, 291:*266.

Galdikas, B. M. F. 1982. Orang-utan tool-use at Tanjung Puting Reserve, Central Indonesian Borneo (Kalimantan Tengah). *Journal of Human Evolution, 10:*19–33.

Galdikas, B. M. F. 1983. The orangutan long call and snag crashing at Tanjung Puting Reserve. *Primates, 24:*371–384.

Galdikas, B. M. F. 1984. Adult female sociality among wild orangutans at Tanjung Puting Reserve. *In* Female Primates: Studies by Women Primatologists, pp. 217–235, ed. M. F. Small. Alan R. Liss, New York.

Galdikas, B. M. F. 1985a. Subadult male orangutan sociality and reproductive behavior at Tanjung Puting. *American Journal of Primatology, 8:*87–99.

Galdikas, B. M. F. 1985b. Orangutan sociality at Tanjung Puting. *American Jouranl of Primatology, 9:*101–119.

Gingerich, P. D., Smith, B. H. and Rosenberg, K. 1982. Allometric scaling in the dentition of primates and prediction of body weight from tooth size in fossils. *American Journal of Physical Anthropology, 58:*81–100.

Greenfield, L. O. 1979. On the adaptive pattern of *"Ra-*

mapithecus". *American Journal of Physical Anthropology, 50*:527–548.

Harrisson, B. 1962. Orang-utan. Collins, London.

Hatley, T. and Kappelman, J. 1980. Bears, pigs, and Plio-Pleistocene hominids: a case for the exploitation of belowground food resources. *Human Ecology, 8*:371–387.

Hildebrand, M. 1967. Symmetrical gaits of primates. *American Journal of Physical Anthropology, 26*:119–130.

Horr, D. A. 1972. The Borneo orang-utan. *Borneo Research Bulletin, 4*:46–50.

Horr, D. A. 1975. The Borneo orang-utan: population structure and dynamics in relationship to ecology and reproductive strategy. *In* Primate Behavior, vol. 4, pp. 307–323, ed. L. A. Rosenblum. Academic Press, New York.

Horr, D. A. 1977. Orang-utan maturation: growing up in a female world. *In* Primate Bio-Social Development, pp. 289–321, ed. S. Chevalier-Skolnikoff and F. E. Poirier. Garland Press, New York.

Ishida, H. 1972. On the muscular composition of lower extremities of apes based on the relative weight. *Journal of the Anthropological Society, Nippon, 80*:125–142.

Jenkins, F. A., Jr. and Fleagle, J. G. 1975. Knuckle-walking and the functional anatomy of the wrists in living apes. *In* Primate Functional Morphology and Evolution, pp. 213–227, ed. R. H. Tuttle. Mouton, The Hague.

Jungers, W. L. 1985. Body size and scaling of limb proportions in primates. *In* Size and Scaling in Primate Biology, pp. 345–381, ed. W. L. Jungers. Plenum Press, New York.

Kay, R. F. 1981. The nut-crackers: a new theory of the adaptations of the Ramapithecinae. *American Journal of Physical Anthropology, 55*:141–151.

Kay, R. F. and Simons, E. L. 1983. A reassessment of the relationship between later Miocene and subsequent Hominoidea. *In* New Interpretations of Ape and Human Ancestry, pp. 577–624, ed. R. L. Ciochon and R. S. Corruccini. Plenum Press, New York.

von Koenigswald, G. H. R. 1982. Distribution and evolution of the orang utan, *Pongo pygmaeus* (Hoppius). *In* The Orang Utan. Its Biology and Conservation, pp. 1–15, ed. L. E. M. de Boer. W. Junk, The Hague.

Landau, M., Pilbeam, D. and Richard, A. 1982. Human origins a century after Darwin. *BioScience, 32*:507–512.

Lewis, O. J. 1965. Evolutionary change in the primate wrist and inferior radio-ulnar joints. *Anatomical Record, 151*:275–286.

Lewis, O. J. 1969. The hominoid wrist joint. *American Journal of Physical Anthropology, 30*:251–268.

Lewis, O. J. 1971. Brachiation and the early evolution of the Hominoidea. *Nature, 230*:577–579.

Lewis, O. J. 1972. Evolution of the hominoid wrist. *In* The Functional and Evolutionary Biology of Primates, pp. 207–222, ed. R. Tuttle. Aldine, Chicago.

Lewis, O. J. 1974. The wrist articulations of the Anthropoidea. *In* Primate Locomotion, pp. 143–169, ed. F. A. Jenkins, Jr. Academic Press, New York.

Lipson, S. and Pilbeam, D. 1982. *Ramapithecus* and hominoid evolution. *Journal of Human Evolution, 11*:545–548.

MacKinnon, J. 1971. The orang-utan in Sabah today. *Oryx, 11*:141–191.

MacKinnon, J. 1974. The behaviour and ecology of wild orang-utans (*Pongo pygmaeus*). *Animal Behaviour, 22*:3–74.

MacKinnon, J. 1979. Reproductive behavior in wild orangutan populations. *In* The Great Apes, pp. 256–273, ed. D. A. Hamburg and E. R. McCown. Benjamin/Cummings, Menlo Park, CA.

McHenry, H. M. and Corruccini, R. S. 1983. The wrist of *Proconsul africanus* and the origin of hominoid postcranial adaptations. *In* New Interpretations of Ape and Human Ancestry, pp. 353–367, ed. R. L. Ciochon and R. S. Corruccini. Plenum Press, New York.

Mitani, J. C. 1985a. Mating behaviour of male orangutans in the Kutai Game Reserve, Indonesia. *Animal Behaviour, 33*:392–402.

Mitani, J. C. 1985b. Sexual selection and adult male orangutan long calls. *Animal Behaviour, 33*:272–283.

Napier, J. R. 1960. Studies on the hands of living primates. *Proceedings of the Zoological Society of London, 134*:647–657.

Napier, J. R. 1964. The evolution of bipedal walking in the hominids. *Archives de Biologie (Liège), 75*:673–708.

Oxnard, C. E. 1967. The functional morphology of the primate shoulder as revealed by comparative anatomical, osteometric and discriminant function techniques. *American Journal of Physical Anthoropology, 26*:219–240.

Pilbeam, D. 1982. New hominoid skull material from the Miocene of Pakistan. *Nature, 295*:232–234.

Pilbeam, D. 1985. Patterns of hominoid evolution. *In* Ancestors: the Hard Evidence, pp. 51–59, ed. E. Delson. Alan R. Liss, New York.

Pilbeam, D. R., Behrensmeyer, A. K., Barry, J. C. and Shah, S. M. I. 1979. Miocene sediments and faunas of Pakistan. *Postilla, 179*:1–45.

Pilbeam, D., Meyer, G. E., Badgley, C., Rose, M. D., Pickford, M. H. L., Behrensmeyer, A. K. and Shah, S. M. I. 1977. New hominoid primates from the Siwaliks of Pakistan and their bearing on hominoid evolution. *Nature, 270*:689–695.

Pilbeam, D. R., Rose, M. D., Badgley, C. and Lipschutz, B. 1980. Miocene hominoids from Pakistan. *Postilla, 181*:1–94.

Pilbeam, D. and Smith, R. 1981. New skull remains of *Sivapithecus* from Pakistan. *Memoir Geological Survey of Pakistan, 2*:1–13.

Preuss, T. M. 1982. The face of *Sivapithecus indicus*: description of a new, relatively complete specimen from the Siwaliks of Pakistan. *Folia Primatologica, 38*:141–157.

Raza, S. M., Barry, J. C., Pilbeam, D., Rose, M. D., Shah, S. M. I. and Ward, S. 1983. New hominoid primates from the middle Miocene Chinji Formation, Potwar Plateau, Pakistan. *Nature, 306*:52–54.

Rijksen, H. D. 1978. A Field Study on Sumatran Orang

Utans *(Pongo pygmaeus abelii* Lesson 1827). H. Veenman & Zonen, Wageningen.

Roberts, D. 1974. Structure and function of the primate scapula, *In* Primate Locomotion, pp. 171–200, ed. F. A. Jenkins, Jr. Academic Press, New York.

Rodman, P. S. 1973. Population composition and adaptive organization among orang-utans of the Kutai Reserve. *In* Comparative Ecology and Behaviour of Primates, pp. 171–209, ed. R. P. Michael and J. H. Crook. Academic Press, London.

Rodman, P. S. 1977. Feeding behaviour of orang-utans of the Kutai Nature Reserve, East Kalimantan. *In* Primate Behaviour: Studies of Feeding and Ranging Behaviour in Lemurs, Monkeys and Apes, pp. 383–413, ed. T. H. Clutton-Brock. Academic Press, London.

Rodman, P. S. 1979. Individual activity pattern and the solitary nature of orangutans. *In* The Great Apes, pp. 234–255, ed. D. A. Hamburg and E. R. McCown. Benjamin/Cummings, Menlo Park, CA.

Rodman, P. S. 1984. Foraging and social systems of orangutans and chimpanzees. *In* Adaptations for Foraging in Nonhuman Primates, pp. 134–160, ed. P. S. Rodman and J. G. Cant. Columbia University Press, New York.

Rose, M. D. 1984. Hominoid postcranial specimens from the Middle Miocene Chinji Formation, Pakistan. *Journal of Human Evolution, 13:*503–516.

Schaller, G. B. 1961. The orang-utan in Sarawak. *Zoologica, 46:*73–82.

Schultz, A. H. 1936. Characters common to higher primates and characters specific to man. *Quarterly Review of Biology, 11:*259–283, 425–455.

Schultz, A. H. 1941. Growth and development of the orang-utan. Carnegie Institute of Washington Publication 525, *Contributions to Embryology, 27:*57–110.

Schultz, A. H. 1956. Postembryonic age changes. *Primatologia I,* pp. 887–964, ed. H. Hofer, A. H. Schultz, and D. Starck. Karger, Basel.

Schultz, A. H. 1963. Relations between the lengths of the main parts of the foot skeleton in primates. *Folia Primatologica, 1:*150–171.

Shea, B. T. 1985. On aspects of skull form in African apes and orangutans, with implications for hominoid evolution. *American Journal of Physical Anthropology, 68:*329–342.

Sigmon, B. A. 1975. Functions and evolution of hominid hip and thigh musculature. *In* Primate Functional Morphology and Evolution, pp. 235–252, ed. R. H. Tuttle. Mouton, The Hague.

Smith, R. J. and Pilbeam, D. R. 1980a. Evolution of the orang-utan. *Nature, 284:*447–448.

Smith, R. J. and Pilbeam, D. R. 1980b. Smith and Pilbeam reply. *Nature, 288:*301.

Smith, R. J. and Pilbeam, D. R. 1981. Smith and Pilbeam reply. *Nature, 291:*266.

Straus, W. L. , Jr. 1942. Rudimentary digits in primates. *Quarterly Review of Biology, 17:*228–243.

Sugardjito, J. 1982. Locomotor behaviour of the orang utan (*Pongo pygmaeus abelii*) at Ketambe, Gunung Leuser National Park. *Malayan Nature Journal, 35:*57–64.

Sugardjito, J. 1983. Selecting nest-sites of Sumatran orang-utans, *Pongo pygmaeus abelii* in the Gunung Leuser National Park, Indonesia. *Primates, 24:*467–474.

Susman, R. L. 1974. Facultative terrestrial hand postures in an orangutan and pongid evolution. *American Journal of Physical Anthropology, 40:*27–38.

Susman, R. L. 1979. Comparative and functional morphology of hominoid fingers. *American Journal of Physical Anthropology, 50:*215–236.

Susman, R. L. and Tuttle, R. H. 1976. Knuckling behavior in captive orangutans and a wounded baboon. *American Journal of Physical Anthropology, 45:*123–124.

Tardieu, C. 1983. L'Articulation du Genou. Centre National de la Recherche Scientifique, Paris.

Temerin, L. A. 1980. Evolution of the orang-utan. *Nature, 288:*301.

Tuttle, R. H. 1967. Knuckle-walking and the evolution of hominoid hands. *American Journal of Physical Anthropology, 26:*171–206.

Tuttle, R. H. 1969a. Quantitative and functional studies on the hands of the Anthropoidea. I. The Hominoidea. *Journal of Morphology, 128:*309–364.

Tuttle, R. H. 1969b. Knuckle-walking and the problem of human origins. *Science, 166:*953–961.

Tuttle, R. H. 1970a. Postural, propulsive, and prehensile capabilities in the cheiridia of chimpanzees and other great apes. *In* The Chimpanzee, vol. 2, pp. 167–253, ed. G. H. Bourne. Karger, Basel.

Tuttle, R. H. 1970b. Propulsive and prehensile capabilities in the hands and feet of the great apes: a preliminary report. *Proceedings of the VIIIth International Congress of Anthropological and Ethnological Sciences, 1:*327–331. Science Council of Japan, Tokyo.

Tuttle, R. 1972. Relative mass of cheiridial muscles in catarrhine primates. *In* The Functional and Evolutionary Biology of Primates, pp. 262–291, ed. R. Tuttle. Aldine, Chicago.

Tuttle, R. 1974. Darwin's apes, dental apes, and the descent of man: normal science in evolutionary anthropology. *Current Anthropology, 15:*389–426.

Tuttle, R. 1975. Parallelism, brachiation and hominoid phylogeny. *In* Phylogeny of the Primates, pp. 447–480, ed. W. P. Luckett and F. S. Szalay. Plenum Press, New York.

Tuttle, R. H. 1977. Naturalistic positional behavior of apes and models of hominid evolution, 1929–1976. *In* Progress in Ape Research, pp. 277–296, ed. G. H. Bourne. Academic Press, New York.

Tuttle, R. H. 1986. Apes of the World. Studies on the Lives of Great Apes and Gibbons, 1929–1985. Noyes, Park Ridge, New Jersey.

Tuttle, R. H. and Basmajian, J. V. 1974. Electromyography of forearm musculature in gorilla and problems related to knuckle-walking. *In* Primate Locomotion, pp. 293–347, ed. F. A. Jenkins, Jr. Academic Press, New York.

Tuttle, R. H. and Basmajian, J. V. 1977. Electromyography of pongid shoulder muscles and hominoid evolution. I. Retractors of the humerus and rotators of the scapula. *Yearbook of Physical Anthropology, 20:*491–497.

Tuttle, R. H. and Basmajian, J. V. 1978a. Electromyography of pongid shoulder muscles, II. Deltoid, rhomboid and "rotator cuff." *American Journal of Physical Anthropology, 49:*47–56.

Tuttle, R. H. and Basmajian, J. V. 1978b. Electromyography of pongid shoulder muscles, III. Quadrupedal positional behavior. *American Journal of Physical Anthropology, 49:*57–70.

Tuttle, R. H., Basmajian, J. V. and Ishida, H. 1978. Electromyography of pongid gluteal muscles and hominid evolution. *In* Recent Advances in Primatology, vol. 3, pp. 463–468, ed. D. J. Chivers and K. A. Joysey. Academic Press, London.

Tuttle, R. H. , Basmajian, J. V. and Ishida, H. 1979. Activities of pongid thigh muscles during bipedal behavior. *American Journal of Physical Anthropology, 50:*123–136.

Tuttle, R. and Beck. B. B. 1972. Knuckle walking hand postures in an orangutan (*Pongo pygmaeus*). *Nature, 236:*33–34.

Tuttle, R. H., Cortright, G. W. and Buxhoeveden, D. P. 1979. Anthropology on the move: progress in experimental studies of nonhuman primate positional behavior. *Yearbook of Physical Anthropology, 22:*187–214.

Tuttle, R. H. and Rogers, C. M. 1966. Genetic and selective factors in reduction of the hallux in *Pongo pygmaeus. American Journal of Physical Anthropology, 24:*191–198.

Tuttle, R. H., Velte, M. J. and Basmajian, J. V. 1983. Electromyography of brachial muscles in *Pan troglodytes* and *Pongo pygmaeus. American Journal of Physical Anthropology, 61:*75–83.

Tuttle, R. H. and Watts, D. P. 1985. The positional behavior and adaptive complexes of *Pan gorilla. In* Primate Morphophysiology, Locomotor Analyses and Human Bipedalism, pp. 261–288, ed. S. Kondo. University of Tokyo Press, Tokyo.

de Vos, J. 1983. The *Pongo* faunas from Java and Sumatra and their significance for biostratigraphical and paleo-ecological interpretations. *Proceedings of the Koninklijke Nederlandse Akademie van Wetenschappen, Series B, 86:*417–425.

de Vos, J. 1984. Reconsideration of Pleistocene cave faunas from South China and their relation to the faunas from Java. *Courier Forschungsinstitut Senkenberg, 69:*259–266.

Ward, S. C. and Pilbeam, D. R. 1983. Maxillofacial morphology of Miocene hominoids from Africa and Indo-Pakistan. *In* New Interpretations of Ape and Human Ancestry, pp. 211–238, ed. R. L. Ciochon and R. S. Corruccini. Plenum Press, New York.

Wu, R., Xu, Q. and Lu, Q. 1983. Morphological features of *Ramapithecus* and *Sivapithecus* and their phylogentic relationships. *Acta Anthropologica Sinica, 2:*1–14.

24

Anomalies and Atavisms in Appendicular Myology

JOHN C. ANDERTON

This contribution results from the dissection of the limbs of an orang-utan, *Pongo pygmaeus.* The specimen was a Sumatran male, born and raised at the Yerkes Regional Primate Center, Emory University, where it died of pneumonia at age 9 years. From the dissections were prepared detailed descriptions of the musculature which were subsequently contrasted with the extant literature on the appendicular anatomy of the orang-utan. During this procedure, a few morphologies emerged as being of particular interest. Thus, given the constraints here of length, I decided to examine in detail these striking characters rather than provide a broad discussion of all the musculature.

The significance of morphologies peculiar to a single specimen is difficult to assess. Unusual characteristics may be due simply to individual variation. Comparisons with earlier anatomical accounts enlarges one's sample and distinguishes those morphologies produced by variation in age, sex, and other factors from those truly abnormal in structure. To illuminate the evolutionary implications of these anomalies, comparisons must be undertaken with the anatomy of organisms within the subject's clade (in this case, the large hominoids) and of "outgroups," related groups whose morphologies designate them as relatively primitive (the Cercopithecoidea and the Prosimii).

Morphological surveys of primitive outgroups establish an "anatomical backdrop" against which one may determine the derived characteristics (synapomorphies) of the advanced group. In turn, structures present in an organism but lacking in the members of its clade and in the outgroup are likely to be evolutionary novelties for that organism. However, while some of the outstanding muscle configurations of the orang-utan discussed here are not found in other hominoids, including other orang-utans, they do resemble the apparently homologous structures in monkeys and particularly prosimians. Such features are atavisms, anatomical arrangements associated with primitive groups which have appeared in a more advanced organism through the suppression or arrest of developmental processes and are assumed to have characterized an ancestor of the advanced organism (Dunlap et al., 1985).

Some of the morphologies discussed here—such as those of the flexors of the pedal digits—are clearly nonatavistic, and as these are shared to an extent with some specimens of African apes they may be interpreted as synapomorphic of a "great ape" group, or alternatively, as features acquired in parallel among the three genera.

THE FORELIMB

This specimen had been used previously for a study of the cervical and cranial anatomy (Chapter 14, this volume) and thus was missing on both sides much of the musculature relating the scapula to the vertebral column (i.e., the rhomboids, trapezius, and levator scapulae) as well as the clavicular muscles. The pectoral muscles were destroyed during the autopsy. Otherwise, both upper limbs were intact and did not exhibit differences bilaterally (unlike the hindlimbs, see below). Of all the musculature associated with the forelimb only two, both in the brachium, displayed noteworthy deviations from the norm in the orang-utan; the peculiarities of these two muscles, the brachialis and the triceps brachii, are clearly interrelated.

The *M. Brachialis*

The brachialis (Fig. 24–1) occupied its usual position on the distal half of the anterior aspect of the humerus and inserted on the coronoid process of the ulna. The origin of the lateral half of the muscle was, however, markedly extended: it passed posterior to the insertion of the deltoid (which was embraced by the brachialis as it is in humans) and covered the dorsum of the humeral shaft lateral to the spiral groove. The spiral

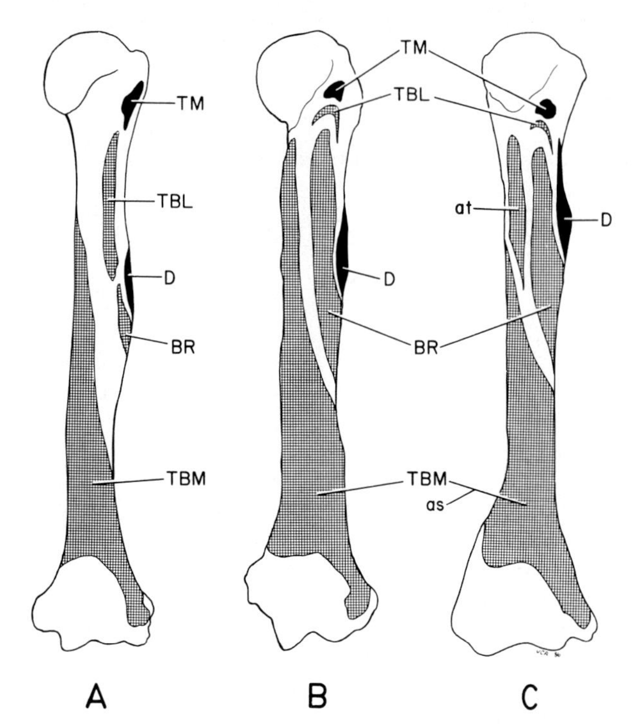

FIG. 24–1. The origins *(shaded)* and insertions *(black)* of muscles on the posterior aspect of the right humerus in (A) humans, (B) the 9-year-old orang-utan discussed in this paper, and (C) the tree shrew *Tupaia.* BR, brachialis; D, deltoid; TBL, triceps brachii lateral head; TBM, triceps brachii medial head; TM, teres minor. These brachial muscles in *Homo* represent the basic arrangement in anthropoids, but this particular specimen of *Pongo* very clearly recalls the more primitive pattern, which characterizes most prosimians. The medial head of the triceps in the tree shrew, as in prosimians, is often subdivided into two parts, the anconeus tertius (at) and the anconeus sextus (as). (A from Woodburne [1978]; C based on George [1977] and Le Gros Clark [1924].)

groove separated the brachialis from the medial head of the triceps. In hominoids this area of the humeral shaft is normally occupied by the origin of the lateral head of the triceps, which in this specimen was displaced and thus arose only from a small region immediately inferior to the insertion of the teres minor. This uppermost origin of the brachialis was thin and largely tendinous.

Howell and Straus (1932) stated that in prosimians the medial and lateral portions of the brachialis are distinct, and that the lateral portion is usually the more strongly developed, arising, in *Galago* and *Tarsius,* quite high on the posterior aspect of the humeral shaft. Howell and Straus equated this prosimian condition with that of rodents and insectivores and, to a degree, the opossum, *Didelphis.* Studies on lemuriforms (Murie and Mivart, 1872), *Tarsius* (Woollard, 1925), and the Tupaiidae (Davis, 1938; George, 1977; Le Gros Clark, 1924, 1926) confirm this configuration as primitive. It may be seen on occasion in monkeys, such as *Macaca* (Howell and Straus, 1933); Swindler (1973) illustrates it as such in *Papio,* but incongruously describes the brachialis as being similar to that in hominoids. In any case, such forms of the brachialis in catarrhines, if not all anthropoids, are probably atavistic. Howell and Straus (1932) asserted that the equal development and fusion of the two halves of the brachialis—as seen in anthropoids generally—is an advanced state, while the embracement by the brachialis of the deltoid insertion is even further advanced, often found in the large apes (such as the present orang-utan) and is also characteristic of humans.

The primitively high lateral origin of the brachialis does not seem to have occurred in any hominoid other than the orang-utan discussed here. As such, there can be little question that this feature in this specimen is an atavism, and it is therefore of further interest to find it in association with a supposedly advanced (for hominoids) feature as the embracement of the deltoid insertion.

The *M. Triceps Brachii*

The long head of the triceps was unremarkable in this orang-utan, in contrast to the lateral and medial heads (Fig. 24–1). The lateral head exhibited a notably restricted origin from the posterior aspect of the humerus, just superior to the surgical neck and between the insertions of the teres minor superiorly, and the extended lateral origin of the brachialis inferiorly. The origin of the lateral head was largely fibrous, although a slight fleshy origin persisted down the lateral intermuscular septum. The origin of the medial head was also unusual, extending very high on the medial aspect of the humeral shaft to the attachment of the articular capsule of the shoulder joint. The medial head in hominoids normally does not extend above the lower limit of the teres major insertion on the medial lip of the bicipital groove. In all other respects the medial head of the triceps in this orang-utan was normal.

A review of the literature revealed that, among the hominoids, a very high origin of the medial head was present only in Sonntag's (1923) chimpanzee. But in contrast to the orang-utan discussed here, this specimen of African ape was even more unusual in that this high origin was confined to the proximal third of the humeral shaft, and thus was lacking the common muscle mass inferior to the spiral groove. The lateral head of Sonntag's chimpanzee also bore a restricted origin similar to that seen in my orang-utan. There are no other references to similar morphologies of the triceps among the hominoids. However, in the prosimians *Lemur, Galago, Daubentonia* (Murie and Mivart, 1872), and *Tarsius* (Burmeister, 1846; Woollard, 1925), and in the Tupaiidae (George, 1977; Le Gros Clark, 1924), the triceps is described as consisting of four parts, as a result of the medial head being divided into two components, one superior (the anconeus tertius) and the other inferior (the anconeus sextus) to the spiral groove; the lateral head (which may also be subdivided [Le Gros Clark, 1924]) arises from a small area inferior to the teres minor.

The Old and New World monkeys seem for the most part to hold to the tricipital organization common to the hominoids. Beattie (1927) found a prosimian-like triceps in a specimen of *Callithrix,* but this is not true of a number of other examinations of platyrrhines (Ashton and Oxnard, 1963; Campbell, 1937; Schön, 1968; Sirena, 1871), and Dunlap et al. (1985), while mentioning peculiarities in the long tricipital head of some New World monkeys, did not discover any morphologies of the other two heads that differed from the rest of the Anthropoidea.

In achieving the configuration normal for anthropoids, the superior portion of the medial head (the prosimian anconeus tertius) has apparently become integrated with the lateral head, thereby (1) extending the origin of the latter down to the spiral groove; (2) reducing the origin of the medial head from the medial aspect of the surgical neck of the humerus; and (3) supplanting the high lateral origin of the brachialis. The orang-utan studied here would therefore be atavistic in its arrangement of the humeral portions of the triceps brachii, since it displays a prosimian-like small origin of the lateral head in addition to a high lateral origin of the brachialis. The medial head of this orang-utan's triceps was not divided by the spiral groove, but this is not a characteristic of all prosimians either (e.g., *Nycticebus;* Murie and Mivart, 1872). Thus the high origin of the medial head of triceps in the orang-utan discussed here may be considered a remnant of the anconeus tertius.

Sonntag's (1923) account gives the impression that his chimpanzee retained a distinct anconeus tertius but was lacking the inferior portion of the medial head, the anconeus sextus. The retention of the former muscle as distinct from the lateral head is an atavism, while the complete absence of the inferior mass is a striking, possibly pathological, anomaly. This chimpanzee exhibited the same characters of the lateral head as seen in my orang-utan, and thus the conclusions reached here as to the significance of this feature would seem to be applicable.

THE HINDLIMB

More characters of interest were found in the musculature of the lower than in the upper limb. Among these was the disparity in size between the musculature of the right and left sides: muscles of the left lower limb were consistently bulkier than their right counterparts. Although this could not be substantiated quantitatively (contraction states and tissue desiccation differed between the muscles of the two sides), it was particularly noticeable in the structure of the muscles. For example, the sartorius on the left side arose from the anterior margin of the iliac blade, whereas on the right side it was anchored only to the tough superficial fascia of the inguinal region; and the left tibialis anterior produced three tendons, each arising from a distinct longitudinal subdivision of the muscle, while, on the right side, the comparatively unified tibialis anterior generated only two tendons, the more common number in the orang-utan. Of interest here is that, when muscle weights were compared, Chibber and Singh (1970) found left-sidedness to be common in humans.

Of the muscles discussed below, three—the flexor digitorum brevis, the peroneus digiti quinti, and the "subgluteus"—displayed differences bilaterally, and only the first of these three favored the right limb. This is the one exception to the general "left-leggedness" of this orang-utan specimen.

The "Subgluteus"

The "subgluteus" (Fig. 24–2) was a mysterious muscle located in the hip, deep to the gluteus superficialis. It arose from the lateral aspect of the superior half of the sacrospinous ligament, passed in parallel with the gluteus medius superficial to the piriformis, and crossed over the tendon of the latter to insert by a very narrow, thin tendon on the lateral aspect of the femur approximately 2 cm proximal to the insertion of the gluteus superficialis. The left subgluteus was much more robust than the right one (2.5 cm compared to 1 cm wide) and bore a heavier tendon. On both sides this muscle was innervated by the inferior gluteal nerve, which also supplied the gluteus superficialis.

It is difficult to assess the significance of the subgluteus because it seems to have no identifiable parallels in the literature on primate anatomy. No muscle is described which either arises from the sacrospinous ligament or inserts upon the femur in just this manner. Judging from its innervation and its general disposition, the subgluteus would appear to have a closer relationship to the gluteus superficialis than to any other muscle of the hip joint. The subgluteus may be a vestige of the caudal division of the gluteus superficialis of prosimians (McArdle, 1981; Murie and Mivart, 1872) and monkeys (Howell and Straus, 1933; Schön, 1968; Sirena, 1871), which, in addition to originating from the gluteal aponeurosis, is formed by fascicles that arise independently from the costal processes of the proximal caudal vertebrae and converge toward insertion. The drastic reduction of the caudal vertebrae to form the coccyx in hominoids has eliminated this latter origin, so that only the origin from the gluteal fascia of the dorsum of the coccyx remains. The structure of the subgluteus seems to recall the caudal muscular bundles of the more primitive primates, and the common source of innervation for the subgluteus and the gluteus superficialis reinforces the impression of the former as an atavistic feature.

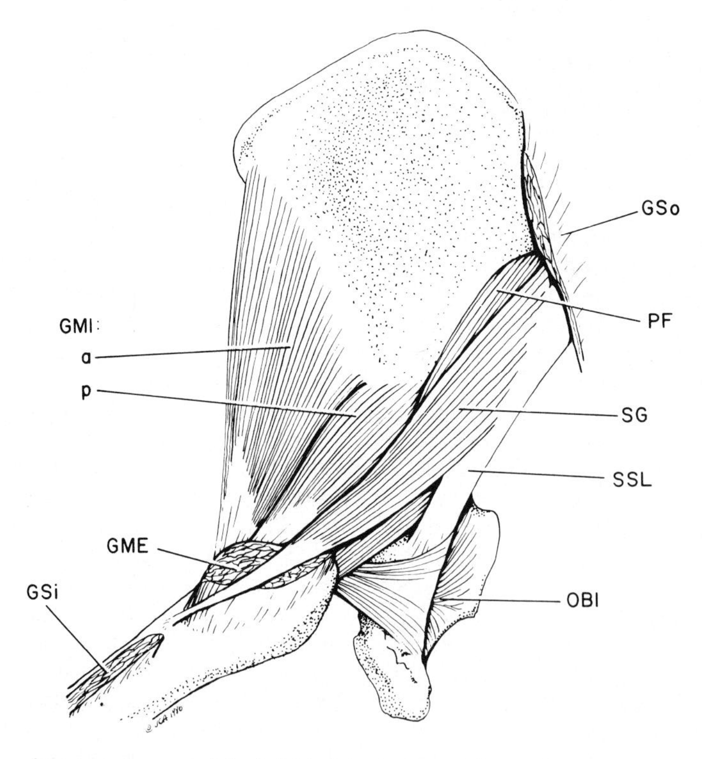

FIG. 24–2. The muscles of the lateral aspect of the left hip joint in the specimen of the orang-utan discussed here. GME, gluteus medius insertion; GMI, gluteus minimus anterior (a, also known as the scansorius) and posterior (p) portions; GS, gluteus superficialis origin (o) and insertion (i); OBI, obturator internus and associated gemelli muscles; PF, piriformis; SG, the aberrant "subgluteus" muscle; and SSL, sacrospinous ligament.

However, it should be noted that, in the left leg, where the subgluteus was more fully developed, the piriformis was weaker and less bulky than on the right side and was partially subdivided into superficial and deep parts. This suggests some relationship between these two muscles.

The *M. Peroneus Digiti Quinti*

In the left leg of this orang-utan, a small muscular slip arose from a 1-cm area of the lateral aspect of the fibula, 2 cm superior to the lateral malleolus; it was deeply contiguous with the peroneus brevis. This fusiform muscular slip produced a very thin fascia-like tendon which expanded to attach along a line on the dorsolateral aspect of the calcaneus, between the peroneus brevis tendon and the most proximal origin of the extensor digitorum brevis. This muscle was not found in the right leg.

Peroneal muscles to the fourth (peroneus digiti quarti) and the fifth (peroneus digiti quinti) pedal digits are common in prosimians and primitive mammals. Both muscles arise from the fibula inferior to the larger peronei, with the quinti superior to the quarti (Straus, 1930). Of the two the quarti is absent more often; in *Loris,* both peroneal muscles may be lacking (Murie and Mivart, 1872). Woollard (1925) described the peroneus digiti quarti in *Tarsius.* Straus (1930) stated that the occurrence of this muscle is "rather regular" only in some monkeys but is unknown in Hominoidea. The peroneus digiti quinti is a very common anatomical feature of prosimians, platyrrhines, and cercopithecoids, but it is usually lacking in hominoids.

Some anatomists (cited in Straus, 1930) nevertheless have identified a complete peroneus digiti quinti in the orang-utan. In these cases the body of the muscle is positioned as is the flimsy distal fascicle in the specimen discussed here, but from it passes forward a narrow filament-like tendon to the dorsum of the fifth proximal phalanx. This fine tendon was present in both feet of my 9-year-old male orang-utan, but it was fused to the peroneus brevis tendon, and not associated with the muscle mass. This fused tendon is frequently encountered in hominoids: Raven (1950) found it in the gorilla, Hepburn (1892) and Sonntag (1924) in the orang-utan, Beddard (1893), Champneys (1872), and Sonntag (1923) in the chimpanzee; Straus (1930) cites cases for all the nonhuman hominoids, including the gibbon. Gray (1942) also lists the complete peroneus digiti quinti as rare but the vestigial tendon as rather common in humans, while Aziz (1981) discovered the complete muscle in four aneuploid human infants—two specimens of trisomy 13 (Patau's syndrome) and two of trisomy 18 (Edwards's syndrome). Judging from the literature, the body and proximal length of the tendon of the peroneus digiti quinti have been generally lost in the hominoids, and the distal tendon is the only remaining common vestige. However, a body and a proximal portion of a tendon inserting, as it does in the present orang-utan, onto the calcaneus (or, less frequently, the cuboid) occurs in about 13% of humans and has been identified as the "peroneus quartus" (Gray, 1942).

It would, therefore, appear that the peroneus digiti quinti of prosimians and monkeys has in hominoids been fragmented into two parts, with the body and proximal tendon being lost and the distal half of the tendon frequently becoming integrated with the tendon of the peroneus brevis. The reappearance of the proximal portion (as the "peroneus quartus") in a hominoid is atavistic in a sense, although it does not constitute the complete anatomical characteristic of a hominoid ancestor.

THE FLEXORS OF THE PEDAL DIGITS

The *M. Flexor Digitorum Tibialis*

The flexor digitorum tibialis (Figs. 24–3A, 24–4, and 24-5) arose from the posterior aspect of the medial half of the tibia, the origin being interposed between the insertion of the popliteus and the origin of the tibialis posterior. An additional and disjunct origin of the flexor digitorum tibialis was located distally on the tibia and the adjacent interosseous membrane, lateral to the tibialis posterior; the flexor digitorum tibialis thus cloaked the latter muscle in this specimen. This small distal origin was not mentioned in any of the references on hominoids used for this study.

The muscle fibers of the tibial flexor dwindled as the muscle's bulky tendon passed around the medial malleolus posteromedial to the ensheathed tibialis posterior tendon. At this point the tibial flexor tendon divided into a stout tendon to digit II and the trunk of the four remaining lateral tendons. Curiously, muscle tissue persisted on the lateral border of the trunk as it cleaved into four uneven tendons. There was a small tendon to digit III, a bulky one to digit IV, and two tendons, one large and one small, to digit V. With the exception of the tendon to the second digit and the large medial tendon to the fifth digit, muscle tissue remained on the mar-

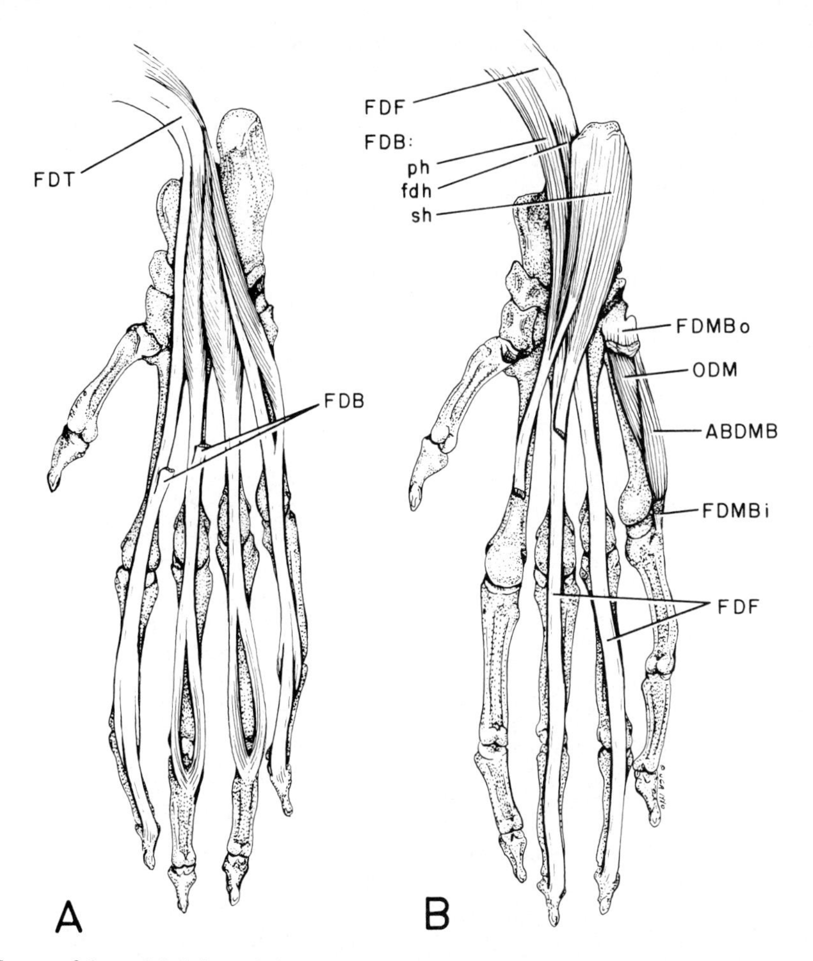

FIG. 24–3. The flexors of the pedal digits and the least toe in the present specimen of the orang-utan. ABDMB, abductor digiti minimi brevis; FDB, flexor digitorum brevis: fdh, the fibular deep head, ph, the proximal head, and sh, the superficial head; FDF, flexor digitorum fibularis; FDMB, flexor digiti minimi brevis origin (o) and insertion (i); FDT, flexor digitorum tibialis; ODM, opponens digiti minimi. In life, the flexor digitorum tibialis tendons are interposed between those of the flexor digitorum brevis and the flexor digitorum fibularis; it has been separated here to show the relationship between the latter two muscles.

gins of the tendons about halfway down the length of the metatarsus. The tendon to digit II inserted on both the middle and terminal phalanges, while the tendons to digits III and IV were perforated by the deeper flexor digitorum fibularis tendons and inserted on the middle phalanges of their respective digits. The two tendons to digit V fused to produce a single structure that inserted on both the middle and distal phalanges.

The *M. Flexor Digitorum Fibularis*

The flexor digitorum fibularis (Figs. 24–3B, 24–4, and 24–5) lay deep to the lateral head of the gastrocnemius. This muscle, which is the homolog of the human flexor hallucis longus, had two distinct origins: the superficial head arose from the posterior aspect of the lateral femoral epicondyle, and did so in common with the lateral head of the gastrocnemius; the deep head of the fibular flexor was inferior to the superficial head, arising from the superior two-thirds of the posterior aspect of the fibular shaft and the lateral intermuscular septum. The heavy tendon of the flexor digitorum fibularis passed deep to the flexor retinaculum and tendo Achilles and lateral to the tendon of the flexor digitorum tibialis. A groove on the posterior aspect of the distal tibia conducted the tendon over the posterior aspect of the talus and through the sulcus on the plantar surface of the sustenaculum tali. The fibularis tendon expanded deep to the tendon of the

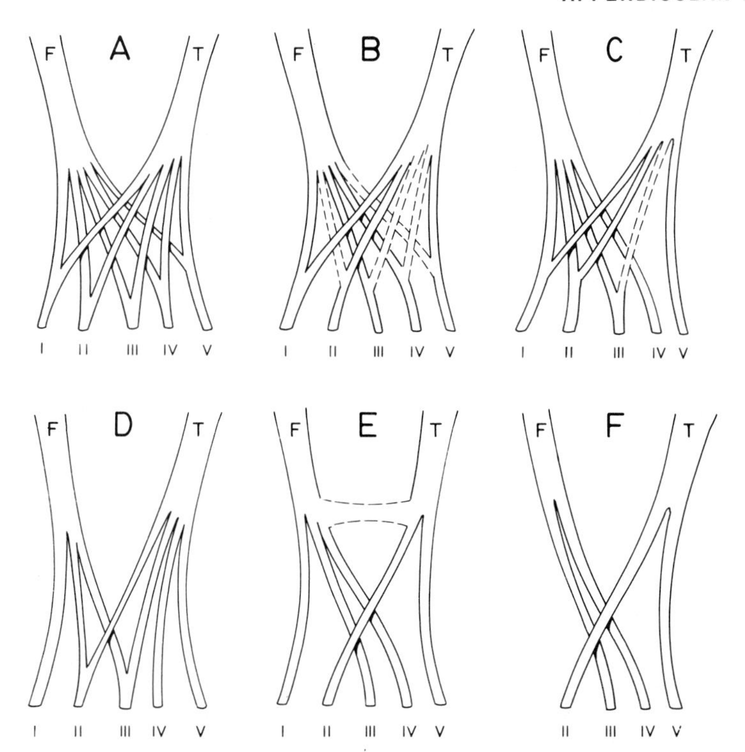

FIG. 24–4. Diagrammatic representation of the arrangement of the tendons of the flexor digitorum fibularis (F) and the flexor digitorum tibialis (T) in primates; modified from Straus (1930). The monkeys (B) often resemble the primitive arrangement of prosimians (A). But variation is extreme, and the reduction or loss of certain tendons of both muscles (in dashed lines), especially frequent in Old World monkeys, creates a pattern that parallels that of the African apes (E), save for the typical double tendon supply to the hallux in the former. Both gibbons (C) and humans (D) remain comparatively primitive, the former undergoing greatest reduction in the tibial flexor, while in humans the fibular flexor is the more reduced. Variation in the African apes is great, and the configuration shown here is a generalization that seems to hold more often for the gorilla than for the chimpanzee, although both may display elements of the prosimian pattern at times. The bridge between the flexors in the African apes indicates that the two muscles are very often fused proximal to the division into individual tendons, and thus elements of either flexor may contribute to the formation of a tendon. The orang-utan (F) is the furthest removed from the prosimian condition and thus the most specialized. All accounts surveyed appeared to conform to the pattern depicted here, which is notable for the absence of a hallucal tendon and for the complete lack of fusion between the two long flexors.

flexor digitorum tibialis and divided, sending very heavy tendons to insert on the distal phalanges of digits III and IV.

The origin and location of the long flexors of the pedal digits are fairly constant among hominoids and, in this regard, the orang-utan examined here did not differ significantly from those of earlier studies. The flexor digitorum fibularis in the orang-utan is unique among hominoids in its origin from the lateral femoral epicondyle.

Dobson (1883) and Glaesmer (1908) maintained that the typical but by no means absolute configuration of the long digital flexors (the flexor digitorum tibialis and the flexor digitorum fibularis) in prosimians represents the primitive condition for primates. Generally, in prosimians, each long digital flexor produces five tendons; there are thus five pairs of tendons with each pair fusing to form the deep (perforating) tendon of each digit. Each digit therefore receives tendinuous components of both the flexor digitorum tibialis and fibularis. These deep tendons insert on the distal phalanges. The superficial (perforated) tendons, which insert on the middle phalanges, are provided by the flexor digitorum brevis, which is an intrinsic pedal muscle.

Monkeys may display pedal flexor arrangements similar to those of prosimians—such as Schön's (1968) howler monkey, *Alouatta*—but they more often possess some features of the great ape pattern—as did Sirena's (1871) specimen of *Alouatta*. Generally, the configuration of the pedal flexor tendons in monkeys, especially

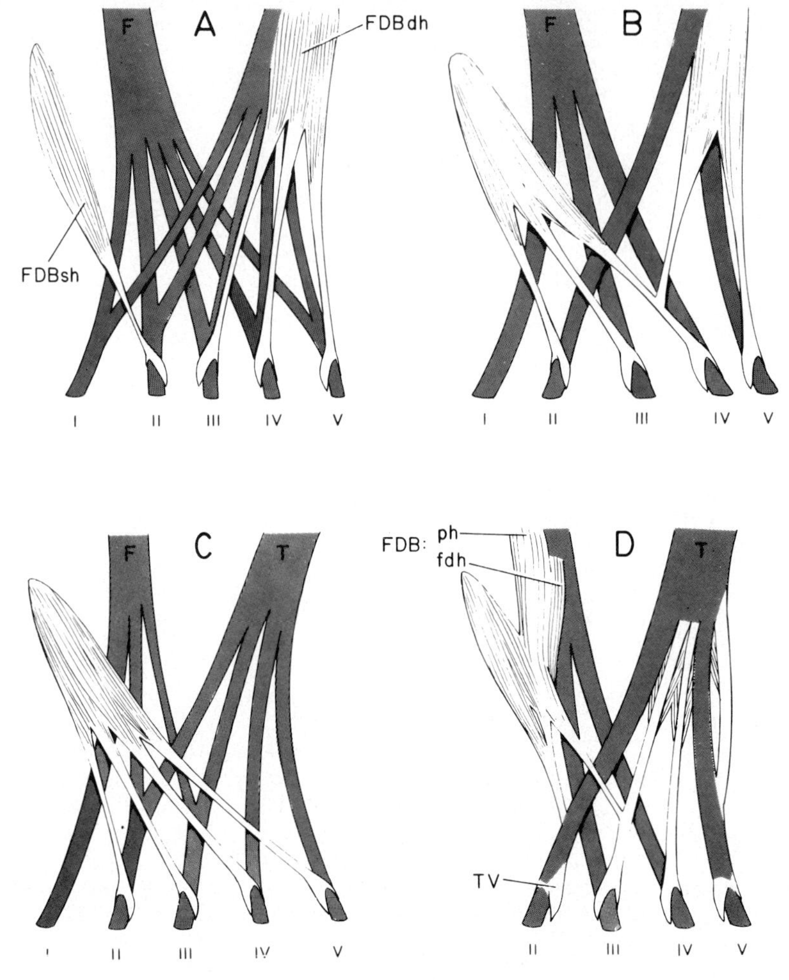

FIG. 24–5. Diagrammatic representation of the structure of the flexor digitorum brevis (FDB), superficial head (sh) and deep head (dh), in primates. The components of the muscle are superimposed on the configurations of the flexor digitorum fibularis and the flexor digitorum tibialis taken from Fig. 24–4. In prosimians (A), the deep head of the short flexor usually supplies tendons to lateral toes. Straus (1930) and others have pointed out that in the transferral of these tendons to the superficial head [seen in fullest development in humans (C)] the large nonhuman hominoids (B), including the orang-utan, remain somewhat intermediate, with some digits receiving tendon components from both heads. The orang-utan discussed here (D) represents an extreme example of another development common in the large apes, in which the tendons of the flexor digitorum brevis deep head are adopted by the flexor digitorum tibialis. In the case of this orang-utan, even the tendons of the superficial head were absorbed, leaving only the insertions distinct, as "triangular vincula" (TV), in digits II and V. The proximal (ph) and fibular deep (fdh) heads in this orang-utan were anomalous.

the Old World monkeys, is intermediate between those of the prosimians and the great apes (Champneys, 1872; Keith, 1894).

The hominoids, particularly the great apes, have tended to reduce the degree of fusion of the long flexor tendons, often limiting the source of the deep tendon of each toe to one rather than both of the long digital flexors. In the great apes, although there is a great deal of variation, the flexor digitorum fibularis usually provides the deep tendons of the hallux and the third and fourth digits, while the flexor digitorum tibialis supplies those of the second and fifth digits (Straus, 1930). The gibbon remains more primitive in its pedal flexor anatomy: the flexor digitorum fibularis often supports the medial four (or all five: Hepburn, 1892) digits with the tibialis fusing to it in the hallux and the indicis, and occasionally in the lateral digits (Huxley, 1864; Straus, 1930, 1949). Humans also retain a less specialized configuration than seen in the great apes: the tibialis flexor (the flexor digitorum longus) supports the lateral toes while the fibular muscle (the flexor hallucis longus) commonly supplies the bulk of the hallucal tendon; in digits II and III, in which the two flexors often fuse, the fibular muscle usually contributes more to the tendons than does the tibial flexor (Gray, 1942; Straus, 1930). Keith (1894) pointed out the similarity between the gibbon and humans in the degree of participation of the fibular flexor in the formation of the deep tendon of the second digit.

Hepburn (1892) and Straus's (1930) chimpanzees were identical in the anatomy of the pedal flexors. The flexor digitorum fibularis reached digits I, III, and IV and the flexor digitorum tibialis supplied digits II–IV, fusing to the fibularis flexor in digits III and IV; such a uniform distribution of the long flexor tendons and extensive fusion between them recalls the more primitive primate condition. Sonntag's (1923) *Pan* was strikingly human-like (and therefore also primitive) in that it had a flexor digitorum fibularis serving only the hallux and thus had a true flexor hallucis longus. In this specimen, the tibial flexor reached digits II–V and the arrangement overall was very much like that in Hepburn's (1892) gorilla—also human-like. Beddard's (1893) chimpanzee, however, displayed the "usual" great ape pattern. Champney's (1872) specimen also conforms to that model, although, as often in the gorilla, the long tendon trunks were fused prior to the division into digital tendons.

Gorilla varies greatly in the flexor anatomy of its foot. Much as in humans, Hepburn's (1892) specimen possessed a flexor digitorum fibularis that supplied only a hallucal tendon, while the tibial flexor distally supplied the four lateral toes; the two long flexor tendon trunks were joined at their bases. Also rather primitive, a gorilla Straus (1930) described was reminiscent of the condition in prosimians: fibularis tendons supplied all the pedal digits, those to digits I–IV fused to tendons of the flexor digitorum tibialis. However, Straus's summary tables of numerous other anatomical studies indicate that, most frequently, the gorilla conforms to the configuration described above typical of the great apes.

The orang-utan is normally one step further removed than the African apes from the primitive configuration of the pedal flexors in that a hallucal tendon is usually absent (Beddard, 1893; Boyer, 1935; Chapman, 1880; Hepburn, 1892: Primrose, 1899; Sonntag, 1924; Straus, 1930); the contributions of the flexor digitorum fibularis are thus typically confined to the third and fourth digits. By all accounts, the tibialis flexor represents the sole deep tendon supply to the second and fifth digits, thus conforming to the "great ape pattern." The fusion of the two long flexors of the pedal digits, a characteristic of prosimians, monkeys, and, to a lesser degree, gibbons and humans, is apparently most reduced among the great apes in *Pongo*. In the orang-utan of the present study there was no connection whatsoever between the two long flexors; Hepburn (1892) concluded similarly for his orang-utan, and this would also appear to be the case for the specimens of Boyer (1935), Chapman (1880), Primrose (1899), and Sonntag (1924). In those African apes in which fusion was reported to be entirely lacking between the individual tendons of the flexor digitorum tibialis and fibularis—Chapman's (1878) and Hepburn's (1892) *Gorilla* and Sonntag's (1923) *Pan*—there was still a juncture between the tendon trunks of the two muscles proximal to their division into separate digital tendons.

There are many reports in hominoids, especially the orang-utan, of the flexor digitorum tibialis tendons also forming connections with the tendons of the flexor digitorum brevis: Beddard (1893), Boyer (1935), Hepburn (1892), Primrose (1899), and Sonntag (1924) found a supernumerary tibialis tendon that fused to a tendon of the flexor digitorum brevis to form a perforated (rather than a deep) tendon. In the 9-year-old male of the present study there was a total of five tibialis tendons (seemingly a primitive condition), but those to digits III and IV constituted superficial, rather than deep, tendons. Only in digit III did a slip of the flexor digitorum brevis fuse to the tibialis tendon. Those tibialis tendons that went to digits II and V, in which perforated tendons were entirely absent, established middle phalangeal insertions simply by producing flanges on both sides of each tendon which inserted in the usual manner of flexor digitorum brevis tendons. The remaining central, bulky portion of these tendons proceeded distally to insert on the terminal phalanx, as is usual for flexor tibialis tendons. Therefore, true perforated tendons did not exist in this orang-utan's second and fifth digits. Sonntag's (1924) chimpanzee also apparently lacked a true perforated tendon to its lateralmost digit; he referred to the middle phalangeal insertion of this digit as a "triangular vinculum" of the tibialis tendon.

Another remarkable feature of the orang-utan I dissected is the muscle tissue that arose on the lateral margin of the undivided flexor digitorum tibialis tendon trunk and, with the exception of the medial of the two tendons to the fifth digit, persisted down both sides of each of the tendons to the lateral three digits. I failed to find this condition described for any primate. However, the muscular deep head of the flexor digitorum brevis in prosimians and monkeys usually arises from the tendon trunk of the flexor digitorum tibialis (see below), and Straus mentioned that he (1930), Turner (1867), and Keith (1899) (all cited in Straus, 1930) had found in humans a small muscle belly arising from the tibialis tendon trunk which produced a superficial tendon that went to digit V in the manner of a flexor digitorum brevis tendon. Partial origins for the short flexor tendons for digits III and IV in humans were also found, representing remnants of the normally absent deep head of the flexor digitorum brevis. Considering the reduced state of the flexor digitorum brevis in this orang-utan, and its deep origin from the tendon of the flexor digitorum tibialis rather than the fibularis (see below), it is probable that the supernumerary tibialis tendons to the third, fourth, and fifth digits—which served as superficial tendons in digits III and IV and in digit V contributed to the formation of a superficial tendon in place of the flexor digitorum brevis—were actually vestiges of the flexor digitorum brevis and not homologs of the tibialis tendons to these digits as seen in prosimians. I must stress that the supernumerary tendons of the flexor digitorum tibialis were not simply muscular slips appended to that muscle (as are the prosimian deep flexor brevis and the structures in humans as described by Straus ([1930]), but structures which were actual divisions of the trunk of the tibial flexor.

The *M. Flexor Digitorum Brevis*

The flexor digitorum brevis (Figs. 24–3B and 24–5) in my orang-utan was composed of three heads in the right foot but only two in the left. The proximal head, missing on the left side, arose from the sheath of the posterior tibial artery and tibial nerve in the distal third of the leg and was visible between the medial head of the gastrocnemius and the flexor digitorum tibialis. This small fusiform muscle passed posterior to the medial malleolus directly behind the tendon of the tibial flexor, where it was joined by the smaller deep head which arose from the tendinous sling of the flexor digitorum fibularis tendon. At this juncture, the deep head of the flexor digitorum brevis was united into a single belly and thereafter was identical in both feet, lying lateral and superficial to the flexor digitorum tibialis tendon. The deep head of the flexor digitorum brevis contributed a slight tendon to digit II tendon of the flexor digitorum tibialis.

The superficial head of the flexor digitorum brevis arose from the plantar rim of the calcaneal tuberosity and was placed lateral to the deep head and directly deep to the origin of the abductor hallucis muscle. This muscle formed three slips; the medial slip fused immediately to the deep head tendon which joined the flexor digitorum tibialis tendon to digit II. The lateral two slips developed tendons, of which the medial one also joined the tendon of the deep head. The lateralmost and largest tendon blended into the tibialis tendon that went to digit III. The flexor digitorum brevis was not associated with the fourth and fifth digits in this specimen.

The origin of the superficial head of the flexor digitorum brevis from the plantar aponeurosis is limited in orang-utans due to the lateral (as compared to the other hominoids) origin of the abductor hallucis, which overlies the former structure; otherwise, the disposition of the muscle is rather consistent among hominoids. But the occurrence and structure of the deep head of the short flexor, which arises from the tendon trunk of the flexor digitorum tibialis, is much more variable. Hepburn (1892) found it only in the chimpanzee and gibbon. Sonntag (1923, 1924) and Beddard (1893) do not mention a deep head of the flexor digitorum brevis in the chimpanzee or the orang-utan. This component, however, is common in the gorilla (Chapman, 1878; Raven, 1950; Straus, 1930) and is a fairly constant characteristic of monkeys and prosimians (Straus, 1930). In all primates (except in the present orang-utan), apparently, when the deep head is present, it distributes perforated tendons to the lateral two or three digits. Concerning the configuration seen in the right foot of the present orang-utan, only Straus's (1930) gorilla, among hominoids, possessed a similar posterior and superior extension of the deep head, causing it to arise in the region of the ankle as well; due to the incompleteness of his specimen, Straus could not provide details of this origin. A proximal head of the flexor digitorum brevis was also described by Howell and Straus (1933) for *Macaca mulatta.* The deep head of Boyer's (1935) orang-utan arose from the medial aspect of the soleus tendon, thus recalling somewhat the unusual proximal head of the short digital flexor in the present specimen. Since McMurrich (1907) described the flexor brevis superficialis as arising as far proximal as the crus in the opposum, *Didelphis,* it is likely that this configuration in higher primates is primitive.

Hepburn (1892) and Straus (1930) suggested that there is a "trend" among the hominoids toward the loss of the deep head of the flexor digitorum brevis with a concurrent incorporation of its tendons to the three lateral digits into the superficial head of that muscle. Although this deep head may fail to develop in any hominoid, it is most frequently absent in humans and orang-utans. Straus's (1930) survey of the literature also delineated the tendency among hominoids toward the loss of the flexor digitorum brevis tendon to the fifth digit. Although its absence is relatively common among the hominoids, it is most frequently lacking in the gorilla.

The flexor digitorum brevis in the orang-utan under discussion is exceptional on several counts. First, the deep head, which appeared to be well developed, arose from the tendon of the flexor digitorum fibularis rather than the flexor digitorum tibialis. The only similar deviations I found in the literature were in reference to specimens of *Gorilla* (Chapman, 1878) and the colobine monkey *Pygathrix* (Straus, 1930); the flexor digitorum brevis deep head in both specimens bore a small and partial origin from the fibular flexor tendon—but it was still anchored in large part to the flexor digitorum tibialis tendon. Second, this orang-utan's short digital flexor appeared to lack tendons not only to digit V—as do many hominoids, as discussed above—but to digit IV as well. Third, the superficial tendons to the second and third digits were feeble and fused to the respective flexor digitorum tibialis tendons rather than forming the perforated tendons independently. Fourth, and most unusual of all, the single tendon of the deep head of the flexor digitorum brevis, which developed with two slips from the superficial head, passed to the second digit rather than to a lateral digit. Such a condition does not appear to have been remarked upon elsewhere in the literature.

Although there were no perforated tendons provided solely by the flexor digitorum brevis in this orang-utan, there still existed a full complement of middle phalangeal insertions. Therefore, despite the impression that the muscle was grossly underdeveloped, it appears to have been functionally present. The flexor digitorum brevis seems to have been fragmented, and many of its components adopted by the flexor digitorum tibialis.

Some authors (Hepburn, 1892; Raven, 1950), while describing the deep head of the flexor digitorum brevis in the great apes, pointed to the continuation of the fleshy fibers of the flexor digitorum tibialis into the body of the former muscle. This condition is reminiscent of the appearance of the unusual fleshy bodies of the flexor digitorum tibialis tendons to digits III and IV and of the lateral tendon to digit V (which are anomalies in themselves) in the orang-utan under examination. I suspect that the deep portions of a flexor digitorum brevis are represented in this orang-utan by these fleshy tibialis flexor components (rather than by the deep head arising from the fibularis tendon) and that the perforated tendons of digits II–V have become fully integrated into the flexor digitorum tibialis, which, being free of any attachments to the fibular flexor, assumed the role of the short digital flexor. This model does not account for the small tendinous slip from the superficial head of the flexor digitorum brevis; however, Straus (1930), after tracing the migration of the short digital flexor tendons to the lateral three digits from the deep head (as, e.g., in various prosimians) to the superficial head (as in *Homo*), found that there were some "intermediate"cases (in certain prosimians and monkeys) in which digits III and IV had tendons composed of elements from both heads (see Fig. 24–4B). This could account for the double tendon in digit III of this male orang-utan: the tendon from the superficial head of the short flexor is present in its usual configuration, but the tendon from the deep head became integrated with the flexor digitorum tibialis. The fourth digit of this orang-utan was similar in this respect.

This pattern of flexor tendon transferral is not without precedence in the orang-utan. Beddard (1893), Boyer (1935), Hepburn (1892), Primrose (1899), and Sonntag (1924) recorded the fusion of a normal flexor digitorum brevis tendon to a flexor digitorum tibialis tendon to produce a perforated tendon to digit IV. Boyer (1935) and Hepburn (1892) also found double tibialis flexor tendons to the fifth digit, as in my specimen. Beddard (1893) and Sonntag (1923) reported that the flexor digitorum tibialis contributed to the superficial tendon of the fourth digit in the chimpanzee. In the gibbon, Hepburn (1892) observed tendinous slips coursing from the flexor digitorum tibialis to the short flexor tendons of the lateral three digits, and Chapman (1878) found the same in the lateral two digits of the gorilla. For nonhominoids, Schön (1968) described the tibial flexor in a single specimen of the platyrrhine *Alouatta* as providing all of the perforated tendon to digit V. Although the foregoing cases demonstrate the transferral of flexor digitorum brevis deep head tendons to the flexor digitorum tibialis, in none has the absorption been as marked as in the orang-utan under discussion here, in which the short flexor of the pedal digits is no more than an accessory muscle which does not produce any perforated tendons independently.

If, in this orang-utan, the tibialis tendons to digits III and IV and the lateral fleshy tendon to digit V are identified as modified flexor digitorum brevis components, then those parts that represent the true flexor digitorum tibialis—the tendon to the second digit and the medial tendon to the fifth—conform to the model presented by Straus (1930) for the great apes: i.e., the flexor digitorum tibialis tends to become restricted to the second and fifth toes. This also explains why both of these tendons (to digits II and V) lack the fleshy bodies of the other tibialis tendons—true tendons of flexor digitorum tibialis are, by all accounts, nonmuscular. Only the development of the middle phalangeal insertions by these true tibialis tendons is remarkable, and it is readily apparent that these "triangular vincula" arise from the tibial flexor's complete absorption of the flexor digitorum brevis tendons to those digits and the retention of the short flexor's distinctive insertions. Therefore, Straus's observation that the flexor digitorum brevis tendon to the fifth digit is often lost in the hominoids is not entirely accurate; rather, this tendon is absorbed by the flexor digitorum tibialis.

The only feature of the pedal flexor anatomy of my orang-utan that remains to be explained is the origin of a deep portion of the flexor digitorum brevis from the tendon trunk of the flexor digitorum fibularis. Since the deep head of the flexor digitorum brevis in primates consistently serves the lateral digits and has been shown in this orang-utan to be incorporated into the flexor digitorum tibialis, the "deep head" observed in this specimen obviously cannot be homologous with the deep head of the flexor digitorum brevis commonly seen in primates. Since the deep slip here reaches only the second digit and receives tendinous contributions from two slips that constitute nearly half the bulk of the superficial head of the flexor digitorum brevis, one must assume that this deep portion is actually part of the superficial head that has "sunk" into a deeper, but adjacent, plane of the foot to meet the tendon of the flexor digitorum fibularis.

THE SHORT FLEXORS OF THE LEAST TOE

The *M. Flexor Digiti Minimi Brevis*

The flexor digiti minimi brevis (Fig. 24–3B) was located deep and medial to the abductor digiti minimi and lateral to the digit V tendon of the flexor digitorum tibialis. The origin of the flexor digiti minimi brevis was from the base and tuberosity of the fifth metatarsal; it also arose from the fascia extending from the plantar tuberosity of the cuboid and the sheath of the peroneus longus tendon. The muscle was multifasciculate and tightly bound with fascia. Fanning out from the origin, the deepest of its fibers, constituting the opponens digiti minimi, inserted on nearly the entire length of the plantolateral border of the fifth metatarsal. The more superficial stratum of the flexor digiti minimi brevis inserted via a very short tendon onto the lateral aspect of the base of the proximal phalanx of the fifth digit. Its insertion was deep and proximal to the insertion of the abductor digiti minimi.

Arising from the entire lateral border of the fifth metatarsal was another muscle bundle. Lateral to the insertion of the opponents digiti minimi, it inserted with the flexor digiti minimi brevis. This small bundle was distinct enough that it may be called the "abductor digiti minimi brevis." It was separated from the short flexor of the digit by the tendon of the abductor digiti minimi.

In hominoids, the component of the flexor digiti minimi brevis that arises from the sheath of the peroneus longus and the fifth metatarsal base—as well as occasionally from the base of the fourth metatarsal, as in Hepburn's (1892) orang-utan and the gorillas of Straus (1930) and Raven (1950)—is always present and always has the same insertion. On occasion it may blend with the third interosseus (Hepburn, 1892; Straus, 1930). However, an opponens of the least digit is frequently absent in primates, and, when it is present, there is little pattern to its occurrence. An opponens was well developed in the gorilla and the chimpanzee studied by Hepburn (1892), Raven's (1950) gorilla, and in Sonntag's (1924) chimpanzee. All of these African apes resembled the orang-utan discussed here in the continuity of the opponens with the flexor digiti minimi brevis at the origin and medial border of the former muscle. An opponens was lacking in Straus's (1930) gorilla, but he did cite cases of its appearance in several prosimians (*Lemur, Daubentonia,* and *Galago*) and monkeys, as well as in the gibbon, chimpanzee, and gorilla. Murie and Mivart (1872) found an opponens only in *Lemur.* Neither Woollard (1925) nor Burmeister (1846) recorded this muscle for *Tarsius.* In humans the opponens digiti minimi is often present, but it usually consists of only a few lateral fibers of the flexor digiti minimi brevis (Gray, 1942); this muscle is not so well developed as in the present orang-utan. Due to the irregularity of its occurrence, it is difficult to attach any evolutionary significance to the development of an opponens digiti minimi. The muscle appears to be a variable proximal extension of the insertion of the flexor digiti minimi brevis onto the shaft of the fifth metatarsal.

Straus (1930) claimed that the "additional superficial layer" of the flexor digiti minimi brevis he found in his gorilla had not been recorded by other authors. However, it would seem that such a layer is actually homologous with the short flexor of the least digit, as that muscle is usually described, and that the discussed origin from the base of the fourth metatarsal was merely a simple, and fairly common, variation. What is remarkable is the origin of the flexor digiti minimi brevis from the fibular aspect of the fifth metatarsal shaft, which appears to be a peculiarity observed only in Straus's gorilla and my orang-utan. This latter muscle, the "abductor digiti minimi brevis," was more developed in Straus's specimen than in my orang-utan, and arose from the plantar aspect of the base as well as the shaft of the fifth metatarsal, thus sharing its origin with the flexor digiti minimi brevis. The relatively larger size of the short abductor in Straus's specimen may be due to the absence of an opponens digiti minimi, the presence of which restricted the area available for origin from the fifth metatarsal in my orang-utan.

In this orang-utan the abductor digiti minimi brevis was likely an extension of the closely associated flexor digiti minimi brevis. As in the primate hand (Lewis, 1965), the pedal flexor digiti minimi brevis is a vestige of the outermost muscle in the series of primitive flexores breves of each digit which now constitute the plantar and much of the dorsal interossei (McMurrich, 1907). In development, the flexor digiti minimi brevis in hominoids normally assumes a plantar and medial position in relation to the metatarsal, but its precursor had a more lateral orientation, similar to that retained by the interossei, and such a muscle mass would resemble the abductor digiti minimi brevis described here.

CONCLUSIONS

Of the variations present in the appendicular anatomy of this 9-year-old male *Pongo pygmaeus,* those of the brachialis, triceps brachii, and peroneus digiti quinti are the most easily interpreted: they are atavisms, appearances of morphologies characteristic of related but comparatively primitive primate groups. All of these configurations closely resemble those found in prosimians, and, to a lesser degree, monkeys. In the case of the brachialis and the triceps, they are of primary interest because, together, they constitute a primitive arrangement of brachial muscles that seems not to have been recorded previously for the orang-utan. With regard to the hindlimb, the peroneus digiti quinti of *Pongo* has been described, but here it is incomplete and in two parts; thus the body, having lost its connection with its tendon (which is now fused to the tendon of the peroneus brevis) and acquired a new fascial connection to the calcaneus, would have acted as a very weak everter of the foot. The "abductor digiti minimi brevis" in this orang-utan was apparently a vestige of a developmental precursor of the flexor digiti minimi.

Evaluation of the remaining variations, all of which were confined to the hindlimb, is much more difficult. The "subgluteus" defies explanation, since a similar muscle apparently has not been found in other primates. As I suggested above, the "subgluteus" is likely to have been derived from the gluteus superficialis, with which it shared innervation. However, the increased size of the "subgluteus" in the left hip, coupled with the decrease in the size of the neighboring piriformis, suggests a relationship with the latter muscle.

The significance of the variations in the pedal flexor anatomy of the orang-utan is entirely different, suggesting adaptive processes rather than evolutionarily "backward" traits, and the trend observed by Hepburn (1892) and Straus (1930, 1949) in hominoids—in which the tendons of the flexor digitorum brevis deep head are acquired by the superficial head of the same muscle—should be reevaluated in light of these variations. To this trend it should be added that, in the great apes, these deep head tendons might alternatively be integrated into the flexor digitorum tibialis. In cases of such a transferral, in digits to which the flexor digitorum tibialis normally distributes deep tendons (to digits II and V), there will appear to be a double tendon, providing insertions on both the middle and distal phalanges. In other instances, the adopted flexor digitorum brevis tendon may be absorbed completely by the flexor tibialis tendon, leaving as a remnant only the middle phalangeal insertion—a "triangular vinculum"—such as was found in the second digit of the orang-utan discussed here and the lateralmost digit of Sonntag's (1924) chimpanzee and Straus's (1930) gorilla. The fifth digit of my orang-utan is "intermediate" between these two possible states, with the double tendon fusing just proximal to the insertion, via a triangular vinculum, on the middle phalanx.

In some cases the two processes of tendon transferral are exhibited simultaneously, as in the third toe of the present orang-utan and in the fourth toes of a number of specimens cited above. The perforated tendon here arises from both the flexor digitorum tibialis and the superficial head of the flexor digitorum brevis. Such instances demonstrate that there is no real "transferral" of the flexor digitorum brevis deep head tendon—it remains as the supernumerary tendon of the tibial flexor—and that the tendon from the superficial head of the short flexor is actually a new structure altogether. This is essentially what one finds in various prosimians and monkeys, in which the perforated tendon finds origin in both heads of the flexor digitorum brevis (which Straus [1930] calls "partial transfer"), except that the deep tendon in the above cases has been overtaken by the tibialis flexor.

The involvement of the flexor digitorum tibialis in the formation of the flexor digitorum brevis tendons has been noted in all the apes but appears to be most frequent and generally pronounced in the orang-utan. Aside from the orang-utan discussed here, the specimens that retained the fewest distinct short flexor tendons (in the literature surveyed) were Sonntag's *Pongo* (1924) and *Pan* (1923): each possessed only two fully independent tendons (to digits II and III). My orang-utan alone had lost all free short flexor tendons; every one of the four was to some degree adopted by the flexor digitorum tibialis.

In general it would seem that, in the great apes, the role of the flexor digitorum brevis as the superficial flexor of the toes is being diminished. This certainly appears to be the case at least with regard to the fifth digit, in which the perforated tendon is frequently "missing." The only instances encountered in the literature of the flexor digitorum fibularis (as opposed to the tibialis) adopting a flexor brevis tendon were in the fifth digit of Chapman's (1878) and Straus's (1930) gorillas. In the latter specimen, the presence of a fibular flexor tendon in the least digit was a primitive characteristic.

It is tempting to suggest that this intriguing alteration of pedal flexor anatomy, as well as Straus's model of long flexor distribution in the great apes, could serve as evidence for the monophyly of the "Pongidae"—the African apes and the orang-utan. But there is such a great degree of individual variation in each of the great apes' pedal flexor anatomy, with some specimens of African apes even being almost "prosimian" in arrangement, that such an argument would not be very substantial. It should not also be forgotten that the gibbon may exhibit a degree of short flexor tendon integration, though not complete transferral, even while retaining a primitive layout of long flexor tendons (Hepburn, 1892). There are also scattered occurrences in various monkeys. It is altogether more likely that the trends in pedal flexor modification are occurring in parallel between the African apes and the orang-utan through some common behavioral factor. Since the great apes use their feet as grasping organs to a greater extent than does the gibbon and far more than humans (Andrews and Groves, 1976), the modifications may have adaptive significance in locomotion. The separation of the long flexors would allow greater freedom of movement in individual digits, while the adoption of deep flexor digitorum brevis tendons by the flexor digitorum tibialis, rather than by the superficial head of the short flexor, would certainly endow these tendons with a much more powerful grip, especially given that the extrinsic flexor digitorum tibialis is a far more massive muscle than is the diminutive intrinsic flexor digitorum brevis. Since the arboreal orang-utan uses its feet in moving about (to the degree that it is often called "quadrumanous") more extensively and efficiently than the more terrestrial gorilla and chimpanzee, this would account for the predominance, both in frequency and degree of development, of the flexor modifications in the orang-utan. Indeed, since none of the orang-utans described in the literature exhibited notably primitive arrangements of the long pedal flexors—as did certain specimens of *Pan* (Hepburn, 1892) and *Gorilla* (Straus, 1930)—and all appeared to display some integration between the short flexor and the tibial flexor, these modifications of flexor anatomy might be included in the suite of characteristics (such as the femoral origin of the flexor digitorum fibularis, the reduction of the hallux, and the loss of the ligamentum teres and a hallucal flexor tendon) that are recognized locomotory adaptations of the pelvic limb of *Pongo.*

ACKNOWLEDGMENTS

This research was made possible by the Yerkes Regional Primate Center, Emory University (NIH RR-00165).

REFERENCES

Andrews, P. and Groves, C. P. 1976. Gibbons and brachiation. *Gibbon and Siamang, 4:*167–218.

Ashton, E. H. and Oxnard, C. E. 1963. The musculature of the primate shoulder. *Transactions of the Zoological Society of London, 29:*553–650.

Aziz, M. A. 1981. Possible "atavistic" structures in human aneuploids. *American Journal of Physical Anthropology, 54:*347–353.

Beattie, J. 1927. The anatomy of the common marmoset (*Hapale jacchus* Khul). *Proceedings of the Zoological Society of London, 1927:*593–718.

Beddard, F. E. 1893. Contributions to the anatomy of the anthropoid apes. *Transactions of the Zoological Society of London, 13:*177–218.

Boyer, E. L. 1935. The musculature of the inferior extremity of the orang-utan *Simia satyrus. American Journal of Anatomy, 56:*193–255.

Burmeister, H. 1846. Beiträge zur näheren Kenntnis der Gattung *Tarsius.* Reimers, Berlin.

Campbell, B. 1937. Shoulder musculature of the platyrrhine monkeys. *Journal of Mammalogy, 18:*66–71.

Champneys, 1872. On the muscles and nerves of a chimpanzee *(Troglodytes niger)* and a *Cynocephalus anubis. Journal of Anatomy and Physiology, London, 6:*176–211.

Chapman, H. C. 1878. On the structure of the gorilla. *Proceedings of the Academy of Natural Sciences, Philadelphia, 30:*385–394.

Chapman, H. C. 1880. On the structure of the orang-outang. *Proceedings of the Academy of Natural Sciences, Philadelphia, 32:*160–175.

Chibber, S. R. and Singh, I. 1970. Asymmetry and one-sided dominance in the human lower limbs. *Journal of Anatomy, 106:*553–556.

Davis, D. D. 1938. Notes on the anatomy of the tree shrew *Dendrogale. Field Museum Publications, Chicago, Zoology, 20:*383–405.

Dobson, G. E. 1883. On the homologies of the long flexor muscles of the feet of Mammalia, with remarks on the value of the leading modifications in classification. *Journal of Anatomy and Physiology, London, 17:*142–179.

Dunlap, S. S., Thorington, R. W. and Aziz, M. A. 1985. Forelimb anatomy of the New World monkeys: myology and the interpretations of primate anthropoid models. *American Journal of Physical Anthropology, 68:*499–517.

George, R. M. 1977. The limb musculature of the Tupaiidae. *Primates, 18:*1–34.

Glaesmer, E. 1908. Untersuchung über die Flexorengruppe am Unterschenkel und Fuss der Säugertiere. *Morphologisches Jahrbuch, 38:*36–90.

Gray, H. 1942. Anatomy of the Human Body, 24th edition. Lea and Febiger, Philadelphia.

Hepburn, D. 1892. The comparative anatomy of the muscles and nerves of the superior and inferior extremities of the anthropoid apes. *Journal of Anatomy and Physiology, London, 26:*149–186, 324–356.

Howell, A. B. and Straus, W. L., Jr. 1932. The brachial flexor muscles in primates. *Proceedings of the U.S. National Museum, 80:*1–31.

Howell, A. B. and Straus, W. L., Jr. 1933. The muscular system. *In* The Anatomy of the Rhesus Monkey, pp. 89–176, ed. C. G. Hartman and W. L. Straus, Jr. Williams and Wilkins, Baltimore.

Huxley, T. 1864. The structure and classification of the mammalia. *Medical Times and Gazette, 1* and *2.*

Keith, A. 1894. Notes on a theory to account for the various arrangements of the flexor digitorum profundus in the hand and foot of Primates. *Journal of Anatomy and Physiology, London, 28:*335–339.

Keith, A. 1899. On the chimpanzees and their relationship to the gorilla. *Transactions of the Zoological Society of London, 1899:*296–312.

Le Gros Clark, W. E. 1924. The myology of the tree shrew *(Tupaia minor). Proceedings of the Zoological Society of London, 1924:*461–497.

Le Gros Clark, W. E. 1926. On the anatomy of the pen-tailed tree shrew *(Ptilocercus lowii). Proceedings of the Zoological Society of London, 1926:*1179–1309.

Lewis, O. J. 1965. The evolution of the mm. interossei in the primate hand. *Anatomical Record, 153:*275–288.

McArdle, J. E. 1981. Functional morphology of the hip and thigh of the Lorisiformes. *Contributions to Primatology, vol. 17.* S. Karger, Basel.

McMurrich, J. P. 1907. The phylogeny of the plantar musculature. *American Journal of Anatomy, 6:*407–437.

Murie, J. and Mivart, St. G. 1872. On the anatomy of the Lemuroidea. *Transactions of the Zoological Society of London, 7:*1–114.

Primrose, A. 1899. The anatomy of the orang-outang. *Transactions of the Canadian Institute, 6:*507–597.

Raven, H. C. 1950. Regional anatomy of the gorilla. *In* The Anatomy of the Gorilla, pp. 15–188, ed. W. K. Gregory. Columbia University Press, New York.

Schön, M. A. 1968. The muscular system of the red howling monkey. *Bulletin of the United States National Museum, 273:*1–185.

Sirena, S. 1871. Richerche sulla miologia del *Mycetes fuscus. Giornale de Scienze Naturali ed Economiche, Palermo, 7:*164–244.

Sonntag, C. F. 1923. On the anatomy, physiology, and pathology of the chimpanzee. *Proceedings of the Zoological Society of London, 1923:*323–429.

Sonntag, C. F. 1924. On the anatomy, physiology, and pathology of the orang-outang. *Proceedings of the Zoological Society of London, 1924:*349–449.

Straus, W. L., Jr. 1930. The foot musculature of the highland gorilla *(Gorilla beringei). Quarterly Review of Biology, 5:*261–317.

Straus, W. L., Jr. 1949. The riddle of man's ancestry. *Quarterly Review of Biology, 24:*200–223.

Swindler, D. R. 1973. An Atlas of Primate Gross Anatomy: Baboon, Chimpanzee, and Man. University of Washington Press, Seattle.

Turner, W. 1867. On variability in human structures, with illustrations, from the flexor muscles of the fingers and the toes. *Transactions of the Royal Society, Edinburgh, 24:*175–189.

Woodburne, R. T. 1978. Essentials of Human Anatomy, Sixth edition. Oxford University Press, London.

Woollard, H. H. 1925. The anatomy of *Tarsius spectrum. Proceedings of the Zoological Society of London, 1925:*1071–1184.

25

Relative Growth of the Locomotor Skeleton in Orang-utans and Other Large-Bodied Hominoids

WILLIAM L. JUNGERS AND STEVE E. HARTMAN

> The higher primates differ very widely in many of their body proportions, of which some change profoundly during growth.
>
> SCHULTZ (1973:338)

Simple visual inspection allows one to discern a variety of apparently significant differences in adult body shape among the five species of large-bodied hominoids (orang-utans, humans, and the African apes). Traditionally, many of these differences in adult postcranial proportions have been characterized through use of simple ratios such as the intermembral, brachial, and crural indices (e.g., Mollison, 1911; Napier and Napier, 1967; Schultz, 1930, 1937). More recently, "allometric" analyses have attempted to dissect these ratios by examining the effect of scale differences on both the numerator and denominator (Aiello, 1981; Biegert and Maurer, 1972; Jungers, 1984, 1985). For example, when compared to empirical allometric gradients in either African apes alone or quadrupedal primates as a group (using body mass as the independent variable), it is obvious that orang-utans are characterized by unusually long forelimbs and relatively short hindlimbs; by contrast, humans possess exceptionally long hindlimbs and relatively short forelimbs (Jungers, 1985). Hence, the orang-utans exhibit the highest intermembral index among large-bodied hominoids and humans have the lowest. As such, these findings are nicely congruent with our biomechanical expectations about how relative limb lengths should vary with size and locomotor differences, and they permit informed speculations about motor function in extinct species such as the australopithecines (Jungers, 1982; Jungers and Stern, 1983; Susman et al., 1984). Such indices, therefore, contain both phylogenetic and functional information in some poorly defined combination.

The approaches to the study of body proportions mentioned above are "static" in design, in that comparisons are limited to fully adult individuals of each species. The study of growth-related metric changes in the locomotor skeleton of primates has also proven to be a useful tool in functional morphology and evolutionary biology. For example, analysis of ontogenetic pathways culminating in adult differences can yield important information on different locomotor adaptations in closely related species (Jungers and Fleagle, 1980). Additionally, a careful analysis of relative growth in different parts of the postcranium can lead to evolutionary hypotheses about the role of extrapolation or truncation of ontogenetic patterns as a source of morphological differences among closely related species (Jungers and Susman, 1984; Lumer, 1939; Shea, 1981, 1983, 1984; also cf. Gould, 1977). Given the clear-cut differences in adult proportions among large-bodied hominoids, it is also possible to investigate whether postnatal growth preserves or alters these relationships. For example, Schultz (1941) compared a variety of anthropometric indices in different age categories in the orang-utan to see if growth modified or maintained the shape of the neonate. Among other findings, he noted that the intermembral index was quite stable throughout postnatal growth in orang-utans, but that the brachial index changed such that adults had consistently longer forearms relative to arm length.

Inspired by Huxley's (1932) work, Lumer (1939) was the first author to successfully apply the bivariate power formula for simple allometry ($y = bx^k$) to the study of relative growth in limb proportions of both lesser and great apes (using a cross-sectional data base supplied by Schultz). Although never acknowledged by Schultz himself, Lumer provided a quantitative foundation for and empirical confirmation of many of Schultz's later observations on relative growth in orang-utans (Schultz, 1941), chimpanzees (Schultz, 1940), and other hominoids (e.g., Schultz, 1956, 1973). Lumer's efforts have recently been supplemented by the addition of comparable data on limb growth in humans (Buschang, 1982a,b) and expanded by a series of papers on relative growth of bodily proportions in African apes (Jungers and Susman, 1984; Shea, 1981, 1983, 1984).

One goal of the present study resembles those that have preceded it; namely, to expand the comparative data base documenting ontogenetic pathways taken by different elements of the locomotor skeleton that result in marked dissimilarities among adults of different species. Using linear measurements collected on long bones and both pectoral and pelvic girdles, the following questions are addressed:

i. Are hominoid shape relationships measured at or near birth preserved throughout postnatal development into adulthood? In other words, how closely does postnatal growth correspond to a null hypothesis of overall isometric growth?
ii. If nonisometric postnatal growth obtains, how do patterns of allometry compare among the large-bodied hominoids? Do any of the species become more similar in proportions with growth, or does growth exaggerate differences already evident at birth?
iii. What is the functional significance of the size-related patterns we find?

Finally,

iv. What is the nature and strength of the phylogenetic signal inherent in patterns of intraspecific growth allometry?

Following the example of Shea (1985), we will begin our analysis with a consideration of within-group (intraspecific) multivariate allometry and proceed to an investigation of multigroup (interspecific) patterns of relative growth. We wish to emphasize that our data base is strictly ontogenetic in composition because intraspecific scaling of adults (another form of "static allometry") rarely resembles true relative growth based on an analysis of individuals of all ages (Cheverud, 1982; Cock, 1966; Jungers and Susman, 1984; Shea, 1981).

MATERIALS AND MEASUREMENTS

The sample sizes of nonadult skeletons used in this study range from 29 (*Pan paniscus,* the pgymy chimpanzee or bonobo) to 81 (*Pongo pygmaeus,* the orang-utan), depending on the availability of such material in U.S. and European museum collections. All nonhuman species are individuals collected in the wild. Animals raised in captivity or which died in zoos were omitted in view of the possible impact of environmental conditions on linear dimensions (e.g., Fooden and Izor, 1983, for orang-utans). Because preliminary analyses uncovered no consistent differences in growth trajectories of males and females of each species, the sexes were pooled and unsexed specimens were added to the samples. Only complete specimens are included in the multivariate analyses, but partial skeletons greatly increased the samples for some of the bivariate analyses. As noted above, 81 nonadult orang-utan skeletons were measured; this sample includes members of both subspecies of *Pongo pygmaeus.* Of the 81 specimens, only 60 were complete enough for all measurements to be taken. The chimpanzee sample consists of 51 specimens of *Pan troglodytes troglodytes* (48 complete individuals, of which the majority were from the Cameroons). The gorilla specimens are all *Gorilla gorilla gorilla,* again primarily from the Cameroons; 40 of the 45 individuals included in this study were complete. Of the 29 *Pan paniscus* skeletons measured, 27 were complete. Thirty complete human skeletons of African descent (drawn from a large cadaver population) comprised the *Homo sapiens* ontogenetic series. Ages were unavailable for the nonhuman samples, but each was represented by a series from infants to young adults, as judged from the degree of epiphyseal fusion with diaphysis. The humans ranged in age from 1 month to 19 years. All groups were treated, therefore, as mixed cross-sectional growth samples (Cock, 1966; Johnston, 1980). We would prefer longitudinal growth data on individuals, but cross-sectional data appear to be the best available and the only type possible for noncaptive animals.

Nine linear measurements were taken in order to assess growth-related changes in overall postcranial proportions (Table 25–1; see Fig. 1 in Jungers and Susman, 1984). Four variables measure length of the main limb elements (arm, forearm, thigh, and leg). Two variables (clavicular length and scapular breadth) assess the overall size of the pectoral girdle, and three the sizes of the pelvic girdle elements (ilium, ischium, and pubis). What we call scapular breadth is sometimes referred to as the morphological "length" of the scapula (Shea, 1986), but we prefer the nomenclature of human osteology (e.g., Bass, 1971), especially when the taxa being considered are limited to large-bodied hominoids. These

TABLE 25–1. Variable List and Abbreviations

Length of humeral diaphysis	HDL
Length of radial diaphysis	RDL
Length of femoral diaphysis	FDL
Length of tibial diaphysis	TDL
Scapular "breadth" along spinal axis	SBSA
Length of clavicle	CL
Length of ilium without crest[1]	IL
Length of pubis[1]	PL
Length of ischial shank[1,2]	ISL

See Jungers and Susman (1984) for an illustration of these measurements.

[1]Measured from approximate center of acetabulum.

[2]Ischial length without tuberosity.

measurements were selected to permit the use of bones lacking epiphyses, thereby extending the sampled size range down to infants.

SPECIES-SPECIFIC MULTIVARIATE GROWTH ALLOMETRY

The possibility that all nine postcranial dimensions grow in such a way that all proportions among them remain constant—the null hypothesis of isometric relative growth—was investigated using the multivariate generalization of the allometry equation formalized by Jolicoeur (1963a,b). This method examines multivariate proportionality via a principal components analysis of the covariance matrix of logarithmically transformed variables. If the first principal axis accounts for the overwhelming majority of the variance (Sprent, 1972) and if all loadings (direction cosines) are of the same sign, this axis can be interpeted as a "vector of relative growth" (Shea, 1985). This procedure "implicitly chooses the geometric mean of all measurements as a size variable" (Mosimann and James, 1979:455). For p variables, multivariate isometry exists when all p dimensions increase at the same rate as the internally defined size variable, such that all p loadings are equal to $p^{-1/2}$. For nine variables $p^{-1/2} = 0.333$, and loadings can be standardized by this value so that adjusted coefficients greater than 1.0 indicate positive allometry and those less than 1.0 denote negative allometry. The allometric relationship between any two variables can also be assessed by the ratio of their coefficients (Davies and Brown, 1972; Thorington, 1972). Furthermore, it is possible to calculate an angle θ between the hypothetical (isometric) "growth vector" and the observed vector of relative growth; the larger the angle, the greater the observed vector diverges from that indicative of overall isometric growth (Jolicoeur, 1963a:18). Jolicoeur (1984) has recently proposed a direction test to evaluate the statistical significance of such departures from isometry; this test is employed here in lieu of that proposed earlier by Anderson (1963).

Table 25–2 summarizes the results of this type of analysis for the five species of living, large-bodied hominoids. Note that the first principal axis accounts for at least 98% of the variance in all five species. Angles of divergence between the observed relative growth vectors and isometry range from 2.8° in gorilla to 6.0° in humans; in a sense, then, humans depart more from the null hypothesis of multivariate isometry than do any of the other four species. However, the departure is highly significant in *all* five species.

The patterns of growth allometry clearly distinguish humans from the great apes. Relative to the other skeletal variables, diaphyseal lengths of all four long bones exhibit positive allometry; conversely, allometries of all measurements on the bony girdles of humans are negative. In other words, limb bones increase in length throughout ontogeny at a faster pace than do either pectoral or pelvic dimensions. This relative increase is especially pronounced in the hindlimb (also cf. Buschang, 1982a,b), a finding also illustrated graphically by Schultz (1925:Fig. 10). We concur with Buschang's (1982b:295) conclusion that "considering the human specialization for bipedality, greater allometric growth of the lower extremities might be expected." For similar reasons, it is not surprising to find that lesser apes are characterized by quite the opposite growth trend; these long-armed brachiators exhibit negative growth allometry of the hindlimbs relative to forelimb length (Lumer, 1939). Our findings are also compatible with Schultz's (1973:351) observation that postnatal changes in intermembral index are "most marked in man" in contrast to the great apes. Brachial and crural indices change much less with growth, but there is a slight tendency for both to increase postnatally in this human data set. Other authors have reported a minor decrease in brachial index during human postnatal growth (e.g., Buschang, 1982b; Schultz, 1973). Within the human pelvic girdle, postnatal shape changes affect relative ilium length most markedly (strong negative allometry). Clavicle length exhibits the strongest negative allometry relative to the internally defined size variable.

Both elements of the forelimb exhibit some

TABLE 25–2. Multivariate Growth Allometry of the Skeleton in Hominoids

Variable	Orang-utan (N = 60)	Gorilla (N = 40)	Chimpanzee (N = 48)	Bonobo (N = 27)	Human (N = 30)
HDL	0.979	0.992	0.969	0.940	1.036
RDL	1.025	0.971	0.984	0.931	1.065
FDL	0.959	0.964	1.013	1.013	1.151
TDL	0.985	0.958	0.984	0.986	1.174
SBSA	0.962	1.005	0.977	1.050	0.903
CL	0.979	0.944	0.896	0.816	0.846
IL	1.039	1.032	1.057	1.023	0.886
PL	1.105	1.100	1.130	1.150	0.949
ISL	0.985	1.026	0.969	1.077	0.944
% Variance	98.4	98.6	98.5	98.0	98.8
Angle of divergence	3.0°	2.8°	4.1°	3.1°	6.0°
Test of isometric null hypothesis*	$F = 8.965$ $p < 0.001$	$F = 5.677$ $p < 0.001$	$F = 8.344$ $p < 0.001$	$F = 3.805$ $p < 0.01$	$F = 11.816$ $p < 0.001$
	Reject	Reject	Reject	Reject	Reject

Measurements defined in Table 25–1.

*Test statistic defined in Jolicoeur (1984).

degree of negative allometry in all three species of African apes. Orang-utans deviate from this trend with respect to relative growth of radius length (slight positive allometry). Tibial diaphyseal length is negatively allometric in all four great apes, but femoral allometry is variable (negative in orang-utan and gorilla but slightly positive in both species of *Pan*). The humerofemoral index tends to increase postnatally in the orang-utan and gorilla, but it decreases in common and pygmy chimpanzees. Postnatal changes in the brachial index clearly distinguish the orang-utan (increasing) from the gorilla (decreasing), whereas this ratio appears to remain fairly stable throughout growth in both species of *Pan.* Such findings correspond well with the bivariate results of Lumer (1939) with respect to the brachial index. Buschang's (1982b) assertion that humerus and radius grow isometrically in all great apes is not supported in this analysis, nor in that by Jungers and Susman (1984, which was limited to African apes). The crural index tends to decrease with age in western common chimpanzees *(troglodytes)* and bonobos (but not significantly so); it remains more or less constant in orang-utans and gorillas. It should be noted that significant negative allometry exists for TDL relative to FDL in the eastern subspecies *(schweinfurthii)* of the common chimpanzee (Jungers and Susman, 1984). It is premature, therefore, to suggest that isometry is the best description of the relationship between these two hindlimb elements for all hominoids (Buschang, 1982b). Only in the orang-utan does relative growth of the clavicle exceed that for scapular breadth; nonetheless, negative allometry obtains for clavicle length in all four nonhuman species. Proportional changes within the innominate are similar for all great apes in that relative pubic length increases most during growth, usually followed by relative growth in iliac length (except in bonobos). Regardless, compared to humans (wherein negative allometry was found in all three elements) the overall proportional changes in the pelvis are relatively similar for the four great apes.

The overall similarity or disparity among these intraspecific, multivariate coefficients will be reexamined shortly in the context of phylogenetic affinities among the five species. The suggestion that "selection may have favored [ontogenetic] shape preservation within species" provides a biological foundation for the null hypothesis of isometry throughout growth (Shea, 1981:191). However, the generality of this proposition is weakened considerably by significant departures from multivariate isometry in all five species of large-bodied hominoids, including the African apes. The proposed connection between raw (non-normalized) first component coefficients and *specific* growth within species also merits serious consideration (Shea, 1985; also cf. Takai, 1977). One expectation associated with this postulate is for the magnitude of axis I eigenvalues (variance explained by each factor) to be correlated with absolute species size. This is indeed the case within African apes (gorilla = 0.195; chimpanzee = 0.170; bonobo = 0.108), but orang-utans and humans greatly complicate such interpretations because both have first component eigenvalues greater than gorillas (orang-utan = 0.203; human = 0.272). The discrepancy may be due in this case to a better sampling of very small infants in the orang-utan and human skeletal series, but additional work on this issue is warranted in any case.

MULTIGROUP COMPONENTS OF RELATIVE GROWTH

Multivariate growth relationships among large-bodied hominoids were further examined through use of a multiple group principal components analysis of the *total* covariance matrix of log-transformed variables. The same nine postcranial variables were utilized, but bonobos were excluded due to their extensive overlap with the common chimpanzee. The total matrix (N = 178) was used rather than the pooled within-group covariance matrix (Boag, 1984; Shea, 1985). The latter matrix has been advocated by Thorpe (1983; Thorpe and Leamy, 1983) and applied by Wiig (1985), but seems best suited to intraspecific or "racial" variation rather than to an interspecific situation. In other words, there is no reason to expect homogeneity of covariance matrices among different species (Reyment, 1969), which renders pooling inappropriate in theory.

In Fig. 25–1 convex polygons bound the scores for individuals of each species. The first two components of this analysis (Table 25–3) account for over 95% of the variance. The first component captures the overwhelming majority of the variance and distributes individuals more or less according to overall size. We have reported the raw loadings in this part of the analysis rather than the transformed "allometric" coefficients such as those of Table 25–2 because it is not clear how closely the interspecific or multigroup case corresponds to the multivariate generalization by Jolicoeur for within-group allometry. Although signs of all the loadings are the same, they vary in magnitude. This suggests more than a simple size gradient without shape changes (Shea, 1985). Pubis length loads most strongly on this axis, whereas clavicle length has the lowest value.

The recent suggestion that the first component in multigroup analyses "summarizes shape variation resulting from the extension of a common trajectory of growth allometry" (Shea, 1985:383) is a potentially significant finding. This inference seems more secure when only closely related species are compared, as in the case of African apes. Inclusion of orang-utans and humans, however, greatly complicates the search for a common growth trajectory among these four species by adding quite divergent, species-specific patterns of relative growth. In such cases, axis I may represent statistical compromise among groups more than biological commonality. For example, the long bone coefficients suggest stability of both brachial and crural indices with growth, but

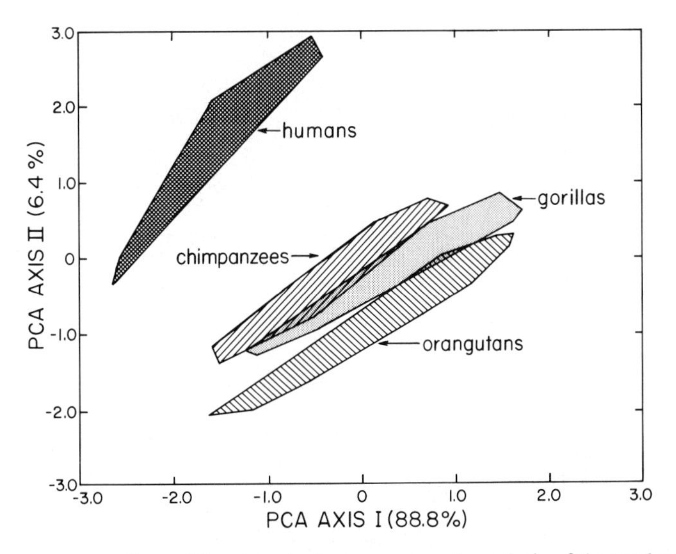

FIG. 25–1. Plot of the first two axes of a multigroup principal components analysis of the total covariance matrix of log-transformed variables. Species-specific convex polygons are shown to facilitate comparing patterns of overall dispersion, degree of overlap, and approximate orientation of relative growth in humans, common chimpanzees, gorillas, and orang-utans.

TABLE 25–3. Principal Components Based on Total Covariance Matrix of Log-Transformed Variables

Variable	Component I loadings	Component II loadings
HDL	0.155	−0.018
RDL	0.154	−0.050
FDL	0.150	0.073
TDL	0.150	0.070
SBSA	0.149	−0.015
CL	0.140	−0.010
IL	0.152	−0.043
PL	0.170	0.002
ISL	0.152	−0.008
Eigenvalue	0.210	0.015
% Variance	88.8	6.48

these ratios actually exhibit quite disparate growth-related changes in the differernt species (Table 25–2). Similarly, one might infer a common pattern of increasing humerofemoral indices from the multigroup coefficients, but this description is clearly misleading for both chimpanzees and humans.

The second principal axis of the multigroup analysis serves to separate humans from the three great apes (which show varying degrees of overlap). The different signs of the variable loadings for this axis are suggestive of a "contrast vector" (Blackith and Reyment, 1971) between FDL and TDL on the one hand, and IL and RDL on the other. The significance of this contrast will be considered below. As Shea (1985) has noted, because the growth trends summarized by the first two components are not strictly parallel to the first axis, the obliquity implies that the second axis also contains growth or size-related changes in shape. Presumably, differences in orientation and location of the convex polygons reflect "divergent growth trajectories" due to some combination of (1) differences in species-specific allometry (slopes in a bivariate case) and (2) in transpositional displacements of "onset signals" (Alberch et al., 1979; Gould, 1982). That the taxa occupy different locations in this specially defined space comes as no surprise given that within-group allometries were shown previously to be different in various ways among the species.

Selected bivariate plots (Figs. 25–2 to 25–6) of log-transformed variables provide additional insights into the presumed divergent growth trajectories. In each case, the best-fitting line is estimated by the major axis (Kuhry and Marcus, 1977; Sokal and Rohlf, 1981). Significant departures from isometry are indicated by a plus sign for positive allometry ($k > 1.0$) and a negative sign for negative allometry ($k < 1.0$) between variables. The major axis solution represents a

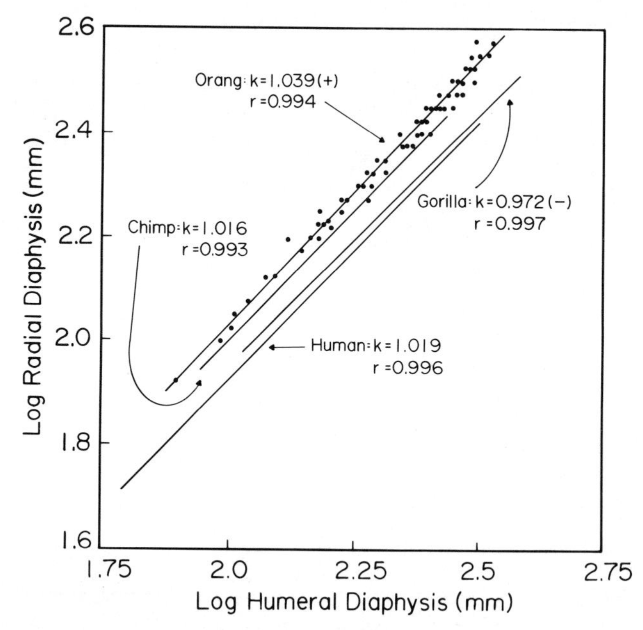

FIG. 25–2. Bivariate log-log plot of radial diaphyseal length vs. humeral diaphyseal length in orang-utans, common chimpanzees, gorillas, and humans. The principal or major axis for each species is indicated by a straight line; k is the slope of the line and r is the correlation coefficient. A plus sign (+) in parentheses indicates that the lower end of the 95% confidence limits for k was greater than 1.0; this represents significant positive allometry. Negative signs (−) denote an upper limit less than 1.0, and, therefore, negative allometry. Slope values without either sign have 95% confidence limits that include 1.0. Individual points (solid circles) are also provided for the orang-utan growth series.

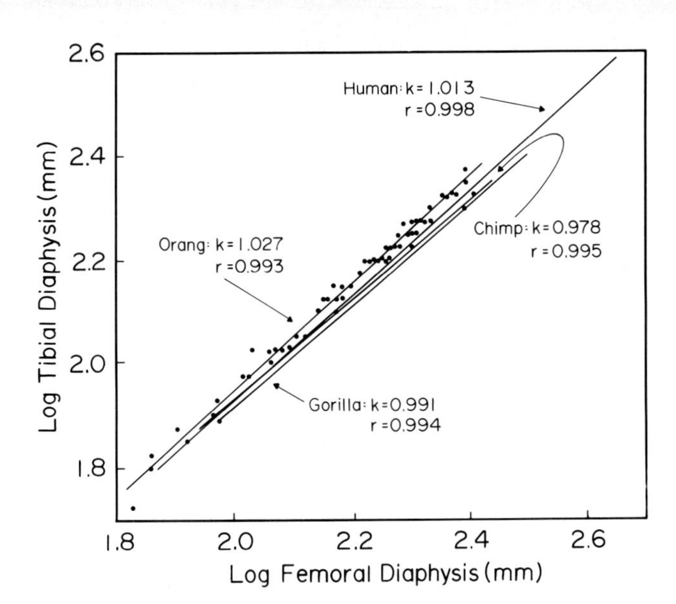

FIG. 25–3. Bivariate log-log plot of tibial diaphyseal length vs. femoral diaphyseal length. Conventions are as in Fig. 25–2.

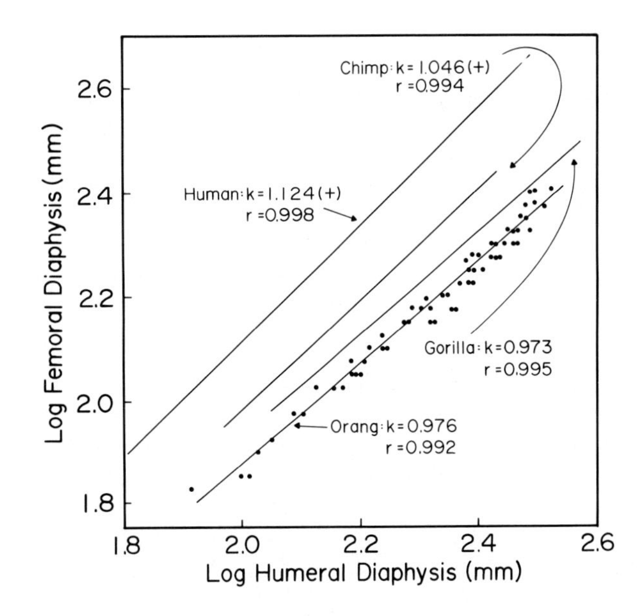

FIG. 25–4. Bivariate log-log plot of femoral diaphyseal length vs. humeral diaphyseal length. Conventions are as in Fig. 25–2.

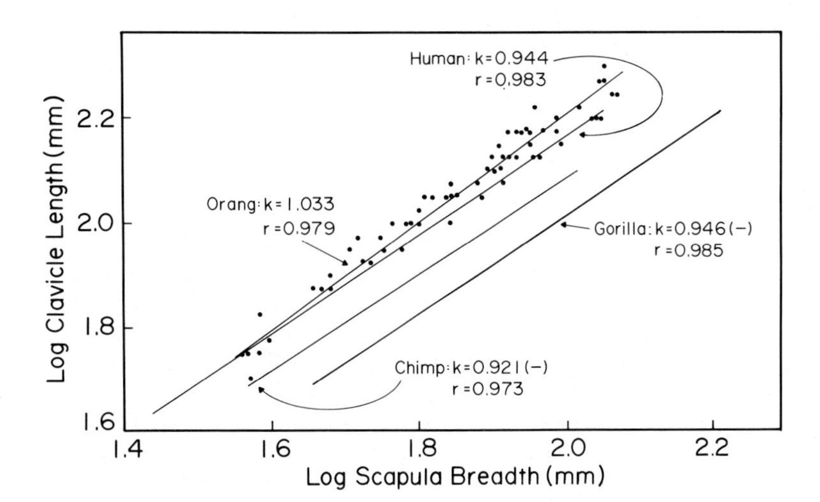

FIG. 25–5. Bivariate log-log plot of clavicle length vs. scapular "breadth" along the spinal axis. Conventions are as in Fig. 25–2.

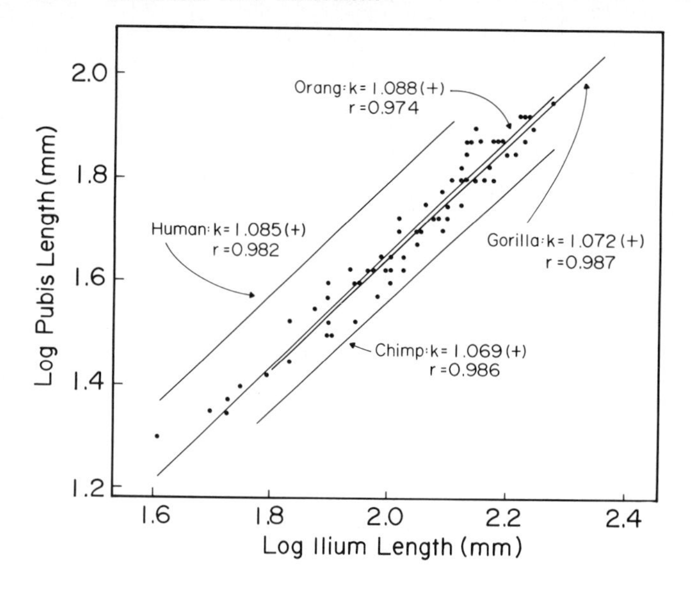

FIG. 25–6 Bivariate log-log plot of pubic length vs. iliac length. Conventions are as in Fig. 25–2.

bivariate case of the multivariate generalization of principal components (Thorington, 1972), but has the advantage of additional positional information (e.g., transpositions of growth trends can be compared). Clearly, the use of least-squares regression to estimate this type of relationship (e.g., Shea, 1986) is inappropriate. Figure 25–2 shows that in orang-utans RDL scales in a positive allometric fashion with HDL, whereas this relationship is significantly negative in gorillas. Moreover, at any humerus length, orang-utans possess the longest radii while humans have the shortest. During growth, the brachial indices of humans and gorillas converge somewhat as a consequence (Schultz, 1973). Figure 25–3 suggests that the crural index should be the highest in orang-utans and lowest in gorillas (this is indeed the case; Jungers, 1984), but overall differences should be relatively small and exhibit only minor growth-related changes. Figure 25–4 indicates major transpositional shifts combined with significant growth allometry in interlimb proportions. At any given humerus length (from neonate to adult), orang-utans possess the shortest femora, whereas humans have by far the longest. The human condition is exaggerated postnatally by significant positive allometry of FDL on HDL. Figure 25–5 provides additional evidence of both vertical shifts and significant departures from isometry. At any clavicle length, gorillas clearly possess the "broadest" scapula; this trend is exaggerated postnatally. Humans and orang-utans exhibit similar proportionalities in the pectoral girdle at birth, but growth tends to reduce this similarity. Finally, Fig. 25–6 depicts significant positive allometry of pubis length relative to iliac length in all four species; this is coupled with marked transpositions that serve to separate humans (short ilia) from chimpanzees (long ilia) at opposite extremes.

Returning to Fig. 25–1, we can conclude that both the overall positions and the differing degrees of obliquity seen in the convex polygons do indeed reflect divergent growth trajectories. The high frequency of marked transpositional differences in the bivariate slopes of Figs. 25–2 to 25–6 suggests that displacements of the onset signals (i.e., those proportional relationships present at birth) have contributed greatly to the observed differences. The contrast vector of principal axis II appears to be related closely to these vertical shifts. Human hindlimbs (FDL and TDL) are long relative to all other dimensions at birth and become relatively even longer during growth. The human ilium is relatively very short at birth and remains so throughout ontogeny. In other words, many of the proportional differences among species created during fetal development are either preserved or exaggerated postnatally. That the human differences are related directly to features associated with a specialization for bipedalism should not be surprising. Long hindlimbs have a direct effect on relative stride length (Jungers, 1982; Reynolds, 1987), and reduced iliac height is correlated with a center of gravity closer to the hip joint (Jungers and Stern, 1983). Not only do the large-bodied orang-utan and gorilla display adult interlimb proportions associated with mechanical competence in climbing (Jungers, 1985), but their postnatal interlimb proportions change in a way compatible with the same biomechanical expectations (e.g., increasing humerofemoral indices with increasing size). It should be noted that the findings of this ontogenetic analysis of postcranial proportions are compatible with the functional inferences drawn from interspecific analyses where "size" was taken as body weight or mass (Jungers, 1984, 1985; contra Aiello, 1984).

PHYLOGENETIC IMPLICATIONS OF POSTCRANIAL GROWTH ALLOMETRY

As with the simple indices discussed at the beginning of this chapter, it seems likely that our data contain both functional and phylogenetic information. Accordingly, we have used both phenetic and cladistic methods to evaluate the strength of the phylogenetic signal in the (within-group) coefficients of multivariate relative growth (i.e., nine allometric variables or characters for each of the five species [Table 25–2]).

Although there is still disagreement regarding the evolutionary relationships among chimpanzees, humans, and gorillas (Andrews, 1986; Ciochon and Corruccini, 1983; Creel, 1986; Goodman and Cronin, 1984; Groves, 1986; Hartman, 1986; Martin, 1986; Sibley and Ahlquist, 1984), their status as a monophyletic group (to the exclusion of the orang-utan) is now supported by most relevant evidence of which we are aware (cf. Schwartz, 1984a,b, and this volume for an opposing view). Likewise, there is no disputing the congeneric status of the two species of chimpanzee. In fact, as an external standard against which we can compare phylogenetic implications of our data, we believe only two branching arrangements warrant consideration: one unites humans and chimpanzees most closely, followed by gorillas, and finally orang-utans (based primarily on biomolecular data [e.g., Hasegawa et al., 1985; Nei et al., 1985; Sibley and Ahlquist, 1984]); the other unites all African apes before the addition of humans, and again, includes the orang-utan as sister to the rest (based primarily on morphological considerations [e.g., Martin, 1986]). Our purpose, then, in performing the phenetic and cladistic analyses described below is twofold: (1) to assess as objectively as possible whether our measures of multivariate, ontogenetic allometry offer phylogenetic implications congruent with either of the likely evolutionary histories outlined above; and (2) to explain, if possible, any lack of congruence.

Several programs within NT-SYS (Rohlf et al., 1986) were used to assess overall phenetic similarity in multivariate growth among the five species of large-bodied hominoids. Initially, a five-by-five symmetric matrix of average taxonomic distances (*d*—Sneath and Sokal, 1973) was computed from the nine allometry coefficients for each species from Table 25–2. A principal coordinates ordination of this matrix was then performed (Gower, 1966). The matrix correlation between the distances implied by the resulting graphic summary (Fig. 25–7) and the average taxonomic distances from which it derived was 0.994; this indicates that two dimensions are ad-

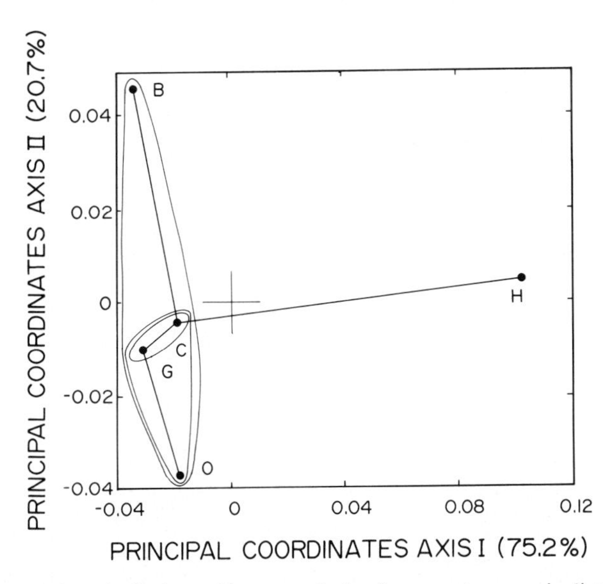

FIG. 25–7 Plot of first two axes of a principal coordinates analysis of average taxonomic distances based on intraspecific, multivariate allometric coefficients of relative growth. The origin is indicated by the cross-hairs. Species abbreviations: **C,** common chimpanzee; **G,** gorilla; **O,** orang-utan; **B,** bonobo or pygmy chimpanzee; **H,** humans. A minimum spanning tree based on the total distance matrix is represented by linear connections between nearest neighbors. Nested subsets of mutually closest taxa (clusters) based on overall similarity are also indicated. Percentages represent proportion of variance explained by each axis.

equate to summarize the phenetic structure embodied by the original interhominoid distances. A minimum spanning tree (uniting "nearest neighbors" in the original distance matrix [Sneath and Sokal, 1973]) and an indication of "mutually closest taxa" (a cluster all members of which are closer to one another than any are to any taxa outside the group) have been superimposed on the diagram. Both aim to reveal any distortion between the ordination summary and the original distances, and none is apparent.

The status of chimpanzees and gorillas as a mutually closest pair (Fig. 25–7) indicates they are more similar in their appendicular growth allometry than either is to any other taxon. They, plus orang-utans, are seen to be more similar than any is to bonobos or humans, and humans are most divergent. The minimum spanning tree suggests human growth is most similar to growth in chimpanzees, but the distance implied is relatively very great. The placement of orang-utans with African apes, to the exclusion of humans, and the relatively wide separation of bonobos and chimpanzees suggests that phenetic resemblances embodied by allometric growth coefficients have relatively little to do with phylogenetic affinity.

Table 25–4 presents correlations between the original allometric coefficients and the first two principal coordinates axes; high correlations are one indication of those variables primarily responsible for separation of taxa along a given axis. On axis I, positive correlations with limb bone allometric coefficients are contrasted with negative correlations with bony girdle coefficients. Again, it is evident that relative growth of the postcranial skeleton in humans is very dissimilar to patterns observed in the other four species. Divergent growth trends within the pectoral girdle are especially important in separating species along axis II.

TABLE 25–4. Correlations Between Raw Variables (Within-Group Coefficients of Multivariate Allometry) and the First Two Principal Coordinates Axes (Based on Average Taxonomic Distances)

Variable	Principal coordinates Axis I	Principal coordinates Axis II
HDL	0.857	−0.377
RDL	0.823	−0.552
FDL	0.928	0.340
TDL	0.990	0.119
SBSA	−0.754	0.624
CL	−0.347	−0.924
IL	−0.958	−0.181
PL	−0.971	0.129
ISL	−0.689	0.585
Eigenvalue	0.0132	0.0036
% Variance	75.2	20.7

We have also applied a numerical approach to cladistic analysis of relative growth within species; again, the characters used were the species-specific coefficients of multivariate allometry. The program WAGPROC (Version 4; Swofford, 1982) was employed to infer phylogenies from continuous characters via a Wagner parsimony algorithm (Farris, 1970).

For five taxa there are only 15 different nondirected (unrooted) trees (Sneath and Sokal, 1973). Lengths for these are plotted as a frequency distribution in Fig. 25–8A. The three shortest, or "most parsimonious," solutions are depicted in Fig. 25–8B. No rooting of the two

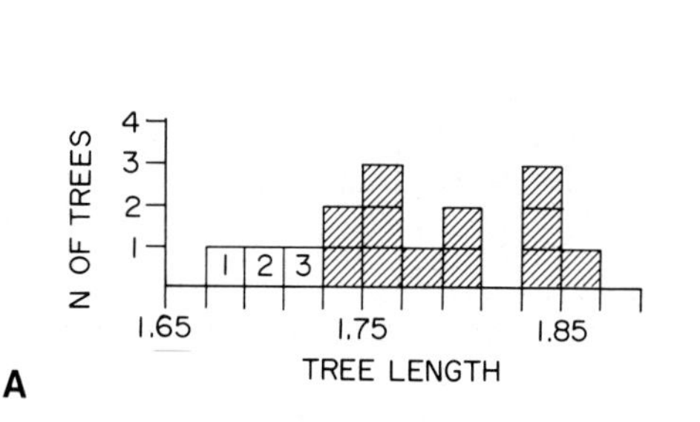

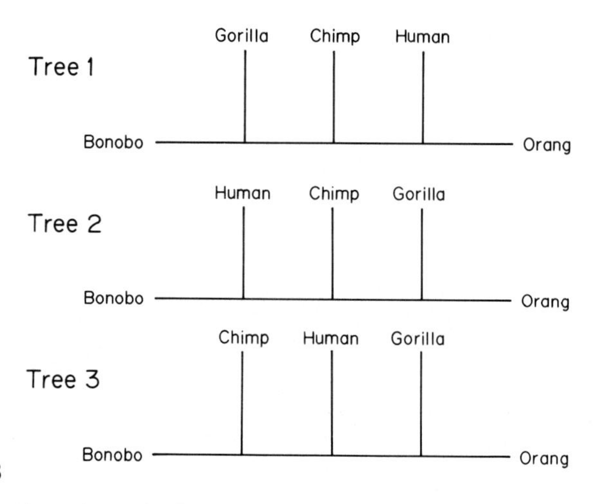

FIG. 25–8 (A) Frequency distribution of all 15 possible nondirected trees for five species of large-bodied hominoids. (B) The three most parsimonious trees (WAGPROC).

shortest nondirected trees supports union of the two species of chimpanzee. Rooting the third shortest tree at the orang-utan edge does give the topology supported by DNA–DNA hybridization data (Felsenstein, 1986; Sibley and Ahlquist, 1984), and may reflect evolutionary reality. Nonetheless, in the absence of other information, the parsimony criterion would have led us to accept an evolutionary topology (the shortest) that is almost certainly not correct.

Thus, neither phenetic nor parsimony analyses of the growth coefficients of this study yield results coincident with the overwhelming body of accumulated biomolecular and morphological evidence on hominoid evolutionary history. Instead, both inferential techniques have given results dominated by the impact of functional specialization in appendicular proportions. The multivariate treatments have, however, helped to expose and clarify differences among living large-bodied hominoids with respect to their species-specific growth allometries. Also, we have seen once again (see also Oxnard, 1981; Hartman, in prep.) that one should be very cautious when drawing phylogenetic inferences based on any limited set of morphological information.

ACKNOWLEDGMENTS

A special thanks is due to the many curators who provided hospitality, assistance, and access to the skeletal material included in this study: Dr. G. Musser at the American Museum of Natural History; Dr. R. Thorington at the National Museum of Natural History; Maria Rutzmoser at the Museum of Comparative Zoology; Dr. W. Kimbel and B. Latimer at the Cleveland Museum of Natural History; D. Howlett at the Powell-Cotton Museum; P. Jenkins at the British Museum (Natural History); Dr. F. Jouffroy at the Muséum National d'Histoire Naturelle; Dr. D. Meirte and Dr. T. van den Audernaerde at the Tervuren Museum; Prof. J. J. Picard at the Laboratoire d'Embryologie (Louvain-la-Neuve); Dr. C. Smeenk at the Rijksmuseum van Natuurlijke Historie; Dr. C. Edelstam at the Naturhistoriska Riksmuseet; and Dr. R. Kraft at the Zoologische Staatssammlung-Munchen. We also wish to thank L. Jungers for the illustrations and J. Kelly for typing the manuscript and tables. This work was supported by NSF grants BNS 82-17635 and BNS 86-06781. Finally, we wish to thank Dr. J. Schwartz for his invitation to contribute to this volume and for his patience with deadlines.

REFERENCES

Aiello, L. C. 1981. The allometry of primate body proportions. *Symposium of the Zoological Society of London, 48:*331–358.

Aiello, L. C. 1984. Applications of allometry: the postcrania of the higher primates. *In* The Lesser Apes: Evolutionary and Behavioral Biology, pp. 170–179, ed. H. Preuschoft, D. Chivers, W. Brockelman, and N. Creel. Edinburgh University Press, Edinburgh.

Alberch, P., Gould, S. J., Oster, G. and Wake, D. 1979. Size and shape in ontogeny and phylogeny. *Paleobiology, 5:*296–317.

Anderson, T. W. 1963. Asymptotic theory for principal component analysis. *Annals of Mathematical Statistics, 34:*122–148.

Andrews, P. 1986. Molecular evidence for catarrhine evolution. *In* Major Topics in Primate and Human Evolution, pp. 107–129, ed. B. Wood, L. Martin, and P. Andrews. Cambridge University Press, Cambridge.

Bass, W. M. 1971. Human Osteology: A Laboratory and Field Manual of the Human Skeleton. Missouri Archeological Society Special Publications, Columbia, MO.

Biegert, J. and Maurer, R. 1972. Rumpfskelettläge, Allometrien und Körperproportionen bei catarrhinen Primaten. *Folia Primatologica, 17:*142–156.

Blackith, R. E. and Reyment, R. A. 1971. Multivariate Morphometrics. Academic Press, New York.

Boag, P. T. 1984. Growth and allometry of external morphology in Darwin's finches *(Geospiza)* on Isla Daphne Major, Galapagos. *Journal of Zoology, 204:*413–441.

Buschang, P. H. 1982a. Differential long bone growth of children between two months and eleven years of age. *American Journal of Physical Anthropology, 58:*291–295.

Buschang, P. H. 1982b. The relative growth of the limb bones for *Homo sapiens*—as compared to anthropoid apes. *Primates, 23:*465–468.

Cheverud, J. 1982. Relationships among ontogenetic, static and evolutionary allometry. *American Journal of Physical Anthropology, 59:*139–149.

Ciochon, R. L. and Corruccini, R. S., editors. 1983. New Interpretations of Ape and Human Ancestry. Plenum Press, New York.

Cock, A. G. 1966. Genetical aspects of metrical growth and form in animals. *Quarterly Review of Biology, 41:*131–190.

Creel, N. 1986. Size and phylogeny in hominoid primates. *Systematic Zoology, 35:*81–99.

Davies, R. G. and Brown, V. 1972. A multivariate analysis of postembryonic growth in two species of *Ectobius* (Dictyoptera: Blattidae). *Journal of Zoology (London), 168:*51–79.

Farris, J. S. 1970. Methods for computing Wagner trees. *Systematic Zoology, 19:*83–92.

Felsenstein, J. 1986. Statistical inference of phylogenies from molecular data. Abstracts of the Twentieth International Numerical Taxonomy Conference.

Fooden, J. and Izor, R. J. 1983. Growth curves, dental emergence norms, and supplementary morphological observations in known-age captive orangutans. *American Journal of Primatology, 5:*285–301.

Goodman, M. and Cronin, J. E. 1984. Molecular anthropology: its development and current directions. *In* A History of Physical Anthropology, 1930–1980, pp. 105–146, ed. F. Spencer. Academic Press, New York.

Gould, S. J. 1977. Ontogeny and Phylogeny. Harvard University Press, Cambridge.

Gould, S. J. 1982. Change in developmental timing as a mechanism of macroevolution. *In* Evolution and Development, pp. 333–346, ed. J. T. Bonner. Springer-Verlag, Berlin.

Gower, J. C. 1966. Some distance properties of latent root and vector methods used in multivariate analysis. *Biometrika, 53:*325–338.

Groves, C. P. 1986. Systematics of great apes. *In* Comparative Primate Biology, vol. 1, Systematics, Evolution and Anatomy, pp. 187–217, ed. D. R. Swindler and J. Erwin. Alan R. Liss, New York.

Hartman, S. E. 1986. A stereophotogrammetric analysis of the occlusal morphology of extant hominoid molars. Doctoral dissertation, State University of New York at Stony Brook.

Hasegawa, M., Kishino, H. and Yano, T. 1985. Dating of the human-ape splitting by a molecular clock of mitochondrial DNA. *Journal of Molecular Evolution, 3:*1–18.

Huxley, J. S. 1932. Problems in Relative Growth. Cambridge University Press, London.

Johnston, F. E. 1980. Research design and sample selection in studies of growth and development. *In* Human Physical Growth and Maturation: Methodologies and Factors, pp. 5–19, ed. F. E. Johnston, A. F. Roche, and C. Susanne. Plenum Press, New York.

Jolicoeur, P. 1963a. The degree of generality of robustness in *Martes americana. Growth, 27:*1–27.

Jolicoeur, P. 1963b. The multivariate generalization of the allometry equation. *Biometrics, 19:*497–499.

Jolicoeur, P. 1984. Principal components, factor analysis, and multivariate allometry: a small sample direction test. *Biometrics, 40:*685–690.

Jungers, W. L. 1982. Lucy's limbs: skeletal allometry and locomotion in *Australopithecus afarensis. Nature, 297:*676–678.

Jungers, W. L. 1984. Scaling of the hominoid locomotor skeleton with special reference to the lesser apes. *In* The Lesser Apes: Evolutionary and Behavioral Biology, pp. 146–169, ed. H. Preuschoft, D. Chivers, W. Brockelman, and N. Creel. Edinburgh University Press, Edinburgh.

Jungers, W. L. 1985. Body size and scaling of limb proportions in primates. *In* Size and Scaling in Primate Biology, pp. 345–381, ed. W. L. Jungers, Plenum Press, New York.

Jungers, W. L. and Fleagle, J. G. 1980. Postnatal growth allometry of the extremities in *Cebus albifrons* and *Cebus apella:* a longitudinal and comparative study. *American Journal of Physical Anthropology, 53:*471–478.

Jungers, W. L. and Stern, J. T., Jr. 1983. Body proportions, skeletal allometry and locomotion in the Hadar hominoids: a reply to Wolpoff. *Journal of Human Evolution, 12:*673–684.

Jungers, W. L. and Susman, R. L. 1984. Body size and skeletal allometry in African apes. *In* The Pygmy Chimpanzee: Evolutionary Morphology and Behavior, pp. 131–177, ed. R. L. Susman. Plenum Press, New York.

Kuhry, B. and Marcus, L. 1977. Bivariate linear models in biometry. *Systematic Zoology, 26:*201–209.

Lumer, H. 1939. Relative growth of the limb bones in the anthropoid apes. *Human Biology, 11:*379–392.

Martin, L. 1986. Relationships among extant and extinct great apes and humans. *In* Major Topics in Primate and Human Evolution, pp. 151–187, ed. B. Wood, L. Martin, and P. Andrews. Cambridge University Press, Cambridge.

Mollison, T. 1911. Die Körperproportionen der Primaten. *Morphologisches Jahrbuch, 42:*79–304.

Mosimann, J. E. and James, F. C. 1979. New statistical methods for allometry with application to Florida red-winged blackbirds. *Evolution, 33:*444–459.

Napier, J. R. and Napier, P. H. 1967. A Handbook of Living Primates. Academic Press, London.

Nei, M., Stephens, J. C. and Saitou, N. 1985. Methods for computing the standard errors of branding points in an evolutionary tree and their application to molecular data from humans and apes. *Molecular Biology and Evolution, 2:*66–85.

Oxnard, C. E. 1981. The place of man among the primates: anatomical, molecular and morphometric evidence. *Homo, 32:*149–176.

Reyment, R. A. 1969. A multivariate paleontological growth problem. *Biometrics, 25:*1–8.

Reynolds, T. R. 1987. Stride length and its determinants in humans, early hominids, primates and mammals. *American Journal of Physical Anthropology, 72:*101–115.

Rohlf, F. J., Kishpaugh, J. and Kirk, D. 1986. NT-SYS: Numerical taxonomic system of multivariate statistical programs. State University of New York at Stony Brook.

Schultz, A. H. 1925. Embryological evidence of the evolution of man. *Journal of the Washington Academy of Sciences, 15:*247–263.

Schultz, A. H. 1930. The skeleton of the trunk and limbs of higher primates. *Human Biology, 2:*303–438.

Schultz, A. H. 1937. Proportions, variability and asymmetries of the long bones of the limbs and the clavicles in man and apes. *Human Biology, 9:*281–328.

Schultz, A. H. 1940. Growth and development of the chimpanzee. Carnegie Institution of Washington Publication 518. *Contributions to Embryology, 28:*1–63.

Schultz, A. H. 1941. Growth and development of the orang-utan. Carnegie Institution of Washington Publication 525. *Contributions to Embryology, 29:*57–110.

Schultz, A. H. 1956. Postembryonic age changes. *Primatologia, 1:*887–964.

Schultz, A. H. 1973. Age changes, variability and generic differences in body proportions of recent hominoids. *Folia Primatologica, 19:*338–359.

Schwartz, J. H. 1984a. The evolutionary relationships of man and orang-utans. *Nature, 308:*501–505.

Schwartz, J. H. 1984b. Hominoid evolution: a review and a reassessment. *Current Anthropology, 25:*655–672.

Shea, B. T. 1981. Relative growth of the limbs and trunk of the African apes. *American Journal of Physical Anthropology, 56:*179–202.

Shea, B. T. 1983. Paedomorphosis and neotony in the pygmy chimpanzee. *Science, 222:*521–522.

Shea, B. T. 1984. An allometric perspective on the morphological and evolutionary relationships between pygmy *(Pan paniscus)* and common *(Pan troglodytes)* chimpanzees. *In* The Pygmy Chimpanzee: Evolutionary Morphology and Behavior, pp. 89–130, ed. R. L. Susman. Plenum Press, New York.

Shea, B. T. 1985. Bivariate and multivariate growth allometry: statistical and biological considerations. *Journal of Zoology (London), 206:*367–390.

Shea, B. T. 1986. Scapula form and locomotion in chimpanzee evolution. *American Journal of Physical Anthropology, 70:*475–488.

Sibley, C. G. and Ahlquist, J. E. 1984. The phylogeny of the hominoid primates, as indicated by DNA-DNA hybridization. *Journal of Molecular Evolution, 20:*2–15.

Sneath, P. H. A. and Sokal, R. R. 1973. Numerical Taxonomy. Freeman, San Francisco.

Sokal, R. R. and Rohlf, F. J. 1981. Biometry, Second edition. Freeman, New York.

Sprent, P. 1972. The mathematics of size and shape. *Biometrics, 28:*23–37.

Susman, R. L., Stern, J. T., Jr., and Jungers, W. L. 1984. Arboreality and bipedality in the Hadar hominids. *Folia Primatoligica, 43:*113–156.

Swofford, D. L. 1982. WAGPROC, version 4: a FORTRAN 77 program for inferring phylogenies via the Wagner method. Illinois Natural History Survey. Champaign, IL.

Takai, S. 1977. Principal component analysis of the elongation of metacarpal and phalangeal bones. *American Journal of Physical Anthropology, 47:*301–304.

Thorington, R. W. 1972. Proportions and allometry in the gray squirrel, *Sciurus carolinensis. Nemouria, 8:*1–17.

Thorpe, R. S. 1983. A review of the numerical methods for recognizing and analyzing racial differentiation. *In* Numerical Taxonomy, pp. 403–423, ed. J. Felsenstein. Springer-Verlag, Berlin.

Thorpe, R. S. and Leamy, L. 1983. Morphometric studies in inbred and hybrid house mice (*Mus* sp.): multivariate analysis of size and shape. *Journal of Zoology (London), 199:*421–432.

Wiig, O. 1985. Multivariate variation in feral American mink *(Mustela vison)* from Southern Norway. *Journal of Zoology (London), 206:*441–452.

Subject Index

Author Index